WAVE AND SCATTERING METHODS FOR NUMERICAL SIMULATION

WAVE AND SCATTERING METHODS FOR NUMERICAL SIMULATION

Stefan Bilbao

John Wiley & Sons, Ltd

Copyright 2004   John Wiley & Sons Ltd, The Atrium, Southern Gate, Chichester,
West Sussex PO19 8SQ, England

Telephone (+44) 1243 779777

Email (for orders and customer service enquiries): cs-books@wiley.co.uk
Visit our Home Page on www.wileyeurope.com or www.wiley.com

All Rights Reserved. No part of this publication may be reproduced, stored in a retrieval system or transmitted in any form or by any means, electronic, mechanical, photocopying, recording, scanning or otherwise, except under the terms of the Copyright, Designs and Patents Act 1988 or under the terms of a licence issued by the Copyright Licensing Agency Ltd, 90 Tottenham Court Road, London W1T 4LP, UK, without the permission in writing of the Publisher. Requests to the Publisher should be addressed to the Permissions Department, John Wiley & Sons Ltd, The Atrium, Southern Gate, Chichester, West Sussex PO19 8SQ, England, or emailed to permreq@wiley.co.uk, or faxed to (+44) 1243 770620.

This publication is designed to provide accurate and authoritative information in regard to the subject matter covered. It is sold on the understanding that the Publisher is not engaged in rendering professional services. If professional advice or other expert assistance is required, the services of a competent professional should be sought.

Other Wiley Editorial Offices

John Wiley & Sons Inc., 111 River Street, Hoboken, NJ 07030, USA

Jossey-Bass, 989 Market Street, San Francisco, CA 94103-1741, USA

Wiley-VCH Verlag GmbH, Boschstr. 12, D-69469 Weinheim, Germany

John Wiley & Sons Australia Ltd, 33 Park Road, Milton, Queensland 4064, Australia

John Wiley & Sons (Asia) Pte Ltd, 2 Clementi Loop #02-01, Jin Xing Distripark, Singapore 129809

John Wiley & Sons Canada Ltd, 22 Worcester Road, Etobicoke, Ontario, Canada M9W 1L1

Wiley also publishes its books in a variety of electronic formats. Some content that appears in print may not be available in electronic books.

Library of Congress Cataloging-in-Publication Data

Bilbao, Stefan D.
  Wave and scattering methods for numerical simulation / Stefan D. Bilbao.
    p. cm.
  Includes bibliographical references and index.
  ISBN 0-470-87017-6 (cloth : alk. paper)
  1. Electric filters, Wave-guide. 2. Electric filters, Digital.
  3. Simulation methods. I. Title.
  TK7872.F5B535 2004
  620′.001′1—dc22
                                               2004004939

British Library Cataloguing in Publication Data

A catalogue record for this book is available from the British Library

ISBN 0-470-87017-6

Produced from LaTeX files supplied by the author, typeset by Laserwords Private Limited, Chennai, India
Printed and bound in Great Britain by Antony Rowe Ltd, Chippenham, Wiltshire
This book is printed on acid-free paper responsibly manufactured from sustainable forestry
in which at least two trees are planted for each one used for paper production.

Contents

Preface   xi

Foreword   xv

1 Introduction   1
     1.1 An Overview of Scattering Methods 3
         1.1.1 Remarks on Passivity 3
         1.1.2 Case Study: The Kelly–Lochbaum Digital Speech Synthesis Model   4
         1.1.3 Digital Waveguide Networks 12
         1.1.4 A General Approach: Multidimensional Circuit Representations and Wave Digital Filters 18
     1.2 Questions 24

2 Wave Digital Filters   25
     2.1 Classical Network Theory 27
         2.1.1 N-ports 27
         2.1.2 Power and Passivity 28
         2.1.3 Kirchhoff's Laws 30
         2.1.4 Circuit Elements 31
     2.2 Wave Digital Elements and Connections 32
         2.2.1 The Bilinear Transform 33
         2.2.2 Wave Variables 35
         2.2.3 Pseudopower and Pseudopassivity 36
         2.2.4 Wave Digital Elements 37
         2.2.5 Adaptors 41
         2.2.6 Signal and Coefficient Quantization 43
         2.2.7 Vector Wave Variables 45
     2.3 Wave Digital Filters and Finite Differences 48

3 Multidimensional Wave Digital Networks   53
     3.1 Symmetric Hyperbolic Systems 55
     3.2 Coordinate Changes and Grid Generation 60
         3.2.1 Structure of Coordinate Changes 61
         3.2.2 Coordinate Changes in $(1+1)$D 61
         3.2.3 Coordinate Changes in Higher Dimensions 62

3.3 MD-passivity 65
3.4 MD Circuit Elements 68
    3.4.1 The MD Inductor 68
    3.4.2 Other MD Elements 70
    3.4.3 Discretization in the Spectral Domain 71
    3.4.4 Other Spectral Mappings 73
3.5 The $(1+1)$D Advection Equation 74
    3.5.1 A Multidimensional Kirchhoff Circuit 75
    3.5.2 Stability 76
    3.5.3 An Upwind Form 77
3.6 The $(1+1)$D Transmission Line 79
    3.6.1 MDKC for the $(1+1)$D Transmission Line Equations 80
    3.6.2 Discussion: The Inductive Lattice Two-port 82
    3.6.3 Energetic Interpretation 83
    3.6.4 An MDWD Network for the $(1+1)$D Transmission Line .. 83
    3.6.5 Simplified Networks 85
3.7 The $(2+1)$D Parallel-plate System 86
    3.7.1 MDKC and MDWD Network 87
3.8 Finite Difference Interpretation 89
    3.8.1 MDWD Networks as Multistep Schemes 90
    3.8.2 Numerical Phase Velocity and Parasitic Modes 93
3.9 Initial Conditions 97
3.10 Boundary Conditions 99
    3.10.1 MDKC Modeling of Boundaries 101
3.11 Balanced Forms 105
3.12 Higher-order Accuracy 108

4 Digital Waveguide Networks   115
4.1 FDTD and TLM 117
4.2 Digital Waveguides 118
    4.2.1 The Bidirectional Delay Line 118
    4.2.2 Impedance 119
    4.2.3 Wave Equation Interpretation 120
    4.2.4 Note on the Different Definitions of Wave Quantities 121
    4.2.5 Scattering Junctions 122
    4.2.6 Vector Waveguides and Scattering Junctions 124
    4.2.7 Transitional Note 126
4.3 The $(1+1)$D Transmission Line 127
    4.3.1 First-order System and the Wave Equation 127
    4.3.2 Centered Difference Schemes and Grid Decimation 127
    4.3.3 A $(1+1)$D Waveguide Network 129
    4.3.4 Waveguide Network and the Wave Equation 131
    4.3.5 An Interleaved Waveguide Network 133
    4.3.6 Varying Coefficients 135
    4.3.7 Incorporating Losses and Sources 141
    4.3.8 Numerical Phase Velocity and Dispersion 143
    4.3.9 Boundary Conditions 144

	4.4	The $(2+1)$D Parallel-plate System 146
		4.4.1 Defining Equations and Centered Differences 146
		4.4.2 The Waveguide Mesh 149
		4.4.3 Reduced Computational Complexity and Memory Requirements in the Standard Form of the Waveguide Mesh 156
		4.4.4 Boundary Conditions 158
	4.5	Initial Conditions 162
	4.6	Music and Audio Applications of Digital Waveguides 164

5 Extensions of Digital Waveguide Networks   169

	5.1	Alternative Grids in $(2+1)$D 169
		5.1.1 Hexagonal and Triangular Grids 170
		5.1.2 The Waveguide Mesh in Radial Coordinates 173
	5.2	The $(3+1)$D Wave Equation and Waveguide Meshes 180
	5.3	The Waveguide Mesh in General Curvilinear Coordinates 182
	5.4	Interfaces between Grids 186
		5.4.1 Doubled Grid Density Across an Interface 187
		5.4.2 Progressive Grid Density Doubling 193
		5.4.3 Grid Density Quadrupling 196
		5.4.4 Connecting Rectilinear and Radial Grids 198
		5.4.5 Grid Density Doubling in $(3+1)$D 202
		5.4.6 Note 203

6 Scattering Methods: A Unified Perspective   205

	6.1	The $(1+1)$D Transmission Line Revisited 206
		6.1.1 Multidimensional Unit Elements 207
		6.1.2 Hybrid Form of the Multidimensional Unit Element 208
		6.1.3 Alternative MDKC for the $(1+1)$D Transmission Line 210
	6.2	Alternative MDKC for the $(2+1)$D Parallel-plate System 212
	6.3	Higher-order Accuracy Revisited 214
	6.4	Maxwell's Equations 217

7 Applications to Vibrating Systems   223

	7.1	Beam Dynamics 224
		7.1.1 MDKC and MDWD network for Timoshenko's System 226
		7.1.2 Waveguide Network for Timoshenko's System 228
		7.1.3 Boundary Conditions in the DWN 230
		7.1.4 Simulation: Timoshenko's System for Beams of Uniform and Varying Cross-sectional Areas 232
		7.1.5 Improved MDKC for Timoshenko's System via Balancing 233
	7.2	Plates 235
		7.2.1 MDKCs and Scattering Networks for Mindlin's System 238
		7.2.2 Boundary Termination of the Mindlin Plate 242
		7.2.3 Simulation: Mindlin's System for Plates of Uniform and Varying Thickness 246

	7.3	Cylindrical Shells	247
		7.3.1	The Membrane Shell	248
		7.3.2	The Naghdi–Cooper System II Formulation	250
	7.4	Elastic Solids	252
		7.4.1	Scattering Networks for the Navier System	255
		7.4.2	Boundary Conditions	258

8 Time-varying and Nonlinear Systems — 261

	8.1	Time-varying and Nonlinear Circuit Elements	262
		8.1.1	Lumped Elements	262
		8.1.2	Distributed Elements	263
	8.2	Linear Time-varying Distributed Systems	264
		8.2.1	A Time-varying Transmission Line Model	266
	8.3	Lumped Nonlinear Systems in Musical Acoustics	267
		8.3.1	Piano Hammers	267
		8.3.2	The Single Reed	270
	8.4	From Wave Digital Principles to Relativity Theory	272
		8.4.1	Origin of the Challenge	272
		8.4.2	The Principle of Newtonian Limit	274
		8.4.3	Newton's Second Law	274
		8.4.4	Newton's Third Law and Some Consequences	276
		8.4.5	Moving Electromagnetic Fields	277
		8.4.6	The Bertozzi Experiment	277
	8.5	Burger's Equation	278
	8.6	The Gas Dynamics Equations	280
		8.6.1	MDKC and MDWD Network for the Gas Dynamics Equations	282
		8.6.2	An Alternate MDKC and Scattering Network	283
		8.6.3	Entropy Variables	285

9 Concluding Remarks — 289

	9.1	Answers	289
	9.2	Questions	293

A Finite Difference Schemes for the Wave Equation — 297

	A.1	Von Neumann Analysis of Difference Schemes	298
		A.1.1	One-step Schemes	299
		A.1.2	Multistep Schemes	300
		A.1.3	Vector Schemes	302
		A.1.4	Numerical Phase Velocity	302
	A.2	Finite Difference Schemes for the $(2+1)$D Wave Equation	303
		A.2.1	The Rectilinear Scheme	304
		A.2.2	The Interpolated Rectilinear Scheme	305
		A.2.3	The Triangular Scheme	309
		A.2.4	The Hexagonal Scheme	311
		A.2.5	Note on Higher-order Accuracy	314

	A.3	Finite Difference Schemes for the $(3+1)$D Wave Equation 315
		A.3.1 The Cubic Rectilinear Scheme 315
		A.3.2 The Octahedral Scheme 317
		A.3.3 The $(3+1)$D Interpolated Rectilinear Scheme 318
		A.3.4 The Tetrahedral Scheme 321

B Eigenvalue and Steady State Problems   **325**
    B.1 Introduction 325
    B.2 Abstract Time Domain Models 326
    B.3 Typical Eigenvalue Distribution of a Discretized PDE 326
    B.4 Excitation and Filtering 327
    B.5 Partial Similarity Transform 327
    B.6 Steady State Problems 329
    B.7 Generalization to Multiple Eigenvalues 330
    B.8 Numerical Example 331

Bibliography   **333**

Index   **355**

Preface

Engineering research is a messy business. This is the rare assertion, expressible in six or fewer words, that does not need to be qualified; it is something that most people, whether inside or outside the engineering world, just seem to know. From my time in graduate school and through a few subsequent years during which this book was written, further refinements of this idea on my part were inevitable—after all, what else is writing a book, besides either cleaning up a mess or making a new one? Two thoughts, in particular, stand out: in any sufficiently large field of study, (a) there is no single person who is an authority, and (b) at any given moment, there will be many people working independently on the same problem. To anyone who has done any type of research, these are self-evident truths; to me they remain only slightly less unsettling than when they dawned upon me. Slightly less, because I also now know that ignorance and uncertainty, as well as a vast and chaotic literature are inevitable by-products of a free research environment. When there is no such freedom, it is easy for one to become an authority and, once an authority, to make sure that everyone in one's charge is working together on the same problem; at that stage, it is only a small step to further ensure, as Lysenko did in Soviet Russia, that everyone uncovers the same results, their truth notwithstanding [236].

If there is any field, though, that has suffered more from the dispersive effects of free research than scattering-based numerical methods, it would be interesting to know about it. Poised on the cusp between electrical engineering and applied mathematics, this area of study has attracted specialists in three domains: electromagnetic simulation, digital filter design, and sound simulation. Each of these three groups has arrived, in almost complete isolation from the others, at an electric circuit-based formalism for numerical methods, especially as applied to simulation. The names given to these approaches by the three groups are, respectively, the transmission line matrix (TLM), multidimensional wave digital (MDWD), and digital waveguide network (DWN) methods. The first two methods are markedly different in many ways (from underlying philosophy to notation); the third is quite similar to the first, though it came into being through a radically different path, and served as my point of entry into this field. The history of the development of these methods is more than a little tangled, so a short chronology is in order here.

TLM first appeared in a publication by P. B. Johns and R. L. Beurle [135] in 1971, which was also, coincidentally, the publication date of Alfred Fettweis's original paper on wave digital filters (WDFs) [62], in the filter design context. Each of these publications generated, independently, a huge literature, and research is active in both fields to this day. For a more detailed summary of recent work in each of these areas, the reader may turn to Section 4.1 and the beginning of Chapter 2, respectively, but the important points are the following: TLM was conceived of as a numerical method for electromagnetic field problems, and wave digital filters were, as the name suggests, digital filter structures. Both techniques employ digital wave quantities, to be propagated numerically and scattered, in a computer program that attempts to mimic wave-like behavior in a real electrical device.

In the mid-1980s, DWNs, the invention of Julius O. Smith III at Stanford, appeared first as an artificial reverberation technique and then as efficient structures for digital musical sound synthesis [228]; inspired, in a sense, by the Karplus–Strong algorithm [141] and grounded in electrical network and signal processing theory just like WDFs, they have since become the method of choice in this arena. When extended to multiple dimensions, as was first done by Smith and Scott Van Duyne [267], they are also very similar (and in some cases identical) to TLM structures, though in an acoustic setting; the subsequent history of DWNs is elaborated in Section 4.6. (It is worth mentioning that similar structures had also appeared earlier in inverse filtering applications to problems in speech and geophysics [35].) In the meantime, WDFs had been extended to multiple dimensions, though still intended for digital filtering purposes (i.e., for image and video processing). In approximately 1990, Fettweis and his students, and in particular Gunnar Nitsche, began to apply WDF techniques to multidimensional time-dependent problems for simulation purposes, within a powerful theoretical framework for attacking a very general class of problems [87]. The numerical structures they arrived at were not the same as TLM or DWN methods but possessed a number of very similar properties, including operation based on the scattering principle.

The multidimensional circuit-based formalism developed by Fettweis and Nitsche in their approach to scattering methods for distributed systems distinguishes the MDWD method from TLM and DWNs; for this reason, I have decided to build this book around it[1]. There is also significant coverage of DWNs because it was my original research interest. I apologize to members of the TLM community if they feel that their discipline has been given short shrift. TLM design is a complex art form in itself and, of the three methods mentioned here, TLM is certainly the most widely known; there exist at least two comprehensive TLM texts already [122, 44] and practitioners will certainly recognize the fundamental similarity with the DWN. However, it is not the purpose of this book to detail problem-specific design techniques, or to catalogue the multitude of existing different forms but rather to provide a unified treatment of numerical methods that make use of circuits, waves, and scattering. Indeed, after some exposure to these methods, it becomes clear that the distinctions between TLM, MDWD networks, and DWNs are quite academic; in a sense to be made precise later in this book, they are all part of the same family.

If one spends enough time working on these scattering methods, one is bound to forget that there already exists a good half-century of literature on numerical techniques, as applied to the solution of time-dependent partial differential equations. Traditional approaches to numerical simulation usually involve the direct discretization of a given set of equations by a variety of techniques, such as finite difference, finite element, and spectral or collocation methods. As mentioned earlier, the methods to be discussed here, however, have their roots elsewhere, in electrical network theory, digital filtering, and scattering theory. It is important then, some 30 years into the lifetime of these methods, to answer or at least pose some questions about how they fit into the larger picture of numerical integration methods as a whole. It is worth pointing out that TLM was developed initially as an alternative

---

[1]This is a good place to mention that the single best reference on the subject of MDWD simulation is Gunnar Nitsche's doctoral dissertation [176]; I have referred to and borrowed from it quite a bit and, in fact, many topics in this book have appeared there, in a somewhat more compact form. Unfortunately, it is only available in German and this has been one of the chief reasons for attempting a comprehensive review in English.

to more standard finite difference time domain (FDTD) methods for electromagnetics and has since been viewed as a distinct numerical method (though there have been a few attempts at making the link to FDTD; see Section 4.1 and the references therein). It is clear enough, though, that TLM methods (or indeed any scattering method) are no more than finite difference schemes themselves and should be treated as such—it is this point of view that is taken in this book and a good deal of time is spent on clarifying the link between scattering methods and finite differences in particular.

As might be expected, this book suffers in certain respects (notation among them) from the mismatch between the points of view of the electrical engineer and the specialist in numerical methods. Needless to say, there is much insight to be gained in the attempt to resolve some of the many outstanding distinctions. There are hundreds of researchers at work in the fields of TLM, MDWD networks, and DWNs—I hope that this book will be of some use to them.

Book Summary

The basic organization of this book is as follows: Chapter 1 is an overview of scattering-based numerical methods for the uninitiated, with an emphasis on the important concept of passivity. At the end of this chapter, several general questions are posed about these methods. The book begins in earnest in Chapter 2, which is a review of the basics of classical electrical network theory and wave digital filters, with filtering applications de-emphasized. It is then shown in Chapter 3 how wave digital filtering techniques have been extended to multiple dimensions in order to approach the simulation of distributed systems, and in particular, the transmission line and parallel-plate test problems. Chapter 4 then revisits the simulation of these same systems, from the point of view of digital waveguide networks, which are very similar (and sometimes identical) to the TLM structures that have been proposed. Chapter 5 is something of an interlude; the structures introduced in Chapter 4 are extended to the more general scenarios that arise in simulation, and in particular, those involving irregular geometries requiring coordinate changes or multiple grids. Chapter 6 is a return to the DWN and MDWD approaches for comparison and an eventual unification of the two methods, under the umbrella of the circuit formalism of Fettweis. Scattering methods are then developed in tandem, in Chapter 7, for a variety of distributed elastic systems and in Chapter 8 for time-varying and nonlinear systems. Some concluding remarks (including answers to the general questions posed in Chapter 1) follow in Chapter 9. Basic material on the spectral analysis of finite difference schemes, especially with regard to DWNs, is given in Appendix A. Appendix B is a self-contained discussion of techniques applicable to steady-state problems.

Detailed summaries appear in the opening remarks of each chapter, except the first, last, and Appendix B.

Acknowledgments

This book would not have been written without the help of several people. Foremost among them are Professor Julius O. Smith III, of the Center for Computer Research in Music and Acoustics (CCRMA) at Stanford University, and Dr. Ivan Linscott, of the Space, Telecommunications and Radioscience Laboratory, also at Stanford; I thank them for their guidance and great patience. I also thank Professor Donard DeCogan, of the University of East Anglia, for his enthusiastic support and persistent encouragement to publish this book. Many others have helped me at different stages in the writing of the manuscript. Scott Van Duyne was extremely helpful, especially at the beginning stages of research. Davide Rocchesso and Federico Fontana, both of the University of Verona, Perry Cook at Princeton, Robert M. Gray at Stanford, and Richard Kronland-Martinet and Julien Bensa of the CNRS-Marseille provided many useful comments after their careful reading. Thanks also to Simone Taylor and her associates at Wiley, for proving to me that publishing need not be the nerve-wracking ordeal it is often made out to be.

I am pleased to be able to include contributions from Alfred Fettweis (Section 8.4) and Gunnar Nitsche (Appendix B). Their work forms the core of this book, and I am very grateful for their input, as well as the extensive help they have offered me over the years.

Thanks to my many colleagues at CCRMA, and the School of Music and the Sonic Arts Research Centre, at the Queen's University, Belfast. I am especially grateful for the generous financial support of the Bosack-Kruger Foundation.

Foreword

At an early stage in my scientific career, I had the need to build equipment to destroy discrete electronic devices using short-duration, high-power electrical pulses. The results of the ensuing experiments suggested that complex thermal interactions were operating, and this demanded a means of numerical modeling that could account for the thermal dependence of the specific heats and thermal conductivities of the component materials of the packaged device. By a stroke of extreme good fortune, I was in the Electrical & Electronic Engineering Department of Nottingham University where the late Peter Johns suggested that I use his recently developed Transmission Line Matrix (TLM) technique. This proved extremely successful, and a wider curiosity meant that the characterization and exploitation of TLM has dominated my subsequent academic career.

In view of the above, it is a great honor to be asked to write a foreword to this extraordinary work. It certainly scared us when its precursor came on the scene. I was just finishing writing a book on the applications of TLM in computational mechanics, when one of my coauthors e-mailed in a panic to say that we had been upstaged and that everything that we were trying to do could now be done using the contents of a PhD thesis from Stanford University that had just appeared on the Web. The author of the thesis, Stefan Bilbao stated

The subject of this thesis is the numerical simulation of physical systems. In particular, we look at systems which are dynamic and distributed...The simulation techniques that we will discuss are based on analogies between the systems mentioned above and electrical networks, and make use of scattering principles...

This is what TLM does. He then went on to say

We will be primarily concerned with two such methods. The first is based on wave digital filtering, a filter design technique which was initially intended as a means of translating a lumped analog electrical filtering network into discrete time, while preserving its topology and energetic properties (passivity in particular)...

These are the keystones of our discipline. It is little wonder that people were worried. However, we soon came to realize the beauty of the techniques that were being proposed and the manner in which they could provide fresh insights into our own work. We found ourselves in awe of what Stefan had achieved, in apparent isolation from us and from others, and I know that he has something to say about this here.

At the start of this book, some fundamental questions are posed. These act as guiding lights throughout the work. These are addressed again in the concluding chapter, but in between we see Stefan carving, apparently effortlessly, through swathes of concepts that have kept other researchers busy for most of their lives. There is much reference to TLM within the book and it is obvious that the author is as fascinated with our approach as we are with his. I, for one, look forward to a long collaborative exchange of ideas, and I hope that readers from all areas of scientific modeling will derive as much benefit from its content as I have.

Donard DeCogan
University of East Anglia, UK.

Chapter 1

Introduction

The subject of this book is the numerical simulation of physical systems. In particular, we look at systems that are dynamic and distributed. By *dynamic*, we mean that the system's state evolves as time progresses, and by *distributed*, that the system is defined over some region in space, called the *problem domain*. The systems of interest here, always described mathematically by sets of time-dependent partial differential equations (PDEs) complemented by initial and boundary conditions and possibly external excitations, span a large range of physical scenarios, including electromagnetics, acoustics, transmission lines, the vibration of elastic systems such as strings, membranes, beams, plates and shells, and even nonlinear fluid dynamics.

The simulation techniques that we will discuss are based on analogies between the systems mentioned above and electrical networks, and make use of *scattering* principles. The time evolution of the state of a system is modeled as the movement of energy as it is reflected, transmitted, and propagated throughout an electrical network; the energy is carried by *waves*. The chief benefit of a network formulation is that there is direct access to a measure of the system energy, which can be used to bound the size of the solution of the system as it evolves over time. Because many physical systems, in the absence of external excitations, are inherently *passive* (i.e., they do not produce energy on their own), a network model for a system of PDEs is useful in that this passivity is reflected in an obvious way; a simple positivity condition on all the circuit element values is all that is required. When such a network model is transferred to a discrete setting in an appropriate way (where it will eventually be implemented as a computer program, operating as a recursion over a numerical grid), this passivity condition becomes a sufficient and trivially verifiable condition for the numerical stability of the resulting simulation. It is interesting that these numerical methods have their roots in digital filter-design techniques[1], which were, in turn, based on discrete physical models of mechanical or electrical circuit elements and the connections between them. In a sense, then, simulation is a more 'natural' use for these structures than filtering. Many important ideas regarding the good behavior of these methods in finite machine arithmetic, however, were first introduced in the filtering context—these also result from passivity in the network model.

---

[1] This is not true of the transmission line matrix method (TLM), which was conceived of directly as a numerical method.

We will be primarily concerned with two such methods. The first is based on *wave digital filters* (WDFs), a filter-design technique that was initially intended as a means of translating a lumped analog electrical filtering network into discrete time, while preserving its topology and energetic properties (passivity in particular). Because the voltages and currents in a closed analog electrical network will evolve according to a set of *ordinary differential equations* (ODEs), these digital filter networks can also be viewed as numerical integration methods. The extension of WDFs to multiple dimensions, in which case they are referred to as *multidimensional wave digital filters* (MDWDFs), is direct and makes use of a distributed network formulation of a given system as a means of arriving at a simulation routine. In the simulation context, we will refer to these structures as multidimensional wave digital (MDWD) networks. Here, there is a compact (though quite abstract) multidimensional circuit representation of the model system of PDEs, just as a lumped network is a representation of a system of ODEs. Despite the sometimes abstruse formalism underlying the construction of MDWD networks for simulation (invoking various coordinate changes, spectral mappings, and the use of nonphysical 'circuit elements' that are distributed, and may have a directional character), these numerical methods always reduce to the scattering of digital signals over a numerical grid of nodes that fills the problem domain, and are straightforward to program.

The second method, though very similar to the first from the standpoint of the programmer in that the basic signal-processing operation is the scattering of wave variables, is of a seemingly different origin. Here, the network is composed of a large number of connected elements, which are essentially transmission lines, or *waveguides*, so as to fill the problem domain. Wave propagation along a given waveguide is modeled, in discrete time, by a pair of digital delay lines that transport wave signals in opposite directions. A *digital waveguide network* (DWN), then, is usually thought of as a large network of *lumped* elements; there is traditionally not a multidimensional representation, as there is for MDWD networks. We will spend some time looking at the relationship between DWNs and the multidimensional wave digital networks mentioned above. Though, as mentioned previously, they are derived from digital filter structures, DWNs are very similar (and indeed identical in some cases) to structures arrived at through the TLM method, which has a long independent history of its own (see Section 4.1 for more discussion).

These network approaches are relative newcomers in the field of numerical simulation. There are, of course, many other, older ways of designing a simulation method; the most well established and straightforward approach makes use of *finite difference approximations* to the model system of PDEs. Partial derivatives are replaced by differences between quantities on a numerical grid, and a recursion (or difference scheme) results. These methods are simpler to program, but the wave/scattering interpretation is lost, and the verification of numerical stability can be very involved, especially in the presence of boundary conditions. Because the electrical network models mentioned above also operate, ultimately, as recursions on grids, it is reasonable to ask how scattering methods fit into the finite difference picture. The eventual identification of scattering methods with standard finite difference methods may come as something of a disappointment to anyone who feels that these methods are completely novel. It is best, however, to think of these methods as a different way of organizing calculation, which leads to more robust numerical behavior.

As mentioned in the Preface, the overall goal of this book is to provide a unified picture of how these scattering methods are related to each other and to finite differences. It is possible to rephrase this goal as an attempt to answer a basic set of questions; we will pose these explicitly in Section 1.2. Before we get to that stage, however, it is useful to outline the basics of these methods in a little more detail.

1.1 An Overview of Scattering Methods

In all of the next chapter and in large parts of Chapters 3 and 4, we will be forced to make a long detour in order to lay out fully the details of how scattering-based numerical simulation methods are designed. In this section, we take a brief and informal look at many of the relevant ideas, while putting aside full development until later. The reader who has some familiarity with WDFs and DWNs or TLM may safely skip this section.

1.1.1 Remarks on Passivity

Passivity is the key property of all the methods that we will discuss in this book. Though it will be defined precisely in Chapters 2 and 3, it can be understood in a rough sense as follows: a physical system is passive if it is incapable of producing energy on its own. For many physical systems of interest, this is indeed the case, and passivity should be reflected in any useful mathematical model.

There is, however, a more subtle distinction between *abstract* and *concrete* passivity [14]. A system that is abstractly passive is passive only in a global sense; its internal description may be unknown. A concretely passive system is one that is made up entirely of simpler elements that are themselves passive. Typically, these elements cannot be further decomposed. For instance, any physically assembled collection of masses, springs, and dashpots is concretely passive; at any given moment, the energy contained in such a collection is bounded by the energy contained initially in the system (i.e., the kinetic energy of moving masses plus the potential energy of compressed or extended springs). The same can be said of any electrical network composed of resistors, capacitors, and inductors.

The distinction is that passivity of a concretely passive system may be deduced by an examination of the constituent elements themselves (and the connections between them)—as these elements are often quite simple, the determination of passivity can be very straightforward, even trivial. This idea is most easily approached in the electrical setting. For instance, consider the two electrical networks shown in Figures 1.1(a) and (b). The first is a series connection of a capacitor of capacitance C and a resistor of resistance R (assumed positive), and the second is a series connection of a capacitor and two resistors, one of

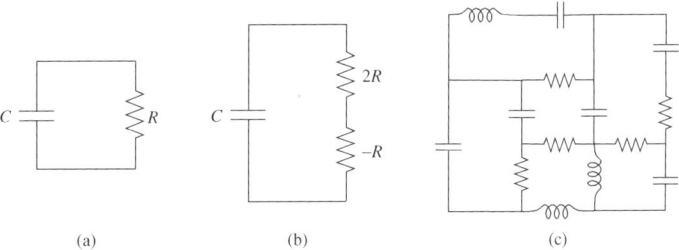

Figure 1.1: (a) *A concretely passive RC circuit,* (b) *an abstractly, but not concretely passive equivalent to the network shown in (a),* (c) *a network for which passivity is not easy to determine if parameter values can be positive or negative.*

resistance $2R$, the other of resistance $-R$. From a global point of view, these two devices are identical, and are both abstractly passive: the electric field energy initially surrounding the capacitor is dissipated exponentially, with time constant RC. In the first case, all this energy is dissipated in the single resistor, but in the second case, an active unphysical 'negative' resistance supplies power, but not enough to make up for that dissipated in the positive resistance. The point here is that in the first case, checking the positivity of the resistances is sufficient for passivity, but in the second, passivity is slightly less apparent: one must check that the sum of the resistances is positive. It should be clear that as the complexity of a system grows, so does the difficulty of performing such a task. Considering the general RLC network shown in Figure 1.1(c), if all resistances, inductances, and capacitances are positive, the network is passive. Otherwise, the check can be much more involved.

A good model of a physical system should exhibit concrete passivity—that is, energetic properties should not rely on cancellation between passive and active contributions (except, of course, when such active components actually form part of the physical system). As it turns out, concrete passivity offers more than just a means of simple passivity verification— indeed, it points the way toward numerical simulation methods that possess a similar transparent passivity property; in this context, it can be thought of as a sufficient condition for numerical stability. Many numerical methods reflect abstract, rather than concrete passivity, and for this reason, stability verification techniques often require a global analysis. For a numerical method that inherits concrete passivity, we might expect that stability can be ascertained through purely local analysis. This is in fact true; we call the reader's attention to these rather obvious observations in order to motivate the material in the rest of this book.

1.1.2 Case Study: The Kelly–Lochbaum Digital Speech Synthesis Model

As we mentioned above, the numerical methods to be discussed in this book have their origin in digital filter design, even though they are intended, ultimately, for use in simulation, and not filtering. (This is not true of TLM, however.) Though these two goals may seem to be at cross purposes, there is a very early instance of an engineering problem that straddles both worlds.

Kelly and Lochbaum [142] developed a digital speech synthesis model by treating the vocal tract as a slowly time-varying one-dimensional acoustic tube of variable (but circular) cross section, excited at one end (periodically by the glottis, or by turbulent noise), and radiating a speech waveform at the other (see Figure 1.2(a)). At any given time t, the shape of the tube as a function of the spatial coordinate x determines the system resonances, or formants [196], which serve as important perceptual cues for the listener in distinguishing among various voiced and unvoiced vocal sounds. The problem, then, is to develop a numerical method, suitable for computer implementation, which somehow simulates the time evolution of the acoustic 'state' of the vocal tract, that is, the *pressure* and *velocity* distributions in the interior. We follow the standard exposition of the Kelly–Lochbaum model here [45, 196].

1.1. AN OVERVIEW OF SCATTERING METHODS

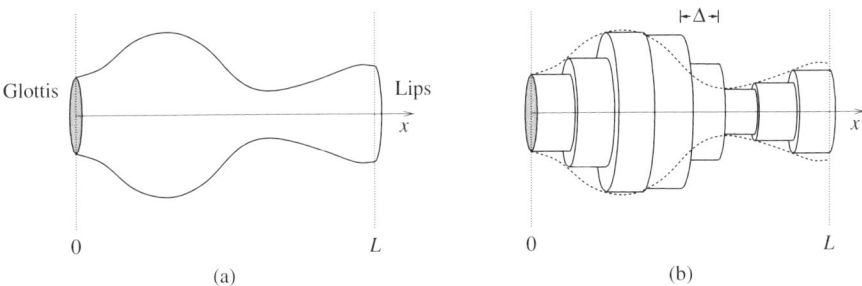

Figure 1.2: (a) *The vocal tract, modeled as a single one-dimensional acoustic tube of varying cross-sectional area and* (b) *an eight tube model suitable for discretization.*

Concatenated Acoustic Tube Model of the Vocal Tract

The first step toward a digital model is in representing the tube as a series of N concatenated tubes of constant cross-sectional areas, as in Figure 1.2(b) (where $N = 8$). The tubes are assumed to be of equal length Δ; if L is the total length of the vocal tract, we have $N\Delta = L$. In the limit as Δ becomes small, the shape of the approximation of the series of tubes will converge to that of a continuous vocal tract shown in Figure 1.2(a).

Wave Propagation in a Tube of Constant Cross-sectional Area

The concatenated tube model is useful because the acoustic behavior of a single tube of constant cross-sectional area A is quite simple to describe, in terms of a volume velocity $u(x, t)$, and a pressure deviation $p(x, t)$ from the mean tube pressure. Provided wavelengths are long in comparison with the tube radius, and that pressures do not become too large (both these requirements are easily satisfied in the speech context), and also that wall losses are negligible (an assumption somewhat less justifiable), the time evolution of the acoustic state of any single tube, such as that shown in Figure 1.3(a), will be described completely by

$$\frac{\rho}{A}\frac{\partial u}{\partial t} + \frac{\partial p}{\partial x} = 0 \qquad (1.1a)$$

$$\frac{A}{\rho\gamma^2}\frac{\partial p}{\partial t} + \frac{\partial u}{\partial x} = 0 \qquad (1.1b)$$

subject, of course, to initial conditions, and the effect of the boundary terminations on adjacent tubes. Given that the cross-sectional tube area A, the air density ρ and the sound-speed γ are constant, the general solution to (1.1) can be written as

$$p(x, t) = p^l(t + x/\gamma) + p^r(t - x/\gamma) \qquad (1.2a)$$

$$u(x, t) = \underbrace{Yp^l(t + x/\gamma)}_{u^l} + \underbrace{(-Yp^r(t - x/\gamma))}_{u^r} \qquad (1.2b)$$

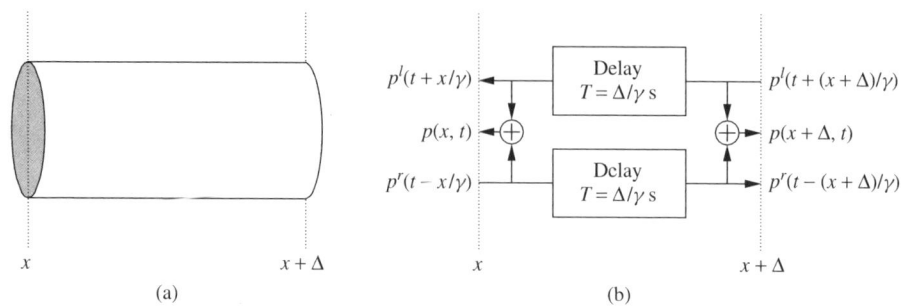

Figure 1.3: (a) *An acoustic tube and* (b) *a representation of the traveling-wave solution; traveling pressure waves can be added together at either end of the tube to give the physical pressure, as per* (1.2a).

Here, the physical pressure p has been decomposed into a sum of a *leftward-traveling wave* p^l and a *rightward-traveling wave* p^r; both are arbitrary functions of one variable. The volume velocity u, which is dual[2] to p in the system (1.1), can be similarly expressed as a sum of leftward- and rightward-traveling velocity waves u^l and u^r. But these velocity waves are simply the pressure waves, scaled by the tube *admittance*, defined by

$$Y \triangleq \frac{A}{\rho\gamma}$$

In addition, the rightward-traveling wave component of the velocity is sign-inverted with respect to the corresponding pressure wave. System (1.1) can be simplified to a single second-order PDE in pressure alone,

$$\frac{\partial^2 p}{\partial t^2} = \gamma^2 \frac{\partial^2 p}{\partial x^2} \qquad (1.3)$$

from which the traveling pressure wave solution may also be extracted. The volume velocity satisfies an identical equation.

Consider one of the tube segments of length Δ from Figure 1.2(b). It should be clear that we can represent the pressure traveling-wave solution to (1.1) by using two delay lines, each of duration Δ/γ; (see Figure 1.3). We can obtain the physical pressure at either end of the tube by summing the leftward- and rightward-traveling components, as per (1.2a). (The physical volume velocity can be obtained, from (1.2b), by taking the difference of p^l and p^r, and scaling the result by Y.) The discrete-time implementation of this single isolated acoustic tube is immediate. Taking

$$T \triangleq \frac{\Delta}{\gamma} \qquad (1.4)$$

as the *unit delay*, or *sampling period* for our discrete-time system, we can see that there is no loss in generality in treating the paired shifts as *digital* delay lines, accepting and shifting discrete-time pressure wave signals, at intervals of T seconds. The discrete-time model of the acoustic tube will still calculate an exact solution to system (1.1), at times

---

[2] By dual, we mean that the product pu has units of power.

1.1. AN OVERVIEW OF SCATTERING METHODS

that are integer multiples of T. (This solution can be considered to be exact at *all* time instants as long as all signals in the network are assumed to be *bandlimited* to half of the sampling rate, $F_s = 1/T$.)

Also note that because the traveling pressure and volume velocity waves are simply related to one another by a scaling, in a computer implementation it is only necessary to propagate one of the two types of waves in a given discrete tube section—we will assume, then, that pressure waves are our signal variables.

Junctions between Two Uniform Acoustic Tubes

Consider now a *junction* between two of the uniform acoustic tubes in the concatenated tube model shown in Figure 1.2(b). The wave speeds in all the tubes (of cross-sectional areas $A_i = A(i\Delta)$, $i = 0, \ldots, N - 1$) are assumed to be constant, and equal to γ. The discrete-time representation of any single tube will thus have the form of the pair of digital delay lines shown in Figure 1.3(b). At the junction between the ith and $(i + 1)$th tubes, we will then have a pressure and a velocity on either side; we will write these pressure/velocity pairs as (p_i, u_i), and (p_{i+1}, u_{i+1}) respectively (see Figure 1.4(a)). Continuity arguments (or conservation laws) dictate that these quantities should remain unchanged as we pass through the boundary between the two tubes, and thus

$$p_i = p_{i+1} \qquad\qquad u_i = u_{i+1} \qquad (1.5)$$

Note that we have dropped the arguments t and x, since the relationships of (1.5) hold instantaneously, and only at the tube boundaries.

As per (1.2a) and (1.2b), the pressures and velocities can be split into leftward- and rightward-traveling waves as

$$p_i = p_i^l + p_i^r \qquad\qquad u_i = Y_i \left(p_i^l - p_i^r \right) \qquad (1.6a)$$

$$p_{i+1} = p_{i+1}^l + p_{i+1}^r \qquad\qquad u_{i+1} = Y_{i+1} \left(p_{i+1}^l - p_{i+1}^r \right) \qquad (1.6b)$$

where Y_i, the admittance of the ith tube, is defined by

$$Y_i \triangleq \frac{A_i}{\rho\gamma} \qquad (1.7)$$

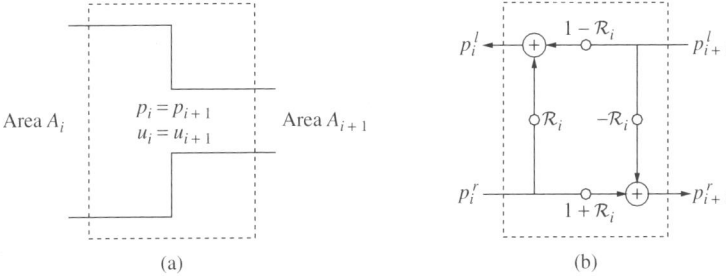

(a)   (b)

Figure 1.4: (a) *The junction between the ith and $(i + 1)$th acoustic tubes in the Kelly–Lochbaum vocal-tract model and* (b) *the resulting scattering junction for pressure waves.*

It is then possible, using (1.6) to rewrite (1.5) purely in terms of the wave variables, as

$$p_i^l = \mathcal{R}_i p_i^r + (1 - \mathcal{R}_i) p_{i+1}^l \tag{1.8a}$$

$$p_{i+1}^r = (1 + \mathcal{R}_i) p_i^r - \mathcal{R}_i p_{i+1}^l \tag{1.8b}$$

where $\mathcal{R}_i$ is defined by

$$\mathcal{R}_i \triangleq \frac{Y_i - Y_{i+1}}{Y_i + Y_{i+1}}$$

Here we have written a formula for calculating the pressure waves p_i^l and p_{i+1}^r leaving the junction in terms of the waves p_i^r and p_{i+1}^l entering the junction (see Figure 1.4(b) for the resulting signal flow diagram). In particular, (1.8) can be viewed as a scattering operation; incident waves on either side of an interface are reflected and transmitted according to the mismatch in the admittances between the two tubes. The mismatch is characterized by the reflection parameter $\mathcal{R}_i$ which is bounded in magnitude by unity, *as long as the admittances of the two tubes are positive*[3]. (If $Y_i = Y_{i+1}$, for instance, then $\mathcal{R}_i = 0$, and there is no reflection at the interface.) As we mentioned before, the calculations (1.8) should be viewed as occurring pointwise at the junction interface itself, which does not occupy physical space.

Suppose that we define a set of *power-normalized* wave variables by

$$\underline{p}_i^l = \sqrt{Y_i}\, p_i^l \qquad\qquad \underline{p}_i^r = \sqrt{Y_i}\, p_i^r \tag{1.9}$$

(so-named since such variables have units of root power). Then the scattering operation (1.8) can be written, in matrix form, as

$$\begin{bmatrix} \underline{p}_i^l \\ \underline{p}_{i+1}^r \end{bmatrix} = \begin{bmatrix} \mathcal{R}_i & \sqrt{1 - \mathcal{R}_i^2} \\ \sqrt{1 - \mathcal{R}_i^2} & -\mathcal{R}_i \end{bmatrix} \begin{bmatrix} \underline{p}_i^r \\ \underline{p}_{i+1}^l \end{bmatrix} \tag{1.10}$$

Because the $\mathcal{R}_i$ are bounded in magnitude by 1, it is easy to see that scattering, in this case, corresponds to an orthogonal matrix transformation applied to the input wave variables.

Power Conservation at Scattering Junctions

At the junction between the ith and $(i+1)$th tubes, the continuity relations (1.5), when multiplied together, imply that

$$p_i u_i = p_{i+1} u_{i+1}$$

This is simply a statement of *conservation of power* at the interface. Using the definitions of traveling-wave variables from (1.6), we have that

$$\left(p_i^l + p_i^r\right) Y_i \left(p_i^l - p_i^r\right) = \left(p_{i+1}^l + p_{i+1}^r\right) Y_{i+1} \left(p_{i+1}^l - p_{i+1}^r\right)$$

---

[3]More generally, the reflection parameter is bounded by unity as long as the admittances are of the same sign; by convention, admittances are usually taken to be positive.

1.1. AN OVERVIEW OF SCATTERING METHODS

or, rearranging terms,

$$Y_i \left(p_i^l\right)^2 + Y_{i+1} \left(p_{i+1}^r\right)^2 = Y_i \left(p_i^r\right)^2 + Y_{i+1} \left(p_{i+1}^l\right)^2$$

In other words, the sum of the squares of the incident waves, weighted by their respective tube admittances, is equal to the same weighted square sum of the reflected waves. Assuming that the Y_i are positive, then, a weighted l_2 measure of the signal variables (pressure waves) is preserved through the scattering operation. This reflects the inherent *losslessness* of the tube interface.

In terms of the power-normalized variables defined by (1.9), and scattered according to (1.10), we will have (due to the orthogonality of the scattering matrix)

$$\left(\underline{p}_i^l\right)^2 + \left(\underline{p}_{i+1}^r\right)^2 = \left(\underline{p}_i^r\right)^2 + \left(\underline{p}_{i+1}^l\right)^2$$

Thus, the Euclidean norms of the incident and the reflected vectors of the power-normalized wave variables are the same.

Discrete-time Vocal-Tract Model

Now that we have discussed both the digital delay line representation of wave propagation within a single acoustic tube as well as the scattering that occurs at any junction between adjacent tubes, we are ready to present the full discrete-time model of the vocal tract. For an N tube model of the vocal tract, then, we will have the digital signal flow graph shown in Figure 1.5. Here, the scattering junctions are indicated by rectangles, marked by $\mathbf{S}_i$ (representing a matrix transformation of the form of (1.8) or (1.10), which is parameterized by $\mathcal{R}_i$, which itself depends on the adjoining tube admittances Y_i and Y_{i+1}).

The structure is driven at the left end by an input waveform (typically an impulse train for voiced speech or white noise for unvoiced speech, or a combination of the two), and an output speech waveform is emitted at the right end. The gray boxes, representing boundary conditions at the glottis and lips, we leave unspecified—such terminations can be modeled in a variety of ways [196].

Leaving aside a discussion of these boundaries, we can see that a single cycle in the recursive structure shown in Figure 1.5 (one pass through the main loop of the computer program that it implies) will involve two distinct steps:

- Wave variables incident on the junctions are *scattered*.

- The output waves are *shifted* to the inputs of the junctions immediately to the left and right.

We have already seen that the scattering operation preserves a weighted l_2 norm of the signal variables; it should be obvious that the shifting operation also does so, trivially (indeed, in the computer program, shifting amounts to no more than a permutation of the set of pressure signals stored in memory). Thus, we have a simple positive definite measure of the state of the tube in terms of signal values stored in the delay registers, which remains

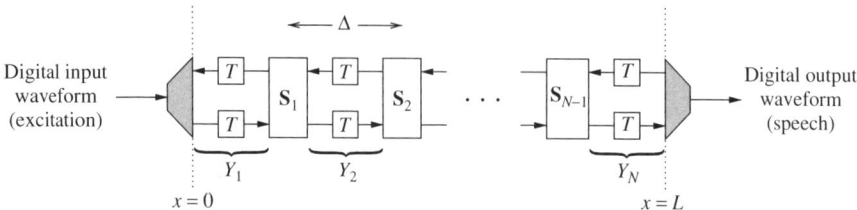

Figure 1.5: *Signal flow graph for an N tube vocal-tract model.*

constant as time progresses (again, excepting the effect of the boundary conditions). What is more, the *numerical stability* property of this structure is very easy to verify; we need only check that all the reflection coefficients $\mathcal{R}_i$ are bounded by 1 in magnitude, or equivalently, that all the admittances are positive. The excitation at the left boundary will, of course, introduce energy into the system, but we can at least be sure that signal energy is not being produced in the problem interior. The energy drain at the radiating (right) boundary is similarly localized.

Several other features are worthy of comment. First, we have treated the vocal tract here as a *static* or *time-invariant linear* (LTI) system. As we mentioned before, however, the configuration of the vocal tract must necessarily change during any utterance—these variations are assumed to be slow with respect to the frequency content of the excitation. The slow variation in the acoustic tube profile will cause shifts in the system resonances (formants), and these shifts will be perceived, by the listener, as phoneme transitions. In our discussion of scattering and energy conservation in the Kelly–Lochbaum model, we have not taken the time variation of the tube cross-sectional areas (and thus the reflection coefficients) into account. It should be clear, though, that if we are using the power-normalized signal variables defined by (1.9), then scattering defined by (1.10) at any junction remains an orthogonal (and thus norm-preserving) operation, even if the $\mathcal{R}_i$ are functions of time[4] [228]. Second, it is also simple to extend the model to include the nasal pathways (necessary for the production of certain vocal sounds, and also modeled as acoustic tubes [45]), without compromising overall losslessness. Third, we note that the stability of this model can be maintained even if the reflection coefficients $\mathcal{R}_i$ are quantized [228]—this will necessarily occur in any finite word-length machine implementation. As long as the quantized coefficients remain bounded by 1, we still have a perfectly lossless system. Signal quantization can also be performed so as to maintain overall stability, though the system will become more generally *passive* and not strictly lossless. Fourth, although the acoustic tube of varying cross-sectional area is often considered to be analogous to a lossless electrical transmission line of spatially varying inductance and capacitance, it is better thought of as a special case of the latter. For the acoustic tube, the local admittance varies directly with the cross-sectional area, but the wave speed γ remains constant; this is important, because for a given tube length of Δ, the time delay is dependent on the wave speed, from (1.4). For a transmission line, both the admittance and the wave speed may vary from point to point

---

[4]It should be said, however, that a time-varying acoustic tube is not, strictly speaking, a lossless system—energy is pumped into the system by the variations themselves. While the lossless time-varying concatenated acoustic tube model may be a useful signal-processing construct, it cannot be said to correspond to the numerical solution of a commonly known system of PDEs. We will revisit the full time-varying system more rigorously in Section 8.2.

1.1. AN OVERVIEW OF SCATTERING METHODS

along its length. We cannot then approximate the full transmission line by concatenated uniform transmission line segments in the same way as for the acoustic tube without losing synchronization of the resulting discrete-time structure (i.e., delay durations in the segments are not all the same). We will show how to approach this problem in Chapter 4.

Relationship to Digital Filters

Discrete-time structures such as that shown in Figure 1.5 are also used in digital filtering applications [180, 190], in which case the notion of a *spatial location* associated with a particular junction or delay element is often lost. For example, consider the digital filter structure shown in Figure 1.6(a). With $x(n)$ as a real discrete-time input sequence indexed by integer n, and $y(n)$ as the output sequence, this structure is called an *all-pole lattice filter* [180] when any of the types of section shown in Figure 1.6(b), (c) or (d) is used. T is the sample period, or unit delay, and the structure is parameterized by the constants k_i, $i = 1, \ldots, N$. It is possible to show that $x(n)$ and $y(n)$ are related by the familiar all-pole difference equation

$$y(n) = x(n) + \sum_{i=1}^{N} a_i y(n-i) \tag{1.11}$$

where the direct-form filter coefficients a_i, $i = 1, \ldots, N$ can be derived from the k_i through simple recursive procedures [180]. While the direct-form filter implementation implied by (1.11) requires fewer arithmetic operations than the lattice forms in Figure 1.6, the lattice implementation may be preferable because (a) stability is guaranteed by the simple condition $|k_i| < 1$, for all i, (determining stability by direct examination of the a_i is difficult, though it can of course be performed by finding the equivalent set of k_i parameters) and (b) pole locations are much less sensitive to coefficient quantization when applied to the k_i rather than the a_i. We also mention that the same structure also doubles as a useful *all-pass* filter design [180] when $x(n)$ is taken as the input and $w(n)$ as the output. It is also possible

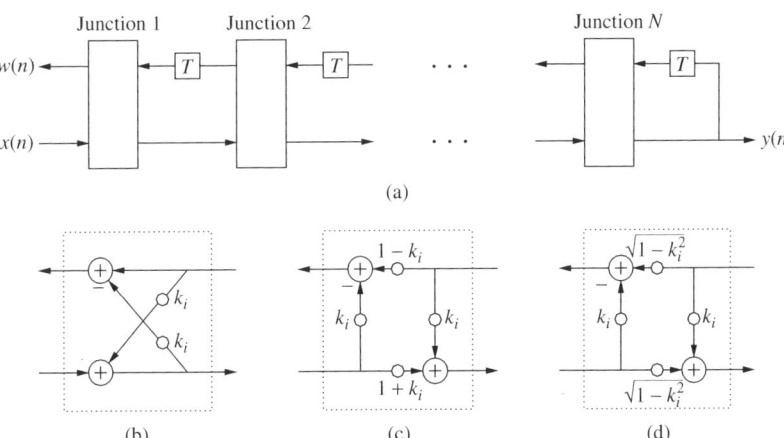

Figure 1.6: (a) *An all-pole lattice filter*, (b) *a standard lattice junction*, (c) *a Kelly–Lochbaum junction*, and (d) *a normalized lattice junction*.

to extend this filter design in order to implement any general stable pole-zero filter by summing readout taps from the leftward signal path into the output [190].

The structure of Figure 1.6(a) is quite similar to the Kelly–Lochbaum discrete-time acoustic tube model, but there are two minor differences. First, the Kelly–Lochbaum structure contains delay elements in both the leftward and rightward signal paths, reflecting the traveling-wave nature of the solution to the physical acoustic tube problem. In the lattice filter structure, however, the delays all occur in the upper (leftward) signal path. It is possible to transform the Kelly–Lochbaum structure into the lattice form by signal flow graph manipulations involving pushing delays through the junctions, combining them, and then downsampling by a factor of two—this can be done provided the acoustic tube model is terminated by a zero or infinite impedance at the right end [228]. (We remark that this downsampling operation can also be applied to DWNs in higher dimensions, in which case we will refer to it as *grid decimation*; we will examine grid decimation for a variety of grid arrangements in Appendix A.) Second, the Kelly–Lochbaum and normalized junctions in our treatment of the acoustic tube model differ slightly from the signal flow graphs shown in Figure 1.6(c) and (d). This difference is due to our choice of pressure waves instead of velocity waves as our signal set. While these quantities behave symmetrically in the one-dimensional acoustic tube, this symmetry is lost when we move to acoustics problems in higher dimensions, and it is more natural to work with pressure variables[5].

The same lattice structure is also arrived at in the analysis context when *linear predictive coding* (LPC) techniques are applied to a speech waveform [165]. The assumption underlying LPC is that speech can be treated as a source signal (such as a glottal waveform), filtered by the vocal tract, and the goal is to design an all-pole filter of the form of (1.11) that models the system resonances (or formant structure). Though this filter is obtained through purely autoregressive (i.e., nonphysical) analysis of a given measured speech signal, the reflection coefficients k_i (also known as *partial correlation* or *PARCOR* coefficients) are calculated as a by-product of the main calculation of the direct-form filter coefficients a_i. The k_i are identical to the $\mathcal{R}_i$ in the acoustic tube model, except for a sign inversion. This is not to say that the filter arrived at through LPC immediately implies a particular vocal-tract shape; it is best thought of as the solution to a filter design or system identification problem, devoid of any physical interpretation [196]. We note, though, that transmission line models such as the concatenated acoustic tube model have long been used for such system identification purposes in the *inverse scattering* context, in which case they are sometimes referred to as 'layer-peeling' or 'layer-adjoining' methods [35, 36, 285]. Provided certain assumptions are made about the glottal waveform and the effects of radiation on the measured speech waveform, it is possible to make some inferences about the vocal-tract shape [45].

1.1.3 Digital Waveguide Networks

The principal components of the Kelly–Lochbaum speech synthesis model, paired delay lines that transport wave signals in opposite directions, and the scattering junctions to which they are connected, are the basic building blocks of DWNs [228], and also the TLM method.

---

[5]Another reason for our choice of pressure waves is that when we transfer DWNs to the electrical framework in Chapter 4, then pressure waves become *voltage waves*, which are also the signal variables in the wave digital filtering literature.

1.1. AN OVERVIEW OF SCATTERING METHODS

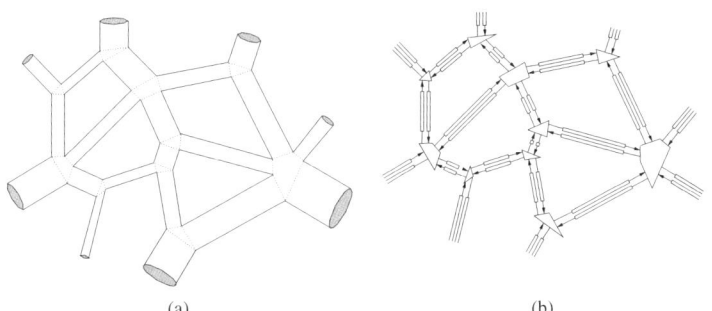

Figure 1.7: (a) *A portion of a general network of one-dimensional acoustic tubes and* (b) *its discrete-time realization using paired bidirectional delay lines and scattering junctions.*

Keeping within the acoustic tube framework, it should be clear that *any* interconnected network of uniform acoustic tubes can be immediately transferred to discrete time by modeling each tube as a pair of digital delay lines (or *digital waveguide*) with an admittance depending on its cross-sectional area[6]. At a junction where several tubes meet, these waves are scattered. See Figure 1.7 for a representation of a portion of a network of acoustic tubes, and its DWN equivalent.

The scattering operation performed on wave variables must be generalized to the case of the junction of M tubes, as shown in Figure 1.8. Though we will cover this operation in more detail in Chapter 4, and in the wave digital context in Chapter 2, we note that as for the case of the junction between two tubes, the scattering equations result from continuity requirements on the pressures and volume velocities at the junction. That is, if the pressures in the M tubes at the junction are p_j, and the velocities are u_j, $j = 1, \ldots, M$ (we now fix the sign of u_j to be positive if velocities are in the direction of the junction), then the relations are

$$p_1 = p_2 = \cdots = p_M \triangleq p_J \tag{1.12a}$$

$$u_1 + u_2 + \cdots + u_M = 0 \tag{1.12b}$$

In other words, the pressures in all the tubes are assumed to be identical and equal to some common pressure p_J at the junction, and the flows must sum to zero, by conservation of mass. These are the acoustic analogues of Kirchhoff's Laws for a parallel connection of M electrical circuit elements, where pressures are interpreted as voltages, and velocities as currents[7].

The pressures and velocities can be split into incident and reflected waves p_j^+ and p_j^- as per (1.2a) and (1.2b), by

$$p_j = p_j^+ + p_j^- \qquad\qquad u_j = Y_j \left(p_j^+ - p_j^- \right)$$

---

[6] In order for the network to be synchronic, or realizable as a recursive computer program, all the delay durations must be integer multiples of a common unit delay (the sampling period). Because the physical length of a digital waveguide is directly proportional to the delay (by a factor of γ, the wave speed), a synchronic DWN always corresponds to a network of acoustic tubes whose lengths are appropriately quantized.

[7] In the electrical setting, there is of course a dual set of laws describing a series connection, but there is no simple acoustic analogues for such a connection.

where Y_j is the admittance of the jth tube. The scattering relation, which can then be derived from (1.12), is

$$p_k^- = -p_k^+ + \frac{2}{\sum_{j=1}^M Y_j} \sum_{j=1}^M Y_j p_j^+ \qquad k = 1, \ldots, M \qquad (1.13)$$

and can be represented graphically as per Figure 1.8(b). It is worth examining this key operation in a little more detail. First, note that the scattering operation can be broken into two steps in the following way. Calculate the junction pressure p_J by

$$p_J = \sum_{j=1}^M \alpha_j p_j^+ \qquad \text{where} \qquad \alpha_j \triangleq \frac{2Y_j}{\sum_{j=1}^M Y_j} \qquad \text{for} \qquad j = 1, \ldots, M \qquad (1.14)$$

Then, calculate the outgoing waves from the incoming waves by

$$p_k^- = -p_k^+ + p_J \qquad k = 1, \ldots, M$$

Although (1.13) produces M output waves from M input waves, and can thus be written as an $M \times M$ matrix multiply, the number of operations is $O(M)$ (M multiplies and $2M - 1$ adds). Also note that the physical junction pressure is calculated, from (1.14), as a natural by-product of the scattering operation; because in a numerical integration setting, this physical variable is always what we are ultimately after, we may immediately suspect some link with standard differencing methods, which operate exclusively using such physical 'grid variables'. In Chapter 4, we examine the relationship between finite difference methods and DWNs in some detail.

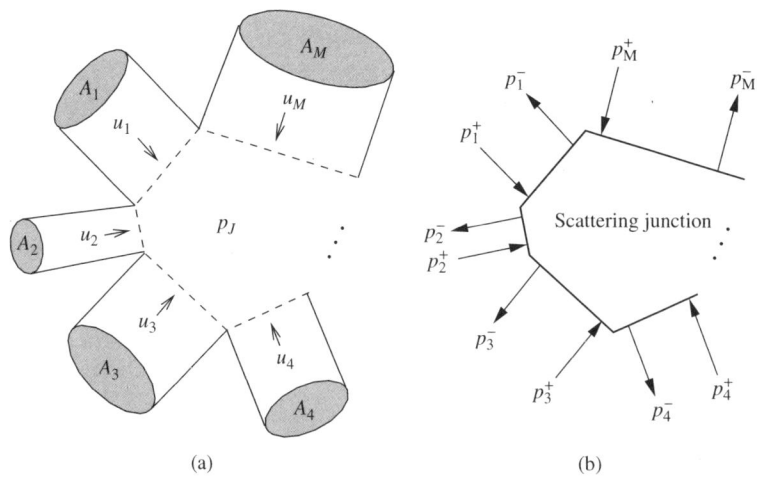

Figure 1.8: (a) A junction of M acoustic tubes, indicating the common pressures p_J and volume velocities u_j in the jth tube, $j = 1, \ldots, M$ at the junction and (b) a scattering junction relating outgoing pressure waves p_j^- to incoming waves p_j^+, $j = 1, \ldots, M$.

1.1. AN OVERVIEW OF SCATTERING METHODS

It is simple to show that the scattering operation also ensures that

$$\sum_{j=1}^{M} Y_j(p_j^+)^2 = \sum_{j=1}^{M} Y_j(p_j^-)^2 \qquad (1.15)$$

which is, again, merely a restatement of the conservation of power at a scattering junction. Notice that if *all the admittances are positive*, then a weighted Euclidean norm of the wave variables is preserved through the scattering operation. (If power-normalized variables are employed, then scattering is again equivalent to an $M \times M$ orthogonal matrix transformation.) The network as a whole will behave losslessly through the scattering and shifting operations, which constitute a single step in the global recursion that such a network implies.

Waveguide Meshes and the Wave Equation

The DWN shown in Figure 1.7(b) is unstructured; though the individual acoustic tubes are assumed to have lengths proportional to the delays in the resulting digital waveguides, they do not fall in any regular arrangement. In fact, although we have drawn what appears to be a network spanning two-dimensional space, we have not associated any physical coordinates with the various tube endpoints; it should be clear that losslessness of the digital structure is maintained regardless. At each step in the computer implementation of the DWN, signals are scattered, then shifted—the notion of 'where the signals are' is unimportant in this abstract setting.

Consider now a regular arrangement, or *mesh* [267] of acoustic tubes, as in Figure 1.9(a). The tubes are all of length Δ and admittance Y, and intersections of four tubes occur at grid points in a Cartesian coordinate system. The resulting DWN is shown in Figure 1.9(b); any scattering junction is linked to its four neighbors to the north, south, east, and west by bidirectional delay lines of delay $T = \Delta/\gamma$. We have indicated the scattering operation by

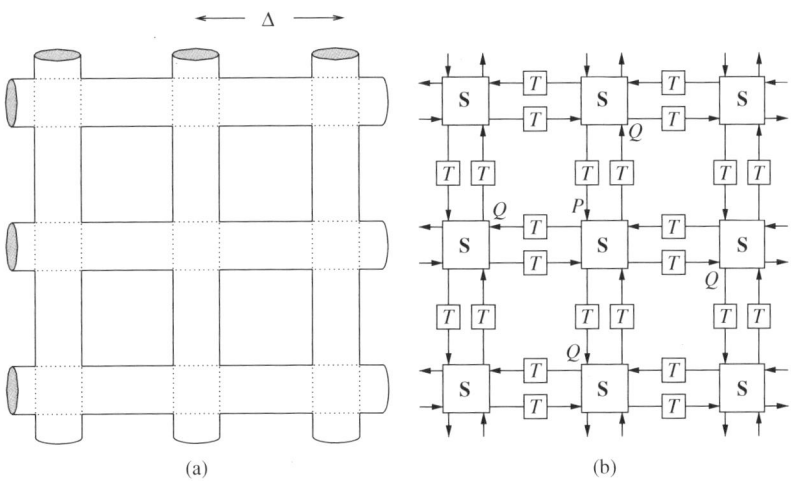

Figure 1.9: (a) *A regular mesh of acoustic tubes of equal admittances and* (b) *the associated digital waveguide network.*

the letter **S**. Because the admittances of all the tubes are identical, this scattering operation at any junction, from (1.13), has a particularly simple form:

$$\begin{bmatrix} p_N^- \\ p_S^- \\ p_E^- \\ p_W^- \end{bmatrix} = \frac{1}{2} \underbrace{\begin{bmatrix} -1 & 1 & 1 & 1 \\ 1 & -1 & 1 & 1 \\ 1 & 1 & -1 & 1 \\ 1 & 1 & 1 & -1 \end{bmatrix}}_{\mathbf{S}} \begin{bmatrix} p_N^+ \\ p_S^+ \\ p_E^+ \\ p_W^+ \end{bmatrix} \quad (1.16)$$

where p_j^+ and p_j^- are the incident and reflected pressure waves from direction j, $j = N, E, S, W$. Because the tube admittances are all identical, the scattering matrix **S** is orthogonal here, even if we are not using power-normalized waves. This is, incidentally, the exact form of the scattering equations that was proposed in two-dimensional TLM [44].

Referring to Figure 1.9(b), suppose we initialize this structure with a single incident pressure wave $p_N^+ = 1$ at location P, the north port of some junction. After scattering, the energy of the incident wave has been distributed among four reflected waves; from (1.16), we will have $p_S^- = p_E^- = p_W^- = 1/2$ and $p_N^- = -1/2$. After a delay of T seconds, these reflected waves are then shifted to the inputs of the four neighboring junctions, at points labeled Q. This process is then repeated, and over many time steps, signals will have propagated far from the original excitation at point P. At any time step, however, it should be clear that the sum of the squares of the signals in all the delay registers will be 1.

It is possible to view this propagation of signal energy (in a very rough sense) as a discrete time and space version of *Huygens' Principle* [53], an early description of diffraction phenomena: the advance of a wave front can be analyzed by considering each point on the wave front to be the generator of a secondary source of waves. A mesh of acoustic tubes, however, is far from a physical medium supporting multidimensional wave propagation, and a basic question then arises: Is this network of one-dimensional acoustic tubes approximating the behavior of a two-dimensional acoustic medium?

DWNs and Numerical Integration

To answer this question, let us consider the two-dimensional waveguide mesh at a junction with coordinates $x = i\Delta$ and $y = j\Delta$, for integer i and j. The discrete-time junction pressure $p_{J,i,j}(n)$ at time $t = nT$, for integer n (recall that our DWN operates with a sampling period of T), can be written in terms of the four incident wave variables at the same location, from (1.14), as

$$p_{J,i,j}(n) = \frac{1}{2}\left(p_{N,i,j}^+(n) + p_{E,i,j}^+(n) + p_{S,i,j}^+(n) + p_{W,i,j}^+(n)\right)$$

By tracing the propagation of the wave variables through the network backward in time through two time steps, it is in fact possible to write a recursion in terms of the junction pressures alone:

$$p_{J,i,j}(n) + p_{J,i,j}(n-2) = \frac{1}{2}\Big(p_{J,i-1,j}(n-1) + p_{J,i+1,j}(n-1) + p_{J,i,j-1}(n-1)$$

$$+ p_{J,i,j+1}(n-1)\Big) \quad (1.17)$$

1.1. AN OVERVIEW OF SCATTERING METHODS

Assume, for the moment, that these discrete time and space junction pressure signals are in fact samples of a continuous function $p(x, y, t)$ of x, y and t. Expanding the terms in the recursion above in Taylor series about the location with coordinates $x = i\Delta$ and $y = j\Delta$, at time $t = (n-1)T$ gives

$$T^2 \left.\frac{\partial^2 p}{\partial t^2}\right|_{x,y,t-T} + O(T^4) = \frac{\Delta^2}{2}\left(\frac{\partial^2 p}{\partial x^2} + \frac{\partial^2 p}{\partial y^2}\right)\bigg|_{x,y,t-T} + O(\Delta^4)$$

Recalling that $\Delta = \gamma T$, where γ is the speed of wave propagation in the one-dimensional tubes, and discarding higher-order terms in T and Δ (they are assumed to be small), we get

$$\frac{\partial^2 p}{\partial t^2} = \tilde{\gamma}^2 \left(\frac{\partial^2 p}{\partial x^2} + \frac{\partial^2 p}{\partial y^2}\right) \qquad (1.18)$$

This is simply the two-dimensional wave equation, with the wave speed $\tilde{\gamma}$ defined by

$$\tilde{\gamma} = \gamma/\sqrt{2} \qquad (1.19)$$

This equation describes wave propagation in a lossless two-dimensional acoustic medium, and the DWN of Figure 1.9(b) can thus be considered to be a numerical integrator of this equation, assuming the wave speeds in the tubes are set according to (1.19); the discrete Huygens' Principle interpretation of the behavior of the mesh is justified, at least in the limit as T and Δ become small[8]. The recursion (1.17) in the junction pressures, however, can be seen as a simple finite difference scheme that could have been derived directly from (1.18) by replacing the partial derivatives by differences between values of a grid function $p_{i,j}(n)$ on a numerical grid. Because the DWN operates using wave variables, we can see that the DWN is simply a different organization of the same calculation—in particular, it has been put into a form for which all operations (scattering, and shifting) rigidly enforce conservation of energy, in a discrete sense.

We can also reconsider the Kelly–Lochbaum model in this light; forgetting, for the moment, about the approximation of the tube by a series of concatenated *uniform* tubes, it is possible to write the equations of motion for the gas in the tube directly [196] as

$$\frac{\rho}{A(x)}\frac{\partial u}{\partial t} + \frac{\partial p}{\partial x} = 0 \qquad (1.20a)$$

$$\frac{A(x)}{\rho\gamma^2}\frac{\partial p}{\partial t} + \frac{\partial u}{\partial x} = 0 \qquad (1.20b)$$

subject to initial conditions and boundary conditions at the glottis and lips. This system is identical in form to (1.1) for a uniform tube, except for the variation in x of the cross-sectional area. It can be condensed to a single second-order equation in the pressure alone,

$$\frac{\partial^2 p}{\partial t^2} = \frac{\gamma^2}{A(x)}\frac{\partial}{\partial x}\left(A(x)\frac{\partial p}{\partial x}\right) \qquad (1.21)$$

---

[8]What we have done, in the jargon of numerical integration methods, is to show the *consistency* [238] of the waveguide mesh with the two-dimensional wave equation.

which is sometimes called *Webster's horn equation* [20, 45, 94]. Owing to the variation in
the cross-sectional area, it is not equivalent to the one-dimensional wave equation (1.3), and
does not possess a simple solution in terms of traveling waves (which is why we needed
a concatenated uniform tube model in the first place). Returning now to the DWN of
Figure 1.5, it can be shown that the junction pressures $p_{J,i}(n)$ (at spatial locations $x = i\Delta$
and at time $t = nT$ for i and n integer) satisfy a recursion of the form

$$p_{J,i}(n) + p_{J,i}(n-2) = \frac{2}{Y_i + Y_{i+1}} \Big(Y_i p_{i-1}(n-1) + Y_{i+1} p_{i+1}(n-1) \Big) \qquad (1.22)$$

where Y_i is the admittance of the ith acoustic tube, running from $x = (i-1)\Delta$ to $x = i\Delta$.
With the Y_i set according to (1.7), it is again possible to show that (1.22) is a finite difference
scheme for (1.21), with $\Delta = \gamma T$.

1.1.4 A General Approach: Multidimensional Circuit Representations and Wave Digital Filters

For the Kelly–Lochbaum vocal-tract model, it is straightforward to arrive at a numerical
scattering formulation of the problem; the approximation of a smoothly varying tube by
a series of concatenated tubes is intuitively satisfying, and leads immediately to a wave
variable numerical solution to Webster's equation. The identification of the mesh of one-
dimensional tubes of Figure 1.9(a) as a numerical solver for the two-dimensional wave
equation is more difficult, because it is by no means clear that such a mesh behaves like
a two-dimensional acoustic medium. Although as we have seen, it is possible to prove
(through a finite difference treatment) that the tube network is indeed solving the right
equation, we have not shown a way of *deducing* such a structure from the original defining
PDE system. If one wants to develop a DWN for a more complex system (such as a stiff
vibrating plate of variable density and thickness, for example), then guesswork and attempts
at invoking Huygens' Principle will be of limited use.

The scattering operation we introduced in Section 1.1.2 and Section 1.1.3 is at the
heart of all the numerical methods we will discuss in this book, whether they are based
on DWNs or WDFs, which we will introduce shortly. A given system of PDEs is numerically solved by filling the problem domain with scattering nodes, or junctions, such as that
shown in Figure 1.8(b), which calculate reflected waves from incident waves according
to (1.13) (or its series dual form). The topology of the network of interconnected junctions will be dependent on the particulars of the system we wish to solve. As we have
seen, these scattering junctions act as power-conserving signal-processing blocks, and in
a DWN, they are linked by discrete-time acoustic tubes, or transmission lines, which are
also power-conserving, and serve to transport energy from one part of the network to
another. The key concept here is the losslessness of the network components, which is
dependent on the positivity of the various element values (admittances); as we have seen,
this positivity condition ensures that some weighted Euclidean norm of the signals in the
discrete-time network will remain constant as time progresses. In other words, the simulation routine that such a network implies is guaranteed to be stable by enforcing this
condition.

Wave digital filters are also based on the idea of preserving losslessness (and more
generally passivity) in a discrete-time simulation of a physical system, though the approach

1.1. AN OVERVIEW OF SCATTERING METHODS

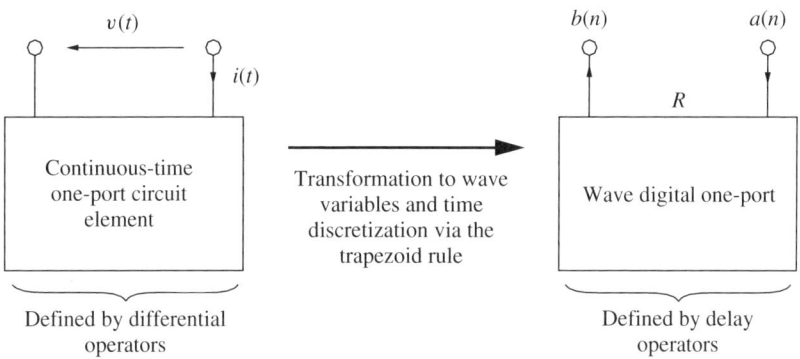

Figure 1.10: *Wave digital discretization of a one-port circuit element.*

is somewhat different from what we have just seen. As they were originally intended to transfer analog electrical filter (RLC) networks to discrete time, it is best to begin by looking briefly at lumped circuit elements. A one-port element [14], such as that shown on the left in Figure 1.10 is characterized by a voltage v, and a current i, both of which are functions of time t. In the time domain, the one-port generally relates $v(t)$ and $i(t)$ through some combination of differential or integral operators. If the one-port (or more generally, N-port) is linear and time-invariant, then there is a simple description of its behavior in the frequency domain, but we will wait until Chapter 2 before entering into the details. An analog filter is simply an interconnected network of such elements; it is operated by applying a voltage at one pair of free terminals, and then reading the filtered output at another pair. In particular, if the network is made up of passive elements such as resistors, capacitors, inductors, and so on, then it must behave as a stable filter.

Fettweis [68] developed a procedure for mimicking the energetic behavior of an analog filtering network in discrete time. The input and filtered output become digital signals, and the filtering network becomes a recursion, to be realized as a computer program. Most importantly, the digital network has the same topology as the analog network, and can be thought of as its discrete-time 'image.' One-ports (or more generally N-ports) are first characterized in terms of *wave variables*,

$$a = v + iR$$
$$b = v - iR$$

where R is some arbitrary positive constant, assigned to the particular one-port, called a *port resistance*. The continuous-time element, described by differential operators, is then replaced by a discrete-time element operating on digital signals, and composed of algebraic operations and delay operators or shifts. The signal $a(n)$ is called the input wave, and $b(n)$ is the output wave; both are discrete-time sequences indexed by integer n. If the discretization procedure is carried out in an appropriate way (to be more precise, if differentiation is approximated by the trapezoid rule of numerical integration), then the resulting *wave digital one-port* has energetic properties very similar to the continuous element from which it is derived. In particular, if the analog element is passive (lossless), then the wave digital element can be considered to be passive (lossless) in a similar sense. In fact, if a wave digital circuit element is composed of delay operators (hence requiring memory), then a weighted sum of the squares of the signal values stored in the element's delay registers is the direct counterpart to the physical energy stored in the electric and magnetic fields surrounding the corresponding analog element. The passivity property is contingent on the positivity of the port resistance; given this constraint, it can often be chosen such that there is no delay-free path from the input $a(n)$ to the output $b(n)$. This, as we will see, has important implications for explicit computability in simulation routines.

Consider a parallel connection of two one-port circuit elements, as shown in Figure 1.11(a). The one-ports are defined by some relationship between their respective voltages and currents, which we will write as v_1, i_1 and v_2, i_2. For such a parallel connection, *Kirchhoff's Laws* dictate that

$$v_1 = v_2 \qquad\qquad i_1 + i_2 = 0 \qquad\qquad (1.23)$$

(We could equally well treat this as a series connection, by reversing the directions of the arrows that define v_2 and i_2 in Figure 1.11(a).) We can now define two sets of wave variables at the two one-ports by

$$a_1 = v_1 + i_1 R_1 \qquad\qquad a_2 = v_2 + i_2 R_2$$
$$b_1 = v_1 - i_1 R_1 \qquad\qquad b_2 = v_2 - i_2 R_2$$

In the scattering formulation, the Kirchhoff connection is treated as a *separate two-port element*, with inputs a'_1 and a'_2 and outputs b'_1 and b'_2. These are simply the outputs and inputs, respectively, of the one-ports, as shown in Figure 1.11(b).

Kirchhoff's Laws for the parallel connection can then be rewritten in terms of the wave variables as

$$b'_1 = \mathcal{R} a'_1 + (1 - \mathcal{R}) a'_2 \qquad\qquad (1.24a)$$
$$b'_2 = (1 + \mathcal{R}) a'_1 - \mathcal{R} a'_2 \qquad\qquad (1.24b)$$

where the reflection coefficient $\mathcal{R}$ is defined by

$$\mathcal{R} \triangleq \frac{R_2 - R_1}{R_2 + R_1}$$

Equations (1.24) define a wave digital two-port *parallel adaptor*. They are identical in form to the equations defining a parallel junction of two acoustic tubes, from (1.8)—this is to be expected, since Kirchhoff's Laws (1.23) are equivalent to the pointwise continuity equations

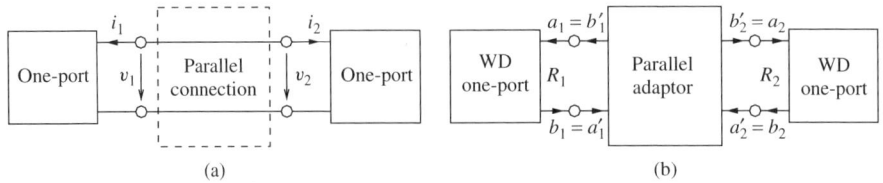

Figure 1.11: (a) *Parallel connection of two continuous-time one-ports and* (b) *its wave digital counterpart.*

1.1. AN OVERVIEW OF SCATTERING METHODS

(1.5) at an acoustic junction. Thus, all comments we made about scattering junctions in Section 1.1.2 hold for the wave digital adaptor as well—in particular, if we define power-normalized waves, then the scattering operation again is equivalent to an orthogonal (i.e., l_2 norm-preserving) transformation, as long as the port resistances R_1 and R_2 are chosen positive (implying, again, that $|\mathcal{R}| < 1$).

WDFs and the Numerical Integration of ODEs

In the closed network of Figure 1.11, we have left the two one-ports unspecified. Suppose we connect an inductor of constant inductance $L > 0$ at the left-hand port and a capacitor of constant capacitance $C > 0$ at the right-hand port, as shown in Figure 1.12(a). Then the voltage–current relations are defined by

$$v_1 = L\frac{di_1}{dt} \qquad i_2 = C\frac{dv_2}{dt} \qquad (1.25)$$

When these relations are closed by Kirchhoff's parallel connection rules (1.23), it is possible to write a single second-order ODE describing the time evolution of the circuit state,

$$\frac{d^2 w}{dt^2} = -\frac{1}{LC} w$$

where $w(t)$ stands for any of the voltages or currents in the network. This network thus behaves as a harmonic oscillator, of frequency $1/\sqrt{LC}$; the voltages and currents, assumed real, evolve according to

$$w(t) = A \cos(t/\sqrt{LC}) + B \sin(t/\sqrt{LC})$$

for some arbitrary constants A and B determined by the initial voltages and currents in the network.

Though we have not explicitly derived the forms of the wave digital inductor and capacitor, this is a good opportunity to see what these elements look like—the wave digital network corresponding to the LC harmonic oscillator circuit is shown in Figure 1.12(b). (The reader may glance ahead to Section 2.2.4 for a glimpse of how these forms are derived.) We have a parallel adaptor, which is a digital signal-processing block defined by equations (1.24), terminated on delay elements (one of which incorporates a sign inversion). Special choices of the port resistances R_1 and R_2 (marked in the figure) were chosen in

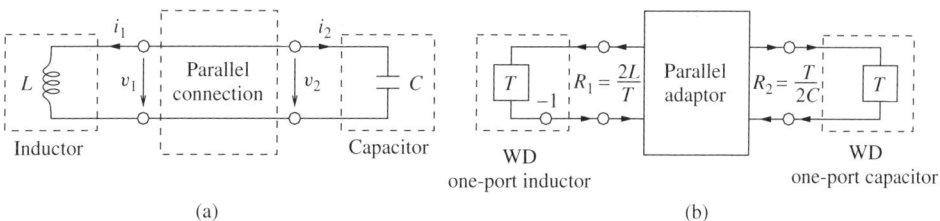

Figure 1.12: *The LC harmonic oscillator—(a) a parallel connection of an inductor, of inductance L and a capacitor of capacitance C and (b) the corresponding wave digital network.*

order to obtain these simple signal flow graphs. This diagram implies a recursion, which, like DWN methods consists of a scattering step, and a delay step (possibly with sign inversion). Because it makes use of only two delay operators, it should be obvious that this simple network must behave as a *two-pole resonator*—the discrete-time counterpart to the continuous-time harmonic oscillator. The wave digital network thus functions as a *numerical integrator*.

Although this example is very simple, the same ideas can be applied to large networks, and the result is always an explicitly recursible structure for which passivity can be simply guaranteed. We will return to the harmonic oscillator for further discussion in Section 2.3.

Multidimensional WDFs as PDE Simulators

Wave digital filter networks are derived from lumped analog circuits, and we have seen that they can be interpreted as numerical ODE integrators. Most importantly, we saw that a given analog circuit immediately implies a corresponding WDF structure; if the original circuit is lossless, then the WDF network, which is its discrete-time image, will be lossless as well. It is easy to extend the maintenance of losslessness to the more general case of passivity (i.e., we allow our networks to dissipate energy, as well as recirculate it).

Fettweis and Nitsche [90] found a way of directly extending this simulation technique to distributed systems. First, it is necessary to generalize the definition of a circuit element to multiple dimensions, in which case it is called an *MD circuit element*; an MD one-port is shown on the left in Figure 1.13. The one-port is still defined in terms of a voltage v and current i across its terminals, but these quantities are now, more generally, functions of an n-dimensional spatial coordinate $\mathbf{x}$ as well as time t. In particular, v and i will be, in general, related by partial differential operators. Though the representation is the same as in the lumped case, this circuit element is itself a *distributed* object, occupying physical space. Such a distributed circuit element is merely a generalization of a lumped one-port circuit element; it should not be conceived of as a physical entity. The rules of classical network theory, however, (and in particular Kirchhoff's connection rules), can still be applied in order to form combinations of such objects.

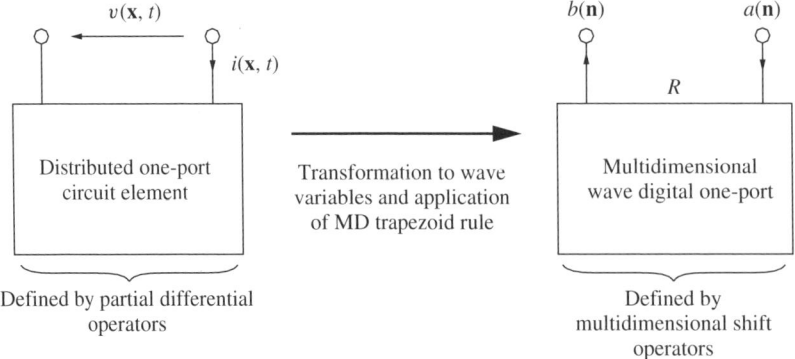

Figure 1.13: *Multidimensional wave digital discretization of a distributed one-port circuit element.*

1.1. AN OVERVIEW OF SCATTERING METHODS

It is also possible to extend the notions of passivity and losslessness to multiple dimensions and to introduce wave variables, which, like the voltages and currents, will also be distributed quantities. Finally, it is possible to discretize these elements in such a way that this passivity is retained in the discrete time and space domain (through the use of the trapezoid rule in multiple dimensions). The result is the MDWD element shown on the right of Figure 1.13. Just as for the lumped case, where differential operators are mapped to *delays*, here partial differential operators are mapped to *shifts* in the discrete multidimensional problem space. We again have an input wave a and an output wave b, which take on values at a discrete set of locations; these are to be interpreted as *grid functions* over a set of points, indexed by an integer-valued vector **n**.

Though we will discuss the MDWD discretization procedure in much more detail in Chapter 3, we outline the basic steps in Figure 1.14. Beginning from a given passive physical system, we first model it with a suitable system of PDEs. It may be possible, then, to interpret the individual equations as *loop equations* in a closed multidimensional Kirchhoff circuit (MDKC) made up of elements of the form shown on the left of Figure 1.13. Typically, the *currents* flowing through the 'wires' in such a network will be the dependent, or state variables describing the physical system; all the partial differential operators are then consolidated in the various elements. For a first-order system of PDEs, it will usually be true that the number of equations is equal to the number of loops in the circuit. It is important, at this stage, to ensure that such a network representation is composed of multidimensional circuit elements that are individually passive—this can generally be determined by a cursory examination of the circuit element values (such as inductances, capacitances, etc., which may be functions of several variables). Once the work of manipulating the system into a suitable circuit form is complete, the discretization step is immediate, and a multidimensional wave digital network results; if the MDKC is made up of passive elements, then the discrete network will be as well. It can then be interpreted as a *stable explicit numerical integration scheme* for the original defining system of PDEs. The basic operations will be, just as for DWNs, the scattering and shifting of wave variables through a numerical grid of nodes. The resulting structures, however, differ markedly from DWNs in many ways, though they can still be viewed as finite difference schemes.

We note that each of the various steps (i.e., the arrows in Figure 1.14) involves a good deal of choice on the part of the algorithm designer. For a given system, there is almost always a variety of PDE systems that could serve as adequate models; not all are suitable for

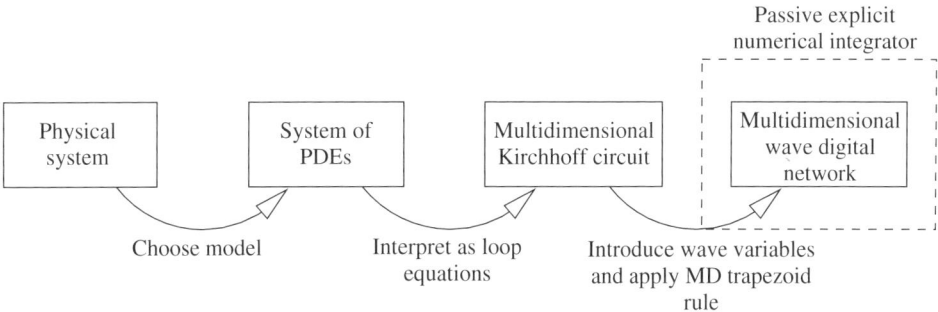

Figure 1.14: *Steps in the construction of a multidimensional wave digital filtering simulation routine.*

circuit-based discretization. It is also true that for a given system of PDEs there is no unique network representation (though they should all be related by equivalence transformations from classical network theory). Finally, though Fettweis et al. make use of the MD trapezoid rule as a means of arriving at a passive discrete network, this is by no means the only way of proceeding—many integration rules possess the desired passivity-preserving properties. We will explore the consequences of these choices extensively throughout the rest of this book.

1.2 Questions

Before proceeding to the main development, it is worthwhile to take a step back and view the underlying motivations for writing this book. The overall goal can be expressed as an attempt to answer, or at least address, several general questions about wave digital filter and digital waveguide network numerical simulation algorithms:

- *To what types of systems can wave digital and digital waveguide network simulation approaches be applied?*

- *What features do these two methods share, and what distinguishes them?*

- *What are their relative advantages?*

- *Can they be unified in a formal way?*

These were the guiding questions that the author had in mind while writing this book. The impatient reader can flip to Section 9.1 for the answers—some clear-cut, some much less so.

Chapter 2

Wave Digital Filters

The entry point, for any study of numerical methods based on wave and scattering ideas, must necessarily be a review of *wave digital filters* (WDFs) [62, 68]. This filter-design technique, proposed by Alfred Fettweis in the early 1970s, was an attempt at translating analog filters into the digital realm, with a pointed emphasis on preserving as much of the underlying physics as possible. In particular, a digital filter structure arrived at through Fettweis's procedure has the same precise network topology and energetic properties as the lumped analog electrical circuit (called the *reference circuit*) from which it is derived. One might also argue that waveguide networks or TLM could also serve as such a starting point for this study, but the WDF framework, with its fundamental emphasis on energetic properties offers a more global and comprehensive perspective (and, as we will see, contains these transmission line based methods as a special case).

The theory is straightforward; analog circuit components (N-port devices or elements), usually defined by a voltage–current relation, are first given an equivalent characterization in terms of *wave variables*. While this is merely a change of variables, it has the advantage of allowing an alternate description of the dynamic behavior of the network: the energy incident on a circuit element (incident from the rest of the network to which it is connected through a *port*) may be reflected back from the element through the same port, transmitted through to another part of the network through a different port, or stored within the element itself. The incident, reflected, transmitted, and stored energies are carried by *waves*. The reflectances and transmittances themselves are determined by arbitrary positive constants called *port resistances* that are assigned to individual wave ports. An important result of using wave variables is that the entire network may then be parameterized by these reflection and transmission coefficients that are, at least for passive networks, bounded independently of the numerical circuit element values themselves (inductances, capacitances, and resistances etc.), which may vary over a wide range.

The true advantage of using wave variables becomes much more tangible when we seek to obtain, from a given analog filter design, a digital filter structure. This is usually done in the WDF context at steady state via a particular type of *bilinear transformation* or *spectral mapping* [62] from continuous to discrete frequency variables. Unless wave variables are employed, the resulting filter structure will usually not be recursively computable, and hence not directly implementable as a computer program without matrix inversion routines.

Wave and Scattering Methods for Numerical Simulation   Stefan D. Bilbao
© 2004 John Wiley & Sons, Ltd   ISBNs: 0-470-87017-6

In addition, because the reflectances and transmittances of the network (which become the filter multiplier values) are bounded in a simple way, a host of desirable filter properties result, which are especially valuable in a fixed-point computer implementation: complete elimination of certain types of limit cycles or parasitic oscillations and very low sensitivity of the filter response to coefficient truncation are the most frequently mentioned [68]. A further advantage stems from the fact that the network topology of the reference circuit has been inherited by the digital filter structure, and we thus have convenient access to a simple energy measure for the discrete dynamical system; this energy, which is a direct analogue of the energy stored in the electrical and magnetic fields surrounding the reference circuit, may be used as a discrete-time Lyapunov function [56, 63] in order to provide further rules for dealing with the inevitable truncation of the filter state in a fixed-point implementation.

Many of the underlying ideas, however, had existed for some time before they coalesced into Fettweis's digital filter-design technique. In fact, it is perhaps best to describe wave digital filtering not as an unprecedented invention, but as the synthesis of two principal preexistent ideas. The crucial wave variable and scattering concepts were borrowed from microwave filter design [13, 14], and digital structures based on the reflection and transmission of waves had appeared previously, especially in 'layer-peeling' and 'layer-adjoining' methods for solving inverse problems that arise in geophysics [35, 36, 285], and in models of the human vocal tract used in the analysis and synthesis of speech [142, 196], as we saw in Section 1.1.2. Many other digital filter structures make use of similar ideas, and have similar useful properties—among these are digital ladder and lattice forms [109], normalized filters [110], and orthogonal filters. This last type of structure has been formally unified with WDFs in [260]. The other cornerstone of wave digital filtering, the concept of a continuous-to-discrete spectral mapping that is, in some sense, energy preserving, was not new to circuit discretization approaches. It appeared in the 1960s in the numerical analysis community that was concerned with the stability of the discretization of sets of ordinary differential equations (ODEs); indeed, wave digital filtering can be thought of as an *A-stable* [49, 93, 106, 166] numerical method that discretizes the defining differential equations of an analog electrical network.

Wave digital filtering has, since its inception, developed in many directions, and has become a large subfield of the vast expanse of digital filter design. Because this book is devoted to the use of wave digital filters for simulation purposes, and not for filtering, this introductory chapter is intended merely to motivate material in the sequel, and to provide enough basic information for the reader to understand the WDF symbology (which is, unfortunately, somewhat idiosyncratic and takes a bit of getting used to). Indeed, many filtering issues do not arise at all in a simulation setting, at least from the point of view of traditional numerical analysis. The single best WDF review paper is certainly that of Fettweis [68], which is filled with practical filter-design information and references. We briefly mention that some lines of development have been in the areas of multirate systems and filter banks [82, 158, 248] and in many areas in time-varying and nonlinear systems (to be discussed in detail in Chapter 8). Another important direction has been the generalization of WDFs to the multidimensional case [90], which are the subject of the next chapter.

Summary

This chapter is intended as a review of lumped wave digital filters, minus any discussion of filtering applications, since we will only be looking at simulation applications in the remainder of the book. Because these concepts are used extensively throughout the rest of the book, the reader is advised to begin here, even though discussion of numerical methods for PDE solving does not begin in earnest until the next chapter. We follow the standard development (as in, say, Fettweis's comprehensive review paper [68], which is the chief reference for this chapter) and begin with a brief introduction to the theory of electrical N-port devices [14], and, in particular, the key concept of *passivity*, which later plays a pivotal role as the stability criterion for multidimensional simulation networks. We then review the basics of the lumped wave digital discretization procedure, involving the use of a passivity-preserving continuous-to-discrete spectral bilinear transformation (the trapezoid rule in the time domain) and the transformation to wave variables. The wave digital counterparts of the standard circuit elements (capacitors, inductors, resistors, transformers, etc.,) are then introduced, as are adaptors, which are simply the wave variable counterparts to Kirchhoff's series and parallel connection rules. We then continue with a brief description of finite word-length arithmetic properties of WDFs, and a look at some specialized vector elements that will come in handy later (and are in fact necessary) for the simulation of some elastic dynamic systems. Finally, we examine the application of WDFs to the numerical integration of a very simple system of ODEs. It is important to keep in mind that though we only discuss lumped elements and networks in Chapter 2, the basic set of construction rules (essentially classical electrical network theory) remains unchanged when we move to a multidimensional setting in the next chapter.

2.1 Classical Network Theory

2.1.1 N-ports

Classical network theory [14] is partly concerned with the properties of connections of N-port devices. In the abstract, an N-port is a mathematical entity whose internal behavior is only accessible through its N ports. With the jth port is associated a *current* i_j, a *voltage* v_j, and two *terminals* (see Figure 2.1). The two terminals of any port must always be connected to the terminals of another port. A *network* is simply a collection of N-ports connected such that no port is left free[1].

For lumped networks, the voltages, currents, and possibly element values in the networks are allowed to be real-valued functions of a sole real parameter t that is usually interpreted

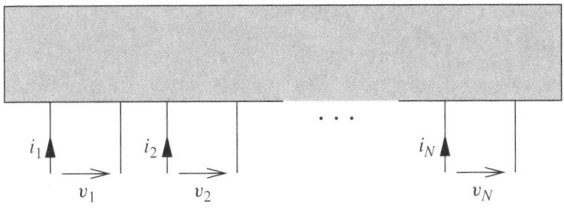

Figure 2.1: *N-port*.

---

[1] More generally, network theory is concerned with the so-called *t-terminal* or *multipole networks* [278], for which terminals are not necessarily associated in pairs.

as physical time. Multidimensional networks [280] are more general in the sense that the voltages, currents, and port resistances may be functions of one or many other parameters, which may represent spatial dimensions. In this introductory chapter, we will be concerned only with lumped networks, but it should be kept in mind that Chapters 3, 6, 7, and 8 are devoted chiefly to a particular class of multidimensional network that can represent the behavior of a distributed physical system.

If an N-port is *linear* and *time-invariant* (LTI), then the port quantities may exhibit a purely exponential time dependence at a single complex frequency s. For such an exponential state [14], it is also useful to define, for any port with voltage $v(t)$ and current $i(t)$, the complex amplitudes $\hat{v}$ and $\hat{i}$. We can then write

$$v(t) = \hat{v}e^{st} \qquad\qquad i(t) = \hat{i}e^{st}$$

Under certain conditions [14], an LTI network will possess an $N \times N$ *impedance matrix*[2] $\mathbf{Z}$, so that the steady state voltages and currents are related by

$$\hat{\mathbf{v}} = \mathbf{Z}\hat{\mathbf{i}} \qquad (2.1)$$

where $\hat{\mathbf{v}}$ and $\hat{\mathbf{i}}$ are the column N-vectors containing the amplitudes $\hat{v}_1, \ldots, \hat{v}_N$ and $\hat{i}_1, \ldots, \hat{i}_N$ respectively. In general, if the N-port contains elements that behave as differential or integral operators, then we will have $\mathbf{Z} = \mathbf{Z}(s)$. The *admittance* of such an N-port is defined as

$$\mathbf{Y} = \mathbf{Z}^{-1}$$

at frequencies s for which $\mathbf{Z}$ is invertible.

In many cases of interest, the entries of $\mathbf{Z}(s)$ will be rational functions of s. An N-port so defined is called *real* if the coefficients of these rational functions are real numbers. In this case, there is no loss in generality [14] in considering the port voltages and currents to be real-valued functions of t, in which case we may write, for an exponential state,

$$v(t) = \operatorname{Re}\left(\hat{v}e^{st}\right) \qquad\qquad i(t) = \operatorname{Re}\left(\hat{i}e^{st}\right)$$

Now $v(t)$ and $i(t)$ are referred to as the real instantaneous port voltage and current respectively.

2.1.2 Power and Passivity

The total instantaneous power absorbed by a real N-port is defined by

$$w_{\text{inst}}(t) \triangleq \sum_{j=1}^{N} v_j(t) i_j(t) \qquad (2.2)$$

---

[2] A linear and time-invariant N-port need not have an impedance; the ideal transformer, for example, does not. In such cases, a more general 'hybrid' matrix [14], from which all relevant properties may be deduced, can be defined. We will make special use of hybrid forms in Chapter 6.

2.1. CLASSICAL NETWORK THEORY

where $v_j(t)$ and $i_j(t)$ are the real instantaneous voltage and current at port j. In general, for an N-port that contains stored energy $E(t)$, which dissipates energy at rate $w_d(t)$, and which contains sources that provide energy at rate $w_s(t)$, the energy balance

$$\int_{t_1}^{t_2} (w_{\text{inst}} + w_s - w_d) \, dt = E(t_2) - E(t_1) \tag{2.3}$$

must hold over any interval $[t_1, t_2]$. Such an N-port is called *passive* if we have

$$\int_{t_1}^{t_2} w_{\text{inst}} \, dt \geq E(t_2) - E(t_1) \tag{2.4}$$

over any time interval; in other words, the increase in stored energy must be less than the energy delivered through the ports. The N-port is called *lossless* if (2.4) holds with equality over any interval.

For an LTI N-port in an exponential state of complex frequency s, we can define the *total complex power* absorbed to be the inner product

$$w = \hat{\mathbf{i}}^* \hat{\mathbf{v}}$$

and the *average* or *active power* as

$$\overline{w} = \text{Re}(\hat{\mathbf{i}}^* \hat{\mathbf{v}})$$

where $*$ denotes transpose conjugation. For an N-port defined by an impedance relationship, we may immediately write, in terms of the voltage and current amplitudes,

$$\overline{w} = \text{Re}\left(\hat{\mathbf{i}}^* \hat{\mathbf{v}}\right) = \frac{1}{2}\left(\hat{\mathbf{i}}^* \hat{\mathbf{v}} + \hat{\mathbf{v}}^* \hat{\mathbf{i}}\right) = \frac{1}{2}\left(\hat{\mathbf{i}}^* \mathbf{Z} \hat{\mathbf{i}} + \hat{\mathbf{i}}^* \mathbf{Z}^* \hat{\mathbf{i}}\right) = \frac{1}{2}\hat{\mathbf{i}}^* \left(\mathbf{Z} + \mathbf{Z}^*\right) \hat{\mathbf{i}}$$

For such a real LTI N-port, passivity may be defined in the following way. If the active power absorbed by an N-port is always greater than or equal to zero for frequencies s such that $\text{Re}(s) \geq 0$, then it is called *passive*. This implies that

$$\mathbf{Z} + \mathbf{Z}^* \geq \mathbf{0} \qquad \text{for} \qquad \text{Re}(s) \geq 0 \tag{2.5}$$

A matrix $\mathbf{Z}$ with such a property is called a positive matrix. In the present case of a real N-port, $\mathbf{Z}$ is called positive real (though in general, positivity is all that is required for passivity). If the average power absorbed is identically zero for $\text{Re}(s) = 0$, or, in terms of impedances, if

$$\mathbf{Z} + \mathbf{Z}^* = \mathbf{0} \qquad \text{for} \qquad \text{Re}(s) = 0$$

then the N-port is called *lossless*.

2.1.3 Kirchhoff's Laws

Connections between individual ports can be made through an appeal to *Kirchhoff's Laws*, which specify two important connection rules. Kirchhoff's Voltage Law (KVL) states that for a series connection, as pictured in Figure 2.2(a), the currents will be equal in all ports to be connected, and that the sum of the voltages at all ports is zero, or, in other words, if we have a series connection of M ports,

$$i_1 = i_2 = \ldots = i_M$$
$$v_1 + v_2 + \ldots + v_M = 0$$

Kirchhoff's Current Law (KCL) specifies the dual relationship among the voltages and currents in the case of a parallel connection of M ports, as per Figure 2.2(b), as

$$v_1 = v_2 = \ldots = v_M$$
$$i_1 + i_2 + \ldots + i_M = 0$$

Both sets of constraints hold instantaneously and can be thought of as M-ports in their own right. In addition, both types of M-port are passive, and in fact lossless. For example, in the case of a series connection of M ports where the currents at every port are the same and equal to i, we have, from (2.2), that

$$w_{\text{inst}} \doteq \sum_{j=1}^{M} v_j i_j = i \sum_{j=1}^{M} v_j = 0$$

Losslessness of the parallel connection can be similarly demonstrated. It is possible to show, through the use of *Tellegen's Theorem* [185], that a network made up of Kirchhoff connections of passive N-ports will behave passively as a whole.

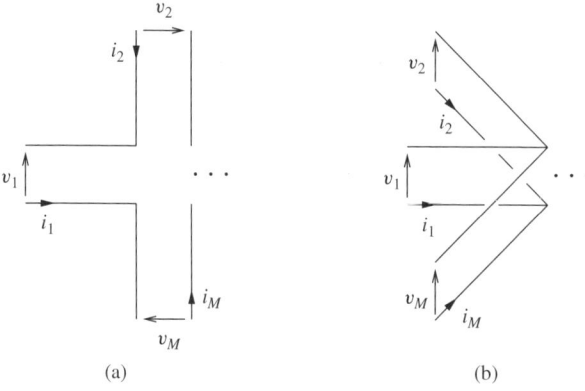

Figure 2.2: *Kirchhoff connections of M ports, in (a) series and (b) parallel.*

2.1.4 Circuit Elements

The most commonly encountered linear *one-ports* are the *inductor* of inductance L, the *resistor* of resistance R_0 and *capacitor* of capacitance C; their schematic representations are shown in Figure 2.3.

The equations relating voltage and current in the three one-ports, as well as their associated impedances are as follows:

$$\text{Inductor:} \quad v = L\frac{di}{dt} \quad Z = Ls \quad (2.6)$$

$$\text{Resistor:} \quad v = R_0 i \quad Z = R_0 \quad (2.7)$$

$$\text{Capacitor:} \quad i = C\frac{dv}{dt} \quad Z = \frac{1}{Cs} \quad (2.8)$$

Each of these circuit elements is passive as long as its element value (L, C or R_0) is positive[3]; the inductor and capacitor are easily shown to be lossless as well. The inductor and capacitor are examples of *reactive* circuit elements—all power instantaneously absorbed by either one will be stored and eventually returned to the network to which it is connected. The resistor is passive but not lossless.

In addition to the one-ports mentioned above, we can also define the *short-circuit*, *open-circuit*, *current source*, and *voltage source* (see Figure 2.4) by

$$\text{Short-circuit:} \quad v = 0$$

$$\text{Open-circuit:} \quad i = 0$$

$$\text{Voltage source:} \quad v = e(t)$$

$$\text{Current source:} \quad i = f(t)$$

The impedances of the short- and open-circuit one-ports are zero and infinity respectively. Both are lossless.

The two-ports that will occur most frequently in the WDF context are the *transformer* and *gyrator*, both shown in Figure 2.5. Each of these two-ports has two voltage/current pairs, one for each port. The transformer has associated with it one free parameter n, called the *turns ratio*, and the gyrator is defined with respect to a parameter $R_G > 0$, called the

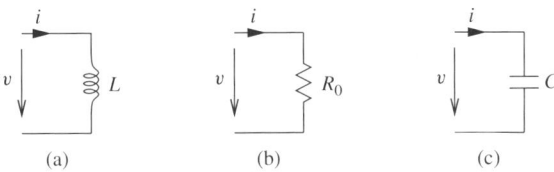

Figure 2.3: *One-port elements*—(a) *an inductor of inductance* L, (b) *a resistor of resistance* R_0, *and* (c) *a capacitor of capacitance* C.

---

[3]More generally, we allow these values to be zero as well. In these cases, the inductor and resistor are interpreted as short-circuits and the capacitor as an open-circuit.

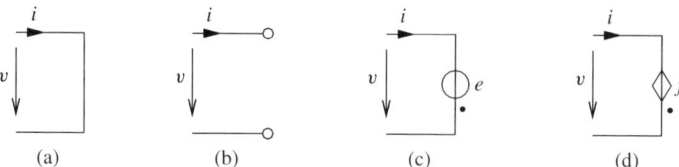

Figure 2.4: *Other one-ports*—(a) *a short-circuit*, (b) *an open-circuit*, (c) *a voltage source*, *and* (d) *a current source. Dots adjacent to the sources indicate polarity.*

gyrator coefficient, as well as a direction, represented graphically by an arrow. The relation among the port variables in each case is given by

$$\text{Transformer:} \quad v_2 = nv_1 \quad\quad i_1 = -ni_2 \quad (2.9)$$

$$\text{Gyrator:} \quad v_1 = -R_G i_2 \quad\quad v_2 = R_G i_1 \quad (2.10)$$

It is easily checked that both the transformer and gyrator are lossless two-ports. The gyrator is the first example we have seen so far of a *non-reciprocal* element—that is, its impedance matrix is not Hermitian; while we will not make nearly as much use of it here as of the other elements, it will find a place in certain parts of this work, especially when dealing with physical systems that have a certain type of asymmetric coupling (see Chapter 7), and in optimizing certain wave digital structures for simulation (see Section 3.11), and will play a pivotal role in linking digital waveguide networks to wave digital networks (see Chapter 6). A gyrator can be thought of as a dualizing element—when terminated on a capacitor, it behaves as an inductor, and vice versa. There are other non-reciprocal circuit elements that appear in the literature, particularly in the WDF context (such as the circulator, for example [68], or the quasi-reciprocal line (QUARL) [64]), but we will not need to make use of them in this book.

2.2 Wave Digital Elements and Connections

Wave digital filters result from the mapping of a lumped analog electrical network (usually made up of the elements mentioned in the previous section, connected using Kirchhoff's Laws, and which is intended for use as a filter) into the discrete-time domain.

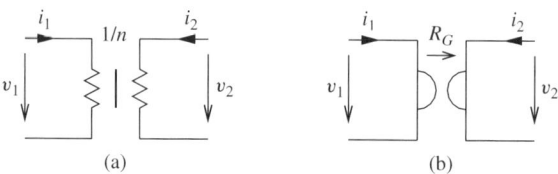

Figure 2.5: *Two-ports*—(a) *a transformer, of turns ratio n and* (b) *a gyrator, of gyrator coefficient* R_G.

2.2.1 The Bilinear Transform

In the linear time-invariant case, discretization is carried out using a particular type of spectral mapping between the analog frequency variable s and a new discrete frequency variable ψ, which will be a rational function of $z^{-1} = e^{-sT}$. (As per standard digital filtering theory, z^{-1} can be interpreted as the frequency-domain equivalent to a unit delay, of duration T.) The mapping affects only reactive N-ports, that is, those whose behavior is frequency-dependent, such as the inductor and capacitor. Memoryless elements, such as the transformer, gyrator, and resistor (as well as the parallel or series connection, interpreted as an N-port) are frequency-independent, and will be unaltered by such a transformation.

The frequency mapping proposed by Fettweis[4] [62] is a particular type of bilinear transform, given by

$$s \to \psi \triangleq \frac{2}{T}\frac{1-e^{-sT}}{1+e^{-sT}} = \frac{2}{T}\frac{1-z^{-1}}{1+z^{-1}} \tag{2.11}$$

We can then write

$$\mathrm{Re}(\psi) = \frac{2}{T}\frac{1-e^{-2\mathrm{Re}(s)T}}{|1+e^{-sT}|^2} = \frac{2}{T}\frac{1-|z|^{-2}}{|1+z^{-1}|^2}$$

and clearly,

$$\mathrm{Re}(s) > 0 \iff \mathrm{Re}(\psi) > 0 \iff |z| > 1$$
$$\mathrm{Re}(s) < 0 \iff \mathrm{Re}(\psi) < 0 \iff |z| < 1$$
$$\mathrm{Re}(s) = 0 \iff \mathrm{Re}(\psi) = 0 \iff |z| = 1$$

This implies that stable, causal transfer functions in s will be mapped to stable causal transfer functions in the discrete variable z^{-1}, and moreover that positive real functions will be mapped to functions that are *positive real in the outer disk* [224] (see Figure 2.6). Such functions are often called *pseudopassive* [63], and have an energetic interpretation similar to that of their counterparts in the analog domain. (Indeed, Fettweis views pseudopassivity as simply passivity using a warped frequency variable ψ [68].)

In particular, for a harmonic state, that is, for real frequencies ω such that $s = j\omega$ and $z = e^{j\omega T}$, we have that

$$\omega \to \frac{2}{T}\tan\left(\frac{\omega T}{2}\right) \tag{2.12}$$

so that the entire analog frequency spectrum is mapped to the discrete frequency spectrum exactly once. In particular, we have that the analog DC frequency $s = 0$ is mapped to discrete DC $z = 1$, and that analog infinite frequency is mapped to the Nyquist frequency. It should be clear that there will be significant warping of the spectrum away from either extreme.

---

[4]Fettweis in fact proposes the mapping $s \to (1-z^{-1})/(1+z^{-1})$, which is similar to (2.11) except for the factor $2/T$. Although this factor is of little importance in filtering applications, it is necessary here for the interpretation of such mapping as an integration rule. This should become clear in Chapter 3.

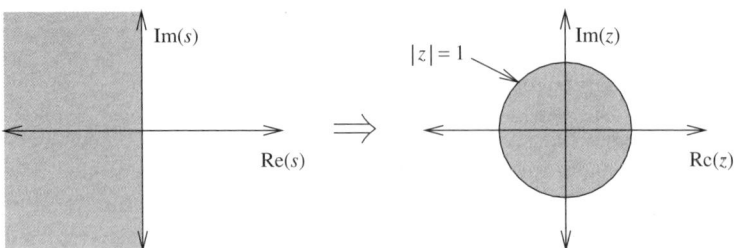

Figure 2.6: *Spectral mapping corresponding to the trapezoid rule.*

It is also worthwhile examining the mapping (2.11) on the unit circle in the low-frequency limit, in which case we can expand the right side of the mapping about $\omega = 0$, to get

$$\frac{2}{T} \tan\left(\frac{\omega T}{2}\right) = \omega - \frac{T^2}{12}\omega^3 + \ldots$$

The mapping (2.12) can be rewritten as

$$\omega \to \omega + O(\omega^3 T^2)$$

The frequency mapping thus becomes more accurate near $\omega = 0$ in the limit as $T \to 0$. The order of this approximation (namely to T^2) will play an important role in numerical integration methods, because it defines the accuracy of a numerical scheme [93, 176, 238].

It is important to mention that the time domain interpretation of the bilinear mapping (2.11) is called the *trapezoid rule* for numerical integration. That is, treating z^{-1} as the unit delay, the right-hand side of (2.11) serves as an approximation to the derivative in a discrete-time setting. In other words, we make the replacement

$$\frac{d}{dt} \to \frac{2}{T}(1 + \delta_t)^{-1}(1 - \delta_t)$$

where δ_t is a unit shift defined by $\delta_t x(t) = x(t - T)$. For example, in the case of the inductor, application of the mapping yields the following difference equation relating the voltage and current:

$$v(n) + v(n-1) = \frac{2L}{T}\left(i(n) - i(n-1)\right) \tag{2.13}$$

It should be understood here that $v(n)$ and $i(n)$ in (2.13) now represent discrete approximations[5] to the voltage and current of (2.6) at time $t = nT$, for integer n. Generalizations of the WDF approach to cases for which the N-port of interest is time-varying or nonlinear

---

[5]Since the two systems are assumed to be the same, modulo a spectral warping, we will not use a special notation to distinguish a discrete variable from a continuous one; the type of variable should be clear from context, and in cases where confusion may arise, we will always explicitly note the argument. In Chapter 4, however, we will use capital letters to distinguish discrete from continuous variables; this notational switch is unfortunate, but is a compromise necessary in order to remain coherent with the different literatures.

are based on this time domain formulation, because in these cases, we no longer have a well-defined notion of frequency.

For the rest of this section, so as to avoid unnecessary extra notation, we will assume that we have discrete-time voltages and currents. Thus v and i now refer to sequences $v(n)$ and $i(n)$, for n integer, and the steady state quantities $\hat{v}$ and $\hat{i}$ are complex amplitudes of a sequence at the discrete frequency z.

2.2.2 Wave Variables

At this point, one may assume that we have finished; indeed, we can derive a discrete-time equivalent to any LTI N-port (graphically represented by a signal flow diagram involving shifts and arithmetic operations), and such elements can be connected using Kirchhoff's Laws, which remain unchanged by the mapping (2.11). In particular, a network consisting of a collection of connected passive N-ports will possess a discrete equivalent of the passivity property, which has been called *pseudopassivity* [63]. The problem, however, is that a simple application of the bilinear transform to a given N-port usually leaves us with port variables that are not related to each other in a strictly causal way. For example, the difference equation (2.13) that results in the case of the inductor relates $v(n)$ to $i(n)$ at every time step n so that if we try to connect such a discrete-time one-port to another that has the same property (using Kirchhoff's Laws, which are memoryless), we necessarily end up with nonrealizable *delay-free loops* [68] in our resulting signal flow diagram. In other words, we will not be able to explicitly update all the port variables in our algorithm using only past values stored in the delay registers.

The problem of these delay-free loops was solved by Fettweis [62] with the introduction of *wave variables*, a concept with a long history borrowed from microwave electronics [13, 14]. For a port with voltage v and current i, *voltage waves* are defined by

$$a = v + iR \qquad \text{Input voltage wave} \qquad (2.14a)$$

$$b = v - iR \qquad \text{Output voltage wave} \qquad (2.14b)$$

a and b are referred to as *wave variables*, and in particular, a is called an *input wave* and b an *output wave*; the significance of these names will become clear in the examples of Section 2.2.4. This definition holds instantaneously and will also be true for continuous v and i, though we will almost never have occasion to refer to analog wave variables in this book. The parameter $R > 0$ is a free parameter known as the *port resistance*—its choice is governed by the character of the element itself. We also can define the *port conductance* G by

$$G = \frac{1}{R} \qquad (2.15)$$

at a port with port resistance R.

It is also possible to define power-normalized waves [68] $\underline{a}$ and $\underline{b}$ at any port with port resistance R by

$$\underline{a} = \frac{v + iR}{2\sqrt{R}} \qquad \text{Input power wave} \qquad (2.16a)$$

$$\underline{b} = \frac{v - iR}{2\sqrt{R}} \qquad \text{Output power wave} \qquad (2.16b)$$

The two types of waves are simply related to each other by

$$a = 2\sqrt{R}\underline{a} \qquad (2.17a)$$

$$b = 2\sqrt{R}\underline{b} \qquad (2.17b)$$

but power-normalized quantities have certain advantages in cases for which a port resistance is time-varying or signal dependent (indeed, in these cases, power-normalized waves *must* be employed if passivity in the digital simulation is to be maintained). In general, however, in view of (2.17), it should be assumed that we are using voltage waves unless otherwise indicated.

The steady state quantities $\hat{a}$ and $\hat{b}$ are defined in a manner identical to (2.14), where we replace v and i by $\hat{v}$ and $\hat{i}$.

2.2.3 Pseudopower and Pseudopassivity

Fettweis [63] defines the instantaneous pseudopower absorbed by a port with port resistance R (real) at time step n in terms of the discrete input and output wave quantities as

$$w_{\text{inst}}(n) = \frac{1}{R}\left(a^2(n) - b^2(n)\right) = 4\left(\underline{a}^2(n) - \underline{b}^2(n)\right) \qquad (2.18)$$

which, when the transformation (2.14) is inverted, gives

$$w_{\text{inst}}(n) = 4v(n)i(n)$$

This discrete power definition coincides with the standard definition of power in classical network theory from (2.2), aside from the factor of 4, which is of no consequence if definition (2.18) is applied consistently throughout a wave digital network.

For a real LTI N-port, in an exponential state of complex frequency z, the steady state average pseudopower may be written in terms of the $N \times 1$ vectors $\underline{\hat{a}}$ and $\underline{\hat{b}}$, which contain the power-normalized complex amplitudes $\underline{\hat{a}}_j$ and $\underline{\hat{b}}_j$, for $j = 1, \ldots, N$ as

$$\overline{w} = 4\left(\underline{\hat{a}}^*\underline{\hat{a}} - \underline{\hat{b}}^*\underline{\hat{b}}\right)$$

The steady state reflectance $\underline{\mathbf{S}}(z^{-1})$ is defined by

$$\underline{\hat{b}} = \underline{\mathbf{S}}\,\underline{\hat{a}}$$

and gives

$$\overline{w} = 4\left(\underline{\hat{a}}^*(\mathbf{I}_N - \underline{\mathbf{S}}^*\underline{\mathbf{S}})\underline{\hat{a}}\right)$$

where $\mathbf{I}_N$ is the $N \times N$ identity matrix. For pseudopassivity [63], we require (recalling that the bilinear transform (2.11) maps the right half s-plane to the exterior of the unit circle in the z plane), that

$$\underline{\mathbf{S}}^*(z^{-1})\underline{\mathbf{S}}(z^{-1}) \leq \mathbf{I}_N \qquad \text{for} \qquad |z| \geq 1 \qquad (2.19)$$

2.2. WAVE DIGITAL ELEMENTS AND CONNECTIONS

$\underline{\mathbf{S}}(z^{-1})$ is sometimes called a bounded real matrix. If (2.19) holds with equality for $|z| = 1$, then it is called *lossless bounded real* (LBR) [261]. In general, to bounded real matrix reflectances there correspond positive real matrix impedances, and vice versa. In terms of voltage wave quantities, we have, for a wave digital N-port, that

$$\hat{\mathbf{a}} = 2\mathbf{R}^{\frac{1}{2}}\underline{\hat{\mathbf{a}}} \qquad \hat{\mathbf{b}} = 2\mathbf{R}^{\frac{1}{2}}\underline{\hat{\mathbf{b}}}$$

where $\mathbf{R}^{\frac{1}{2}}$ is the diagonal square root of the matrix containing the N port resistances $R_1, \ldots, R_N$ on its diagonal. We then have

$$\mathbf{S} = \mathbf{R}^{\frac{1}{2}}\underline{\mathbf{S}}\mathbf{R}^{-\frac{1}{2}}$$

for the voltage wave scattering matrix $\mathbf{S}$ and thus we require

$$\mathbf{S}^*(z^{-1})\mathbf{R}^{-1}\mathbf{S}(z^{-1}) \leq \mathbf{R}^{-1} \qquad \text{for} \qquad |z| \geq 1 \qquad (2.20)$$

for passivity. For one-ports, the requirements (2.20) and (2.19) are the same.

Also note that we have, by applying the power wave variable definitions (2.16), and the discrete impedance relation $\hat{\mathbf{v}} = \mathbf{Z}\hat{\mathbf{i}}$ (which is identical to the analog relation from (2.1), except that we now have $\mathbf{Z} = \mathbf{Z}(z^{-1})$), that

$$\underline{\mathbf{S}} = (\mathbf{Z}\mathbf{R}^{-1} + \mathbf{I}_N)^{-1}(\mathbf{Z}\mathbf{R}^{-1} - \mathbf{I}_N) \qquad (2.21)$$

If the N-port is not LTI, then it is possible to apply a similar idea to the expression for the instantaneous pseudopower from (2.18) in order to derive a passivity condition [68]; in this case, pseudopassivity has also been called *incremental pseudopassivity* [167].

2.2.4 Wave Digital Elements

We will now present the wave digital equivalents of all the circuit elements mentioned in Section 2.1.4.

Under the bilinear transform (2.11), the steady state equation for an inductor becomes

$$\hat{v} = \frac{2L}{T}\left(\frac{1-z^{-1}}{1+z^{-1}}\right)\hat{i}$$

or, in the discrete-time domain,

$$v(n) + v(n-1) = \frac{2L}{T}\Big(i(n) - i(n-1)\Big)$$

Applying the definition of wave variables (2.14), we get, in the time domain,

$$a(n) + b(n) + a(n-1) + b(n-1) = \frac{2L}{RT}\Big(a(n) - b(n) - a(n-1) + b(n-1)\Big) \qquad (2.22)$$

If we make the choice

$$R = \frac{2L}{T}$$

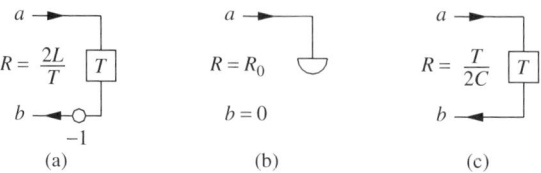

Figure 2.7: *Wave digital one-ports corresponding to the classical one-ports of Figure 2.3—* (a) *the wave digital inductor,* (b) *resistor, and* (c) *capacitor.*

then (2.22) simplifies to

$$b(n) = -a(n-1) \tag{2.23}$$

Thus, the input wave a must undergo a delay and sign-inversion before it is output as b. In terms of steady state quantities, we have

$$\hat{b} = -z^{-1}\hat{a} \quad \Rightarrow \quad S(z^{-1}) = -z^{-1} \tag{2.24}$$

The reflectance $S(z^{-1})$ is, as expected, LBR (see previous section). The resulting *wave digital one-port* is shown in Figure 2.7(a). The construction of the wave digital one-ports corresponding to the resistor and capacitor is similar; their signal flow graphs also appear in Figure 2.7. We note that the same choice of the port resistance R should be made in the case of power-normalized wave variables. We also note in passing that we have used here the symbol T to represent a unit delay in a wave digital filter.

The short-circuit and open-circuit one-ports are, for any choice of the port resistance R, perfectly reflecting (with or without sign inversion, respectively). The appearance of the factor R in the definition of the wave digital current source results from our choice of using voltage waves (as opposed to current waves). In all the wave digital one-ports of Figure 2.8, there is an instantaneous dependence of the output wave b on the input wave a, and we may expect delay-free loops to appear when these elements are connected with others. On the other hand, the form of these one-ports does not depend on a particular choice of the port resistance R (except in a very minor way for a current source), and remains a free parameter, which can be used, in many cases, to remove delay-free loops.

It is also possible to combine resistances and sources [68]; a resistive voltage source, shown in Figure 2.9(a), consists of a voltage source e in series with a resistor of resistance R_0. If the port resistance of the combined one-port is chosen to be R_0, then the wave digital one-port [68] is as shown in Figure 2.9(b). Note that in this case there is no longer any

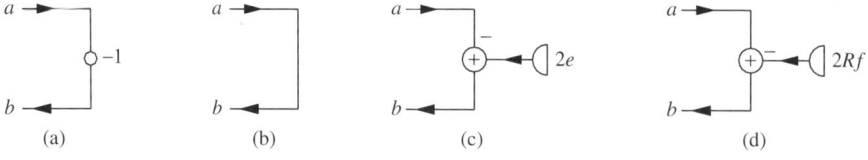

Figure 2.8: *Wave digital one-ports corresponding to the classical one-ports of Figure 2.4—* (a) *the short-circuit,* (b) *open-circuit,* (c) *voltage source, and* (d) *current source.*

2.2. WAVE DIGITAL ELEMENTS AND CONNECTIONS

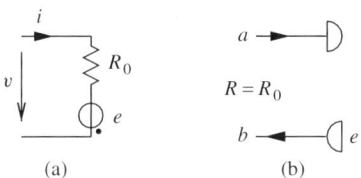

Figure 2.9: (a) *A resistive voltage source and* (b) *the associated wave digital one-port.*

instantaneous dependence of the output wave on the input wave. A wave digital resistive current source can be similarly defined.

The classical transformer and gyrator two-ports can be treated in the same way. For example, the gyrator accepts two input waves a_1 and a_2 and yields two output waves b_1 and b_2. There are two port resistances, R_1 and R_2. The instantaneous equations (2.10) relating the voltages and currents in a gyrator become, upon substitution of wave variables,

$$\begin{bmatrix} b_1 \\ b_2 \end{bmatrix} = \frac{1}{R_G^2 + R_1 R_2} \begin{bmatrix} R_G^2 - R_1 R_2 & -2R_G R_1 \\ 2R_G R_2 & R_G^2 - R_1 R_2 \end{bmatrix} \begin{bmatrix} a_1 \\ a_2 \end{bmatrix} \quad (2.25)$$

which simplify, under the choice of $R_1 = R_G$ and $R_2 = R_G$ to

$$b_1 = -a_2 \qquad b_2 = a_1 \quad (2.26)$$

If we are using power-normalized wave variables, then the scattering equation for the gyrator becomes

$$\begin{bmatrix} \underline{b}_1 \\ \underline{b}_2 \end{bmatrix} = \frac{1}{R_G^2 + R_1 R_2} \begin{bmatrix} R_G^2 - R_1 R_2 & -2R_G \sqrt{R_1 R_2} \\ 2R_G \sqrt{R_1 R_2} & R_G^2 - R_1 R_2 \end{bmatrix} \begin{bmatrix} \underline{a}_1 \\ \underline{a}_2 \end{bmatrix} \quad (2.27)$$

In this case, any choice of the port resistances such that $R_1 R_2 = R_G^2$ gives

$$\underline{b}_1 = -\underline{a}_2 \qquad \underline{b}_2 = \underline{a}_1 \quad (2.28)$$

The ideal transformer also can take on various forms, depending on the choices of the port resistances and on the type of wave variable employed. Under a choice of port resistances R_1 and R_2 such that $R_2 = n^2 R_1$, the equations (2.9) for the ideal transformer of turns ratio n become

$$b_2 = n a_1 \qquad b_1 = \frac{1}{n} a_2 \quad (2.29)$$

For the transformer and gyrator wave digital two-ports, we adopt general symbols that do not reflect a particular choice of the port resistances. If simplifying choices can be made in either case, than we can write the signal flow graph explicitly (see Figure 2.10). There may be occasions when it is not possible to make these simplifying choices of the port resistances that yield (2.26) and (2.29). For example, when we approach the numerical integration of beam and plate systems in Chapter 7, as well as certain *balanced forms* (see Section 3.11) the wave digital networks contain gyrators whose port resistances are constrained, forcing us to use (2.25).

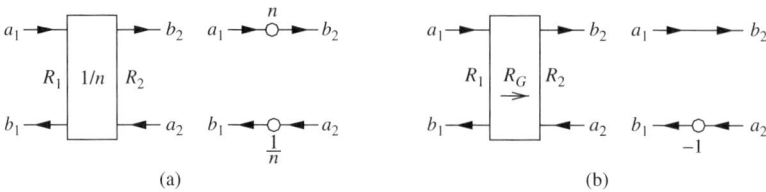

Figure 2.10: *Wave digital two-ports—(a) a transformer with turns ratio n and its simpler form for $R_2 = n^2 R_1$ and (b) a gyrator of gyrator coefficient R_G and its simpler form for $R_1 = R_2 = R_G$.*

We also mention that these two-ports are both lossless and in fact *non-energic* [63] (i.e., we have $w_{\text{inst}}(n) = 0$, for all n).

Numerous other wave digital elements have been proposed, namely circulators, QUARLS, as well as *unit elements* [68]. All have been applied fruitfully to filter-design problems, but the unit element deserves a special mention.

The Unit Element

One wave digital two-port, called the unit element, is usually defined in the discrete-time domain, without reference to an analog counterpart; this element, shown in Figure 2.11, is defined by

$$b_1(n) = a_2(n-1) \qquad\qquad b_2(n) = a_1(n-1)$$

It was considered by Fettweis to be the 'most important two-port element' [68], and was used extensively for realizability reasons in early WDF designs, especially before the appearance of reflection-free ports [85]. It behaves exactly like a transmission line, and is in fact identical to the waveguide or bidirectional delay line, which is the key component[6] of the digital waveguide network (DWN) [228], as we saw in Section 1.1.3. The unit element is time-invariant, and obviously lossless, though it is reactive (able to store energy). Its scattering relation is given by

$$\begin{bmatrix} \hat{b}_1 \\ \hat{b}_2 \end{bmatrix} = \begin{bmatrix} 0 & z^{-1} \\ z^{-1} & 0 \end{bmatrix} \begin{bmatrix} \hat{a}_1 \\ \hat{a}_2 \end{bmatrix}$$

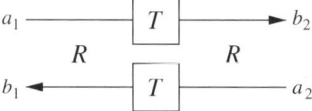

Figure 2.11: *A unit element, with port resistances R and delays T.*

---

[6]The question of which point of view is more fundamental, that of wave digital filtering or digital waveguides, is in evidence here. A wave digital (WD) enthusiast would consider the unit element to be simply a WD two-port, but a DWN or TLM person would likely view one-ports such as the wave digital capacitor and inductor as terminated transmission line segments.

2.2. WAVE DIGITAL ELEMENTS AND CONNECTIONS

2.2.5 Adaptors

Consider now a series connection of M ports, where we have a port resistance $R_j > 0$, $j = 1, \ldots, M$, associated with each port. In terms of instantaneous quantities, we have

$$\sum_{j=1}^{M} v_j = 0$$

or, in terms of wave variables, using the inverse of the transformation (2.14),

$$\sum_{j=1}^{M} (a_j + b_j) = 0$$

Since the currents at all ports are all equal to i, this implies, using $b_j = a_j - 2R_j i$, that

$$\sum_{j=1}^{M} (-2R_j i + 2a_j) = 0$$

and thus

$$i = \frac{1}{\sum_{j=1}^{M} R_j} \sum_{j=1}^{M} a_j \tag{2.30}$$

By applying similar manipulations in the case of a parallel connection of M ports, we can then write down the equations relating the input and output wave variables at the kth port for both types of connection as

$$b_k = a_k - \frac{2R_k}{\sum_{j=1}^{M} R_j} \sum_{j=1}^{M} a_j, \qquad k = 1, \ldots, M \qquad \text{Series connection} \tag{2.31}$$

$$b_k = -a_k + \frac{2}{\sum_{j=1}^{M} G_j} \sum_{j=1}^{M} G_j a_j, \qquad k = 1, \ldots, M \qquad \text{Parallel connection} \tag{2.32}$$

where we recall from (2.15) that G_j is defined as the reciprocal of the port resistance R_j.

For power-normalized wave variables, we thus have, applying (2.17),

$$\underline{b}_k = \underline{a}_k - \frac{2\sqrt{R_k}}{\sum_{j=1}^{M} R_j} \sum_{j=1}^{M} \sqrt{R_j} \underline{a}_j, \qquad k = 1, \ldots, M \qquad \text{Series connection} \tag{2.33}$$

$$\underline{b}_k = -\underline{a}_k + \frac{2\sqrt{G_k}}{\sum_{j=1}^{M} G_j} \sum_{j=1}^{M} \sqrt{G_j} \underline{a}_j, \qquad k = 1, \ldots, M \qquad \text{Parallel connection} \tag{2.34}$$

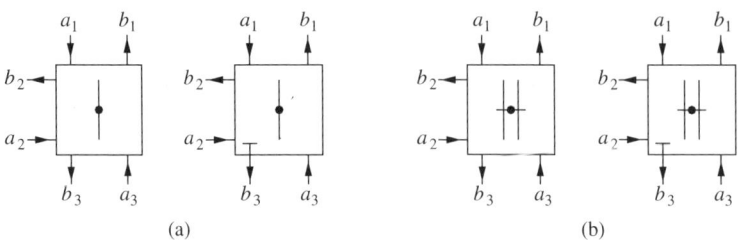

Figure 2.12: *Three-port adaptors*—(a) *a general three-port series adaptor and one for which port three is reflection-free and* (b) *a general three-port parallel adaptor and one for which port three is reflection-free.*

The operator that performs this calculation on the wave variables is called a *series adaptor* or a *parallel adaptor* [68], depending on the type of connection. The graphical representations of three-port adaptors, for either voltage or power-normalized waves, are shown in Figure 2.12.

A useful simplification occurs when we can choose, for a particular port q (called a *reflection-free* port [85]) of an M-port adaptor,

$$R_q = \sum_{j=1, j \neq q}^{M} R_j \qquad \text{Series reflection-free port resistance} \qquad (2.35\text{a})$$

$$G_q = \sum_{j=1, j \neq q}^{M} G_j \qquad \text{Parallel reflection-free port conductance} \qquad (2.35\text{b})$$

in which case the scattering equations (2.31) and (2.32) yield, for the output wave at port q,

$$b_q = - \sum_{j=1, j \neq q}^{M} a_j \qquad \text{Series reflection-free port} \qquad (2.36\text{a})$$

$$b_q = \frac{2}{\sum_{j=1}^{M} G_j} \sum_{j=1, j \neq q}^{M} G_j a_j \qquad \text{Parallel reflection-free port} \qquad (2.36\text{b})$$

Thus, at a reflection-free port q, the output wave b_q is independent of the input wave a_q; such a port can be connected to any other without risk of a resulting delay-free loop. The same choices of port resistances (2.35) will also give a reflection-free port if power wave variables are employed.

Scattering Matrices for Adaptors

The adaptor equations for a connection of M ports, in either the series (2.31) or parallel (2.32) case, may be written as

$$\mathbf{b} = \mathbf{Sa} \qquad (2.37)$$

2.2. WAVE DIGITAL ELEMENTS AND CONNECTIONS

where $\mathbf{b} = [b_1, \ldots, b_M]^T$ and $\mathbf{a} = [a_1, \ldots, a_M]^T$, and where we have

$$\mathbf{S} = \mathbf{I}_M - \boldsymbol{\alpha}_s \mathbf{1}^T \qquad \text{Series adaptor (voltage waves)} \qquad (2.38)$$

$$\mathbf{S} = -\mathbf{I}_M + \mathbf{1}\boldsymbol{\alpha}_p^T \qquad \text{Parallel adaptor (voltage waves)} \qquad (2.39)$$

Here $\mathbf{1}$ is an $M \times 1$ vector containing all ones, $\mathbf{I}_M$ is the $M \times M$ identity matrix, and $\boldsymbol{\alpha}_s$ and $\boldsymbol{\alpha}_p$ are defined by

$$\boldsymbol{\alpha}_s = \frac{2}{\sum_{j=1}^M R_j}[R_1, \ldots, R_M]^T \qquad \boldsymbol{\alpha}_p = \frac{2}{\sum_{j=1}^M G_j}[G_1, \ldots, G_M]^T$$

The sum of the elements of either $\boldsymbol{\alpha}_s$ or $\boldsymbol{\alpha}_p$ is 2. For power-normalized wave variables, we have a similar relationship,

$$\underline{\mathbf{b}} = \underline{\mathbf{S}}\,\underline{\mathbf{a}} \qquad (2.40)$$

where

$$\underline{\mathbf{S}} = \mathbf{I}_M - \sqrt{\boldsymbol{\alpha}_s}\sqrt{\boldsymbol{\alpha}_s}^T \qquad \text{Series adaptor (power-normalized waves)}$$

$$\underline{\mathbf{S}} = -\mathbf{I}_M + \sqrt{\boldsymbol{\alpha}_p}\sqrt{\boldsymbol{\alpha}_p}^T \qquad \text{Parallel adaptor (power-normalized waves)}$$

Here the square root sign indicates an entry-by-entry square root of a vector (all entries of $\boldsymbol{\alpha}_s$ and $\boldsymbol{\alpha}_p$ are nonnegative). The scattering matrices in the power-normalized form are also known as *Householder reflections*.

Defining the Euclidean norm of a column vector $\mathbf{x}$ as $\|\mathbf{x}\|_2 = \sqrt{\mathbf{x}^T\mathbf{x}}$, it is easy to show that a power-normalized scattering matrix $\underline{\mathbf{S}}$ is orthogonal, and thus, l_2 norm-preserving in either the series or parallel case, that is, we have

$$\|\underline{\mathbf{b}}\|_2 = \|\underline{\mathbf{a}}\|_2 \qquad (2.41)$$

For voltage waves, we have the preservation of a weighted l_2 norm, that is,

$$\|\mathbf{b}\|_{\mathbf{P},2} = \|\mathbf{a}\|_{\mathbf{P},2} \qquad (2.42)$$

where $\|\cdot\|_{\mathbf{P},2} = \sqrt{(\cdot)^T \mathbf{P}(\cdot)}$; in this case, $\mathbf{P}$ is an $M \times M$ positive-definite diagonal matrix simply given by $\text{diag}(G_1, \ldots, G_M)$. It should be clear that (2.41) and (2.42) are merely restatements of power conservation at a memoryless, lossless M-port.

Note that multiplying $\mathbf{S}$ or $\underline{\mathbf{S}}$ by a vector requires, in either the series or parallel case, $O(M)$ adds and multiplies—in particular, it is cheaper than a full $M \times M$ matrix multiply.

2.2.6 Signal and Coefficient Quantization

In a machine implementation of a wave digital filter, the signals and coefficients must necessarily be represented with a finite number of bits. As such, it is not immediately obvious that the passivity properties for a given WDF, which are framed in terms of unquantized signals (waves) and filter multipliers (related to the port resistances) will hold in a finite word-length computer implementation. All digital filter implementations are vulnerable to a host of undesirable effects that result from signal and coefficient quantization; among them are parasitic oscillations and high sensitivity of filter pole and zero locations (and thus the frequency response). WDFs, however, offer a number of means of combating these problems. The exploration of these means has produced a large body of literature [65, 68, 86, 167, 242, 276]. We give only a brief outline here, for completeness sake.

From the discussion of wave digital elements, it is easy to see that, in most cases, the only arithmetic operations in a WDF will occur as signals are scattered from adaptors[7]; the wave digital inductor, capacitor, and unit element involve only shifts and possibly sign inversion, and the wave digital resistor, which behaves as a sink, can essentially be ignored by the programmer once its port resistance has been absorbed into the adaptor to which it is connected. Simple quantization procedures [84, 270] were first proposed, and later the concept of incremental pseudopassivity [167] was developed for ensuring that a finite word-length implementation of a wave digital adaptor behaves passively under signal truncation. The most straightforward scheme appears in Figure 2.13, for the case of a three-port adaptor (either series or parallel).

$\tilde{a}_j$ are the input waves (assumed voltage waves) to the junction, for $j = 1, \ldots, M$ (we have $M = 3$ in Figure 2.13), and are assumed to be of some finite word-length. Extended precision is used within the adaptor in order to exactly calculate the output waves b_j, from (2.37). We have assumed that the multiplier coefficients within the junction are of finite word-length as well—we will discuss this presently. The output waves b_j thus satisfy (2.42), where $\mathbf{a}$ is replaced by $\tilde{\mathbf{a}}$, with $\tilde{\mathbf{a}} = [\tilde{a}_1, \ldots, \tilde{a}_M]^T$. Scattering is lossless. In general, however, the number of bits required to represent b_j will now be greater than the number required for $\tilde{a}_j$; in order to reduce the size of the output word-length, we may apply *magnitude truncation* (represented graphically in Figure 2.13 by boxes labeled 'Q', which are *not* wave digital one-ports. Magnitude truncation may be incorporated formally into the

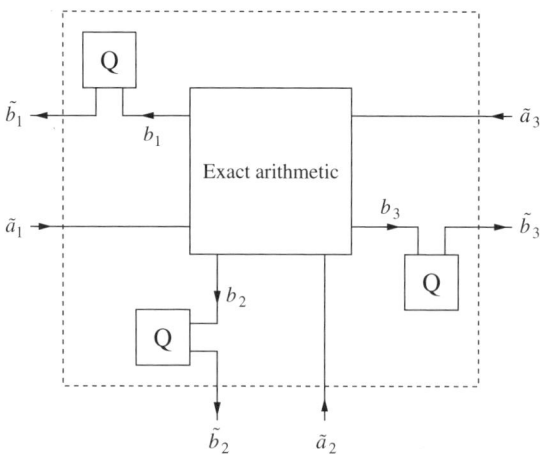

Figure 2.13: *Signal truncation at a three-port adaptor.*

---

[7]Referring to Figure 2.10, it is evident that the transformer with $n = \pm 1$ and $R_1 = R_2$, and the gyrator with $R_1 = R_2 = R_G$ (their most common forms), also are arithmetic-free. Otherwise, a more detailed treatment is required.

2.2. WAVE DIGITAL ELEMENTS AND CONNECTIONS

scattering picture through the use of *circulators* [167]). A reduced word-length wave $\tilde{b}_j$ is obtained from b_j by truncating it in any way as long as the magnitude is decreased. In other words, for any port j,

$$\tilde{b}_j^2 \leq b_j^2$$

This implies, then, that

$$\|\tilde{\mathbf{b}}\|_{\mathbf{P},2} \leq \|\mathbf{b}\|_{\mathbf{P},2} = \|\tilde{\mathbf{a}}\|_{\mathbf{P},2}$$

so that passivity is maintained even considering the finite word-length wave variables. In this way (by ensuring a decrease in the overall energy measure of the WD network), both large- and small-scale parasitic oscillations can be completely eliminated, at least in the zero-input case [68]. Various types of overflow characteristics have been examined [84, 167]. Such a quantization rule has also appeared in other contexts [261], and applies equally well to DWNs [228], which are introduced in Chapter 4.

The quantization of coefficients in WDFs [63, 65, 68, 152] as well as other similar filter structures [261] has been shown to have a minimal effect on the filter response. That is, in many lossless configurations [68], variations in the values of the multiplier coefficients (which are usually the reflection and transmission parameters α_s or α_p at an adaptor) can be shown to have a second-order effect on the filter response. In contrast, when such variations occur in direct-form filter structures, large changes in pole locations can result, and a stable filter may even become unstable [179]. This robustness property of scattering-based filter structures is sometimes called *structural passivity* [198, 203, 261]. As a simple example, consider the scattering equations (2.38) for a series adaptor; as mentioned above, the parameters in the vector α_s are the filter multiplier coefficients, and recall also that the sum of the elements in α_s is exactly 2, in infinite-precision arithmetic. Suppose that the elements of α_s are truncated to some finite word-length values, which can be written as the vector $\tilde{\alpha}_s$. If they are truncated such that all elements of $\tilde{\alpha}_s$ are positive, and their sum is still exactly 2, then it is easy to show that there must correspond a set of nonnegative port resistances, and thus the quantized adaptor can still be considered as exactly lossless. More generally, it is possible to ensure passivity if the sum of the elements of $\tilde{\alpha}_s$ is less than or equal to 2; this has also been discussed in the waveguide filter context [203].

While most of the approaches to quantization have been concerned with fixed-point implementations, many of the same ideas can be applied in floating-point as well. Floating-point signal truncation rules were proposed [51], and an early study of coefficient sensitivity and roundoff noise also appeared [152]. More recent developments include a generalized WDF, which is simply realized using multiply/accumulate operations [81], and a description of passive coefficient-truncation rules [162] based on scattering matrix factorization.

2.2.7 Vector Wave Variables

It is straightforward to extend wave digital filtering principles to the vector case (this has been outlined by Nitsche [176]; the same idea has also appeared in the context of DWNs [228, 203]). For a q-component vector one-port element with voltage $\mathbf{v} = [v_1, \ldots, v_q]^T$, and current $\mathbf{i} = [i_1, \ldots, i_q]^T$, it is possible to define wave variables $\mathbf{a}$ and $\mathbf{b}$ by

$$\mathbf{a} = \mathbf{v} + \mathbf{R}\mathbf{i} \qquad (2.43a)$$

$$\mathbf{b} = \mathbf{v} - \mathbf{R}\mathbf{i} \qquad (2.43b)$$

for a $q \times q$ symmetric positive-definite matrix $\mathbf{R}$; power-normalized quantities may be defined by

$$\underline{\mathbf{a}} = \frac{1}{2}\left(\mathbf{R}^{-T/2}\mathbf{v} + \mathbf{R}^{1/2}\mathbf{i}\right) \qquad (2.44a)$$

$$\underline{\mathbf{b}} = \frac{1}{2}\left(\mathbf{R}^{-T/2}\mathbf{v} - \mathbf{R}^{1/2}\mathbf{i}\right) \qquad (2.44b)$$

where $\mathbf{R}^{1/2}$ is some right square root of $\mathbf{R}$, and $\mathbf{R}^{T/2}$ is its transpose. The power absorbed by the vector one-port will be

$$w_{\text{inst}} = \left(\mathbf{a}^T \mathbf{R}^{-1} \mathbf{a} - \mathbf{b}^T \mathbf{R}^{-1} \mathbf{b}\right) = 4\left(\underline{\mathbf{a}}^T \underline{\mathbf{a}} - \underline{\mathbf{b}}^T \underline{\mathbf{b}}\right) = 4\mathbf{v}^T \mathbf{i} \qquad (2.45)$$

Kirchhoff's Laws, for a series or parallel connection of M q-component vector elements with voltages $\mathbf{v}_j$ and $\mathbf{i}_j$, $j = 1, \ldots, M$ can be written as

$$\mathbf{i}_1 = \mathbf{i}_2 = \cdots = \mathbf{i}_M \qquad \mathbf{v}_1 + \mathbf{v}_2 + \cdots + \mathbf{v}_M = \mathbf{0} \qquad \text{Series connection} \qquad (2.46a)$$

$$\mathbf{v}_1 = \mathbf{v}_2 = \cdots = \mathbf{v}_M \qquad \mathbf{i}_1 + \mathbf{i}_2 + \cdots + \mathbf{i}_M = \mathbf{0} \qquad \text{Parallel connection} \qquad (2.46b)$$

and the resulting scattering equations will be

$$\mathbf{b}_k = \mathbf{a}_k - 2\mathbf{R}_k \left(\sum_{j=1}^{M} \mathbf{R}_j\right)^{-1} \sum_{j=1}^{M} \mathbf{a}_j, \qquad k = 1, \ldots, M \quad \text{Series connection} \qquad (2.47a)$$

$$\mathbf{b}_k = -\mathbf{a}_k + 2 \left(\sum_{j=1}^{M} \mathbf{R}_j^{-1}\right)^{-1} \sum_{j=1}^{M} \mathbf{R}_j^{-1} \mathbf{a}_j, \qquad k = 1, \ldots, M \quad \text{Parallel connection} \qquad (2.47b)$$

in terms of the wave variables $\mathbf{a}_k$, $\mathbf{b}_k$ defined as per (2.43) and the port resistance matrices $\mathbf{R}_k$, $k = 1, \ldots, M$. These are the defining equations of a vector adaptor; their schematics are essentially the same as those of Figure 2.12, except that they are drawn in bold—see Figure 2.14. As before, we use the same representation for power-normalized waves.

Coupled Inductances and Capacitances

Coupled inductances and capacitances defined, in vector form, by

$$\mathbf{v} = \mathbf{L}\frac{d\mathbf{i}}{dt} \qquad \mathbf{i} = \mathbf{C}\frac{d\mathbf{v}}{dt} \qquad (2.48)$$

for symmetric positive-definite matrices $\mathbf{L}$ and $\mathbf{C}$ were first introduced in the WDF context by Nitsche [176]; they turn out to be essential to the construction of WDF-based numerical simulation algorithms for stiff distributed systems such as plates (see Section 7.2) and

2.2. WAVE DIGITAL ELEMENTS AND CONNECTIONS

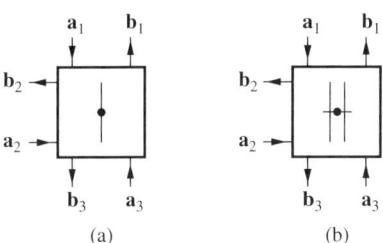

Figure 2.14: *Three-port vector adaptors—(a) a vector series adaptor and (b) a vector parallel adaptor.*

shells (see Section 7.3), as well as for full three-dimensional elastic solid dynamics (see Section 7.4). Though these are best thought of as vector elements, they appear within larger scalar circuits, and it is convenient to have a representation for which the vectors of port quantities are separated out into scalar portwise components.

We show an inductive coupling of q loops in Figure 2.15(a); self-inductances are indicated by directed arrows, accompanied by an inductance L_{jj}, $j = 1, \ldots, q$ (these are the diagonal elements of $\mathbf{L}$), and a mutual inductance between loops j and k, $j \neq k$ is represented by an arrow and the associated inductance L_{kj} (which is the (k, j)th or (j, k)th element of $\mathbf{L}$, and is not constrained to be positive). A coupled capacitance is shown in Figure 2.15(b).

A coupled inductance can be discretized through the use of the trapezoid rule applied directly to the vector equations of (2.48); in terms of wave variables defined by (2.43), we get

$$\mathbf{b}(n) = -\mathbf{a}(n-1) \qquad \mathbf{R} = 2\mathbf{L}/T$$

which is a direct vector generalization of (2.23). Similarly, for a capacitor, we get

$$\mathbf{b}(n) = \mathbf{a}(n-1) \qquad \mathbf{R} = T(2\mathbf{C})^{-1}$$

In practice, if a coupled inductance (capacitance) appears in a circuit that is to be discretized using WDFs, we may treat it as a q-vector two-port made up of a series (parallel) adaptor terminated on a vector wave digital inductor (capacitor) of port resistance $2\mathbf{L}/T$ ($T(2\mathbf{C})^{-1}$). See Figure 2.16 for the signal flow diagrams for these objects and the simplified

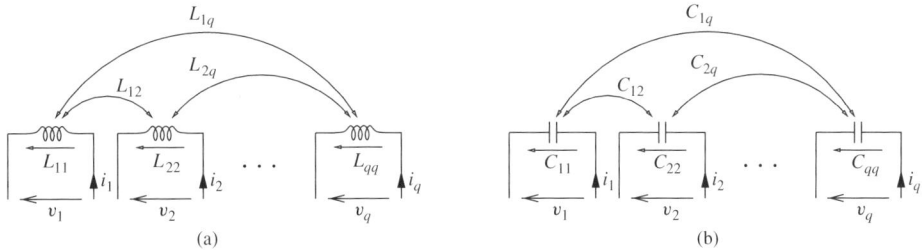

Figure 2.15: (a) *q coupled inductances and* (b) *q coupled capacitances.*

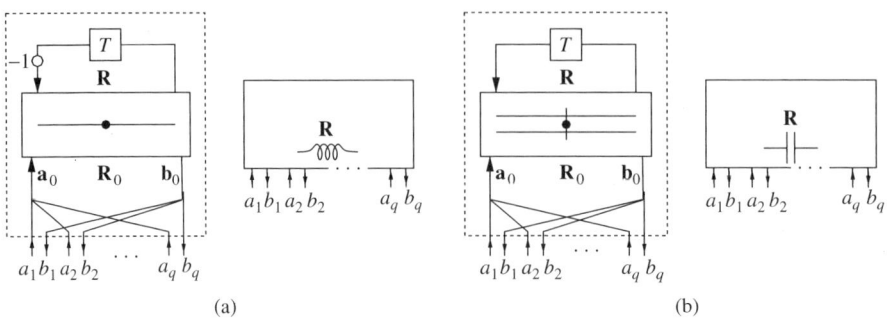

Figure 2.16: (a) *Signal flow graph for a wave digital coupled inductance and a simplified representation. Here the $q \times q$ port resistance $\mathbf{R} = 2\mathbf{L}/T$, and $\mathbf{R}_0$ is a $q \times q$ diagonal matrix; the diagonal entries specify the port resistances at the q scalar ports to which the element is connected.* (b) *The signal flow graph for a wave digital coupled capacitance (vector port resistance $\mathbf{R} = T(2\mathbf{C})^{-1}$) and its simplified representation. In either case, the wave variables at the lower port of the vector adaptor are simply defined by $\mathbf{a}_0 = [a_1, \ldots, a_q]^T$ and $\mathbf{b}_0 = [b_1, \ldots, b_q]^T$.*

representations that we will use. The port resistance at the opposing port will, in general, be diagonal, so that the vector wave variables entering and leaving the adaptor may be decomposed into scalar wave variables; this diagonal port resistance will be determined by the rest of the network to which the q-vector two-port is connected. See Section 4.2.6 for more information on this decomposition in the DWN context; we will return to vector/scalar connections in Chapter 7. We note that in the representations in Figure 2.16, we have not explicitly indicated the order in which the q scalar incoming and outgoing vectors should be 'packed' and 'unpacked' from the vector wave variables at the lower ports of the vector adaptors. In the applications in Chapter 7, for a given coupled inductance (say), self-inductances will all be identical, as will all mutual inductances; thus any ordering will do, as long as the jth elements of both $\mathbf{a}$ and $\mathbf{b}$ correspond to wave variables at the jth scalar port.

2.3 Wave Digital Filters and Finite Differences

We took a brief look at wave digital filters for simulation purposes beginning on page 21. In particular, we examined a wave digital filtering network that numerically integrates the equations defining an electrical harmonic LC oscillator, shown in Figure 1.12(a). It should be clear that, as wave digital elements are derived from continuous-time counterparts via a discretization rule, the network can be viewed as a particular type of finite difference scheme (and in the case of the LC oscillator, a well-known one). Let us now examine this system in more detail.

Clearly, the defining equations of the oscillator (1.25) and Kirchhoff's Laws (1.23) imply that

$$v = L\frac{di}{dt} \qquad\qquad i = -C\frac{dv}{dt} \qquad (2.49)$$

2.3. WAVE DIGITAL FILTERS AND FINITE DIFFERENCES

where we have set, for simplicity, $v = v_1$, and $i = i_1$. This is a parallel connection of an inductor and a capacitor, so we can immediately make the replacement as shown in Figure 1.12(b), with port resistances $R_1 = 2L/T$, and $R_2 = T/2C$. Supposing power-normalized waves are employed, we can then write the update of the values in the delay registers as

$$\underline{\mathbf{a}}(n) = \begin{bmatrix} -1 & 0 \\ 0 & 1 \end{bmatrix} \underline{\mathbf{b}}(n-1) = \begin{bmatrix} -1 & 0 \\ 0 & 1 \end{bmatrix} \underline{\mathbf{S}} \underline{\mathbf{a}}(n-1) \qquad (2.50)$$

Here, $\underline{\mathbf{a}}(n) = [\underline{a}_1(n), \underline{a}_2(n)]^T$ and $\underline{\mathbf{b}}(n) = [\underline{b}_1(n), \underline{b}_2(n)]^T$ are the vectors of wave variables incident on and reflected from the adaptor, respectively, at time step n, and $\underline{\mathbf{S}}$ is the orthogonal scattering matrix for a power-normalized parallel adaptor, as defined on page 43.

The same discrete-time system can, of course, be written in terms of the discrete-time instantaneous voltage and current. Applying the trapezoid rule to (2.49) gives the system

$$\begin{bmatrix} 1 & -R_1 \\ 1 & R_2 \end{bmatrix} \begin{bmatrix} v(n+1) \\ i(n+1) \end{bmatrix} = \begin{bmatrix} -1 & -R_1 \\ 1 & -R_2 \end{bmatrix} \begin{bmatrix} v(n) \\ i(n) \end{bmatrix}$$

Notice that voltage and current are instantaneously coupled at each time step. Though in this simple example, it is not difficult to arrive at an explicit update equation, in the WD setting, this is automatic.

Energy and Stability

It is useful to examine the defining equations (2.49) for the LC oscillator from an energetic point of view. Multiplying the first equation by i and the second by v gives

$$vi = Li\frac{di}{dt} = \frac{d}{dt}\left(\frac{1}{2}Li^2\right) \qquad vi = -Cv\frac{dv}{dt} = -\frac{d}{dt}\left(\frac{1}{2}Cv^2\right)$$

Subtracting the second equation from the first gives an energy relationship

$$\frac{dE}{dt} = 0$$

where

$$E(t) \triangleq \underbrace{\frac{1}{2}Li^2}_{\text{Energy stored in magnetic field surrounding inductor}} + \underbrace{\frac{1}{2}Cv^2}_{\text{Energy stored in electric field surrounding capacitor}} \qquad (2.51)$$

In other words, energy is traded back and forth between the two circuit elements, but is not dissipated.

Such a positive-definite quadratic error measure becomes extremely useful in the discrete domain, as a means of bounding the growth of a numerical solution, or, in other words, ensuring stability. For example, for the LC oscillator discretized using the trapezoid rule, we have the equations

$$\frac{T}{2}(v(n+1) + v(n)) = L(i(n+1) - i(n-1))$$

$$\frac{T}{2}(i(n+1) + i(n)) = -C(v(n+1) - v(n-1))$$

Similar to the continuous-time case, we may now multiply the first equation by $(i(n) + i(n-1))/2$ and the second by $(v(n) + v(n-1))/2$, and then subtract the second equation from the first to get

$$E_{FD}(n+1) = E_{FD}(n)$$

where we define $E_{FD}(n)$ by

$$E_{FD}(n) \triangleq \frac{L}{2}(i(n))^2 + \frac{C}{2}(v(n))^2$$

Thus we have, in this case, a discrete-time analogue of the energy, which is preserved throughout the recursion. As mentioned above, this is indeed a simple quadratic function of the state variables themselves. $E_{FD}(n)$ remains constant and thus the sizes of the state variables themselves may be bounded in terms of the energy initially present in the system, $E_{FD}(0)$. We must have, for example, that

$$|v(n)| \leq \sqrt{\frac{2E_{FD}(0)}{C}}$$

It should be clear that as the wave variable implementation relies on the same discretization, it must preserve a similar energy. In this case, it is immediately apparent from (2.50); due to the orthogonality of the scattering matrix, we must have

$$(a_1(n+1))^2 + (a_2(n+1))^2 = (a_1(n))^2 + (a_2(n))^2$$

and thus

$$E_{WD}(n) \triangleq T\left((a_1(n+1))^2 + (a_2(n+1))^2\right) = \text{constant}$$

From the definition of the wave quantities,

$$a_1(n) = \frac{v(n) + R_1 i(n)}{2\sqrt{R_1}} \qquad a_2(n) = \frac{v(n) - R_2 i(n)}{2\sqrt{R_2}}$$

where again, we have $R_1 = 2L/T$ and $R_2 = T/(2C)$, it is simple to show that

$$E_{WD}(n) = \left(1 + \frac{T^2}{4LC}\right) E_{FD}(n)$$

In other words, the wave digital energy is the same as the energy of the finite difference scheme up to a scaling factor (which approaches 1 in the limit as T approaches zero).

As expected, the wave digital implementation is behaving just as a finite difference scheme does, but note that the preservation of energy is now viewed in terms of the orthogonality of the scattering matrix[8]. We could equally well have carried out all analysis in the frequency domain in this case, but it is worth calling attention to time domain

---

[8] In this simple example (and indeed for any linear and time-invariant system), voltage wave variables could be employed instead of power-normalized quantities, without affecting our energy measure. The scattering matrix would no longer be orthogonal, but it would preserve a weighted l_2 norm of the wave variables through the scattering operation, according to (2.42), which is again sufficient for stability.

energetic analysis of finite difference schemes; when applied to numerical methods for time-dependent partial differential equations, these techniques are often referred to as the *energy method* [112], and serve as a direct means of proving stability when spectral analysis tools may not be easily applied (as is the case for nonlinear or time-varying systems). It is important to note that all the scattering methods to be discussed in this book are finite difference schemes, but are constructed from energy-preserving blocks or modules.

Chapter 3

Multidimensional Wave Digital Networks

The last chapter was concerned with techniques for deriving a digital filter design from an analog network. As mentioned briefly in Section 2.3, the resulting digital filter structure can also be considered to be an explicit numerical solver for the system of ordinary differential equations (ODEs) defined by the analog network that performs the filtering on continuous-time signals. This is perhaps an obvious point but was apparently first noted in the literature in [93]. It is interesting that this link was not made immediately in the multidimensional case, which is the subject of this chapter.

A multidimensional generalization of wave digital filters (MDWDFs) first appeared rather early [66], and most of the initial work involved applications to 2D filter design [159, 160]. It was itself an outgrowth of earlier work in the area of multidimensional circuits and systems [32, 144, 182] where the emphasis was on the synthesis of so-called variable networks (i.e., lumped networks with variable elements). The procedure for deriving a wave digital filter is largely the same in multiple dimensions (MD) as for the lumped case: to a given reference circuit made up of elements connected either in series or parallel, various transformations are applied, specifically a change to wave variables, and spectral mappings. The end product is a multidimensional wave digital (MDWD) network that has nearly all of the same desirable properties as lumped WD networks, especially recursive computability, and insensitivity to signal and coefficient truncation. The difference in multiple dimensions, however, is that the reference circuit, usually called a multidimensional Kirchhoff circuit, or MDKC, is now far more of a mathematical abstraction than a lumped circuit; the circuit state is a function of several variables that may or may not include time, and the circuit elements (as well as the connections between them) must be interpreted in a distributed sense. In particular, it is not a circuit that can be 'built' (except in the case of the variable networks mentioned previously). As such, the major problem confronting the designer of an MDWDF is the construction of this reference circuit [67, 66]. Various techniques were put forth, some involving MD circuits obtained by rotation of a known lumped reference network [68]. A good deal of work went into the related synthesis problem for general multidimensional reactance two-ports [10, 11, 80], which is much more difficult than in the lumped case (and not solvable in general).

The first paper to consider MDWDFs from a simulation perspective (in which case we will refer to them as MDWD networks) appeared in 1990 [87], though it was foreshadowed much earlier in the work of Strube [240]. That is to say, in analogy with the lumped case, a closed MDWD network could be considered to be a simulator of a distributed system that is defined by a system of partial differential equations (PDEs) and represented by an MDKC. Here, unlike in filtering, there is a clear interpretation of the reference circuit that is simply a symbolic restatement of the defining equations of a particular model system. The wave digital numerical integration approach is applicable to a wide variety of physical systems, including electromagnetics [72], coupled transmission lines [91, 145], linear acoustics [217], and elastic solid and beam dynamics [176]. Most surprisingly, the method can be applied to highly nonlinear systems [169], such as those of fluid dynamics [21, 71, 100], as well as even more complex hybrids, such as the magnetohydrodynamic system [259]. The method requires that the propagation speeds in the problem to be modeled be bounded; this is equivalent to saying that the system should be of hyperbolic type[1][238]. This requirement is important because numerical methods derived through this approach can be interpreted as explicit finite difference schemes [112]; as such, they must obey a requirement (the Courant–Friedrichs–Lewy condition [238]) relating the physical region of dependence for the model problem to a similar region on a numerical grid. (We would also like to note that a related approach to numerical integration, based on a transfer function formulation has been taken by Rabenstein et al. [147, 192, 193, 194].)

Although this chapter is intended in part as an extended review and compendium of the work to date in the field of numerical integration through the use of wave digital filters, the subtext is certainly that these methods can and should be treated as a particular class of finite difference methods endowed with a special property, namely, passivity. This point has not been explored in any depth in the literature, except in the lumped case [176]. Such a treatment will also make it easier to compare wave digital methods to digital waveguide networks (DWNs), which can also be used for numerical integration purposes in a very similar way. We will return, in Chapter 6, to the subject of MDWD networks for such a comparison, and eventually, a unification of the two methods. In Chapter 7, we will apply the concepts discussed here to a variety of more complex systems, and in particular, to those describing the vibration of beams, plates, shells, and elastic solids, and in Chapter 8, to nonlinear and time-varying systems.

Summary

This chapter begins with a review of some of the basics of symmetric hyperbolic systems of PDEs, and then proceeds to the generalization of electrical network passivity into

---

[1] It is possible to extend the MDWD approach to cover parabolic systems [238] as well; parabolic systems may not have a bounded propagation velocity but they can be approximated by hyperbolic systems. This is essentially the path taken by Fettweis in the modeling of the full *Navier–Stokes equations* [153] that describe the behavior of a general viscous fluid [71]; we remark that a similar idea, termed 'second-sound theory,' [40, 277] has been used to hyperbolicize parabolic problems (indeed, all time-dependent systems obeying the laws of classical physics *must* be hyperbolic, even if certain models do not reflect this). Elliptic problems that typically occur in describing steady state potential distributions in both electrostatics and fluid dynamics can be dealt with using MDWD networks via a relaxation-type approach [69]. Nitsche [177] has gone further to develop a class of fast eigenvalue solvers for steady-state problems. See Appendix B.

multiple dimensions, where it has been called *MD-passivity* [70, 116]; this can be done in a straightforward way through the application of coordinate transformations [90, 163]. Next, we introduce *multidimensional circuit elements* [67], which are similar to their lumped counterparts, except that they are distributed objects and may have particular directions associated with them. The transformation to wave variables and discretization proceeds as in the lumped case but now the trapezoid rule must be interpreted in a directional sense, as must be the associated MD spectral bilinear transformations. We then proceed through some treatments of typical model problems, namely the *advection equation* [238], the *transmission line system* [146], and its extension to two spatial dimensions, in which case it is called the *parallel-plate system* [88, 89]. We write down MDKC representations and show the corresponding MDWD networks for all these systems. We then spend some time in Section 3.8.1 examining MDWD structures as finite difference schemes and make some comments about frequency-domain behavior, paying particular attention in Section 3.8.2 to *parasitic modes*. In Section 3.9 we discuss the initialization of MDWD methods and then give a brief overview of methods for setting boundary conditions. *Balanced forms* are introduced in Section 3.11 as a means of increasing the computational efficiency of MDWD methods; to date they have been notoriously suboptimal in that the maximum allowable time step can be a great deal smaller than that of conventional finite difference methods (such as, for example, the *finite difference time domain* (FDTD) method [246, 286] and, by extension, DWNs). Finally, in Section 3.12, we turn to a means of incorporating higher-order spatially accurate [238] methods into a circuit framework; this is surprising because it had long been assumed that MDWD methods, traditionally based on the use of the trapezoid rule, could be no better than second-order accurate[2]. We circumvent this problem by applying an alternative integration rule, which is also passivity-preserving (and which will also serve as a 'back door' into the realm of DWNs and also of TLM).

3.1 Symmetric Hyperbolic Systems

In the previous chapter, we examined the discretization of lumped analog circuits; by lumped, we mean that the voltages and currents in these circuits are functions of only one independent variable: time. Although the described procedure was originally intended as a means of developing robust digital filtering structures, an equivalent point of view is that such structures in fact *numerically integrate* the set of ODEs describing the time evolution of these currents and voltages. We examined the numerical integration of a simple ODE system in Section 2.3.

In a distributed problem, the dependent variables are functions not only of time t but also of location within an n-dimensional spatial domain $\mathcal{D}$, with coordinates $\mathbf{x} = [x_1, \ldots, x_n]^T$. Such a problem is referred to as an $(n+1)$D problem in the WDF literature [176]. If the equations that define the problem include differential operators, we are faced with solving a set of PDEs.

A particularly important family of PDE systems is the set of *symmetric hyperbolic systems* [104, 112] of the form

$$\mathbf{P}\frac{\partial \mathbf{w}}{\partial t} + \sum_{k=1}^{n} \mathbf{A}_k \frac{\partial \mathbf{w}}{\partial x_k} + \mathbf{B}\mathbf{w} + \mathbf{f} = \mathbf{0} \tag{3.1}$$

---
[2]Gunnar Nitsche, private communication, 1999.

Here, **w**, the state, is a q-element column vector defined over coordinates $\mathbf{x} \in \mathcal{D} \subset \mathbb{R}^n$ and $t \geq 0$. **P** and $\mathbf{A}_k$, $k = 1, \ldots, n$, are real symmetric $q \times q$ matrices[3]; and in particular, **P** is assumed to be positive definite. **B** is a real $q \times q$ matrix (not necessarily symmetric) whose symmetric part models energy loss or growth, and the q-element real column vector **f** is a *forcing function* or excitation. The matrices $\mathbf{A}_k$ are assumed to be constant, though **P** and **B** are allowed to depend on **x**. These systems are thus linear and time-invariant but not generally shift-invariant, so that we cannot apply spatial Fourier transforms directly to analyze them. System (3.1) must be complemented by initial and boundary conditions [112] in order for the solution to exist and be unique.

Though it is possible to extend this definition to include cases in which the matrices $\mathbf{A}_k$ may depend on **x**, t or even **w** (in which case system (3.1) is nonlinear), this simpler form describes a wide variety of physical systems, from electromagnetics to string, membrane, beam, plate, shell, and elastic solid dynamics, to transmission line systems, to linear acoustics, and so on. Symmetric hyperbolic systems are important because they form a subclass of *strongly hyperbolic* systems, for which the initial-value problem is well posed [238]. Roughly speaking, to say that a system is well posed is to say that the growth of its solution is bounded in a well-defined way; growth in an L_2 norm cannot be faster than exponential. This concept is elaborated in detail in various standard texts [112, 238]. We can examine this growth in the present case as follows.

First, assume that the problem is defined over an unbounded spatial domain $\mathcal{D} = \mathbb{R}^n$, so that we can drop any consideration of boundary conditions, and also that of the forcing function $\mathbf{f} = \mathbf{0}$. We now take the inner product of **w** with (3.1) to get

$$\mathbf{w}^T \mathbf{P} \frac{\partial \mathbf{w}}{\partial t} + \sum_{k=1}^{n} \mathbf{w}^T \mathbf{A}_k \frac{\partial \mathbf{w}}{\partial x_k} + \frac{1}{2} \mathbf{w}^T (\mathbf{B} + \mathbf{B}^T) \mathbf{w} = 0 \tag{3.2}$$

where we have replaced **B** by its symmetric part $\frac{1}{2}(\mathbf{B} + \mathbf{B}^T)$, and where ^{T} denotes transposition. Due to the symmetry of **P** and the $\mathbf{A}_k$, we can then write

$$\frac{\partial}{\partial t} \left(\frac{1}{2} \mathbf{w}^T \mathbf{P} \mathbf{w} \right) + \frac{1}{2} \sum_{k=1}^{n} \frac{\partial}{\partial x_k} \left(\mathbf{w}^T \mathbf{A}_k \mathbf{w} \right) + \frac{1}{2} \mathbf{w}^T \left(\mathbf{B} + \mathbf{B}^T \right) \mathbf{w} = 0 \tag{3.3}$$

Now, integrate (3.3) over $\mathbb{R}^n$, to get

$$\frac{d}{dt} \int_{\mathbb{R}^n} \frac{1}{2} \left(\mathbf{w}^T \mathbf{P} \mathbf{w} \right) dV + \frac{1}{2} \int_{\mathbb{R}^n} \sum_{k=1}^{n} \frac{\partial}{\partial x_k} \left(\mathbf{w}^T \mathbf{A}_k \mathbf{w} \right) dV + \int_{\mathbb{R}^n} \frac{1}{2} \mathbf{w}^T (\mathbf{B} + \mathbf{B}^T) \mathbf{w} \, dV = 0 \tag{3.4}$$

where $dV = dx_1 dx_2 \ldots dx_n$ is the nD differential volume element. The expression $\sum_{k=1}^{n} \frac{\partial}{\partial x_k} \left(\mathbf{w}^T \mathbf{A}_k \mathbf{w} \right)$ is easily seen to be the divergence of a vector field, and by the Divergence Theorem, the integral of this quantity can be replaced by a surface integral over the problem boundary—because we have assumed no boundary, this integral vanishes

---

[3] We will always choose $x_1 = x$, $x_2 = y$, $x_3 = z$, so the matrices $\mathbf{A}_1$, $\mathbf{A}_2$, and $\mathbf{A}_3$ in (3.1) will refer to the matrix coefficients of the partial derivatives in these three directions.

3.1. SYMMETRIC HYPERBOLIC SYSTEMS

and we are left with

$$\frac{d}{dt}\int_{\mathbb{R}^n}\frac{1}{2}\left(\mathbf{w}^T\mathbf{P}\mathbf{w}\right)dV + \frac{1}{2}\int_{\mathbb{R}^n}\mathbf{w}^T(\mathbf{B}+\mathbf{B}^T)\mathbf{w}\,dV = 0 \qquad (3.5)$$

The quantity

$$E(t) \triangleq \int_{\mathbb{R}^n}\frac{1}{2}\left(\mathbf{w}^T\mathbf{P}\mathbf{w}\right)dV \qquad (3.6)$$

can be interpreted as the total energy of system (3.1) at time t. Note that due to the positivity requirement on $\mathbf{P}$, it is a positive definite function of the state, $\mathbf{w}$. If $\mathbf{B}+\mathbf{B}^T$ is positive-semidefinite, then we must have, from (3.5), that

$$\frac{dE}{dt} \leq 0$$

which implies that

$$E(t_2) \leq E(t_1) \qquad \text{for} \qquad t_2 \geq t_1 \qquad (3.7)$$

In other words, the energy of the system must decrease as time progresses.

In the MD circuit models that we will discuss, what we will be doing, in essence, is dividing this energy up among various reactive MD circuit elements. We will elaborate on this in the sections on the $(1+1)$D transmission line and $(2+1)$D parallel-plate system. The passivity condition is essentially equivalent to (3.7). Also, the symmetric nature of the systems will be reflected in the circuit models by the use of mainly *reciprocal* [14] circuit elements, though non-reciprocal elements (gyrators) will come into play if $\mathbf{B}$ is not symmetric (it is not required to be, and note that system (3.1) is well posed regardless of the form of $\mathbf{B}$ [112]). The application of passive circuit methods to systems that are more generally strongly hyperbolic (for which energy estimates such as (3.7) can also be derived [112]) has not been explored in the literature. This would appear to be a worthy direction for future research.

Note on Boundary Conditions

In the analysis above, the spatial domain is assumed unbounded (i.e., we took $\mathcal{D}=\mathbb{R}^n$). It is useful to examine the energetic behavior of (3.1) if this is not the case. Integrating (3.3) over $\mathcal{D}$, we get

$$\frac{d}{dt}\int_{\mathcal{D}}\frac{1}{2}\left(\mathbf{w}^T\mathbf{P}\mathbf{w}\right)dV + \int_{\mathcal{D}}\nabla\cdot\mathbf{b}\,dV + \frac{1}{2}\int_{\mathcal{D}}\mathbf{w}^T(\mathbf{B}+\mathbf{B}^T)\mathbf{w}\,dV = 0$$

where $\nabla \triangleq [\frac{\partial}{\partial x_1},\ldots,\frac{\partial}{\partial x_n}]$, and where we have defined

$$\mathbf{b} \triangleq \tfrac{1}{2}[\mathbf{w}^T\mathbf{A}_1\mathbf{w},\ldots,\mathbf{w}^T\mathbf{A}_n\mathbf{w}]^T$$

If the boundary of $\mathcal{D}$ is sufficiently smooth, then upon applying the Divergence Theorem, we get

$$\frac{d}{dt}\int_\mathcal{D} \frac{1}{2}\left(\mathbf{w}^T\mathbf{P}\mathbf{w}\right)dV + \int_{\partial\mathcal{D}}\mathbf{b}^T\mathbf{n}_\mathcal{D}\,d\sigma + \frac{1}{2}\int_\mathcal{D}\mathbf{w}^T(\mathbf{B}+\mathbf{B}^T)\mathbf{w}\,dV = 0$$

where $\partial\mathcal{D}$ is the boundary of $\mathcal{D}$, $\mathbf{n}_\mathcal{D}$ is defined as the unit outward normal (assumed unique everywhere on $\mathcal{D}$), and $d\sigma$ is a surface element of $\mathcal{D}$. If we define the total energy by

$$E(t) \triangleq \int_\mathcal{D} \frac{1}{2}\left(\mathbf{w}^T\mathbf{P}\mathbf{w}\right)dV$$

then we have

$$\frac{dE}{dt} = -\int_{\partial\mathcal{D}}\mathbf{b}^T\mathbf{n}_\mathcal{D}\,d\sigma - \frac{1}{2}\int_\mathcal{D}\mathbf{w}^T(\mathbf{B}+\mathbf{B}^T)\mathbf{w}\,dV$$

If $\mathbf{B}+\mathbf{B}^T$ is positive-semidefinite, then a simple condition for passivity is

$$\mathbf{b}^T\mathbf{n}_\mathcal{D} \geq 0 \tag{3.8}$$

and the system is lossless if $\mathbf{B}$ is antisymmetric and (3.8) holds with equality.

This analysis is grossly incomplete, however, because we have not said anything about which boundary conditions ensure the existence and uniqueness of a solution; this analysis is rather involved and we refer the reader to standard texts [112] for an introduction. The basic issue is the over- or underspecification of $\mathbf{b}$ on the boundary. We will consider only lossless, memoryless boundary conditions in this book.

Phase and Group Velocity

Since the stability of an explicit numerical method (such as those that we will examine subsequently) that solves a system of hyperbolic equations is dependent on propagation velocities, it is worthwhile to spend a few moments here to define *phase* and *group velocities* [53, 136] for a system such as (3.1). As discussed by Trefethen [252, 253], the concept of group velocity turns out to be crucial when it comes to stability analysis for numerical schemes.

Let us return to the unbounded domain problem with $\mathcal{D} = \mathcal{R}^n$. Suppose that the matrices $\mathbf{P}$, $\mathbf{B}$, and $\mathbf{A}_k$, $k = 1, \ldots, n$ that define system (3.1) are real constants; and in particular, we assume that the driving term $\mathbf{f}$ is zero, and that $\mathbf{B}$ is antisymmetric, so that system (3.1) is lossless. This is then a linear and shift-invariant system and the solution can be written as a superposition of plane wave solutions of the form

$$\mathbf{w}(\mathbf{x}, t) = \mathbf{w}_0 e^{j\omega t + \boldsymbol{\beta}\cdot\mathbf{x}}$$

where $\mathbf{w}_0$ is a constant vector, ω is a real frequency variable, and $\boldsymbol{\beta} = [\beta_1, \ldots, \beta_n]^T$ is the n-component vector wavenumber defining the direction of propagation of the plane wave. Substituting this plane wave solution into the constant-coefficient system (3.1) gives

$$\left(j\omega\mathbf{P} + \sum_{k=1}^n j\beta_k\mathbf{A}_k + \mathbf{B}\right)\mathbf{w}_0 = \mathbf{0} \tag{3.9}$$

3.1. SYMMETRIC HYPERBOLIC SYSTEMS

Nontrivial solutions to (3.9) can only occur when

$$\det \left(j\omega \mathbf{P} + \sum_{k=1}^{n} j\beta_k \mathbf{A}_k + \mathbf{B} \right) = 0 \qquad (3.10)$$

The q solutions to this equation,

$$\omega_l(\boldsymbol{\beta}), \qquad l = 1, \ldots, q \qquad (3.11)$$

(which are not necessarily distinct) define dispersion relations, from which we can derive much useful information.

All the linear systems to be examined in this book are isotropic; propagation characteristics are independent of direction (though not necessarily of location, or frequency). For linear and shift-invariant (LSI) systems, this implies that the dispersion relations (3.11) can be written as functions of $\|\boldsymbol{\beta}\|_2$ alone, where $\|\boldsymbol{\beta}\|_2$ is simply the Euclidean norm of the vector $\boldsymbol{\beta}$. In this case, we may define the *phase* and *group velocities* for the lth relation by

$$\gamma_l^p \triangleq \frac{\omega_l}{\|\boldsymbol{\beta}\|_2} \qquad \gamma_l^g \triangleq \frac{d\omega_l}{d\|\boldsymbol{\beta}\|_2} \qquad (3.12)$$

(For nonisotropic systems, we will need to resort to vector generalizations of these quantities [136].) The phase velocity defines the speed of a single sinusoidal plane wave solution, and the group velocity can be interpreted as the speed of propagation of a wave packet; from the point of view of the stability of numerical methods, it is the group velocities that are of most importance because they define the speeds of information or energy transfer [53]. It is interesting to note that if $\mathbf{B}$ is nonzero, phase velocities may become unbounded in the limit as $\boldsymbol{\beta}$ becomes small—this occurs in several of the systems that will be discussed in Chapter 7, though for all these systems the group velocities will be bounded. This is related to the fact that the system characteristics [104] are independent of $\mathbf{B}$.

In the interest of extending these ideas to spatially inhomogeneous systems (of the form of (3.1) where $\mathbf{P}$ and $\mathbf{B}$ may exhibit a smooth functional dependence on $\mathbf{x} \in \mathcal{D}$), we note that at any location $\mathbf{x} = \mathbf{x}_0 \in \mathcal{D}$, solutions to system (3.1) behave locally as solutions to the *frozen-coefficient* system [112] defined by $\mathbf{P}(\mathbf{x}_0)$ and $\mathbf{B}(\mathbf{x}_0)$. We may then define local group velocities $\gamma_l^g(\|\boldsymbol{\beta}\|_2, \mathbf{x}_0)$, $l = 1, \ldots, q$ in the same way as in (3.12). A quantity that will appear frequently in our subsequent treatment of the stability of numerical methods for these systems will be the maximum global group velocity, defined as

$$\gamma_{\max}^g \triangleq \max_{\substack{l = 1, \ldots, q \\ \|\boldsymbol{\beta}\|_2 \geq 0 \\ \mathbf{x}_0 \in \mathcal{D}}} \gamma_l^g(\|\boldsymbol{\beta}\|_2, \mathbf{x}_0) \qquad (3.13)$$

which, more simply stated, is the maximum propagation velocity over all system modes, wavenumbers, and throughout the entire spatial problem domain.

We briefly note that for problems in which there is some spatial dependence of one or more material parameter values, for analysis purposes we will often need to make use of the maximum or minimum of such a quantity over the problem domain. For example, for a quantity $\chi(\mathbf{x})$, representing perhaps a spatially varying mass density, or a dielectric constant, and so on the maximum and minimum are denoted by

$$\chi_{\min} \triangleq \min_{\mathbf{x} \in \mathcal{D}} \chi \qquad \chi_{\max} \triangleq \max_{\mathbf{x} \in \mathcal{D}} \chi \qquad (3.14)$$

3.2 Coordinate Changes and Grid Generation

Before looking directly at circuits and signal flow diagrams in multiple dimensions, it is useful to introduce coordinate changes, which were first applied in the context of multidimensional wave digital filters by Liu and Fettweis [163]. One might add that it is *useful*, but not strictly *necessary*, since it is possible to develop numerical integration algorithms along the same lines without any explicit reference to new coordinates [89]. It is, however, a very convenient way of understanding causality and grid generation issues, as well as of generalizing the passivity concept to multiple dimensions [70, 116, 176].

Some of the lumped circuit elements we have discussed so far are *passive*—that is, they dissipate (or at least do not create) energy as time progresses, as do Kirchhoff networks composed of connections of such elements (by *Tellegen's Theorem* [185]). In the multidimensional setting, many systems possess a similar property; some measure of energy decreases as a function of time. For example, the amplitude of the vibrations in a struck string or membrane will gradually decrease as a function of time. We have also seen that, for lumped circuits, application of the trapezoid rule translates this passivity property to a discrete equivalent. When attempting a discretization of a set of PDEs, however, we have to cope not only with the time direction but also with spatial ones as well, and passivity (usually a result of the conservative nature of the laws from which a system of equations is derived) does not in general hold with respect to space [73].

The idea of Fettweis and Nitsche [90] was to perform a coordinate transformation such that the new coordinates, generally a mixture of time and space, all contain a part of the physical time variable. Traveling in the positive direction along any of the new coordinates implies forward motion in time (as well as in some spatial direction). More specifically, if

$$(t_1, \ldots, t_{n+1}) = f(x_1, \ldots, x_n, t)$$

are the new coordinates, the authors provide the following conditions:

Any positive change Δt in the variable t must be reflected by a similar positive change Δt_j in all the new coordinates t_j, $j = 1, \ldots, n+1$. (3.15a)

Conversely, any positive change Δt_j in any of the new coordinates must produce a positive change in the old variable t. (3.15b)

As a result, *all* the new coordinates have a timelike character; we will refer to such coordinates as *MD-causal*. The practical implications of this will become apparent in the next section when we introduce multidimensional circuit elements.

3.2.1 Structure of Coordinate Changes

These same authors provide some more detailed guidelines as to what types of coordinate changes are of interest [90]. In particular, they describe transformations of the form,

$$\mathbf{u} = \mathbf{V}^{-1}\mathbf{H}\mathbf{t} \tag{3.16a}$$

$$\mathbf{t} = \mathbf{H}^{-1}\mathbf{V}\mathbf{u} \tag{3.16b}$$

where $\mathbf{t} = [t_1, \ldots, t_{n+1}]^T$ are the new coordinates and $\mathbf{u} = [x_1, \ldots, x_n, t]^T$ are the old. $\mathbf{V}$ is prescribed to be $\mathrm{diag}(1, 1, \ldots, 1, v_0)$ and can be thought of as a simple scaling of the original coordinates $\mathbf{u}$ to a nondimensional (or rather, 'all-spatial') form. v_0 thus plays an important role, as we shall see later in a discrete setting, as the space step/time step ratio on a numerical grid. Its magnitude will be governed by a stability bound [238], sometimes called the *Courant–Friedrichs–Lewy* (CFL) condition, as in conventional explicit finite difference methods (although the manifestation of the condition in the networks we will derive is of a quite different character). The invertible matrix $\mathbf{H}$ is usually chosen to be orthogonal [90].

Here, we can see that the requirement (3.15a) will be satisfied if the elements in the rightmost column of $\mathbf{H}^{-1}$ are positive; if $\mathbf{H}$ is orthogonal, we have $\mathbf{H}^{-1} = \mathbf{H}^T$. The bottom row of $\mathbf{H}$ then consists of positive elements (often chosen equal, so as to give equal contributions from all components t_j to t), and satisfies requirement (3.15b).

The differential operators $\nabla_\mathbf{t} = [\frac{\partial}{\partial t_1}, \ldots, \frac{\partial}{\partial t_k}]^T$ and $\nabla_\mathbf{u} = [\frac{\partial}{\partial x_1}, \ldots, \frac{\partial}{\partial x_n}, \frac{\partial}{\partial t}]^T$ are related by

$$\nabla_\mathbf{t} = \mathbf{H}^T \mathbf{V}^{-1} \nabla_\mathbf{u} \qquad \nabla_\mathbf{u} = \mathbf{V}\mathbf{H}^{-T} \nabla_\mathbf{t} \tag{3.17}$$

Also, we introduce the scaled time variable

$$t' = v_0 t \tag{3.18}$$

which will necessitate a special treatment in the circuit models to follow. See Section 3.4.1 for more details.

3.2.2 Coordinate Changes in (1 + 1)D

Solving a set of PDEs numerically nearly always involves sampling the problem domain and attempting to approximate the solution to the problem at the finite collection of points. Coordinate sampling in the MDWDF context was first examined by Liu and Fettweis [158], and later with regard to MDWD networks by Fettweis and Nitsche [86] and Bass [9]. In (1 + 1)D, there is essentially only one useful type of regular grid; it is shown in Figure 3.1(a), where the grid spacings or step-sizes are assumed equal to Δ in the scaled time (i.e., $t' = v_0 t$) and space directions. Note that the use of the scaled coordinates allows this uniform sampling, without implying any restriction on the relative grid spacings in the unstretched coordinates, since we have introduced the (as yet) free parameter v_0.

Suppose we now change coordinates by

$$t_1 = \frac{1}{\sqrt{2}}(v_0 t + x) \qquad t_2 = \frac{1}{\sqrt{2}}(v_0 t - x) \tag{3.19}$$

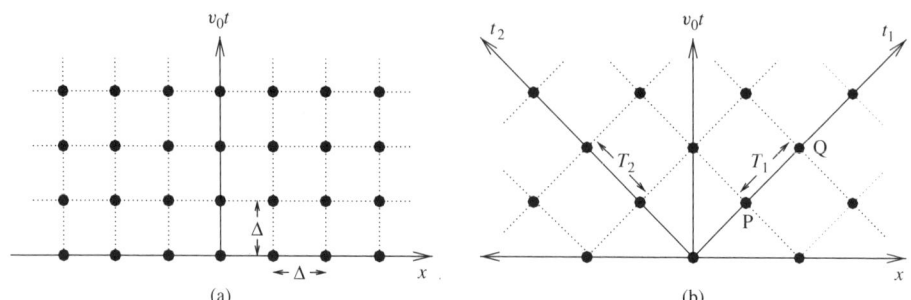

Figure 3.1: *Sampling grids,* (a) *in rectangular coordinates* $(x, v_0 t)$ *and* (b) *in the new coordinates* (t_1, t_2) *defined by* (3.19).

which corresponds to a transformation of type (3.16) with

$$\mathbf{H} = \frac{1}{\sqrt{2}} \begin{bmatrix} 1 & -1 \\ 1 & 1 \end{bmatrix} \quad (3.20)$$

If we now sample the plane in the (t_1, t_2) coordinates, with equal spacings along the two axes, using a step-size of $T_1 = T_2$, we obtain the grid in Figure 3.1(b). Notice that if we choose $T_1 = T_2 = \sqrt{2}\Delta$, then our grid aligns perfectly with exactly half of the grid points sampled uniformly along the $(x, v_0 t)$ axes, as in Figure 3.1(a). In fact, the grid of Figure 3.1(a) can be decomposed into two grids of the form in Figure 3.1(b), where one of the grids is shifted by (Δ, Δ) with respect to the other, in the $(x, v_0 t)$ plane. It will be possible in some instances to exploit this decomposition so as to achieve a gain in computational efficiency; the key idea here is that if we begin with a grid such as shown in Figure 3.1(a), and then are able to develop an algorithm such that only one of the two subdomains is used, then we will have halved the amount of computation, at the expense of a decrease in accuracy by a factor of $\sqrt{2}$ (the step-size in the (t_1, t_2) plane is $T_1 = T_2 = \sqrt{2}\Delta$ versus Δ in the $(x, v_0 t)$ plane). We will mention this *offset* sampling [89, 284] when we look at the (1 + 1)D transmission line problem in Section 3.6, and will examine subgrid decompositions extensively in Section 4.4.3 and Appendix A. It is important to point out that regardless of the coordinate change, updating in any of the WDF-based algorithms that will subsequently be developed will be done with respect to the *time* variable alone (the direction of data flow is still in the time direction), as per standard explicit finite difference methods for hyperbolic problems.

3.2.3 Coordinate Changes in Higher Dimensions

There are more choices for the type of coordinate transformation (and hence the type of grid) that are available when we move to higher-dimensional problems. As an example, let us look at the transformation to 'hexagonal' coordinates [90, 284] defined by

$$\mathbf{H} = \begin{bmatrix} \frac{1}{\sqrt{2}} & -\frac{1}{\sqrt{2}} & 0 \\ \frac{1}{\sqrt{6}} & \frac{1}{\sqrt{6}} & -\sqrt{\frac{2}{3}} \\ \frac{1}{\sqrt{3}} & \frac{1}{\sqrt{3}} & \frac{1}{\sqrt{3}} \end{bmatrix} \qquad \mathbf{V} = \begin{bmatrix} 1 & 0 & 0 \\ 0 & 1 & 0 \\ 0 & 0 & v_0 \end{bmatrix} \quad (3.21)$$

3.2. COORDINATE CHANGES AND GRID GENERATION

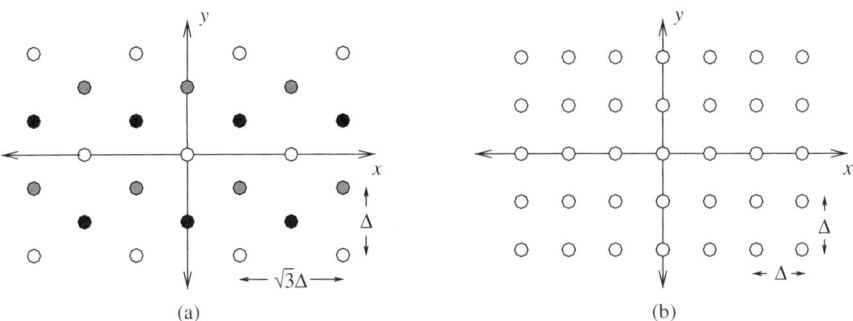

Figure 3.2: $(2+1)$D sampling grids in (a) hexagonal and (b) rectangular coordinates.

Uniform sampling in the (t_1, t_2, t_3) coordinates yields the grid arrangement shown in Figure 3.2(a), flattened onto the (x, y) plane. This is, effectively, a cubic lattice of points viewed along its main diagonal.

At any given time step, one of three different grids (in Figure 3.2(a) the different grids are marked by gray, white or black points), which are simple translations of each other, is used. In a discrete setting, it is sometimes possible (depending on the system at hand) to design an algorithm such that they are used cyclically—grid variables at white grid points can be updated with reference to variables at the gray points, which in turn were updated using stored variables at the black points, and so on. If the separation of the points is as indicated in Figure 3.2(a), then we have used sample steps of $T_1 = T_2 = T_3 = \sqrt{(3/2)}\Delta$.

Embeddings

In order to obtain a standard rectilinear grid in higher dimensions, it is possible to proceed in the same fashion but it is in fact more convenient to extend the class of coordinate transformations so as to embed the problem domain in a higher-dimensional space. In [90], the following generalization of (3.16) has been put forth:

$$\mathbf{u} = \mathbf{V}^{-1}\mathbf{H}\mathbf{t} \qquad (3.22a)$$

$$\mathbf{t} = \mathbf{H}^{-R}\mathbf{V}\mathbf{u} \qquad (3.22b)$$

Here, $\mathbf{u}$ is still the $(n+1)$-dimensional vector $[x_1, \ldots, x_n, t]^T$, but $\mathbf{t}$ is k-dimensional, with $k \geq n+1$. $\mathbf{H}$ must be chosen such that the elements in its bottom row are positive. $\mathbf{H}^{-R}$ is a $k \times (n+1)$ right pseudoinverse [124] of $\mathbf{H}$—in order to satisfy a generalization of the first of conditions (3.15), it must be chosen so that the elements in its rightmost column are all positive. For example, for a $(2+1)$D problem with $\mathbf{u} = [x, y, t]^T$, in order to generate a rectilinear grid, the following choice is usually made

$$\mathbf{H} = \begin{bmatrix} 1 & 0 & -1 & 0 & 0 \\ 0 & 1 & 0 & -1 & 0 \\ 1 & 1 & 1 & 1 & 1 \end{bmatrix} \qquad (3.23)$$

H projects five-dimensional coordinates $\mathbf{t} = [t_1, t_2, t_3, t_4, t_5]^T$ back to the three-dimensional space of **u**. One choice [90] for this right pseudoinverse is

$$\mathbf{H}^{-R} = \frac{1}{2}\mathbf{H}^T \mathrm{diag}\left(1, 1, \frac{2}{5}\right) \tag{3.24}$$

Uniform sampling in the **t** coordinates, with step-sizes of $T_j = \Delta$, $j = 1, \ldots, 5$ yields the standard rectangular grid shown in Figure 3.2(b), which is a pattern equivalent to what one would get by sampling uniformly (see comment below) in the $(x, y, v_0 t)$ coordinates, with a spacing of Δ in all three untransformed variables. It should be clear that to every grid point in the **u** coordinates corresponds a two-parameter family of points in the **t** coordinates; this fact will not influence the resulting difference schemes. This embedding of the problem domain in a higher-dimensional space is simply a means to an end; and in particular, we will not be solving a system numerically over a higher-dimensional grid (which would be computationally infeasible). The new coordinate directions are chosen so that they define a grid, and they will also serve as directions of *energy flow* for the MD circuit elements that we will define presently. In effect, the total energy flow in a physical system is broken up among these new coordinate directions; it will sometimes be true (as in the case of a rectilinear grid in $(2 + 1)$D) that energy can approach a particular grid location from a number of neighbors that is greater than the dimensionality of the problem (for the $(2 + 1)$D parallel-plate problem on a rectilinear grid, at least four: north, south, east, and west). We will take a closer look at this particular transformation and its suitability for calculation on a rectilinear grid in Section 3.7.

In $(3 + 1)$D, in order to obtain a standard rectilinear sampling pattern, Nitsche has proposed seven-dimensional coordinates [90] defined by

$$\mathbf{H} = \begin{bmatrix} 1 & 0 & 0 & -1 & 0 & 0 & 0 \\ 0 & 1 & 0 & 0 & -1 & 0 & 0 \\ 0 & 0 & 1 & 0 & 0 & -1 & 0 \\ 1 & 1 & 1 & 1 & 1 & 1 & 1 \end{bmatrix} \tag{3.25}$$

It is easy to verify that shifts of distance Δ along the coordinates t_j, $j = 1, \ldots, 6$ correspond to shifts of Δ along the positive and negative x, y, z, $-x$, $-y$, and $-z$ directions accompanied by a shift of $T = \Delta/v_0$ in the time direction. We will make use of this coordinate transformation when developing scattering methods for Maxwell's equations (see Section 6.4) and for the system describing elastic solid dynamics (see Section 7.4).

The embedding technique has some tricky aspects. We will make some comments here, in order to complement the information provided in the literature [90]. The two relationships given in (3.22) are not equivalent for general rectangular matrices **H**. Equation (3.22a) serves to define **t**, but the definition of directional derivatives in the **t** coordinates will be given by

$$\nabla_\mathbf{t} = \mathbf{H}^T \mathbf{V}^{-1} \nabla_\mathbf{u} \tag{3.26}$$

and depends only on **H**. The question of how sampling in the new coordinates is to be carried out is not well addressed in the literature. Suppose, for example, that we wish to use embedding (3.23). Grid definition proceeds by letting $\mathbf{t} = \Delta[n_1, n_2, n_3, n_4, n_5]^T$, where n_j, $j = 1, \ldots, 5$ are integers. Clearly, then, using (3.22a), grid points in the original coordinates are given by $\mathbf{u} = [\Delta(n_1 - n_3), \Delta(n_2 - n_4), \Delta/v_0(n_1 + n_2 + n_3 + n_4 + n_5)]^T$, and thus any point of the form $\mathbf{u} = [\Delta m_1, \Delta m_2, \Delta/v_0 m_3]^T$, for integer m_1, m_2, and m_3

3.3. MD-PASSIVITY

is in the range of $\mathbf{V}^{-1}\mathbf{H}$ for some choice of the n_j. This defines the rectilinear grid in the untransformed coordinates. Not all of these points, however, can be mapped back to some $\mathbf{t}$ with $\mathbf{t} = \Delta[n_1, n_2, n_3, n_4, n_5]^T$ under (3.22b). This is worthy of note but will not influence the numerical methods that will depend on discretizing directional derivatives in the $\mathbf{t}$ coordinates, which, as mentioned above, are defined in terms of $\mathbf{H}$ and not $\mathbf{H}^{-R}$. We remark that the inverse relationship for (3.26) will be given by

$$\nabla_{\mathbf{u}} = \mathbf{V}\mathbf{H}^{-RT} \nabla_{\mathbf{t}} \qquad (3.27)$$

where $\mathbf{H}^{-RT}$ is the transpose of $\mathbf{H}^{-R}$.

We do not wish to go too much into the formalism of these coordinate transformations here; it seems excessive since the associated circuit manipulations that we will review are quite straightforward. As mentioned earlier, the coordinate changes in this section are introduced in order to aid in understanding the method and MD-passivity and are not necessary for deriving WDF-based algorithms for numerical integration, though it would appear that some types of reference circuits can only be derived via the transformation approach[4].

3.3 MD-passivity

In dealing with networks and circuit elements in multiple dimensions, we must have a means of generalizing their energetic properties accordingly. In particular, the notion of passivity, which in the lumped case played an important role in developing stable digital filters directly from an analog network, must be expanded to include the distributed character of the system to be modeled. The definition of MD-passivity was first given by Fettweis [70], and more details are provided by Nitsche [176] and Hemetsberger [116]. The idea is nearly the same as in the lumped case—a passive N-port cannot produce energy on its own and hence a well-defined [14] network made up of Kirchhoff connections of such passive N-ports recirculates and possibly dissipates energy. The difference is that in multiple dimensions, we would like to be able to take into account that for most physical systems, conservation of energy is a property holding with respect to time alone. We will need to make use of the coordinates defined in Section 3.2, so as to ensure that passivity holds with respect to all coordinates in the problem. In this section, we recap the main points of the definitions and derivations which appear in the literature mentioned above.

We begin by defining a domain G in the vector space defined by the new coordinates $\mathbf{t} = [t_1, \ldots, t_k]^T$ under a transformation of the type (3.22) (which may be an embedding). Consider an N-port defined over the domain G, with port voltages $v_j(\mathbf{t})$ and currents $i_j(\mathbf{t})$, for $j = 1, \ldots, N$. The instantaneous absorbed power density, at any point in the interior of G is defined by

$$w_{\text{inst}}(\mathbf{t}) = \sum_{j=1}^{N} v_j i_j$$

and the *stored energy flow* as a vector field

$$\mathbf{E} = [E_1, \ldots, E_k]^T$$

---

[4]Gunnar Nitsche, private communication, 1999.

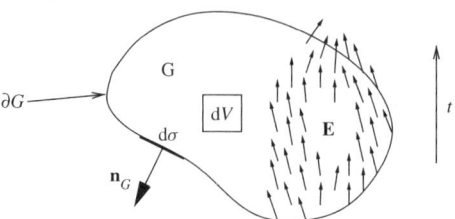

Figure 3.3: *k-dimensional domain G*.

In addition, we can define the source and dissipated power densities within G to be $w_\text{s}(\mathbf{t})$ and $w_\text{d}(\mathbf{t})$. The energy balance of the N-port can then be generalized directly from (2.3):

$$\int_G (w_\text{inst} + w_\text{s} - w_\text{d})\, \mathrm{d}V = \int_{\partial G} \mathbf{n}_G \mathbf{E}\, \mathrm{d}\sigma \qquad (3.28)$$

where $\mathbf{n}_G$ is the k-element row vector outward unit normal to the surface of G, $\mathrm{d}\sigma$ is a surface element of G, and $\mathrm{d}V$ is a volume element internal to G. See Figure 3.3 for a graphical representation of some of the relevant quantities[5]. The N-port is called *MD-passive* if there is a stored energy vector field $\mathbf{E}$ that is a positive-semidefinite function of the state of the N-port (i.e., all components of $\mathbf{E}$ are nonnegative, everywhere in G) such that

$$\int_G w_\text{inst}\, \mathrm{d}V \geq \int_{\partial G} \mathbf{n}_G \mathbf{E}\, \mathrm{d}\sigma \qquad (3.29)$$

and *MD-lossless* if (3.29) holds with equality. The total stored energy lost through the boundary of G must be less than the energy supplied through the ports in G; this is equivalent, from (3.28), to saying that the energy dissipated in G must be greater than the energy coming from the source. The previous definition of MD-passivity has been more precisely called *integral* MD-passivity (with respect to a domain G) [116]. A corresponding differential (pointwise) definition is

$$w_\text{inst} \geq \nabla_\mathbf{t} \cdot \mathbf{E} \qquad \text{in } G \qquad (3.30)$$

An N-port that is differentially MD-passive everywhere throughout a domain G will also be integrally MD-passive with respect to G. The converse is not necessarily true.

It is also useful to define, for an N-port, a scalar total energy [116] by

$$\mathcal{E}(t) = \int_{G_t} \mathbf{e}_t^T \mathbf{E}\, \mathrm{d}x_1 \mathrm{d}x_2 \ldots \mathrm{d}x_n \qquad (3.31)$$

Here G_t a spatial region defined as the cross section of G at time t, and $\mathbf{e}_t$ is a column unit vector in the time direction; note that this definition is framed in terms of the untransformed coordinates $\mathbf{u}$, and $\mathbf{E}$ has been projected onto these coordinates under (3.22a). It can also be used as a measure of the total energy at time t in a given circuit, as we will see in Section 3.6.3.

Fettweis [66] looks at an extension of the idea of positive realness (see Section 2.1.2) to two dimensions, for the case of a real linear and shift-invariant N-port. This idea generalizes

---

[5] Adapted from Figure 2 of [116] and Figure 1 of [70].

3.3. MD-PASSIVITY

easily to higher dimensions, as per some very early work in MD system theory [182]. Consider a real linear and shift-invariant k-dimensional N-port, where the port quantities are in an exponential state of frequency $\mathbf{s_t}$, where

$$\mathbf{s_t} = [s_1, \ldots, s_k]^T$$

are the frequency variables conjugate to $\mathbf{t}$. Thus we have the real instantaneous voltages and currents

$$v_j(\mathbf{t}) = \mathrm{Re}\left(\hat{v}_j e^{\mathbf{s_t}^T \mathbf{t}}\right) \qquad i_j(\mathbf{t}) = \mathrm{Re}\left(\hat{i}_j e^{\mathbf{s_t}^T \mathbf{t}}\right) \qquad j = 1, \ldots, N$$

where $\hat{v}_j$ and $\hat{i}_j$ are complex amplitudes. If there is an impedance relation between the voltages and currents, then we can write

$$\hat{\mathbf{v}} = \mathbf{Z}(\mathbf{s_t})\hat{\mathbf{i}}$$

where $\hat{\mathbf{v}} = [\hat{v}_1, \ldots, \hat{v}_N]^T$ and $\hat{\mathbf{i}} = [\hat{i}_1, \ldots, \hat{i}_N]^T$. The total complex MD power density at frequency $\mathbf{s}_t$ can be defined as

$$w(\mathbf{s_t}) = \hat{\mathbf{i}}^* \hat{\mathbf{v}}$$

and the average or active power density as

$$\overline{w}(\mathbf{s_t}) = \mathrm{Re}\left(\hat{\mathbf{i}}^* \hat{\mathbf{v}}\right)$$

The positive realness condition on $\mathbf{Z}$ for MD-passivity follows immediately and is similar to (2.5), except that we now must have

$$\mathbf{Z}(\mathbf{s_t}) + \mathbf{Z}^*(\mathbf{s_t}) \geq 0 \qquad \text{for} \qquad \mathrm{Re}(s_j) \geq 0 \quad j = 1, \ldots, k \qquad (3.32)$$

Thus the impedance must be positive real in *all* the new coordinates. The N-port is MD-lossless if (3.32) holds with equality for $\mathrm{Re}(s_j) = 0$, $j = 1, \ldots, k$. It is important to note that because of (3.27) and (3.26), we have

$$\mathbf{s_u} = \mathbf{V}\mathbf{H}^{-RT}\mathbf{s_t} \qquad \mathbf{s_t} = \mathbf{H}^T \mathbf{V}^{-1}\mathbf{s_u} \qquad (3.33)$$

where $\mathbf{s_u} = [s_{x_1}, \ldots, s_{x_n}, s_t]^T$ is the vector of frequencies in the untransformed coordinates $\mathbf{u}$. Thus, due to the positivity condition on the elements of the last row of $\mathbf{H}^{-RT}$ and the last column of $\mathbf{H}^T$, we will have

$$\mathrm{Re}(s_j) \geq 0 \quad \text{for} \quad j = 1, \ldots, k \quad \Longleftrightarrow \quad \mathrm{Re}(s_t) \geq 0$$

so that for an MD-passive N-port,

$$\mathbf{Z} + \mathbf{Z}^* \geq 0 \qquad \text{for} \qquad \mathrm{Re}(s_t) \geq 0 \qquad (3.34)$$

It is thus seen that MD-passivity can be interpreted as passivity but spread over a new system of coordinates (regardless of whether the new coordinates number more than the old).

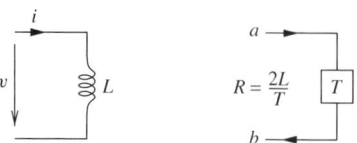

Figure 3.4: *Inductor and its wave digital one-port.*

3.4 MD Circuit Elements

Beginning from the perspective of the one-port circuit elements described in the last chapter, it is not difficult to see how such elements can be generalized to a multidimensional setting. Consider again the inductance and its WD one-port equivalent, shown in Figure 3.4. The voltage across the inductor is integrated and scaled by a factor $1/L$ to yield a current. The fact that the signal flow graph for the WD one-port equivalent is causal indicates that the inductor one-port is associated with the forward time direction. In the multidimensional case, the concepts of a direction associated with a circuit element and causality become crucial.

3.4.1 The MD Inductor

Consider the following (partial) differential equation:

$$v = L \frac{\partial i}{\partial t_j} \tag{3.35}$$

where L is a positive constant, and t_j, for any j, $j = 1, \ldots, k$ is a coordinate defined by transformation (3.22). We now have $v = v(\mathbf{t})$ and $i = i(\mathbf{t})$. Considered as an MD one-port, the instantaneous applied power will be

$$w_{\text{inst}} = vi = \frac{\partial}{\partial t_j}\left(\frac{1}{2}Li^2\right) = \nabla_{\mathbf{t}} \cdot \mathbf{E}$$

if we also define the stored energy flux $\mathbf{E}$ to be

$$\mathbf{E} = \frac{1}{2}Li^2\mathbf{e}_j \tag{3.36}$$

where $\mathbf{e}_j$ is a column unit vector in the t_j direction. If $L \geq 0$, this is indeed a positive-semidefinite vector function of the current across the one-port. Equation (3.35) defines a passive (in fact lossless) element in the sense of (3.30), henceforth called an MD inductor, of inductance L and *direction* t_j.

This equation resembles the voltage/current relation in an inductor with inductance L, the exception being that the integration variable is no longer time but t_j, a mixed space–time variable. In fact, the discretization procedure is identical to that of the lumped case, as we shall see, but for illustrative purposes, we will derive the multidimensional wave digital one-port. The important thing to note here is that even though (3.35) is defined over a k-dimensional domain with coordinates $\mathbf{t}$, it is solved (given v, say) as a series of one-dimensional integrations (since (3.35) must hold for all values of $\mathbf{t}$).

3.4. MD CIRCUIT ELEMENTS

We can immediately approximate (3.35) by the MD trapezoid rule [90] as

$$\frac{v(\mathbf{t} - \mathbf{T}_j) + v(\mathbf{t})}{2} = \frac{L}{T_j}\left(i(\mathbf{t}) - i(\mathbf{t} - \mathbf{T}_j)\right) + O(T_j^2) \tag{3.37}$$

where $\mathbf{T}_j = T_j \mathbf{e}_j$. T_j is interpreted as the step size. Assuming that we have uniformly sampled the $\mathbf{t}$ plane as in Section 3.2, with grid spacings $T_1, \ldots, T_k$, we now define the grid functions $v(\mathbf{n})$ and $i(\mathbf{n})$ where $\mathbf{n} = [n_1, \ldots, n_k]^T$ is an integer-valued vector. We intend to use them to approximate $v(\mathbf{t} = [n_1 T_1, \ldots, n_k T_k]^T)$ and $i(\mathbf{t} = [n_1 T_1, \ldots n_k T_k]^T)$, so we can immediately write the recursion

$$\frac{v(\mathbf{n}) + v(\mathbf{n} - \mathbf{e}_j)}{2} = \frac{L}{T_j}\left(i(\mathbf{n}) - i(\mathbf{n} - \mathbf{e}_j)\right) \tag{3.38}$$

which approximates (3.35) to $O(T_j^2)$.

We can now introduce the wave variables,

$$a(\mathbf{n}) = v(\mathbf{n}) + Ri(\mathbf{n})$$

$$b(\mathbf{n}) = v(\mathbf{n}) - Ri(\mathbf{n})$$

which are also grid functions defined over $\mathbf{n}$. As in the lumped case, R is an arbitrary positive number (here assumed constant). Inserting these wave variables into (3.38) yields, with the choice $R = 2L/T_1$,

$$b(\mathbf{n}) = -a(\mathbf{n} - \mathbf{e}_j) \tag{3.39}$$

In terms of the untransformed coordinates (where we will perform the updating in a simulation), (3.37) becomes

$$\frac{v(\mathbf{u} - \mathbf{V}^{-1}\mathbf{HT}_j) + v(\mathbf{u})}{2} = \frac{L}{T_j}\left(i(\mathbf{u}) - i(\mathbf{u} - \mathbf{V}^{-1}\mathbf{HT}_j)\right) \tag{3.40}$$

again to second order in the transformed spacing. The quantity $\mathbf{V}^{-1}\mathbf{HT}_j$ is the vector corresponding to the same shift, in the untransformed coordinates.

Take, for example, the case of an inductor of direction t_1 under the coordinate change defined by (3.20). A shift of $\mathbf{T}_1 = T_1\mathbf{e}_1$ of T_1 in direction t_1 corresponds to a shift in the old coordinates of $(T_1/\sqrt{2})[1, 1/v_0]^T$. Referring to Figure 3.1(b), where we have chosen $T_1 = \sqrt{2}\Delta$, the *instance* of the wave variable a entering the MD inductor at point P exits, sign-inverted, as b at point Q. From this standpoint, MD-losslessness is obvious since the MD inductor merely shifts and sign-inverts an array of numbers.

One point requires some clarification: the MD inductor as defined by (3.35) is MD-passive for constant $L \geq 0$, and for a transformed coordinate t_j, $j = 1, \ldots, k$. In problems for which material parameters have some spatial variation, some of the MD circuit elements that we will require will, as a rule, have some spatial dependence. If L in (3.35) is a function of $\mathbf{t}$, then in general the equation does not describe an MD-passive one-port. More precisely, if L does not commute with $\partial/\partial t_j$, then the application of the trapezoid rule to (3.35) does not yield the simple wave relationship (3.39). This begs the question then, of how the trapezoid rule can be applied to circuit elements that are not LSI (which we will require in order to numerically integrate systems with spatial material parameter variation).

For almost all the systems of interest in this book, it will be possible to consolidate any material parameter variation in circuit elements defined with respect to the pure time direction (recall that in our general symmetric hyperbolic system (3.1), such variation is confined to the coefficients of the time derivative term). For example, consider an inductor described by

$$v = L \frac{\partial i}{\partial t'} \quad (3.41)$$

in the $(1+1)$D coordinates defined by (3.19). Here, L is strictly positive but may be a function of x, and note that t' is not among the new coordinates defined by (3.19). Since L does commute with $\frac{\partial}{\partial t'}$, it is still possible to apply the trapezoid rule, in the time direction, in order to get a wave relationship of the form of (3.39). The directional shift will then be along the time direction and we need to be sure that the shift does in fact refer to another grid point—from Figure 3.1(b), we can see that this is in fact true (it is true for any of the coordinate systems discussed in Section 3.2). It is of course possible to include a pure time derivative among the new coordinates; this is done, for example, in the case of the embedding defined by coordinates (3.23), for which t_5 is simply t multiplied by a scaling factor. Nitsche [90] has called this the *generalized trapezoid rule*. Note, however, that if we write (3.41) as

$$v = \frac{L}{\sqrt{2}} \left(\frac{\partial i}{\partial t_1} + \frac{\partial i}{\partial t_2} \right) \quad (3.42)$$

it is *not* permissible to treat this as a series connection of two MD inductors—neither one is MD-passive, because L does not commute with either of the two directional derivatives.

A more general definition of an inductor, suitable for use in time-varying or nonlinear problems is

$$v = \sqrt{L} \frac{\partial}{\partial t_j} \left(\sqrt{L} i \right) = \frac{1}{2} \left(L \frac{\partial i}{\partial t_j} + \frac{\partial L i}{\partial t_j} \right) \quad (3.43)$$

for any transformed coordinate t_j. In this case, L can depend on $\mathbf{t}$ or even on v or i; as long as we have $L \geq 0$ and use power-normalized waves, the MD inductor defined by (3.43) is MD-passive [70, 116]. For constant L, (3.43) reduces to (3.35). Circuit elements of this type appear in circuit networks for fluid dynamical systems [21, 71, 100, 259], as well as in a vector-matrix context when dealing with the linearized Euler equations [117]. We also note that passivity under time-varying conditions can be enforced as it has been done in DWNs [228]; it would appear that waveguide networks (to be discussed in-depth in Chapter 4) could be generalized to include the nonlinear case in the same manner (see Section 8.6.2 for an interesting application of these ideas).

3.4.2 Other MD Elements

The inductor and capacitor are the only circuit elements that need a more involved treatment in the MD case[6]. The capacitor is treated as the dual to the inductor, replacing v by i and L by C, and needs no further comment, other than that, as with the lumped capacitor, there

---

[6]We will return to the multidimensional generalization of the unit element in Section 6.1.1.

3.4. MD CIRCUIT ELEMENTS

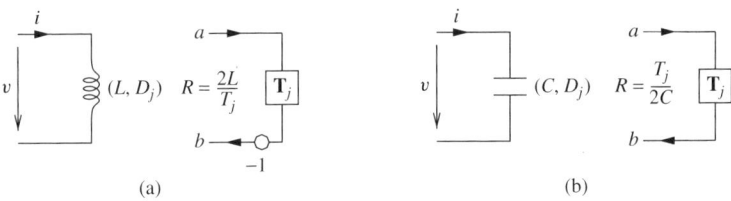

Figure 3.5: *MDWD one-ports*—(a) *an MD inductor, with inductance L, direction t_j and its MDWD counterpart, for step size T_j and $R = 2L/T_j$ and* (b) *an MD capacitor, of capacitance C, direction t_j and its MDWD one-port, with step size T_j and port resistance $R = T_j/2C$.*

is no sign inversion in the resulting MD wave one-port. The graphical representations of these MD one-ports and their MDWD equivalents are shown in Figure 3.5. Note that for the sake of compactness, in the circuit diagrams that will follow, we will use the derivative notation of the MDWD literature [176] where we have

$$D_j \triangleq \frac{\partial}{\partial t_j}$$

for some transformed coordinate t_j. In some instances, derivatives with respect to the original untransformed variables appear and we will write

$$D_t \triangleq \frac{\partial}{\partial t} \quad , \quad D_x \triangleq \frac{\partial}{\partial x} \quad , \quad D_y \triangleq \frac{\partial}{\partial y} \quad \ldots$$

We will also use the notation

$$D_{t'} \triangleq \frac{1}{v_0} \frac{\partial}{\partial t}$$

to refer to the dimensionless time derivative that appears frequently. Also, in a signal flow graph, we represent the operation of shifting by T_j in direction t_j by the symbol $\mathbf{T}_j$. In cases in which the system or N-port is linear and shift-invariant, we will be able to replace $\mathbf{T}_j$ by z_j^{-1}, the transmittance of a shift in direction t_j (see the next section).

All the other elements for which we will have a use, namely, the resistor, transformer, and gyrator, as well as scattering junctions are *memoryless* and hence their pointwise behavior in MD is identical to that of their lumped counterparts. Their graphical representations are also identical (see Section 2.1.4). We must keep in mind, however, that these are still distributed elements. For example, a resistor of resistance $R(t)$ in an MDKC represents some resistivity at every point in the domain of the problem.

A network made up of Kirchhoff connections of N-ports that are individually MD-passive can be shown (through the use of Tellegen's Theorem [185], which is unchanged in multiple dimensions) to be MD-passive as a whole [66].

3.4.3 Discretization in the Spectral Domain

If our network or N-port is linear and shift-invariant, it is also possible to view the discretization procedure as a *spectral mapping*, just as in the last chapter. Consider now the case for which the problem domain is some n-dimensional space, with coordinates

$\mathbf{u} = [x_1, \ldots, x_n, t]^T$, and where we have changed coordinates to $\mathbf{t} = [t_1, \ldots, t_k]^T$, with $k \geq n + 1$ via a transformation of type (3.22). The defining equation of an MD inductor of direction t_j for any $j = 1, \ldots, k$ is

$$v = L \frac{\partial i}{\partial t_j}$$

and for an exponential state at frequencies $\mathbf{s_t}$, we have

$$\hat{v} = L s_j \hat{i}$$

where $v = \hat{v} e^{\mathbf{s_t}^T \mathbf{t}}$ and $i = \hat{i} e^{\mathbf{s_t}^T \mathbf{t}}$. The 'impedance' is here $Z = L s_j$ and clearly satisfies the MD positive realness criteria given in (3.32) (and furthermore is MD-lossless) if $L \geq 0$. As in the lumped case, the trapezoid rule, now applied in the t_j direction, can be interpreted as a spectral mapping

$$s_j \to \psi_j \triangleq \frac{2}{T_j} \frac{1 - e^{-s_j T_j}}{1 + e^{-s_j T_j}} = \frac{2}{T_j} \frac{1 - z_j^{-1}}{1 + z_j^{-1}} \tag{3.44}$$

where T_j is some arbitrary step size in the t_j direction. For notational purposes, we have used

$$z_j^{-1} = e^{-s_j T_j}$$

to represent the frequency domain equivalent of a unit shift in the t_j direction. In complete analogy with the lumped case, (3.44) implies that

$$\text{Re}(s_j) \gtreqless 0 \iff \text{Re}(\psi_j) \gtreqless 0 \iff |z_j| \gtreqless 1$$

This shift can of course also be written in terms of delays and shifts in the $\mathbf{u}$ coordinates. For example, consider the coordinate transformation defined in (3.19). In this case we have, in the frequency domain,

$$s_1 = \frac{1}{\sqrt{2} v_0} s_t + \frac{1}{\sqrt{2}} s_x$$

$$s_2 = \frac{1}{\sqrt{2} v_0} s_t - \frac{1}{\sqrt{2}} s_x$$

where s_t and s_x are the frequency variables conjugate to t and x respectively. (We assume that our spatial domain is of infinite extent, so that s_x corresponds to an imaginary Fourier transform variable.) Suppose we have also chosen the step-sizes in the two coordinates such that the grids overlap—that is, $T_1 = \sqrt{2} \Delta = \sqrt{2} v_0 T$, where T is the shift in the pure time direction. Then, for a shift of T_1 in the t_1 direction, we can write

$$e^{-s_1 T_1} = e^{-\frac{1}{\sqrt{2} v_0} s_t T_1 - \frac{1}{\sqrt{2}} s_x T_1} = e^{-s_t T - s_x \Delta}$$

or

$$z_1^{-1} = z^{-1} w^{-1} \tag{3.45}$$

3.4. MD CIRCUIT ELEMENTS

where z^{-1} represents a delay of duration T in the time direction and w^{-1} corresponds to a shift over distance Δ in the positive space direction. Similarly, we can write

$$z_2^{-1} = z^{-1} w \qquad (3.46)$$

For a more complex example, consider again the transformation defined by

$$\mathbf{H} = \begin{bmatrix} 1 & 0 & -1 & 0 & 0 \\ 0 & 1 & 0 & -1 & 0 \\ 1 & 1 & 1 & 1 & 1 \end{bmatrix}$$

which maps coordinates $[x, y, t]^T$ to a five-dimensional coordinates $[t_1, t_2, t_3, t_4, t_5]^T$. A shift of $T_1 = \Delta$ in direction t_1 corresponds to a transmittance of the form

$$z_1^{-1} = e^{-s_1 T_1} = e^{-(s_x + \frac{1}{v_0} s_t)\Delta} = e^{-(s_x \Delta + s_t T)} = z^{-1} w_x^{-1}$$

where w_x represents a unit shift (of length Δ) in the x-direction, and as before, z^{-1} corresponds to a unit delay of $T = \Delta/v_0$. The other shifts can be written as

$$z_2^{-1} = z^{-1} w_y^{-1} \qquad z_3^{-1} = z^{-1} w_x \qquad z_4^{-1} = z^{-1} w_y \qquad z_5^{-1} = z^{-1}$$

where w_y represents a unit shift (of length Δ) in the y direction. At a given grid point in the old coordinates, the unit delays $z_1^{-1}, \ldots, z_4^{-1}$, interpreted as directional shifts, refer to points on the grid at the previous time step, and located one grid point away in the $-x, -y, x$, and y directions, respectively. The unit delay z_5^{-1} is simply a unit time delay.

It is important to note the manner in which the special character of the coordinate transformation manifests itself here. Owing to the positivity requirement on the elements of the last column of $\mathbf{H}$, a unit delay in *any* of the directions t_j will always include some delay in the pure time direction. By means of this requirement, and the introduction of wave variables, MDWD networks can, in the same way as their lumped counterparts, be designed in which delay-free loops do not appear. Such networks when used for simulation will give rise, in general, to *explicit* numerical schemes [238].

3.4.4 Other Spectral Mappings

One could well ask whether the spectral mappings of the form (3.44), which correspond to an application of the trapezoid rule, are the only means of deriving an MD-passive discrete system from a continuous one. A passivity-preserving mapping is simply one which takes multidimensional positive real functions (i.e., functions whose real parts are positive when the real parts of *all* of their arguments are positive) to functions that have the same property in a generalized multidimensional outer disk.

In a brief section of one of the original papers on the subject of MDWD integration [89], a different type of mapping is proposed, in a discussion of boundary conditions for the transmission line equations. Suppose that, in the $(1 + 1)$D case, our transformed coordinates are given by (3.19). The alternative mapping can be written as

$$s_1 \to \phi_1 \qquad s_2 \to \phi_2$$

where the frequency variables ϕ_1 and ϕ_2 are defined in terms of the variables ψ_1 and ψ_2 from (3.44) by

$$\phi_1 \triangleq \frac{\psi_1}{1 + T_1 T_2 \psi_1 \psi_2 / 4} = \frac{1}{T_1} \frac{(1 - z_1^{-1})(1 + z_2^{-1})}{1 + z_1^{-1} z_2^{-1}} \tag{3.47a}$$

$$\phi_2 \triangleq \frac{\psi_2}{1 + T_1 T_2 \psi_1 \psi_2 / 4} = \frac{1}{T_2} \frac{(1 - z_2^{-1})(1 + z_1^{-1})}{1 + z_1^{-1} z_2^{-1}} \tag{3.47b}$$

We then have that

$$\text{Re}(\phi_1) = \frac{\text{Re}(\psi_1) + T_1 T_2 |\psi_1|^2 \text{Re}(\psi_2)/4}{|1 + T_1 T_2 \psi_1 \psi_2 / 4|^2} \qquad \text{Re}(\phi_2) = \frac{\text{Re}(\psi_2) + T_1 T_2 |\psi_2|^2 \text{Re}(\psi_1)/4}{|1 + T_1 T_2 \psi_1 \psi_2 / 4|^2}$$

from which we can conclude that

$$\text{Re}(\psi_1) \gtreqless 0 \text{ and } \text{Re}(\psi_2) \gtreqless 0 \implies \text{Re}(\phi_1) \gtreqless 0 \text{ and } \text{Re}(\phi_2) \gtreqless 0$$

Another simple way of seeing positive realness is by rewriting (3.47) as

$$\phi_1 \triangleq \frac{1}{1/\psi_1 + T_1 T_2 \psi_2 / 4} \tag{3.48a}$$

$$\phi_2 \triangleq \frac{1}{1/\psi_2 + T_1 T_2 \psi_1 / 4} \tag{3.48b}$$

in which case ϕ_1 and ϕ_2 can be viewed as impedances of parallel combinations of passive (indeed, lossless) elements. For example, ϕ_1 is equivalent to the impedance of a parallel combination of an inductor of impedance ψ_1 and a capacitor of impedance $4/(T_1 T_2 \psi_2)$. Second-order accuracy is also obtained under these mappings; this should be clear from (3.48) as well. This spectral mapping differs from the trapezoid rule in that the discrete spectral images of the two continuous frequency variables s_1 and s_2 are now mixtures of the two discrete frequency variables z_1^{-1} and z_2^{-1}. In addition, the transformation does not have a unique inverse, but this is of little consequence because we will never have any occasion to invert such a mapping. We mention this particular mapping because it will serve as the bridge between MDWD networks and DWNs (to be discussed in Chapter 4). We will spend some time in Chapter 6 elaborating this link. It will also allow us to introduce higher-order accurate methods, which we will discuss in Section 3.12.

3.5 The (1 + 1)D Advection Equation

Perhaps the simplest hyperbolic partial differential equation imaginable is the so-called scalar *advection* or *one-way wave equation* in $(1 + 1)$D, defined by

$$\frac{\partial i}{\partial t} + \alpha \frac{\partial i}{\partial x} = 0 \tag{3.49}$$

where α is a real constant [238]. It is complemented by the initial condition

$$i(x, 0) = i_0(x), \qquad -\infty < x < \infty \tag{3.50}$$

3.5. THE (1 + 1)D ADVECTION EQUATION

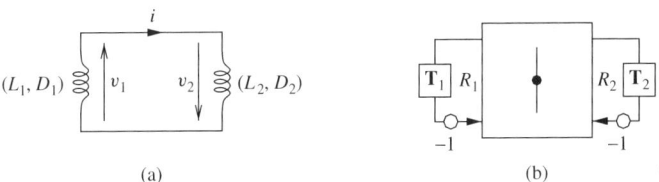

Figure 3.6: *The (1 + 1)D advection equation*—(a) MDKC and (b) MDWD network.

Here, the solution $i(x, t)$ is assumed continuously differentiable (though it need not be[7]), and is defined over the entire x-axis, and for $t \geq 0$. The solution is simply

$$i(x, t) = i_0(x - \alpha t) \tag{3.51}$$

That is, the initial data travels to the left or right (depending on the sign of α) with speed $|\alpha|$. Despite its simplicity, it is often used as a model for numerical schemes [112].

3.5.1 A Multidimensional Kirchhoff Circuit

We first change coordinates via transformation (3.19), which gives

$$\underbrace{\frac{v_0 + \alpha}{\sqrt{2}} \frac{\partial i}{\partial t_1}}_{v_1} + \underbrace{\frac{v_0 - \alpha}{\sqrt{2}} \frac{\partial i}{\partial t_2}}_{v_2} = 0$$

The basis of the MDWD integration approach is to view this equation as a *loop* equation for a multidimensional circuit—that is, a circuit in which voltages and currents may depend not only on time but on space as well. The equation above is to be interpreted as describing a series connection of two inductors, in which the dependent variable i is considered to be the current passing through them. The MD inductors have inductances

$$L_1 \triangleq \frac{(v_0 + \alpha)}{\sqrt{2}} \qquad L_2 \triangleq \frac{(v_0 - \alpha)}{\sqrt{2}}$$

As explained in Section 3.2, these two inductors are associated with the directions t_1 and t_2. The circuit representation of the connection is shown in Figure 3.6(a).

It is important to note that the circuit pictured here is merely a graphical representation of (3.49)—in particular, it represents the pointwise or differential behavior of (3.49) anywhere in the (t_1, t_2) plane. It does, however, permit an immediate discretization via wave digital filters, in exactly the same manner as described in the previous chapter on lumped networks. That is, we can replace the circuit by two MDWD inductor one-ports connected through a series adaptor. This complete wave digital network is shown in Figure 3.6(b), where we have defined the port resistances to be

$$R_1 = \frac{2L_1}{T_1} = \frac{v_0 + \alpha}{\Delta} \qquad R_2 = \frac{2L_2}{T_2} = \frac{v_0 - \alpha}{\Delta} \tag{3.52}$$

---

[7]Given the solution (3.51) to (3.49), it is easy to see that it remains unchanged even if i_0 is not differentiable everywhere. In this case, i must be considered to be a solution to the integral form of (3.49).

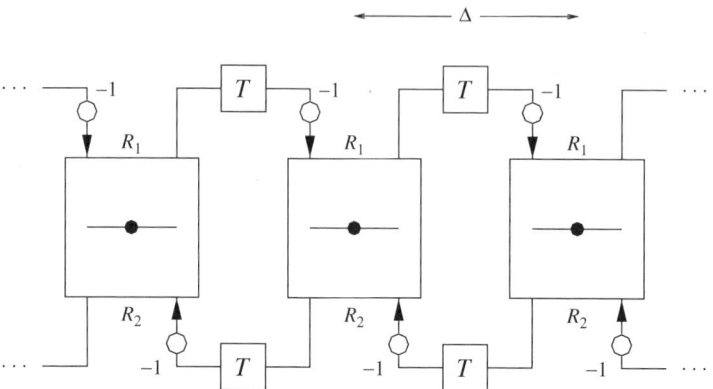

Figure 3.7: *Signal flow graph for Figure 3.6(b).*

Here, we have used $T_1 = T_2 = \sqrt{2}\Delta$. Figure 3.6(b) is an abbreviated notation for a numerical integration routine. We can expand out the spatial dependence into a full signal flow graph in order to better perceive the flow of data. This is shown in Figure 3.7, where we have indicated unit time delays by T; series scattering junctions are separated by a distance Δ.

This signal flow graph can be interpreted as follows: at every grid point in the domain, and at every time step, there are two computational steps, just as discussed on page 9:

- Given wave variable inputs, perform the scattering operation at every adaptor.

- Shift the output wave variables to the inputs of the appropriate adaptors. Referring to Figure 3.6(b), this means that for the port of port resistance R_1 that accepts a wave variable shifted by $\mathbf{T}_1$, we must use the sign-inverted wave quantity output from the corresponding port, one-time step earlier, and one grid point to the *left*. This shifting operation is to be applied at every grid point, as per Figure 3.7. Similarly, the input to the port of port resistance R_2 takes the sign-inverted output of its corresponding port one-time step earlier and one grid point to the *right*.

If $v_0 = |\alpha|$, then either R_1 or R_2 is zero, depending on the sign of α. In this case, the associated inductor can be dropped from the network entirely (i.e., we can treat it as a short-circuit). For example, if $\alpha < 0$, then $v_0 = |\alpha|$ implies that $R_1 = 0$, and we get the simplified network of Figure 3.8. Here, in fact, we have an exact solution to (3.49); the signals in the delay registers are shifted repeatedly to the left and directly implement the traveling wave solution given by (3.51). Note that the sign inversion of the inductor is canceled by that of the reflection from the port.

3.5.2 Stability

It is easy to see that the MDKC of Figure 3.6(a) will be MD-passive if the inductances L_1 and L_2, and consequently the port resistances R_1 and R_2 of the MDWD network in Figure 3.6(b), are nonnegative. From (3.52), this gives a constraint on v_0, the space step/time

3.5. THE (1+1)D ADVECTION EQUATION

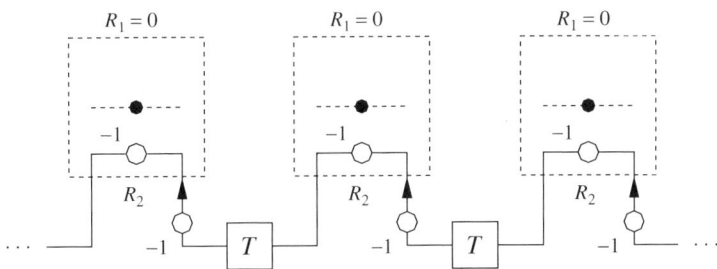

Figure 3.8: *Simplified signal flow graph for Figure 3.6(b), for $v_0 = |\alpha|$, $\alpha < 0$.*

step ratio, namely, that we must have

$$v_0 = \frac{\Delta}{T} \geq |\alpha| \qquad \text{(for passivity)}$$

Any such value of v_0 yields a passive and thus stable algorithm.

It is important to mention, however, that the *instances* of the MDWD network, sampled at every grid point as in Figure 3.7 are *not* connected portwise, as must be true for a traditional lumped WD-network. (Recall that everything discussed in Chapter 2 regarding network passivity depended on portwise connectivity.) The output wave at the bottom port at spatial location $x = i\Delta$ is sign-inverted and then sent as input to the same port, at location $x = (i-1)\Delta$ at the next time step. Thus the realization of Figure 3.7 cannot be analyzed directly as a chain of lumped elements; passivity follows from the multidimensional representations shown in Figure 3.6. This is in direct contrast to digital waveguide methods, for which a lumped representation is always available (see the discussion of the Kelly–Lochbaum model in Section 1.1.2).

3.5.3 An Upwind Form

One of the interesting (and only briefly mentioned [117]) features of the MDKC representation is that it can easily be manipulated to yield what are known as *upwind difference methods*; such methods are usually applied to problems for which there is a directional bias in the propagation speed and are heavily used in fluid dynamical calculations [120].

We can rewrite the advection equation (3.49), where we assume, without loss of generality, that $\alpha > 0$, as

$$\underbrace{\sqrt{2}\alpha \frac{\partial i}{\partial t_1}}_{v_1} + \underbrace{(v_0 - \alpha) \frac{\partial i}{\partial t'}}_{v_2} = 0$$

which can be written as the MDKC shown in Figure 3.9(a). We now have

$$L_1 = \sqrt{2}\alpha \qquad\qquad L_2 = v_0 - \alpha$$

In this case, we have left a directional derivative in the pure time (or scaled time) direction in the MDKC; for this inductor, we apply the generalized trapezoid rule discussed

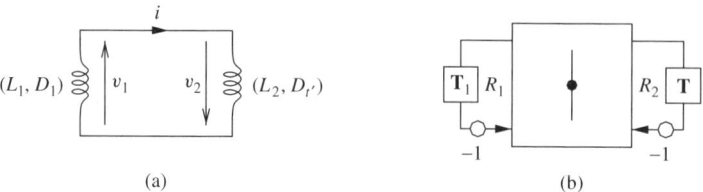

Figure 3.9: *An upwind-differencing form for the advection equation—(a) MDKC and (b) MDWD network.*

in Section 3.4.1, with a step size of $T' = \Delta$. The resulting MDWD network is shown in Figure 3.9(b), with port resistances given by

$$R_1 = \frac{2\alpha}{\Delta} \qquad R_2 = \frac{2}{\Delta}(v_0 - \alpha) \qquad (3.53)$$

(Note that a directional shift of length Δ in the scaled time direction $t' = v_0 t$ corresponds to a pure time shift of duration $\Delta/v_0 = T$, and so we have indicated this shift in Figure 3.9(b) by a **T**). The signal flow graph, with spatial dependence expanded out, is shown in Figure 3.10.

This structure is in a sense a better model for the advection system; recall that for $\alpha > 0$, the solution at any future time instant $t \geq 0$ will simply be the initial distribution shifted to the right by an amount αt. By using upwind differencing, we have dispensed with the unphysical leftward traveling wave that appears in the signal flow diagram in Figure 3.7. As before, the network will be MD-passive for $v_0 \geq \alpha$. It also degenerates to a simple delay line when $v_0 = \alpha$ (in which case we will have $R_2 = 0$, and the right-hand inductor in Figure 3.9(b) can be dropped from the network).

Since all the systems that we will subsequently examine do not have any directional disparities in the wave speed, we will not pursue the subject of upwind differencing further here. We do mention, though, that DWNs [228, 267], which are intimately related to MDWD networks, are incapable of performing upwind differencing for the simple reason that they are constructed from bidirectional delay lines (or unit elements), which carry information symmetrically in opposite directions. (This is presumably true of TLM methods as well.)

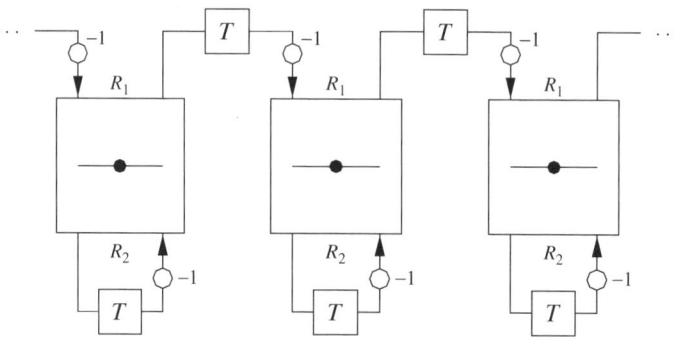

Figure 3.10: *Signal flow graph for Figure 3.9(b), for $\alpha > 0$.*

3.6 The (1 + 1)D Transmission Line

As a slightly more involved example that highlights some of the issues that typically arise in the construction of these algorithms, consider the $(1+1)$D *transmission line* or *telegrapher's* equations [91]:

$$l\frac{\partial i}{\partial t} + \frac{\partial u}{\partial x} + ri + e = 0 \tag{3.54a}$$

$$c\frac{\partial u}{\partial t} + \frac{\partial i}{\partial x} + gu + h = 0 \tag{3.54b}$$

Here, $i(x, t)$ and $u(x, t)$ are the current and voltage in the transmission line, l, c, r, and g are inductance, capacitance, resistance, and shunt conductance per unit length respectively, and are all nonnegative functions of x (l and c are strictly positive[8]). $e(x, t)$ and $h(x, t)$ represent distributed voltage and current source terms. System (3.54) is symmetric hyperbolic; it has the form of (3.1), with $\mathbf{w} = [i, u]^T$, and

$$\mathbf{P} = \begin{bmatrix} l & 0 \\ 0 & c \end{bmatrix} \quad \mathbf{A}_1 = \begin{bmatrix} 0 & 1 \\ 1 & 0 \end{bmatrix} \quad \mathbf{B} = \begin{bmatrix} r & 0 \\ 0 & g \end{bmatrix} \quad \mathbf{f} = \begin{bmatrix} e \\ h \end{bmatrix} \tag{3.55}$$

Phase and Group Velocity

In the constant-coefficient case where $r = g = 0$, the dispersion relation, defined in (3.10), will be

$$-\omega^2 lc + \beta^2 = 0$$

in terms of real frequencies ω and wavenumbers β, and has solutions

$$\omega = \pm\frac{\beta}{\sqrt{lc}}$$

The phase and group velocities, from (3.12) are then

$$\gamma_{TL}^p = \gamma_{TL}^g = \pm\frac{1}{\sqrt{lc}} \tag{3.56}$$

and if l and c are functions of x, the maximal group velocity will be

$$\gamma_{TL,\max}^g = \frac{1}{\sqrt{(lc)_{\min}}} \tag{3.57}$$

where $(lc)_{\min} = \min_{x \in \mathcal{D}}(lc)$ (see (3.14) and the accompanying commentary).

---

[8] In fact, l and c should be bounded away from zero, so that the local wave speed (given by $1/\sqrt{lc}$) remains finite everywhere.

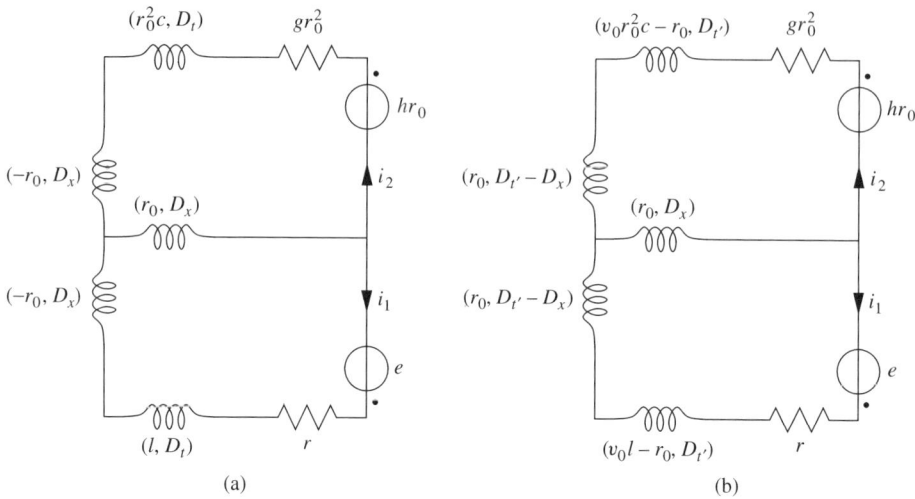

Figure 3.11: *MDKCs for the* $(1+1)$D *transmission line system* (3.59)—(a) *a direct representation and* (b) *after splitting and shifting inductances.*

3.6.1 MDKC for the $(1+1)$D Transmission Line Equations

In order to put this system into the form of an MDKC, let us first change dependent variables by

$$i_1 \triangleq i \qquad i_2 \triangleq \frac{u}{r_0} \qquad (3.58)$$

where $r_0 > 0$ is a free constant parameter that has dimensions of resistance. The primary reason for introducing this parameter is that the numerical algorithm may later be tuned to be optimally efficient (in terms of the largest allowable time step for a given grid spacing). After changing variables, and multiplying the second equation by r_0, we obtain

$$l\frac{\partial i_1}{\partial t} + r_0\frac{\partial i_2}{\partial x} + ri_1 + e = 0 \qquad (3.59a)$$

$$cr_0^2\frac{\partial i_2}{\partial t} + r_0\frac{\partial i_1}{\partial x} + gr_0^2 i_2 + r_0 h = 0 \qquad (3.59b)$$

At this point, it is already possible to write the above system in the form of an MDKC, which is shown in Figure 3.11(a).

Kirchhoff's Current Law gives us the current in the common branch, which is $i_1 + i_2$, and summing voltages around the two loops yields system (3.59). This representation, however, cannot give an explicit algorithm because of the purely spatial MD inductors that form a T-junction between the two loops; that is, if one tries to treat these as one-ports, their WD counterparts will be found to contain delay-free paths from input to output; in other words, the algorithm will be implicit. Nor can it be considered to be MD-passive, since there are negative inductances. By performing a few network theoretic transformations to this MDKC, we can obtain a representation that is MD-passive, and which gives rise to an explicit numerical method. The idea here, grossly speaking, is to make sure that each

3.6. THE (1 + 1)D TRANSMISSION LINE

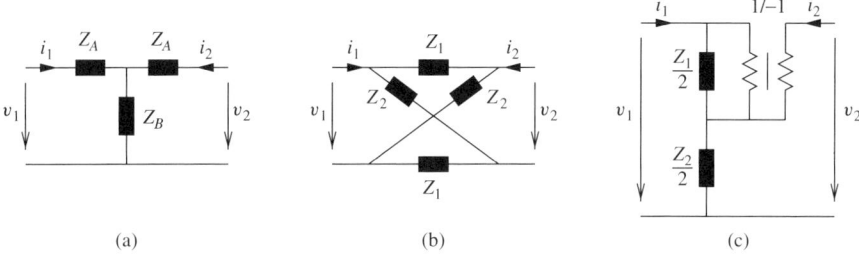

Figure 3.12: *Equivalent two-ports*—(a) T-junction, with impedances Z_A and Z_B and (b) and (c), lattice and Jaumann equivalent two-ports, both with $Z_1 = Z_A$ and $Z_2 = Z_A + 2Z_B$.

inductance is positive and that every inductor 'points' in the direction of a transformed coordinate, as per conditions (3.15).

First note that we can split and shift the differential operators around at will, as long as the loop equations remain unchanged. In particular, we can redraw the circuit as in Figure 3.11(b), where we have introduced the scaled time coordinate $t' = v_0 t$ and its associated derivative $D_{t'}$. Note that some of the time inductance in both loops has been shifted and combined with the purely spatial inductances in the T-junction. Now, examine the three inductors that form this T-junction connecting the two loops. If we are planning to use coordinates defined by (3.20), then the two inductors on the vertical rail can be identified as MD-passive—we have $D_{t'} - D_x = \sqrt{2} D_2$. The inductance in the common branch, however, is not yet in proper form. It is now possible to apply transformations from classical network theory so as to ensure that the resulting equivalent two-port is composed of only MD-passive elements. Although the system as a whole does not change under these manipulations, we would like it to be *concretely* passive[9], so that it may be decomposed into a connection of simpler passive blocks; this is, incidentally, a convenient check for stability of the system as a whole. Since the two-port containing the T-junction will always be, by itself, linear and shift-invariant (i.e., shift-invariant with respect to any coordinate, because the inductances are constant), we are justified in describing it by means of impedances and applying spectral transformations. When it is connected to the other components that are not shift-invariant, the spectrally transformed two-port may be interpreted in terms of difference formulae.

The symmetric T-junction, and its lattice [83, 176] and Jaumann [178] equivalents are shown in Figure 3.12, for arbitrary impedances Z_A and Z_B. Replacement of the T-junction in Figure 3.11(b) by either of the two-ports in Figure 3.12(b) and (c) gives an MDKC that is indeed concretely MD-passive; this circuit is shown in Figure 3.14(a). Note that in this representation, we have left inductors with symbols $D_{t'}$ in the circuit, instead of rewriting them as $D_{t'} = (D_1 + D_2)/\sqrt{2}$. In this case, we must proceed as such because their inductances are possibly spatially varying (note that they depend on l and c); for this reason these elements cannot be split into inductances acting along directions t_1 and t_2 without giving up passivity. For these inductances, we will apply the generalized trapezoid rule that was discussed in Section 3.4.1.

---

[9] By concretely passive, we simply mean, as discussed in Section 1.1.1, that all elements in the network should be individually passive. A network (or N-port [14]) may be abstractly passive, but not concretely passive; the MDKCs shown in Figure 3.11 are of this type.

3.6.2 Digression: The Inductive Lattice Two-port

We have derived the WD equivalents for all the standard circuit elements but the two-ports pictured in Figure 3.12 need a special treatment. Fettweis and Nitsche find the lattice form to be the most straightforward to analyze but we would especially like to call attention to the fact that the resultant WD two-port is the same regardless of which of the equivalent structures we choose; use of a concretely MD-passive two-port, however, makes the passivity of the resulting circuit obvious. We will continue to use the Jaumann equivalent in all future diagrams (though we could equally well use the lattice form). Basu and Zerzghi [12], as well as Fettweis [68] make the point that an electrical network equivalent is a convenient formalism for developing MD-passive discrete networks but it is by no means necessary.

Beginning from any of the equivalent two-port structures in Figure 3.12, we can immediately write down the impedance matrix, which we will denote by $\mathbf{Z}$:

$$\mathbf{Z} = \frac{1}{2}\begin{bmatrix} Z_2+Z_1 & Z_2-Z_1 \\ Z_2-Z_1 & Z_2+Z_1 \end{bmatrix} = \frac{1}{2}\begin{bmatrix} 1 & 1 \\ 1 & -1 \end{bmatrix}\begin{bmatrix} Z_2 & 0 \\ 0 & Z_1 \end{bmatrix}\begin{bmatrix} 1 & 1 \\ 1 & -1 \end{bmatrix} = \mathbf{N}^{-1}\mathbf{\Lambda}\mathbf{N}$$

where we have set $\mathbf{N} = \begin{bmatrix} 1 & 1 \\ 1 & -1 \end{bmatrix}$ and $\mathbf{\Lambda} = \begin{bmatrix} Z_2 & 0 \\ 0 & Z_1 \end{bmatrix}$.

We now introduce a port resistance matrix $\mathbf{R} = \begin{bmatrix} R & 0 \\ 0 & R \end{bmatrix} > \mathbf{0}$, where we can choose equal port resistances because the two-port is symmetric. The scattering matrix is then, from (2.21),

$$\mathbf{S} = \left(\mathbf{N}^{-1}\mathbf{\Lambda}\mathbf{N}\mathbf{R}^{-1} + \mathbf{I}_2\right)^{-1}\left(\mathbf{N}^{-1}\mathbf{\Lambda}\mathbf{N}\mathbf{R}^{-1} - \mathbf{I}_2\right)$$

where $\mathbf{I}_2$ is the 2×2 identity matrix. This is easily rearranged to become

$$\mathbf{S} = \mathbf{N}^{-1}\left(\mathbf{\Lambda}\mathbf{R}^{-1} + \mathbf{I}_2\right)^{-1}\left(\mathbf{\Lambda}\mathbf{R}^{-1} - \mathbf{I}_2\right)\mathbf{N}$$

Returning to the problem of the $(1+1)$D transmission line, for the two-port in question in Figure 3.14(a), we have $Z_1 = \sqrt{2}r_0 s_2$ and $Z_2 = \sqrt{2}r_0 s_1$. Notice, in particular, that for these choices of impedance, any of the two ports of Figure 3.12 are described by MD positive real matrices. We now discretize using the trapezoid rule, that is,

$$s_j \to \psi_j = \frac{2}{T_j}\frac{1-z_j^{-1}}{1+z_j^{-1}} \qquad j=1,2$$

where we also will set $T_1 = T_2 = \sqrt{2}\Delta$. If we make the choice $R = 2r_0/\Delta$, we obtain the discrete-time, causal scattering matrix

$$\mathbf{S}(z_1^{-1}, z_2^{-1}) = \mathbf{N}^{-1}\begin{bmatrix} -z_1^{-1} & 0 \\ 0 & -z_2^{-1} \end{bmatrix}\mathbf{N} \qquad (3.60)$$

The resulting MD two-port is shown in Figure 3.13.

It should be clear that the same procedure can be used for arbitrary impedances Z_1 and Z_2; it should be remarked, however, that we cannot always get a simple form without an instantaneous through path like that pictured in Figure 3.13. Since this two-port is strictly causal, it may easily be connected to other ports without the risk of the appearance of a delay-free loop.

3.6. THE (1+1)D TRANSMISSION LINE

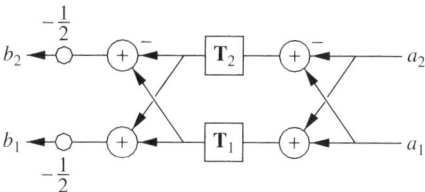

Figure 3.13: *Signal flow diagram for the MDWD lattice or Jaumann two-port.*

3.6.3 Energetic Interpretation

Let us now reexamine the passive MDKC in Figure 3.14(a). The total stored energy flux in the network is contained in the four inductors and will be, from (3.36),

$$\mathbf{E}_{\text{total}} = \frac{1}{2}L_1 i_1^2 \mathbf{e}_{t'} + \frac{1}{2}L_2 i_2^2 \mathbf{e}_{t'} + \frac{1}{2}L_0(i_1 + i_2)^2 \mathbf{e}_1 + \frac{1}{2}L_0(i_1 - i_2)^2 \mathbf{e}_2$$

where the $\mathbf{e}_{t'}$ is a unit vector in direction t', and $\mathbf{e}_1$ and $\mathbf{e}_2$ are unit vectors in directions t_1 and t_2 respectively. Applying the definitions of the inductances, from (3.61), the current definitions from (3.58), and the fact that $\mathbf{e}_1 = (\mathbf{e}_{t'} + \mathbf{e}_x)/\sqrt{2}$ and $\mathbf{e}_2 = (\mathbf{e}_{t'} - \mathbf{e}_x)/\sqrt{2}$, then this total energy flux can be rewritten as

$$\mathbf{E}_{\text{total}} = \frac{1}{2}v_0 l i^2 \mathbf{e}_{t'} + \frac{1}{2}v_0 c u^2 \mathbf{e}_{t'} + ui\mathbf{e}_x = \frac{1}{2}l i^2 \mathbf{e}_t + \frac{1}{2}c u^2 \mathbf{e}_t + ui\mathbf{e}_x$$

Here $\mathbf{e}_t = v_0 \mathbf{e}_{t'}$ is a unit vector in the time direction. The total scalar energy at time t of this network will, from (3.31), be

$$\mathcal{E}(t) = \int_{-\infty}^{+\infty} \mathbf{e}_t^T \mathbf{E}_{\text{total}}\, dx = \int_{-\infty}^{+\infty} \frac{1}{2}\left(l i^2 + cu^2\right) dx = \int_{-\infty}^{+\infty} \frac{1}{2}\mathbf{w}^T \mathbf{P}\mathbf{w}\, dx$$

and thus coincides with the energy definition of the symmetric hyperbolic system, from (3.6). This is certainly not surprising, but the important point here is that in an MDKC such as that of Figure 3.14(a), the scalar energy has been broken down into contributions from several interacting components (the inductors), each of which is passive individually; this useful energy subdivision has been exploited here as a means of developing passive numerical methods.

3.6.4 An MDWD Network for the (1+1)D Transmission Line

Returning to Figure 3.14(a), and making use of the discrete two-port derived in the last section, we can now write the complete wave digital network. It is shown in Figure 3.14(b).
The inductances in the MDKC of Figure 3.14(a) are

$$L_1 = v_0 l - r_0 \qquad L_2 = v_0 r_0^2 c - r_0 \qquad L_0 = \frac{r_0}{\sqrt{2}} \qquad (3.61)$$

and the losses and sources may be grouped as discussed in Section 2.2.4. The port resistances of the MDWD network of Figure 3.14(b) are

$$R_0 = \frac{2r_0}{\Delta} \qquad R_1 = \frac{2L_1}{\Delta} \qquad R_2 = \frac{2L_2}{\Delta} \qquad R_{gh} = gr_0^2 \qquad R_{er} = r \qquad (3.62)$$

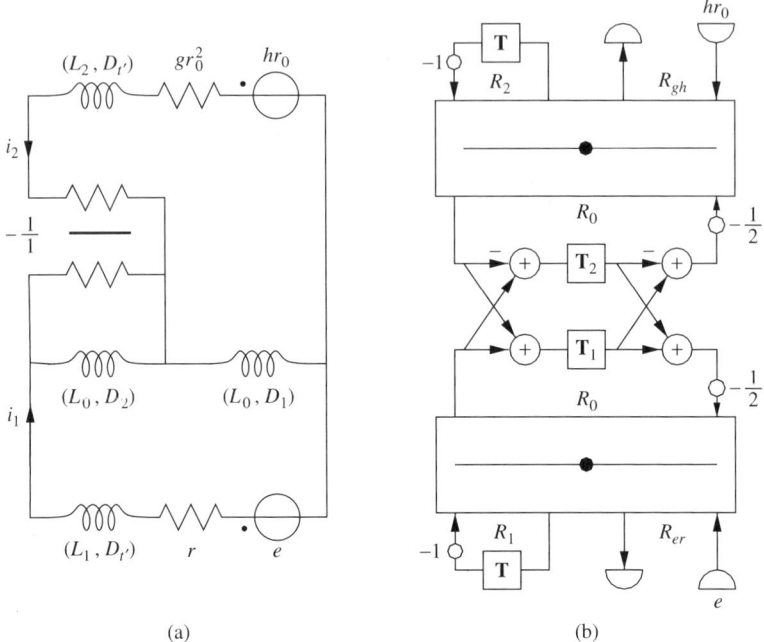

Figure 3.14: (a) *MD-passive network for the* $(1+1)$D *transmission line equations and* (b) *its associated MDWD network.*

The MDWD network is MD-passive if all the port resistances are nonnegative over the entire spatial domain; from (3.61) and (3.62), the only port resistances that are possibly negative are R_1 and R_2. Requiring their positivity gives the constraints

$$v_0 \geq \frac{r_0}{l_{\min}} \qquad v_0 \geq \frac{1}{r_0 c_{\min}}$$

A judicious choice of $r_0 = \sqrt{\frac{l_{\min}}{c_{\min}}}$ [176] allows the largest possible time step for a given grid spacing; the condition is then

$$v_0 \geq \sqrt{\frac{1}{l_{\min} c_{\min}}} \geq \gamma_{\text{TL,max}}^g \qquad (3.63)$$

where $\gamma_{\text{TL,max}}^g$ is defined by (3.57).

If l and c are constant, and (3.63) holds with equality so that we have

$$v_0 = \sqrt{\frac{1}{lc}} = \gamma_{\text{TL,max}}^g \qquad (3.64)$$

then the MDWD numerical scheme is said to be operating at the *CFL bound* [238]. For varying coefficients, however, v_0 is bounded away from $\gamma_{\text{TL,max}}^g$, so the time step will have to be chosen smaller than might be expected; we will look at how to improve upon this bound in Section 3.11.

3.6. THE (1 + 1)D TRANSMISSION LINE

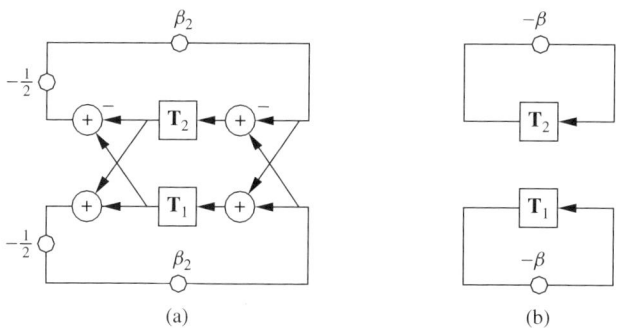

Figure 3.15: *Simplified MDWD network for the (1 + 1)D transmission line equations—* (a) *for constant l and c and* (b) *further simplified in the distortionless case.*

3.6.5 Simplified Networks

In the particular case for which l and c are constants, and where we do not have sources, the MDWD network shown in Figure 3.14(b) can be simplified considerably. If we pick $v_0 = 1/\sqrt{lc}$ and $r_0 = \sqrt{l/c}$, then R_1 and R_2 become zero and their associated inductors may be dropped from the network (that is, they can be treated as short circuits). The two series adaptors then reduce to simple multiplies of the signals output by the lattice two-port as in Figure 3.15(a) where we have written

$$\beta_1 = \frac{r\Delta - 2r_0}{r\Delta + 2r_0} \qquad \beta_2 = \frac{gr_0\Delta - 2}{gr_0\Delta + 2}$$

If, in addition, the transmission line is *distortionless* [43], so that we have $lg = cr$ for all values of x (though as mentioned above, we require l and c to be constant), then the network can be simplified further giving Figure 3.15(b), where $\beta_1 = \beta_2 = \beta$. Now the MDWD network has decoupled into two independent loops, each made up of an MD shift and a scaling. Examine the expanded signal flow graph of Figure 3.16, where the value of the multiplier coefficient, $-\beta$ at location $x = i\Delta$, is written as $-\beta_i$. Values input into the upper array will be shifted repeatedly to the left and attenuated by the factor $-\beta$, and similarly, those in the lower array are shifted to the right and attenuated by the same factor. (Note that $|\beta| \leq 1$.) We thus have a traveling wave formulation of the solution to the transmission line equations, to be compared with the digital waveguide implementation to be discussed in Chapter 4.

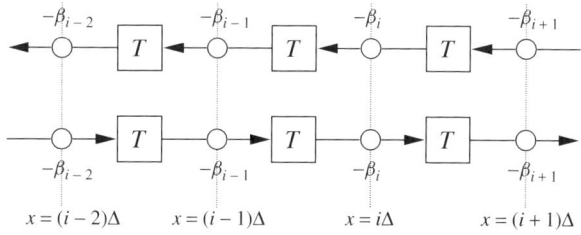

Figure 3.16: *Signal flow graph for the MDWD network of Figure 3.15(b).*

Two special cases are of note here. If the transmission line is lossless so that $r = g = 0$, then $\beta = -1$. The initial values in the storage registers are shifted without attenuation. We would like to note, however, that if $\Delta > 2/gr_0$, then $\beta > 0$, and the traveling waves will be oscillatory and the solution is thus nonphysical. More disturbing is the case $\Delta = 2/gr_0$, in which case we have $\beta = 0$, and all energy leaves the network immediately! Though these examples would seem to indicate that the MDWD network is not behaving correctly, it should be kept in mind that by construction it is stable and consistent with the continuous-time/space transmission line equations and is convergent in the limit as $\Delta \to 0$, by the Lax–Richtmyer Equivalence Theorem [238]. Δ can always be chosen small enough so that β is negative, and thus the solution will be well behaved (i.e., nonoscillatory). This important extra restriction on the grid spacing, which is independent of the time step, is purely a result of the use of the trapezoid rule as our integration method. The lesson here is that passivity, while providing a guarantee of stable numerical methods, does not ensure that we necessarily get a physically acceptable solution in all cases.

3.7 The (2 + 1)D Parallel-plate System

Generalizing the above procedure to several dimensions is straightforward. We examine here, as a practical example, the (2 + 1)D *parallel-plate system*, which is written as

$$l\frac{\partial i_x}{\partial t} + \frac{\partial u}{\partial x} + ri_x + e = 0 \tag{3.65a}$$

$$l\frac{\partial i_y}{\partial t} + \frac{\partial u}{\partial y} + ri_y + f = 0 \tag{3.65b}$$

$$c\frac{\partial u}{\partial t} + \frac{\partial i_x}{\partial x} + \frac{\partial i_y}{\partial y} + gu + h = 0 \tag{3.65c}$$

This system was used early on as a test problem for MDWD methods [90, 284]. Now the dependent variables are a voltage u, and current density components i_x and i_y; these and the sources e, f, and h are functions of time t and two spatial variables, x and y. l, c, r, and g are arbitrary smooth positive functions of x, and y (l and c are strictly positive). It is worth mentioning that the same equations can be used in the contexts of (2 + 1)D linear acoustics, the vibration of a membrane, and, with a trivial modification, (2 + 1)D electromagnetic field problems (involving TE or TM modes).

System (3.65) is symmetric hyperbolic, and thus has the form of (3.1), where $\mathbf{w} = [i_x, i_y, u]^T$, and with

$$\mathbf{P} = \begin{bmatrix} l & 0 & 0 \\ 0 & l & 0 \\ 0 & 0 & c \end{bmatrix} \quad \mathbf{A}_1 = \begin{bmatrix} 0 & 0 & 1 \\ 0 & 0 & 0 \\ 1 & 0 & 0 \end{bmatrix} \quad \mathbf{A}_2 = \begin{bmatrix} 0 & 0 & 0 \\ 0 & 0 & 1 \\ 0 & 1 & 0 \end{bmatrix} \quad \mathbf{B} = \begin{bmatrix} r & 0 & 0 \\ 0 & r & 0 \\ 0 & 0 & g \end{bmatrix} \quad \mathbf{f} = \begin{bmatrix} e \\ f \\ h \end{bmatrix}$$

It will follow, as in the case of the (1 + 1)D transmission line system (see Section 3.6.3), that the total energy of the MDKC that we will derive in the next section will be equal to the energy of system (3.65), as per (3.6).

Phase and Group Velocity

For the constant-coefficient, lossless, and source-free case (i.e., $r = g = e = f = h = 0$), the dispersion relation in terms of the frequency ω and wavenumber magnitude $\|\boldsymbol{\beta}\|_2 = \sqrt{\beta_x^2 + \beta_y^2}$ will be, from (3.10),

$$\omega \left(\omega^2 - \frac{1}{lc} \|\boldsymbol{\beta}\|_2^2 \right) = 0$$

which has roots

$$\omega = 0 \qquad \omega = \pm \sqrt{\frac{1}{lc}} \|\boldsymbol{\beta}\|_2$$

Discounting the stationary mode with $\omega = 0$, the phase and group velocities are then, from (3.12),

$$v_{\text{PP}}^p = v_{\text{PP}}^g = \pm \frac{1}{\sqrt{lc}}$$

and if l and c are functions of x and y, the maximal group velocity will be

$$v_{\text{PP,max}}^g = \frac{1}{\sqrt{(lc)_{\min}}} \tag{3.66}$$

This value is the same as for the $(1 + 1)$D transmission line equations.

3.7.1 MDKC and MDWD Network

The circuit can be constructed along the same lines as for the $(1 + 1)$D case; we deal here with the discretization on a rectilinear grid and will thus apply coordinate transformation defined by the $\mathbf{H}$ of (3.23). Rewriting system (3.65) in terms of the new coordinates $[t_1, \ldots, t_5]^T$ using $\nabla_\mathbf{u} = \mathbf{V}\mathbf{H}^{-R}\nabla_\mathbf{t}$, with the pseudoinverse (3.24) gives

$$(v_0 l - r_0) D_5 i_1 + \frac{r_0}{2} D_1 (i_1 + i_3) + \frac{r_0}{2} D_3 (i_1 - i_3) + r i_1 + e = 0 \tag{3.67a}$$

$$(v_0 l - r_0) D_5 i_2 + \frac{r_0}{2} D_2 (i_2 + i_3) + \frac{r_0}{2} D_4 (i_2 - i_3) + r i_2 + f = 0 \tag{3.67b}$$

$$\left(v_0 r_0^2 c - 2 r_0 \right) D_5 i_3 + \frac{r_0}{2} D_1 (i_3 + i_1) + \frac{r_0}{2} D_3 (i_3 - i_1)$$

$$+ \frac{r_0}{2} D_2 (i_3 + i_2) + \frac{r_0}{2} D_4 (i_3 - i_2) + g r_0^2 i_3 + h r_0 = 0 \tag{3.67c}$$

where we have used the new current-like variables

$$i_1 \triangleq i_x \qquad i_2 \triangleq i_y \qquad i_3 \triangleq \frac{u}{r_0}$$

and r_0 is, as in the $(1 + 1)$D case, an arbitrary positive constant (which has also been used to scale (3.67c)). $D_5 = D_{t'}$ will be treated as a simple time derivative, according to the generalized trapezoid rule discussed in Section 3.4.1. Figure 3.17 shows the MDKC

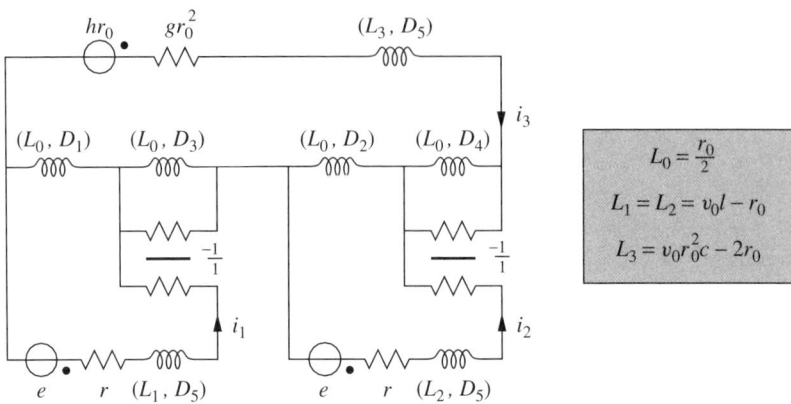

Figure 3.17: *MDKC for the* (2 + 1)D *parallel-plate system in rectangular coordinates.*

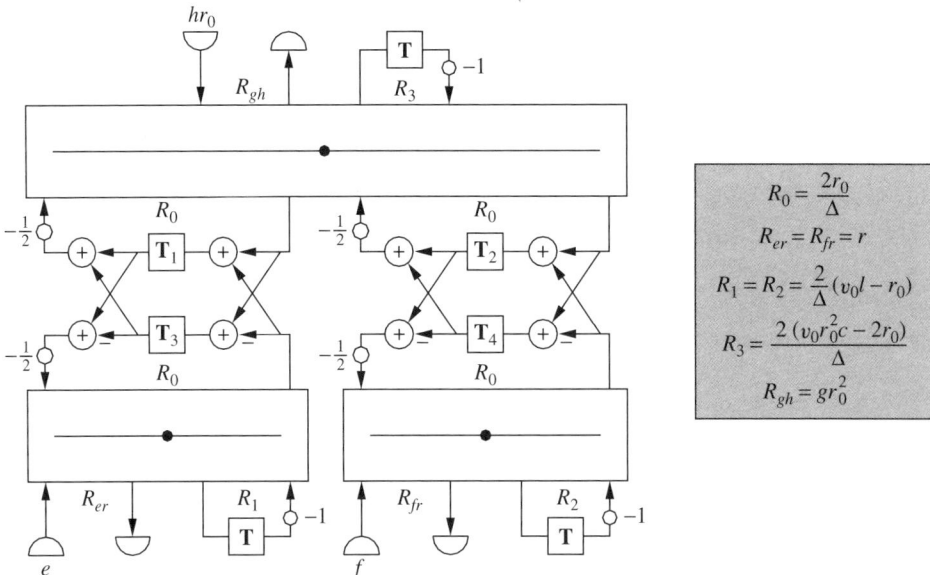

Figure 3.18: *MDWD network for the* (2 + 1)D *parallel-plate system, in rectangular coordinates.*

that results from the transformed set of equations (3.67). The MDWD network corresponding to the MDKC is shown in Figure 3.18, where we have used step-sizes $T_j = \Delta$, $j = 1, \ldots, 5$.

Passivity follows from a positivity condition on the network inductances, and in particular, L_1, L_2, and L_3 (the values of which are given in Figure 3.17). These conditions are

$$v_0 \geq \frac{r_0}{l_{\min}} \qquad v_0 \geq \frac{2}{r_0 c_{\min}} \qquad (3.68)$$

3.8. FINITE DIFFERENCE INTERPRETATION

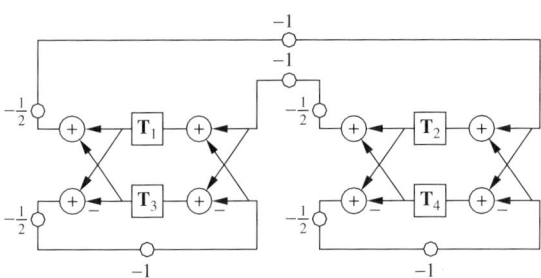

Figure 3.19: *Simplified MDWD network for the $(2+1)$D parallel-plate system, in the lossless, source-free and constant parameter case.*

The choice of $r_0 = \sqrt{\frac{2l_{min}}{c_{min}}}$ gives a stability bound of

$$v_0 \geq \sqrt{\frac{2}{l_{min} c_{min}}} \geq \sqrt{2} \gamma_{PP,max}^g \tag{3.69}$$

which is the best possible bound for this network [89]. Note that v_0 is again bounded away from the maximum group velocity, even taking into account the scaling factor ($\sqrt{2}$ in this case), which is a consistent feature of explicit numerical methods in multiple spatial dimensions.

If l and c are constant, and in addition r, g, e, f, and h are zero, and (3.69) holds with equality, that is, we have

$$v_0 = \sqrt{2} \gamma_{PP,max}^g \tag{3.70}$$

then the network of Figure 3.18 simplifies to the structure shown in Figure 3.19. This particular structure bears a very strong resemblance to the $(2+1)$D waveguide mesh [213, 267] that we saw briefly in Section 1.1.3, and will examine in detail in Chapter 4.

3.8 Finite Difference Interpretation

It should be clear that an MDWD network corresponding to a particular MDKC (and thus to a given set of PDEs) is no more than a particular type of finite difference method and can be analyzed as such. We will do so here for the case of the $(1+1)$D transmission line, in order to compare the schemes that arise from the MDWD approach to the simple centered difference schemes that will be introduced in the next chapter in the waveguide context and can also be put into a scattering form.

3.8.1 MDWD Networks as Multistep Schemes

Recall that the discretization step discussed in Section 3.4.3 consisted of the application of a spectral transformation of the form

$$s_j \to \frac{2}{T_j} \frac{1 - z_j^{-1}}{1 + z_j^{-1}}$$

where s_j is the frequency-domain transform variable corresponding to any MD-causal coordinate t_j, and z_j^{-1} is the frequency-domain unit shift in the same direction.

For spatially inhomogeneous problems, this spectral mapping is equivalent to the application of the trapezoid rule in direction t_j. We can thus write, using operator notation,

$$\frac{\partial}{\partial t_j} \rightarrow \frac{2}{T_j} (1+\delta_j)^{-1} (1-\delta_j) \tag{3.71}$$

where δ_j is a shift operator defined by

$$\delta_j p(t_1, \ldots, t_j, \ldots, t_k) = p(t_1, \ldots, t_j - T_j, \ldots, t_k)$$

when applied to any continuous function $p(\mathbf{t})$. Consider again the lossless $(1+1)$D transmission line equations,

$$l\frac{\partial i}{\partial t} + \frac{\partial u}{\partial x} = 0 \tag{3.72a}$$

$$c\frac{\partial u}{\partial t} + \frac{\partial i}{\partial x} = 0 \tag{3.72b}$$

which can be written as

$$(v_0 l - r_0)\frac{\partial i_1}{\partial t'} + \frac{r_0}{\sqrt{2}}\frac{\partial}{\partial t_1}(i_1 + i_2) + \frac{r_0}{\sqrt{2}}\frac{\partial}{\partial t_2}(i_1 - i_2) = 0 \tag{3.73a}$$

$$\left(v_0 c r_0^2 - r_0\right)\frac{\partial i_2}{\partial t'} + \frac{r_0}{\sqrt{2}}\frac{\partial}{\partial t_1}(i_2 + i_1) + \frac{r_0}{\sqrt{2}}\frac{\partial}{\partial t_2}(i_2 - i_1) = 0 \tag{3.73b}$$

under the application of coordinate transformation (3.20) and using scaled variables $i_1 = i$ and $i_2 = u/r_0$, as well as the scaled time variable $t' = v_0 t$. Under the substitution of (3.71), for $j = 1, 2$, and using the generalized trapezoid rule in time, defined by

$$\frac{\partial}{\partial t'} \rightarrow \frac{2}{T'} (1+\delta_{t'})^{-1} (1-\delta_{t'})$$

where $\delta_{t'}$ is a shift in the scaled time direction t' of duration T', we get

$$(v_0 l - r_0) \frac{2}{T'} (1+\delta_{t'})^{-1} (1-\delta_{t'}) i_1 + \frac{\sqrt{2 r_0}}{T_1} (1+\delta_1)^{-1} (1-\delta_1)(i_1 + i_2)$$

$$+ \frac{\sqrt{2 r_0}}{T_2} (1+\delta_2)^{-1} (1-\delta_2)(i_1 - i_2) = 0$$

$$\left(v_0 c r_0^2 - r_0\right) \frac{2}{T'} (1+\delta_{t'})^{-1} (1-\delta_{t'}) i_2 + \frac{\sqrt{2 r_0}}{T_1} (1+\delta_1)^{-1} (1-\delta_1)(i_2 + i_1)$$

$$+ \frac{\sqrt{2 r_0}}{T_2} (1+\delta_2)^{-1} (1-\delta_2)(i_2 - i_1) = 0$$

3.8. FINITE DIFFERENCE INTERPRETATION

to second order in Δ. This can be rewritten[10] as

$$(1+\delta_1)(1+\delta_2)\frac{R_1}{R_0}(1-\delta_{t'})i_1 + (1+\delta_{t'})\left(1-\delta_{t'}^2\right)i_1 + (1+\delta_{t'})(\delta_2-\delta_1)i_2 = 0 \quad (3.74a)$$

$$(1+\delta_1)(1+\delta_2)\frac{R_2}{R_0}(1-\delta_{t'})i_2 + (1+\delta_{t'})\left(1-\delta_{t'}^2\right)i_2 + (1+\delta_{t'})(\delta_2-\delta_1)i_1 = 0 \quad (3.74b)$$

where we have used $T' = v_0 T = \Delta$, $T_1 = T_2 = \sqrt{2}\Delta$, the fact that $\delta_1\delta_2 = \delta_{t'}^2$, and also the definitions of the port resistances of the MDWD network of Figure 3.14(b), given in (3.62). Upon replacing the quantities i_1 and i_2 by their respective grid functions $I_{1,i}(n)$ and $I_{2,i}(n)$ that take on values for n and i integer, (3.74a) and (3.74b) define recursions on a regular grid, of spacing Δ. (3.74a) can be written as

$$\alpha_i I_{1,i}(n) + \beta_{i+1} I_{1,i+1}(n-1) + \beta_{i-1} I_{1,i-1}(n-1) + \gamma_i I_{1,i}(n-1)$$
$$- I_{2,i-1}(n-1) + I_{2,i+1}(n-1)$$
$$- \beta_{i+1} I_{1,i+1}(n-2) - \beta_{i-1} I_{1,i-1}(n-2) - \gamma_i I_{1,i}(n-2)$$
$$- I_{2,i-1}(n-2) + I_{2,i+1}(n-2)$$
$$- \alpha_i I_{1,i}(n-3) = 0 \quad (3.75)$$

with

$$\alpha_i \triangleq 1 + \frac{R_1(i\Delta)}{R_0} \qquad \beta_i \triangleq \frac{R_1(i\Delta)}{R_0} \qquad \gamma_i \triangleq 1 - \frac{R_1(i\Delta)}{R_0} \quad (3.76)$$

The recursion corresponding to (3.74b) is very similar under the interchange of I_1 and I_2. Note that if l and c are constants and if the difference scheme is operating at the CFL bound (so that $R_1 = R_2 = 0$), then (3.75) can be simplified to

$$I_{1,i}(n) - I_{1,i}(n-2) + I_{2,i+1}(n-1) - I_{2,i-1}(n-1) = 0 \quad (3.77)$$

which is a simple centered difference approximation to (3.72a) and which we will see again in the DWN context in Section 4.3.2. Unlike the case of the DWN, however, away from the passivity bound we have a *multistep scheme* [238] that involves three steps of 'look-back' in order to update a grid variable at a particular location. The introduction of wave variables, then, can be considered a means of expanding the state of the system so that using the new state, the recursion (now in the form of the MDWDF of Figure 3.14) requires access only to wave quantities at the time step immediately preceding the current one.

In order to generate a scheme that operates on alternating interleaved grids (often called *offset sampling* [89]), it is possible to use a doubled time step of $T' = 2\Delta$ in order to implement the generalized trapezoid rule applied to the time derivatives in (3.73a) and (3.73b), that is,

$$\frac{\partial}{\partial t'} \rightarrow \frac{2}{T'}\left(1-\delta_{t'}^2\right)^{-1}\left(1+\delta_{t'}^2\right) = \frac{1}{\Delta}\left(1-\delta_{t'}^2\right)^{-1}\left(1+\delta_{t'}^2\right)$$

---

[10] Because this system is linear and time-invariant (though not shift-invariant), time-shifting operators such as $\delta_{t'}$ commute with purely spatially varying quantities such as R_1.

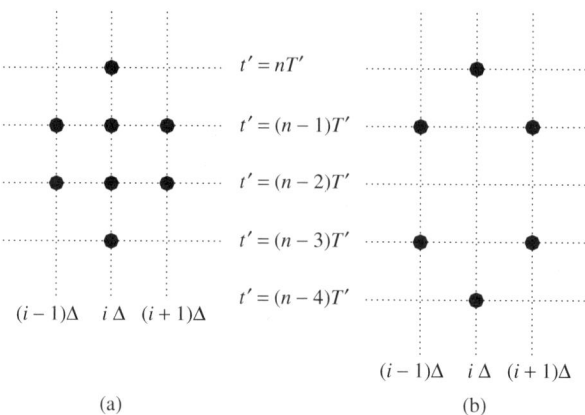

Figure 3.20: *Computational stencils of the equivalent multistep schemes of MDWD networks for the $(1+1)$D transmission line equations—(a) scheme (3.75) and (b) 'offset' scheme (3.78).*

in which case we get, as an approximation to (3.72a),

$$\alpha_i I_{i,1}(n) + \beta_{i+1} I_{1,i+1}(n-1) + \beta_{i-1} I_{1,i-1}(n-1) + I_{2,i+1}(n-1) - I_{2,i-1}(n-1)$$
$$- \beta_{i+1} I_{1,i+1}(n-3) - \beta_{i-1} I_{1,i-1}(n-3) + I_{2,i+1}(n-3) - I_{2,i-1}(n-3)$$
$$- \alpha_i I_{1,i}(n-4) = 0 \tag{3.78}$$

where α and β are defined as per (3.76) but where R_1 is now equal to $\frac{1}{\Delta}(v_0 l - r_0)$. This form also reduces to simple centered differences when l and c are constant, and when we are operating at the CFL bound.

The computational stencils corresponding to the two different schemes are shown in Figure 3.20; the top black dot in either picture represents the location of the grid variable currently being updated (either I_1 or I_2), and the other dots cover the discrete region of influence of the difference scheme. Notice, in particular, that each scheme has a width of only three grid points, corresponding to nearest-neighbor-only updating. Also, because these are multistep methods, one might expect that we will have to take special care when initializing the scheme; we discuss this issue in Section 3.9. For the offset scheme of Figure 3.20(b), the stencil can be shifted one step to the left or right without any overlapping; thus, such a scheme can be subdivided into two mutually exclusive subschemes (operating only for $n+i$ always even or always odd), one of which may be dropped from the calculating scheme entirely. This behavior appears in many of the difference schemes that we will come across subsequently; we will pay particular attention to such schemes in Appendix Λ.

One of the interesting features of the MDKC representation of a set of PDEs is that the same circuit can give rise to an entire family of MDWD networks or, in other words, of difference methods, all of which are consistent with the original set of PDEs. In the case of the MDKC for the transmission line equations derived previously, although we have defined the directions of the various inductors (along which we will be integrating), at the circuit stage we have not as yet specified any spectral mapping that will determine the type of differencing to be applied. Any passivity-preserving mapping that is correct in the low-frequency limit will give rise to a passive, consistent MDWD algorithm. We will examine the important implications of a more exotic type of mapping in Chapter 6, but it is also interesting to note that we can apply the trapezoid rule using different step sizes for all the reactive elements. The constraint on our choices of these step sizes is that all shifting operations refer, ultimately, to another grid point (for computability).

Finally, we note that in general, the determination of stability for a multistep scheme can be quite difficult; even in the constant-coefficient case it will in general be necessary to perform *von Neumann analysis* [238] (see Appendix A for such an analysis applied to difference schemes for the wave equation in $(2+1)$D and $(3+1)$D), which can be quite formidable. For variable coefficients, the *energy method* [112, 199] is a popular avenue of approach. Here, however, we are ensured stability through the passivity condition on the network.

3.8.2 Numerical Phase Velocity and Parasitic Modes

Since, in general, the image MDWD network of a given MDKC for a system of PDEs is a multistep numerical integration scheme, it is reasonable to expect that *parasitic modes* [238] will be present in the solution. Energy in such modes often travels at speeds other than the desired wave speed in the medium and may be highly oscillatory. If the scheme is consistent with the original system of PDEs, and stable, as is an MDWD network derived from the equivalent MDKC under the application of the trapezoid rule, then these parasitic modes must disappear in the limit as the time step is decreased (by the Lax–Richtmyer Equivalence Theorem [238]). They have not as yet been addressed in the wave digital theory, and the subject is related to how initial conditions should be set in an MDWD network. The subject of initialization has been touched upon by Krauss [145].

Analysis of parasitic modes is easiest in the constant-coefficient case. We will examine the simplest possible nontrivial MDWD network, namely that of the constant-coefficient lossless source-free $(1+1)$D transmission line. Since at the stability limit this scheme becomes equivalent to simple centered differences (see previous section), for which we do not have parasitic modes at all, we will look at the MDWD network of Figure 3.14(b) away from this limit[11]. We have chosen $r_0 = \sqrt{l/c}$. The MDWD network is redrawn in Figure 3.21, where we have

$$R_1 = R_2 = \frac{2}{\Delta}(v_0 l - r_0) \qquad R_0 = \frac{2r_0}{\Delta}$$

for some $v_0 > 1/\sqrt{lc}$.

Note that because the system is now linear and shift-invariant, we have replaced the shifts $\mathbf{T}_1$ and $\mathbf{T}_2$ in the two directions t_1 and t_2 by their frequency-domain counterparts z_1^{-1} and z_2^{-1}. Recall also that we have, from (3.45) and (3.46), that

$$z_1^{-1} = z^{-1}w^{-1} \qquad z_2^{-1} = z^{-1}w$$

---

[11] Analysis of a numerical method away from its stability limit is useful because it can give some indication of how the scheme will behave in the presence of material variations; if the system does exhibit such variations, then, locally speaking, we will necessarily be operating away from this limit in at least part of the problem domain.

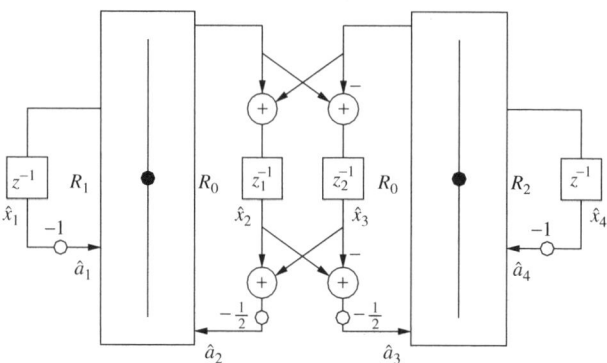

Figure 3.21: *Steady state MDWD network for the lossless, source-free, constant-coefficient (1 + 1)D transmission line system.*

where z^{-1} represents a unit delay in the time direction, and w, a unit shift in the x direction. We have written the outputs of the delay registers, in an exponential state, as

$$x_j(k\Delta, nT) = \hat{x}_j z^n w^k, \quad \text{for} \quad j = 1, \ldots, 4$$

where the $\hat{x}_j$ are complex amplitudes. The updating of the values in the delay registers can be written, in terms of these amplitudes, as

$$\begin{bmatrix} \hat{x}_1 \\ \hat{x}_2 \\ \hat{x}_3 \\ \hat{x}_4 \end{bmatrix} = z^{-1} \begin{bmatrix} -\alpha & \frac{1-\alpha}{2} & \frac{1-\alpha}{2} & 0 \\ \frac{(1+\alpha)}{2}w & \alpha w & 0 & \frac{(1+\alpha)}{2}w \\ \frac{(1+\alpha)}{2}w^{-1} & 0 & \alpha w^{-1} & -\frac{(1+\alpha)}{2}w^{-1} \\ 0 & \frac{1-\alpha}{2} & \frac{\alpha-1}{2} & -\alpha \end{bmatrix} \begin{bmatrix} \hat{x}_1 \\ \hat{x}_2 \\ \hat{x}_3 \\ \hat{x}_4 \end{bmatrix}$$

which is parameterized by a reflectance

$$\alpha = \frac{R_0 - R_1}{R_0 + R_1} \tag{3.79}$$

If we introduce the variables

$$y_1 = x_1 + x_4 \qquad\qquad y_4 = x_1 - x_4$$

the updating decouples into two subsystems, namely,

$$\begin{bmatrix} \hat{x}_2 \\ \hat{y}_1 \end{bmatrix} = z^{-1} \begin{bmatrix} \alpha w & (1+\alpha)w \\ (1-\alpha) & -\alpha \end{bmatrix} \begin{bmatrix} \hat{x}_2 \\ \hat{y}_1 \end{bmatrix} = z^{-1} \mathbf{A}_{21} \begin{bmatrix} \hat{x}_2 \\ \hat{y}_1 \end{bmatrix} \tag{3.80a}$$

$$\begin{bmatrix} \hat{x}_3 \\ \hat{y}_4 \end{bmatrix} = z^{-1} \begin{bmatrix} \alpha w^{-1} & (1+\alpha)w^{-1} \\ (1-\alpha) & -\alpha \end{bmatrix} \begin{bmatrix} \hat{x}_3 \\ \hat{y}_4 \end{bmatrix} = z^{-1} \mathbf{A}_{34} \begin{bmatrix} \hat{x}_3 \\ \hat{y}_4 \end{bmatrix} \tag{3.80b}$$

$\mathbf{A}_{21}$ and $\mathbf{A}_{34}$ are known as *spectral amplification* matrices (see Appendix A). The *symbols* [238] of the two subsystems, $\mathbf{Q}_{24}$ and $\mathbf{Q}_{31}$ are defined by

$$\mathbf{Q}_{21} = \mathbf{I}_2 - z^{-1}\mathbf{A}_{21} \qquad\qquad \mathbf{Q}_{34} = \mathbf{I}_2 - z^{-1}\mathbf{A}_{34}$$

3.8. FINITE DIFFERENCE INTERPRETATION

where $\mathbf{I}_2$ is the 2×2 identity matrix. Nontrivial solutions to the update equations (3.80) occur when the determinants of the symbols vanish. In the absence of boundary conditions, we may assume $w = e^{j\beta\Delta}$, where β is a real wavenumber, in which case we have four solutions in terms of z given by

$$z_{21,\pm} = e^{\frac{j\beta\Delta}{2}} \left(j\alpha \sin\left(\frac{\beta\Delta}{2}\right) \pm \sqrt{1 - \alpha^2 \sin^2\left(\frac{\beta\Delta}{2}\right)} \right) \quad (3.81a)$$

$$z_{34,\pm} = e^{-\frac{j\beta\Delta}{2}} \left(-j\alpha \sin\left(\frac{\beta\Delta}{2}\right) \pm \sqrt{1 - \alpha^2 \sin^2\left(\frac{\beta\Delta}{2}\right)} \right) \quad (3.81b)$$

which are simply the eigenvalues of the spectral amplification matrices. The corresponding eigenvectors of these same matrices are

$$\mathbf{u}_{21,\pm} = \begin{bmatrix} \alpha \cos\left(\frac{\beta\Delta}{2}\right) \pm \sqrt{1 - \alpha^2 \sin^2\left(\frac{\beta\Delta}{2}\right)} \\ (1-\alpha)e^{-\frac{\beta\Delta}{2}} \end{bmatrix} \quad \mathbf{u}_{34,\pm} = \begin{bmatrix} \alpha \cos\left(\frac{\beta\Delta}{2}\right) \pm \sqrt{1 - \alpha^2 \sin^2\left(\frac{\beta\Delta}{2}\right)} \\ (1-\alpha)e^{\frac{\beta\Delta}{2}} \end{bmatrix}$$

All four eigenvalues are of unit magnitude, and thus, using $z = e^{j\omega T}$, we can rewrite solutions (3.81) as

$$e^{j\omega_{21\pm}T} = \pm e^{\frac{j\Delta}{2}(\beta\pm\nu)} \qquad e^{j\omega_{34\pm}T} = \pm e^{-\frac{j\Delta}{2}(\beta\pm\nu)} \quad (3.82)$$

for some real ν defined by $\sin(\nu\Delta/2) = \alpha \sin(\beta\Delta/2)$. ($\nu$ always exists because we have $|\alpha| \leq 1$, from (3.79).) For small wavenumbers, we have

$$\nu \approx \alpha\beta$$

and we thus have in this limit, for the roots subscripted with $+$ in (3.82),

$$e^{j\omega_{21+}T} \approx e^{\frac{j\Delta(1+\alpha)}{2}\beta} \qquad \Rightarrow \qquad \frac{\omega_{21+}}{\beta} \approx \frac{1}{\sqrt{lc}}$$

$$e^{j\omega_{34+}T} \approx e^{-\frac{j\Delta(1+\alpha)}{2}\beta} \qquad \Rightarrow \qquad \frac{\omega_{34+}}{\beta} \approx -\frac{1}{\sqrt{lc}}$$

where we have used the fact that $(1+\alpha)/2 = 1/(v_0\sqrt{lc})$, which follows from (3.79) and the definitions of the port resistances in (3.62) as well as $v_0 = \Delta/T$. The quantities ω_{21+}/β and ω_{34+}/β are called *numerical phase velocities* [238]; they approach the propagation speed in the medium, from (3.56), and these two solutions are to be interpreted as approximations to the traveling wave solution to the transmission line equations. The other modes, however, are *parasitic*, in that they do not propagate near the physical velocity. They are not problematic, provided initial conditions are set properly; indeed, in the limit as Δ becomes small, any reasonable initial conditions tend to align the system with the dominant traveling modes of the system.

Clearly, if we are at the passivity limit, where $v_0 = 1/\sqrt{lc}$, then $R_1 = 0$, and thus $\alpha = 1$, which implies, finally, that $\nu = \beta$, so that we have, from (3.82), that $\omega_{21+}/\beta = 1/\sqrt{lc}$ and $\omega_{34+}/\beta = -1/\sqrt{lc}$; wave propagation is thus dispersionless. As mentioned in the previous section, at this limit, the MDWD network reduces to an exact digital traveling wave solution (this was also noted in Section 3.6.5). It is also interesting to note that when $R_1 = R_2 = R_0$, so that α and ν are zero, then (3.82) implies that wave propagation is also dispersionless in this case as well. It is easy to see here, from Figure 3.21, that because $R_1 = R_2 = R_0$, there will be no scattering through the adaptors; the pure time delays may thus be shifted directly into the lattice two-port and we can perform a manipulation similar to that of Section 3.6.5 to give a simplified digital 'traveling wave' network with doubled time delays. Here, we are in effect implementing a traveling wave solution on a different grid but the implication is that for the corresponding problem with material variation, the MDWD network gives a good approximation to the numerical phase velocity even for certain values of v_0 that are far from the local physical wave speed. This is not true for DWNs, where the numerical phase velocities degrade considerably away from the passivity limit. We will return to these expressions (which provide complete information regarding the *numerical dispersion* properties of the scheme) in Section 4.3.8 in a comparison with the DWN for the same system. In anticipation of the discussion in Section 3.9, we mention that for constant v_0, we have for the eigenvectors corresponding to the dominant modes that

$$\lim_{\Delta \to 0} \mathbf{u}_{21+} = \lim_{\Delta \to 0} \mathbf{u}_{34,+} = \begin{bmatrix} \alpha + 1 \\ 1 - \alpha \end{bmatrix}$$

Since we also have, from Figure 3.21, that $\hat{a}_1 = -\hat{x}_1 = -(\hat{y}_1 + \hat{y}_4)/2$, and $\hat{a}_2 = -(\hat{x}_2 + \hat{x}_3)/2$, we can also write, for the dominant mode,

$$\lim_{\Delta \to 0} \begin{bmatrix} \hat{a}_1 \\ \hat{a}_2 \end{bmatrix} = -\frac{1}{2} \lim_{\Delta \to 0} (\mathbf{u}_{21+} + \mathbf{u}_{34,+}) = -\begin{bmatrix} \alpha + 1 \\ 1 - \alpha \end{bmatrix} = -\frac{2}{R_1 + R_0} \begin{bmatrix} R_1 \\ R_0 \end{bmatrix}$$

Thus, in this limit, the wave variables incident on the left adaptor occur in the same ratio as the port resistances, and are in fact aligned with an eigenvector of the scattering matrix corresponding to the adaptor. A similar statement holds for the quantities incident on the right adaptor. We will return to this observation in the next section.

3.9 Initial Conditions

Numerical simulations for time-dependent systems of PDEs must necessarily be initialized; while this is a relatively straightforward matter for WDF-based integration schemes, it has only been addressed in passing in the literature [145].

We will examine here the initialization of the MDWD network for the source-free transmission line system (3.54) with $e = h = 0$. This system requires initial distributions for both the current and voltage, which we will call $i_0(x)$ and $u_0(x)$, respectively. For the initialization of the MDWD network for this system (shown in Figure 3.22), we will also need their spatial derivatives (assuming they exist), which we will write as $i'_0(x)$ and $u'_0(x)$.

For the MDWD network of Figure 3.22, we must initialize all the wave variables incident upon the scattering junctions, written as a_j, $j = 1, \ldots, 4$; because we have assumed no sources, no wave enters through the loss/source port. This circuit is an MD representation and each of these wave variables refers to an array. Assuming that the spatial grid spacing is

3.9. INITIAL CONDITIONS

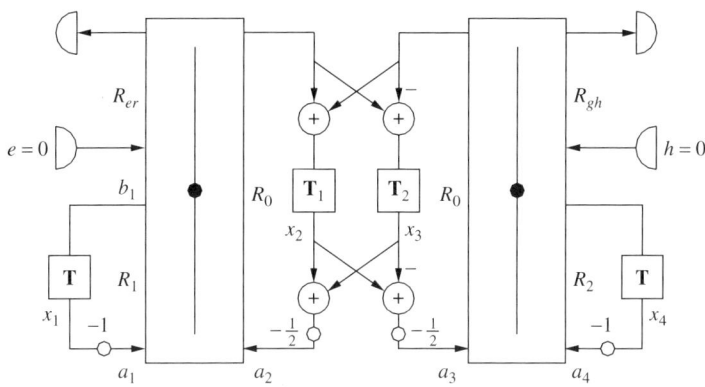

Figure 3.22: *MDWD network for the source-free* $(1+1)$D *transmission line equations.*

Δ, and the time step is T, we can index the elements of these arrays as $a_{jm}(n)$, for m and n integer; this represents an instance of the MD wave variable a_j at grid location $x = m\Delta$, and at time $t = nT$. For initialization, we must thus set $a_{jm}(0)$, over all grid locations $x = m\Delta$ included in the domain of the problem, in terms of the quantities $i_0(m\Delta)$, $u_0(m\Delta)$ and their spatial derivatives.

We will consider only the settings for the wave variables in the left-hand adaptor; one proceeds in the same way for the right adaptor. We recall that the port resistances are defined by

$$R_1 = \frac{2}{\Delta}(v_0 l - r_0) \qquad R_0 = \frac{2r_0}{\Delta} \qquad R_{er} = r$$

Since the port resistances R_1 and R_{er} are functions of position, we will write $R_{1m} \triangleq R_1(m\Delta)$ and $R_{erm} \triangleq R_{er}(m\Delta)$. R_0 is independent of x. It is easy to see, from (2.30), that the initial values $a_{1m}(0)$ and $a_{2m}(0)$ must be arranged such that the initial current $i_0(m\Delta)$ is reproduced. Thus, we need

$$i_0(m\Delta) = \frac{1}{R_0 + R_{1m} + R_{erm}}(a_{1m}(0) + a_{2m}(0)) \tag{3.83}$$

Another condition is required to fully specify the wave variable initial values. Referring to the generating MDKC for this MDWD network in Figure 3.14(a), we can see that the voltage across the inductor of inductance L_1 will be $L_1 \frac{\partial i}{\partial t'}$. We intend to relate this voltage to the associated digital voltage across the inductor of port resistance R_1 in Figure 3.22. We have

$$L_1 \frac{\partial i}{\partial t'} \stackrel{(3.18)}{=} \frac{L_1}{v_0} \frac{\partial i}{\partial t}$$

$$\stackrel{(3.61)}{=} \frac{v_0 l - r_0}{v_0} \frac{\partial i}{\partial t}$$

$$\stackrel{(3.54)}{=} \left(\frac{r_0}{v_0 l} - 1\right)\left(\frac{\partial u}{\partial x} + ri\right) \tag{3.84}$$

At time $t = 0$, and at location $x = m\Delta$, we may write this voltage as

$$L_1 \frac{\partial i}{\partial t'}\bigg|_{x=m\Delta, t=0} \stackrel{(3.62)}{=} -\frac{R_{1m}}{R_0 + R_{1m}}\left(u'_0(m\Delta) + r_m i_0(m\Delta)\right)$$

The wave digital voltage across the same port, at location $x = m\Delta$, is defined by

$$v_m = \frac{1}{2}(a_{1m} + b_{1m}) = \frac{1}{R_0 + R_{1m} + R_{erm}}\left((R_0 + R_{erm})a_{1m} - R_{1m}a_{2m}\right) \quad (3.85)$$

Thus, for initialization, equating the voltages in (3.85) and (3.84), we must have

$$-\frac{R_{1m}(R_{1m} + R_0 + R_{erm})}{R_{1m} + R_0}\left(u'_0(m\Delta) + R_{erm}i_0(m\Delta)\right) = (R_0 + R_{er})a_{1m}(0) - R_1 a_{2m}(0)$$

This requirement, along with (3.83), fully specifies the initial values of the wave variables at the left adaptor. We thus have

$$a_{1m}(0) = \left(R_{1m} - \frac{R_{erm}R_{1m}}{R_{1m} + R_0}\right)i_0(m\Delta) - \frac{R_{1m}}{R_{1m} + R_0}u'_0(m\Delta) \quad (3.86a)$$

$$a_{2m}(0) = \left(R_0 + R_{erm} + \frac{R_{erm}R_{1m}}{R_{1m} + R_0}\right)i_0(m\Delta) + \frac{R_{1m}}{R_{1m} + R_0}u'_0(m\Delta) \quad (3.86b)$$

We note that $u'_0(m\Delta)$ may be obtained from the initial voltage distribution $u_0(x)$ by any reasonable (i.e., consistent) approximation to the spatial derivative.

It is important to recognize that for constant v_0, we have

$$\lim_{\Delta \to 0} a_{1m}(0) = R_{1m}i_0(m\Delta) \qquad \lim_{\Delta \to 0} a_{2m}(0) = R_0 i_0(m\Delta)$$

These values occur in the same ratio as those of an *eigenvector* of the scattering matrix for the left series adaptor. In particular, they follow the distribution of the principal eigenvector (i.e., the unique eigenvector whose elements are all of the same sign) of the scattering matrix. Thus the proper setting for the initial conditions (except at the loss port) should be aligned with the dominant mode of the numerical scheme in this limit (and the fraction of the initial energy injected into the parasitic modes will vanish) (see Section 3.8.2 for a discussion of parasitic modes in this particular system)[12]. We also suggest the following very simple 'rule of thumb' for setting initial conditions:

---

[12] We would like to note here that the eigenvectors of the scattering matrix will give extremal values for the quantity $\mathbf{a}^T \mathbf{P} \mathbf{b}$ for an N-port with N-vector input waves $\mathbf{a}$ and output waves $\mathbf{b}$ and a diagonal weighting matrix $\mathbf{P}$ containing port conductances. This quantity, at least for some simple lumped systems, can be identified with the *Lagrangian*. Many physics problems (indeed, all the systems treated in this book) can be recast as variational problems involving finding an extreme value of the Lagrangian integrated over all possible system states. Though we have not worked out all the details in the distributed case, it would appear that the alignment of the discrete system with an eigenvector of the scattering matrix is the scheme's 'attempt' to conform to Lagrangian mechanics (which is to be expected). This is interesting for two reasons; first, the wave digital Lagrangian could form the basis for a new set of quantization rules, which, in addition to (or perhaps instead of) ensuring passivity, nudge the system toward a preferred state. Second, the variational or Lagrangian formulation of a physics problem is a necessary first step in developing what are known as *finite-element methods* (FEM); these methods are not restricted to operating on regular grids, as MDWD methods are. Could MDWD networks operating on unstructured grids be arrived at through such a formulation? This is, admittedly, a very vague notion.

3.10. BOUNDARY CONDITIONS

> For a given series M-port MD-adaptor in an MDWD network, with associated current i, port resistances R_j and input wave variables a_j, $j = 1, \ldots, M$, if the initial value of i at the adaptor is to be i_0, we should set the initial values of the wave variables to be
>
> $$a_j(0) = 0 \qquad \text{at a loss/source port}$$
> $$a_j(0) = R_j i_0 \qquad \text{otherwise}$$

This rule is to be interpreted in a distributed sense, that is, it holds for every instance of an adaptor on the numerical grid. A similar rule holds for a parallel adaptor. These settings ignore spatial derivative information but give a simple way of proceeding in general, especially during the first stages of programming and debugging, and are correct (to first order) in the limit as Δ approaches 0. If losses are large, though, one may prefer to use exact conditions like (3.86). This rule applies regardless of the number of dimensions of the problem (but may need to be amended if sources or reflection-free ports are present).

3.10 Boundary Conditions

Boundary termination of MDWD networks has been discussed by various authors [89, 146, 176, 284]; it is still an open research problem but we would like to mention the several disparate approaches that have been proposed. The problem of general passive termination of an MD network is very involved and would probably merit a long treatment in a separate work; termination of a $(1 + 1)$D MDWD network, which is all we will be able to discuss here, is simpler and the ideas can be extended to cover certain important cases in higher dimensions. The most straightforward method was put forth by Krauss and Rabenstein [146]. We will refer here to Figure 3.22, the MDWD network for the source-free transmission line equations. This network represents the signal behavior at any grid point in the domain. In particular, the signals x_2 and x_3 are obtained at each time step from signals input into the shift registers at grid locations immediately to the left and right, respectively. Suppose now that we have a left boundary termination at $x = 0$, and that the domain has been sampled such that a grid point coincides with this point. Then, at this location, x_2 cannot be directly obtained because there is no grid point to the left to which the shift register refers.

Let us now examine some simple lossless boundary conditions of the form of (3.8). Suppose that we would like the boundary condition for the transmission line to be that of an *open-circuit* at the left termination, so that we have

$$i(0, t) = 0 \qquad \text{Open-circuit termination}$$

The wave digital approximation to the physical current at any grid point is calculated from the wave variables incident on the left series adaptor, so that we have

$$i = \frac{1}{R_0 + R_{er} + R_1}(a_1 + a_2) = \frac{1}{R_0 + R_{er} + R_1}\left(-x_1 - \frac{1}{2}(x_2 + x_3)\right)$$

If we set, at the left-most grid point,

$$x_2 = -x_3 - 2x_1$$

then the calculated current will be identically zero, and x_2 is easily obtained from x_3 and x_1, both of which are available. A short-circuited termination, that is,

$$u(0, t) = 0 \qquad \text{Short-circuit termination}$$

can be accomplished by treating the right-hand adaptor in a similar manner. It is possible to mix these conditions and to introduce loss and a lumped terminating source as well [146]. This idea is also easily extended to multiple dimensions for rectilinearly sampled grids if the boundary is parallel to one of the grid axes. We would like to add, however, that such a termination has *never been shown to be passive*, and that there is no general theory applicable to boundary termination of MDWD networks (though we will provide a possible foundation for such an approach in Section 3.10.1). While it is easy to prove that the simple cases above correspond to passive lumped terminations, there are situations in higher dimensions when this approach becomes difficult to apply reliably; in several instances, (see Chapter 7 for some added discussion), this approach has failed in simulation. The difficulty with approaching boundary termination in this way is that the physics of the problem (in particular, the passivity at the boundary) is not being taken into account; this method, though easy to apply, is essentially no different from what is done using conventional finite difference methods. Fettweis and Nitsche [89] provided an alternative method that is more satisfying from a physical point of view; in this case, the region beyond the boundary is modeled as a material with extreme parameter values (typically $r = \infty$ or $g = \infty$, for the transmission line or parallel-plate problem). These regions are still passive, though it may now be necessary to employ a 'layer' of this material that will incur extra calculation costs. Other recent work has involved more general lumped boundary terminations [6, 284, 283], as well as the termination of the $(2 + 1)$D parallel-plate problem in hexagonal coordinates; we mention that these approaches are unwieldy in the extreme; in at least one case [282], the proposed modeling of a passive boundary condition requires active elements!

The problem with the termination of MDWD networks is that when spatial dependence is expanded out to get a signal flow graph, we do not end up with a lumped network of portwise-connected elements; see, for example, the flow graph for the simple advective system, shown in Figure 3.7. Such is not the case for DWNs, which can be formulated from the outset, like TLM methods, as large lumped networks. For this reason, boundary termination is much simpler in a DWN. In Chapter 4, which is devoted to DWNs, we will discuss boundary termination for the $(1 + 1)$D transmission line problem in Section 4.3.9, and for the parallel-plate problem in Section 4.4.4. Boundary termination for vibrating beam and plate systems is discussed in detail in Chapter 7.

3.10.1 MDKC Modeling of Boundaries

One of the big hurdles yet to be overcome in the MDWD simulation method is the implementation of boundary conditions. As we mentioned briefly in the previous section, this is a very tricky business, and the approaches in the literature for simple model problems do not generalize to more complex systems. The problem is that there is not, as yet, a

3.10. BOUNDARY CONDITIONS

general theory of boundary conditions for MDWD simulation methods[13]. In this section, we briefly mention a possible foundation for such a theory that is based on earlier work on MD-passivity [70, 116, 176], as outlined in Section 3.3.

Suppose that the problem of interest is $(n+1)$D, and defined with respect to coordinates $\mathbf{u} = [x_1, \ldots, x_n, t]^T$, or equivalently, to k transformed coordinates $\mathbf{t} = [t_1, \ldots, t_k]^T$, with $k \geq n+1$. We will assume that the problem has one spatial boundary, namely, the hyperplane $x_1 = 0$, and is defined over a time interval $[0, t_f]$. As such, the problem domain G is then

$$G = \{\mathbf{u}|\ 0 \leq t \leq t_f, x_1 \geq 0\}$$

or its equivalent in the $\mathbf{t}$ coordinates, obtained under a transformation of the form (3.22). We restate the energy balance for an N-port defined over G, which is

$$\int_G (w_{\text{inst}} + w_s)\, dV_\mathbf{t} = \int_G (w_\text{d} + \nabla_\mathbf{t} \cdot \mathbf{E})\, dV_\mathbf{t}$$

where w_{inst} is the instantaneous applied power at the ports, w_s is an internal source power, w_d is dissipated power, and $\mathbf{E}$ is a column k-vector representing stored energy flux; $dV_\mathbf{t}$ is the differential volume element. As mentioned previously, the N-port is *integrally MD-passive* over G if

$$\int_G w_{\text{inst}}\, dV_\mathbf{t} \geq \int_G \nabla_\mathbf{t} \cdot \mathbf{E}\, dV_\mathbf{t} = \int_{\partial G} \mathbf{n}_G \cdot \mathbf{E}\, d\sigma_G \qquad (3.87)$$

for some $\mathbf{E}$, all of whose components in the $\mathbf{t}$ coordinates are positive everywhere in G. Here, ∂G is the boundary of G, $\mathbf{n}_G$ is the unit outward normal, and $d\sigma_G$ is a differential surface element on the boundary.

Note that ∂G consists of the union of three sets of points, that is,

$$\partial G = \partial G_0 \cup \partial G_f \cup \partial G_b$$

where, in terms of the physical $\mathbf{u}$ coordinates,

$$\partial G_0 = \{\mathbf{u}|\ t = 0, x_1 \geq 0\}$$
$$\partial G_f = \{\mathbf{u}|\ t = t_f, x_1 \geq 0\}$$
$$\partial G_b = \{\mathbf{u}|\ 0 \leq t \leq t_f, x_1 = 0\}$$

We can thus rewrite (3.87) as

$$\int_G w_{\text{inst}}\, dV_\mathbf{t} \geq \int_{\partial G_0} \mathbf{n}_G \cdot \mathbf{E}\, d\sigma_G + \int_{\partial G_f} \mathbf{n}_G \cdot \mathbf{E}\, d\sigma_G + \int_{\partial G_b} \mathbf{n}_G \cdot \mathbf{E}\, d\sigma_G$$

For a *closed network*—that is, an N-port with no free terminals (corresponding to a complete system of PDEs)—the instantaneous applied power is zero, so we are left with

$$0 \geq \int_{\partial G_0} \mathbf{n}_G \cdot \mathbf{E}\, d\sigma_G + \int_{\partial G_f} \mathbf{n}_G \cdot \mathbf{E}\, d\sigma_G + \int_{\partial G_b} \mathbf{n}_G \cdot \mathbf{E}\, d\sigma_G \qquad (3.88)$$

---

[13]Rudolf Rabenstein, private communication, 2000.

In other words, the stored power flux leaving the boundary must be negative (the N-port is passive).

Suppose, now, that there is an nD N-port defined on the spatial boundary ∂G_b of G. Renaming this region $G^{(b)}$, we have another energy balance

$$\int_{G^{(b)}} \left(w_{\text{inst}}^{(b)} + w_{\text{s}}^{(b)} \right) \mathrm{d}V_{\mathbf{t}^{(b)}} = \int_{G^{(b)}} \left(w_{\text{d}}^{(b)} + \nabla_{\mathbf{t}^{(b)}} \cdot \mathbf{E}^{(b)} \right) \mathrm{d}V_{\mathbf{t}^{(b)}}$$

over coordinates $\mathbf{t}^{(b)}$ derived from physical coordinates $\mathbf{u} = [x_2, \ldots, x_n, t]^T$ on $G^{(b)}$. The quantities $w_{\text{inst}}^{(b)}$, $w_{\text{s}}^{(b)}$, $w_{\text{d}}^{(b)}$ and $\mathbf{E}^{(b)}$ are the applied power, source power, dissipated power, and stored energy flux in the boundary network. Again, if the boundary network is passive, we have

$$\int_{G^{(b)}} w_{\text{inst}}^{(b)} \mathrm{d}V_{\mathbf{t}^{(b)}} \geq \int_{G^{(b)}} \nabla_{\mathbf{t}^{(b)}} \cdot \mathbf{E}^{(b)} \mathrm{d}V_{\mathbf{t}^{(b)}} = \int_{\partial G^{(b)}} \mathbf{n}_{G^{(b)}} \cdot \mathbf{E}^{(b)} \mathrm{d}\sigma_{G^{(b)}} \qquad (3.89)$$

where $\partial G^{(b)}$ is the boundary of the region $G^{(b)}$, and consists of the union of the two regions

$$\partial G_0^{(b)} = \{\mathbf{t}^{(b)} | \, t = 0\}$$
$$\partial G_f^{(b)} = \{\mathbf{t}^{(b)} | \, t = t_f\}$$

so that we have, finally,

$$\int_{G^{(b)}} w_{\text{inst}}^{(b)} \mathrm{d}V_{\mathbf{t}^{(b)}} \geq \int_{\partial G_0^{(b)}} \mathbf{n}_{G^{(b)}} \cdot \mathbf{E}^{(b)} \mathrm{d}\sigma_{G^{(b)}} + \int_{\partial G_f^{(b)}} \mathbf{n}_{G^{(b)}} \cdot \mathbf{E}^{(b)} \mathrm{d}\sigma_{G^{(b)}} \qquad (3.90)$$

The boundary network is intended to model a passive distributed termination to the problem defined over the region G. It should be clear that if both networks are passive and if the transfer of energy between them is passive, then the terminated system as a whole will be passive. See Figure 3.23 for a representation of the relevant regions.

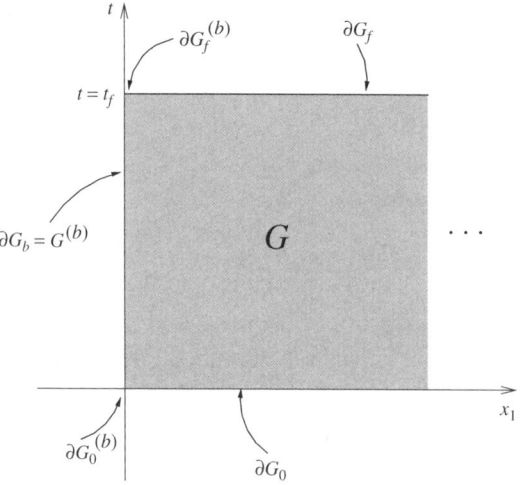

Figure 3.23: *A region with one spatial boundary.*

3.10. BOUNDARY CONDITIONS

We can ensure this by requiring that the power applied through the ports of the boundary network over the region $G^{(b)}$ is equal to the stored energy flux of the interior network leaving through its spatial boundary (recall that we have set $G^{(b)} = \partial G_b$). In other words, we require

$$w_{\text{inst}}^{(b)} = \mathbf{n}_G \cdot \mathbf{E} \quad \text{on} \quad \partial G_b = G^{(b)}$$

Inequality (3.88) can then be rewritten as

$$0 \geq \int_{\partial G_0} \mathbf{n}_G \cdot \mathbf{E} \, d\sigma_G + \int_{\partial G_f} \mathbf{n}_G \cdot \mathbf{E} \, d\sigma_G + \int_{G^{(b)}} w_{\text{inst}}^{(b)} \, dV_{G^{(b)}}$$

or, by employing (3.90), as

$$0 \geq \int_{\partial G_0} \mathbf{n}_G \cdot \mathbf{E} \, d\sigma_G + \int_{\partial G_f} \mathbf{n}_G \cdot \mathbf{E} \, d\sigma_G + \int_{\partial G_0^{(b)}} \mathbf{n}_{G^{(b)}} \cdot \mathbf{E}^{(b)} \, d\sigma_{G^{(b)}}$$

$$+ \int_{\partial G_f^{(b)}} \mathbf{n}_{G^{(b)}} \cdot \mathbf{E}^{(b)} \, d\sigma_{G^{(b)}} \quad (3.91)$$

From (3.31), the quantities in (3.91) have the following interpretation:

$$\mathcal{E}(0) \triangleq -\int_{\partial G_0} \mathbf{n}_G \cdot \mathbf{E} \, d\sigma_G \quad = \quad \text{Energy of interior network at time } t = 0$$

$$\mathcal{E}^{(b)}(0) \triangleq -\int_{\partial G_0^{(b)}} \mathbf{n}_{G^{(b)}} \cdot \mathbf{E}^{(b)} \, d\sigma_{G^{(b)}} = \text{Energy of boundary network at time } t = 0$$

$$\mathcal{E}(t_f) \triangleq \int_{\partial G_f} \mathbf{n}_G \cdot \mathbf{E} \, d\sigma_G \quad = \quad \text{Energy of interior network at time } t = t_f$$

$$\mathcal{E}^{(b)}(t_f) \triangleq \int_{\partial G_f^{(b)}} \mathbf{n}_{G^{(b)}} \cdot \mathbf{E}^{(b)} \, d\sigma_{G^{(b)}} = \text{Energy of boundary network at time } t = t_f$$

The negative signs in the definitions of the initial energies result from the fact that the outward normal to ∂G_0 and $\partial G_0^{(b)}$ points in the negative time direction. As such, (3.91) can be restated simply as

$$\mathcal{E}(t_f) + \mathcal{E}^{(b)}(t_f) \leq \mathcal{E}(0) + \mathcal{E}^{(b)}(0)$$

or, in other words: the total energy stored in the interior and boundary networks must not increase as time progresses.

It is straightforward to extend this idea to more complex boundaries. For example, if the region G were to be defined by

$$G = \{\mathbf{u} | x_1 \geq 0, x_2 \geq 0, 0 \leq t \leq t_f\}$$

so that there is a corner at $x_1 = x_2 = 0$, we could model passive boundary conditions using four networks: an $(n+1)$D network for the interior of G, two nD networks for the two 'faces' and a $(n-1)$D network for the corner itself; an energy inequality similar to (3.91) results.

Here we have said absolutely nothing about discretization. We have, however, indicated the possibility for arbitrary *distributed* passive boundary termination of a given MDKC; only lumped conditions have been examined so far in the literature.

Note on Perfectly Matched Layers

An interesting and related direction in current research into boundary termination involves the use of the so-called *perfectly matched layers* (PMLs) [18, 19] as boundary terminations in problems to be solved over an unbounded spatial domain. The idea, generally speaking, is to surround a numerical problem domain with a layer of a material that creates as little numerical reflection as possible, while also attenuating waves that enter from the problem interior.

Absorbing boundary conditions (ABCs) [247] were long used for this purpose in $(2 + 1)$D and $(3 + 1)$D electromagnetic problems; in terms of the $(2 + 1)$D parallel-plate problem (which is equivalent to $(2 + 1)$D TE or TM mode electromagnetics), the layer is chosen to be matched to the characteristic impedance of the plates, namely, $\sqrt{l/c}$. As such, it can be thought of as an extension to $(2 + 1)$D of the reflectionless matched termination that can be applied to a $(1 + 1)$D transmission line. Unfortunately, in higher dimensions, such a termination is reflection-free only for waves at normal incidence, and there will be significant backscatter into the problem interior at oblique incidence; furthermore, the amount of reflection is frequency-dependent.

Berenger [18] solved this problem, at least in theory, by proposing a new unphysical medium as an absorbing material. For the parallel-plate problem, the dependent variables in this new medium are the current density and two orthogonal ('split') voltage components; if the layer is infinitely thick, then it indeed absorbs and attenuates waves of any frequency or angle of incidence. The problem here, as has been pointed out by various authors [1, 2, 258], is that the proposed medium can be described by a system which, though hyperbolic, is not symmetric hyperbolic and thus not of the form of (3.1). Worse, it is not even strongly hyperbolic [112]; strong hyperbolicity is the necessary requirement for the initial-value problem to be well posed. As a result, lower-order perturbations such as those that might result from numerical discretization, can render such a system ill-posed, and susceptible to numerical instability. (It is worth noting that the MDKC representations that we have discussed in this chapter have only been applied to symmetric hyperbolic systems of the particular form of (3.1). An MDKC representation of a $(3 + 1)$D PML medium has been proposed [173], but in this case the asymmetries in the system were lumped into dependent source terms and MD passivity does not immediately follow.) Other more physical reformulations of the PML in terms of an anisotropic frequency-dependent medium [206, 288] and stretched complex coordinates [247] do not alleviate this problem significantly and other similar approaches, such as sponge layers [187] the transparent absorbing boundary [186] and Lorentz materials [292] appear to lead to similar difficulties.

New PML-type media that can be described by symmetric hyperbolic systems were put forth in [2, 258]; they are of the form of (3.1), but for these media the symmetric part of the **B** matrix is not positive-semidefinite, so an energy estimate of the form of (3.7) is not available. In particular, though one can indeed develop MDKC representations for these systems, the nonpositivity of the symmetric part of **B** leads to active (though purely resistive) coupling between the various circuit loops. This is somewhat curious because it has been shown [2] that field quantities in the absorbing medium decay as a function of distance from the boundary in any direction, so it would be expected that these media should indeed be passive. Several questions arise here that are related to the general issue of when a passive MDKC representation can be derived from a physically passive system. In particular: what kind of symmetries are required of the various system matrices? Is it possible to represent

3.11. BALANCED FORMS

systems that are not symmetric hyperbolic but only strongly hyperbolic (in which case non-reciprocal reactive elements would be necessary)?

Finally, we mention that although these absorbing layers have been proposed for use in electromagnetic field simulation problems, they apply equally well to the associated mechanical and acoustic systems; a version of the layer intended for use in fluid dynamic problems has been proposed [125]. Applications in musical and room acoustics would seem to be manifold (the calculation of the sound field radiating from the open end of a musical instrument, and open-air architectural acoustics problems come to mind as possible examples).

3.11 Balanced Forms

Consider again the $(1+1)$D transmission line, with spatially varying coefficients. It has been noted in the past [176] that the restriction on the time step, namely

$$v_0 \geq \frac{1}{\sqrt{l_{\min} c_{\min}}}.$$

is rather unsatisfying; the local group velocity at any point in the problem domain is given by $\pm 1/\sqrt{lc}$, so we would hope that a more physically meaningful bound such as

$$v_0 \geq \gamma_{TL,\max}^g = \sqrt{\frac{1}{(lc)_{\min}}} \quad (3.92)$$

(which is obtained in using, for example, DWNs, which will be discussed in Chapter 4) could be attainable. Depending on the variation in l and c, the new bound can allow a substantially larger time step. We will show that this is in fact possible via an MDWD approach.

The transmission line equations given in (3.54) can be transformed in the following way: first introduce new dependent variables

$$\tilde{i}_1 = \sqrt{Z} i \qquad \tilde{i}_2 = u/\sqrt{Z}$$

where Z, the local line impedance, is defined by

$$Z(x) \triangleq \sqrt{\frac{l}{c}}$$

Such a transformation in fact changes to variables that both have units of *root power*. After a few elementary manipulations (namely, scaling (3.54a) by $1/\sqrt{Z}$ and (3.54b) by $\sqrt{Z}$), we have

$$\sqrt{lc}\frac{\partial \tilde{i}_1}{\partial t} + \frac{\partial \tilde{i}_2}{\partial x} + \frac{\partial}{\partial x}(\ln(\sqrt{Z}))\tilde{i}_2 + r\tilde{i}_1/Z + e/\sqrt{Z} = 0 \quad (3.93a)$$

$$\sqrt{lc}\frac{\partial \tilde{i}_2}{\partial t} + \frac{\partial \tilde{i}_1}{\partial x} - \frac{\partial}{\partial x}(\ln(\sqrt{Z}))\tilde{i}_1 + g Z\tilde{i}_1 + h\sqrt{Z} = 0 \quad (3.93b)$$

System (3.93) is still symmetric hyperbolic; referring to the general system from (3.1), for $\mathbf{w} = [\tilde{i}_1, \tilde{i}_2]^T$, we now have

$$\mathbf{P} = \begin{bmatrix} \sqrt{lc} & 0 \\ 0 & \sqrt{lc} \end{bmatrix} \quad \mathbf{A}_1 = \begin{bmatrix} 0 & 1 \\ 1 & 0 \end{bmatrix} \quad \mathbf{B} = \begin{bmatrix} r/Z & \frac{\partial}{\partial x}(\ln(\sqrt{Z})) \\ -\frac{\partial}{\partial x}(\ln(\sqrt{Z})) & gZ \end{bmatrix} \quad (3.94)$$

Note that because $\mathbf{P}$ is now a multiple of the identity matrix, there is near complete symmetry between the variables $\tilde{i}_1$ and $\tilde{i}_2$. We use the term 'balanced' to describe such a system. Note also that new off-diagonal terms have appeared in $\mathbf{B}$ (compare (3.94) with (3.55)), but they occur *antisymmetrically*[14], and thus do not give rise to loss—in other words these terms do not appear in $(\mathbf{B} + \mathbf{B})^T$, the term which determines the growth or decay of the solution, as per (3.5). In fact, these off-diagonal terms yield a lossless (but non-reciprocal) gyrator in the circuit setting.

In terms of the coordinates defined by (3.19), we can then rewrite (3.93) as

$$L_1 D_{t'} \tilde{i}_1 + L_0 D_1 (\tilde{i}_1 + \tilde{i}_2) + L_0 D_2 (\tilde{i}_1 - \tilde{i}_2) + R_G \tilde{i}_2 + \tilde{r} \tilde{i}_1 + \tilde{e} = 0$$

$$L_2 D_{t'} \tilde{i}_2 + L_0 D_1 (\tilde{i}_2 + \tilde{i}_1) + L_0 D_2 (\tilde{i}_2 - \tilde{i}_1) - R_G \tilde{i}_1 + \tilde{g} \tilde{i}_2 + \tilde{h} = 0$$

where

$$L_1 = L_2 = v_0 \sqrt{lc} - 1 \quad L_0 = \frac{1}{\sqrt{2}} \quad R_G = \frac{\partial}{\partial x}\left(\ln(\sqrt{Z})\right)$$

(which should be compared with (3.61), for the standard form), and

$$\tilde{r} = r/Z \quad \tilde{e} = e/\sqrt{Z} \quad \tilde{g} = gZ \quad \tilde{h} = h\sqrt{Z}$$

As mentioned previously, in an MDKC setting, the terms with coefficient R_G can be treated as a *gyrator*. The network and its wave digital counterpart are shown in Figure 3.24. The port resistances are given by

$$R_1 = R_2 = \frac{2}{\Delta}\left(v_0 \sqrt{lc} - 1\right) \quad R_0 = \frac{2}{\Delta} \quad R_{\tilde{e}\tilde{r}} = \tilde{r} \quad R_{\tilde{g}\tilde{h}} = \tilde{g}$$

In order to accommodate the gyrator, we have been forced, in order to avoid delay-free loops, to set one of the ports to which it is connected to be *reflection-free* (see Section 2.2.5). In $(1+1)D$, we can choose either of these ports but picking the bottom port in Figure 3.24(b) allows us to extend the idea to $(2+1)D$ easily. This port resistance is then constrained to be

$$R_{G1} = R_1 + R_{\tilde{e}\tilde{r}} + R_0$$

We have two simplifying choices for R_{G2}; either we can choose it to be reflection-free as well, so that we will have a general gyrator described by (2.25), or we can choose

$$R_{G2} = \frac{R_G^2}{R_{G1}},$$

---

[14] Recall from Section 3.1 that the term 'symmetric hyperbolic' does not refer to the coefficient $\mathbf{B}$ of the constant-proportional term, which is not constrained to be of any particular form.

3.11. BALANCED FORMS

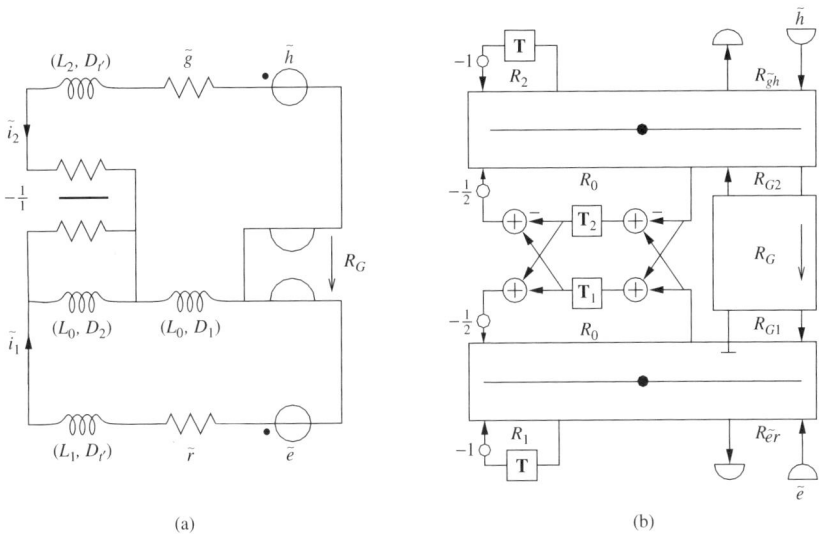

Figure 3.24: (a) *Balanced MD-passive network for the* $(1+1)$D *transmission line equations and* (b) *its associated MDWD network*.

in which case the gyrator equations (2.25) reduce to a pair of throughs, scaled individually by R_G/R_{G1} and its inverse; this latter choice may be problematic if R_G approaches zero, because one of the multipliers becomes unbounded. If R_G is small over some part of the problem domain, however, it is probably wiser to remove the coupling from the network altogether over these regions (it can be replaced by a simple two-port short-circuit). We have assumed, throughout this development, that $l(x)$ and $c(x)$ (or rather, the local characteristic line impedance $Z(x) = \sqrt{l(x)/c(x)}$) are differentiable. An *offset-sampled* version of this network is also possible if we halve the port resistances R_1 and R_2 and double the delays at the same ports.

The stability bound, from a requirement on the positivity of R_1 and R_2, will be exactly (3.92). In a computer implementation, there will be of course the slight additional costs due to the extra gyrator and the rescaling of the new dependent variables $\tilde{i}_1$ and $\tilde{i}_2$ at every time step in order to obtain i and u. We note that this scaling can be fully incorporated into the MDKC by treating the scaling coefficients as transformer turns ratios, though there is no advantage in doing so (other than putting one's mind at ease regarding whether such a scaling is a passive operation).

We will examine how this same technique can be applied to more complex systems when we approach the Timoshenko beam equations in Section 7.1.5; in that case the maximum allowable time step can be radically increased for a system with only mild material parameter variation.

Extension to $(2+1)$D

We briefly note that the same approach can be easily extended to the parallel-plate problem as well; beginning from system (3.65), we can introduce new variables

$$\tilde{i}_1 = \sqrt{Z} i_x \qquad \tilde{i}_2 = \sqrt{Z} i_y \qquad \tilde{i}_3 = u/\sqrt{Z}$$

where $Z(x, y) \triangleq \sqrt{l(x,y)/c(x,y)}$, and then multiply (3.65a) and (3.65b) by $1/\sqrt{Z}$ and (3.65c) by $\sqrt{Z}$. The new system is again symmetric hyperbolic. We do not show the network here but we mention that we will require two gyrators: one linking the series adaptors with associated currents $\tilde{i}_1$ and $\tilde{i}_3$, the other between the adaptors for $\tilde{i}_2$ and $\tilde{i}_3$. One reflection-free port must be chosen for each gyrator; choosing both reflection-free ports at the adaptor with current $\tilde{i}_3$ must be ruled out but other configurations are acceptable.

The stability bound for the balanced $(2+1)$D network will be

$$v_0 \geq \sqrt{\frac{2}{(lc)_{\min}}} = \sqrt{2} \gamma_{\text{PP,max}}^g$$

which is superior to (3.69), the bound for the standard form.

3.12 Higher-order Accuracy

WDF-based numerical methods are, in general, second-order accurate in both the time step and the grid spacing. In all the schemes that have been examined in the literature, these quantities occur in a fixed ratio (usually written as v_0), so we can say that such schemes are accurate to second order in either one (or of any of the shift lengths in the new coordinates). A numerical approximation to a system of PDEs obtained using an MDWD network will converge to the solution to the model problem with a truncation error [238] proportional to the square of any of these spacings.

While this is true in general, in this section we would like to point out that it is indeed possible to devise MD circuit-based schemes that exhibit a higher-order *spatial* accuracy. Temporal accuracy, however, remains fixed at second order[15]; for this reason, such schemes must operate using a small time step; this somewhat limits their usefulness. Even more importantly, however, we note that the schemes we will develop here can be rewritten as very simple finite difference schemes of the form corresponding to DWNs (to be discussed in Chapter 4). We include this section merely to show that higher-order spatial accuracy is not incommensurate with MD passivity, and to indicate a possible direction for future research.

Consider again the lossless source-free transmission line problem, defined by

$$l\frac{\partial i}{\partial t} + \frac{\partial u}{\partial x} = 0 \qquad (3.95a)$$

$$c\frac{\partial u}{\partial t} + \frac{\partial i}{\partial x} = 0 \qquad (3.95b)$$

(Losses and sources may be reintroduced at a later stage in the schemes we will develop here without any difficulty.) Because higher-order spatially accurate explicit methods will

---

[15] We remark here that this restriction may be fundamental. It should be recalled that MD systems are passive with respect to the time coordinate—coordinate transformations are simply a means of distributing this passivity property among all the independent variables of the problem. It is well-known [93] that lumped passive systems of first order can be approximated by passive numerical methods that are at best second-order accurate. This restriction would appear to carry over in MD, though we have not attempted to prove this. Passivity does not hold, however, with respect to the spatial coordinates and it may be this distinction that we are able to exploit in this section.

3.12. HIGHER-ORDER ACCURACY

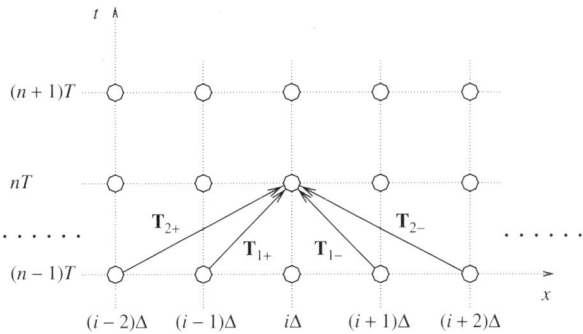

Figure 3.25: *Unit shifts in the coordinates defined by (3.96).*

require access to grid points other than nearest neighbors, we introduce the following coordinate transformation,

$$\mathbf{H} = \begin{bmatrix} 1 & -1 & 2 & -2 & \cdots & q & -q \\ 1 & 1 & 1 & 1 & \cdots & 1 & 1 \end{bmatrix} \qquad \mathbf{H}^{-R} = \mathbf{H}^T \begin{bmatrix} \frac{3}{q(2q+1)(q+1)} & 0 \\ 0 & \frac{1}{2q} \end{bmatrix} \qquad (3.96)$$

for some positive integer q (if $q = 1$, then we get the coordinate transformation defined by (3.19), scaled by a constant factor). $2q$ will shortly be shown to be the order of spatial accuracy of the resulting difference scheme. As before, we have

$$\mathbf{u} = \mathbf{V}^{-1}\mathbf{H}\mathbf{t} \qquad\qquad \mathbf{t} = \mathbf{H}^{-R}\mathbf{V}\mathbf{u}$$

with $\mathbf{u} = [x, t]^T$, and $\mathbf{t} = [t_{1+}, t_{1-}, t_{2+}, t_{2-}, \ldots, t_{q+}, t_{q-}]^T$; the coordinate transformation defined by $\mathbf{H}$ thus describes an embedding of the $(1 + 1)$D problem in a $2q$-dimensional space. A uniform sampling of the new coordinates with spacings $T_{1+} = T_{1-} = \ldots = T_{q+} = T_{q-} = \Delta$ merely regenerates a uniform grid with spacing Δ. The first two pairs of unit shifts are as shown in Figure 3.25.

We now rewrite system (3.95) as

$$v_0 l \frac{\partial i_1}{\partial t'} + r_0 \sum_{j=1}^{q} \alpha_{qj} \frac{\partial i_2}{\partial x} = 0$$

$$v_0 c r_0^2 \frac{\partial i_2}{\partial t'} + r_0 \sum_{j=1}^{q} \alpha_{qj} \frac{\partial i_1}{\partial x} = 0$$

where, as before, we have $i_1 = i$, $i_2 = u/r_0$ for some positive constant r_0, and $t' = v_0 t$. The α_{qj}, $j = 1, \ldots, q$, are constants that satisfy

$$\sum_{j=1}^{q} \alpha_{qj} = 1 \qquad (3.97)$$

We may continue and write

$$\left(v_0 l - r_0 \sum_{j=1}^{q} \frac{|\alpha_{qj}|}{j}\right) \frac{\partial i_1}{\partial t'} + r_0 \sum_{j=1}^{q} \frac{|\alpha_{qj}|}{j} \left(\frac{\partial i_1}{\partial t'} + j\,\text{sgn}(\alpha_{qj})\frac{\partial i_2}{\partial x}\right) = 0$$

$$\left(v_0 c r_0^2 - r_0 \sum_{j=1}^{q} \frac{|\alpha_{qj}|}{j}\right) \frac{\partial i_2}{\partial t'} + r_0 \sum_{j=1}^{q} \frac{|\alpha_{qj}|}{j} \left(\frac{\partial i_2}{\partial t'} + j\,\text{sgn}(\alpha_{qj})\frac{\partial i_1}{\partial x}\right) = 0$$

Since from (3.26), we have that

$$D_{j+} \triangleq \frac{\partial}{\partial t_{j+}} = \frac{\partial}{\partial t'} + j\frac{\partial}{\partial x} \qquad D_{j-} \triangleq \frac{\partial}{\partial t_{j-}} = \frac{\partial}{\partial t'} - j\frac{\partial}{\partial x} \qquad j = 1, \ldots, q$$

we can immediately write

$$L_{1q} D_{t'} i_1 + \sum_{j=1}^{q} M_{qj} \left(D_{j+}(i_1 + \beta_{qj} i_2) + D_{j-}(i_1 - \beta_{qj} i_2)\right) = 0 \tag{3.98a}$$

$$L_{2q} D_{t'} i_2 + \sum_{j=1}^{q} M_{qj} \left(D_{j+}(i_2 + \beta_{qj} i_1) + D_{j-}(i_2 - \beta_{qj} i_1)\right) = 0 \tag{3.98b}$$

with

$$L_{1q} = v_0 l - r_0 \sum_{j=1}^{q} \frac{|\alpha_{qj}|}{j} \qquad L_{2q} = v_0 c r_0^2 - \sum_{j=1}^{q} \frac{r_0 |\alpha_{qj}|}{j}$$

$$M_{qj} = \frac{r_0 |\alpha_{qj}|}{2j} \qquad \beta_{qj} = \text{sgn}(\alpha_{qj}) \tag{3.99}$$

The system (3.98) can immediately be identified with an MDKC, as in Figure 3.26.

Each of the Jaumann two-ports can be discretized according to the trapezoid rule; as long as our choice of the constants α_{qj} satisfies the constraint (3.97) and L_{1q} and L_{2q} remain

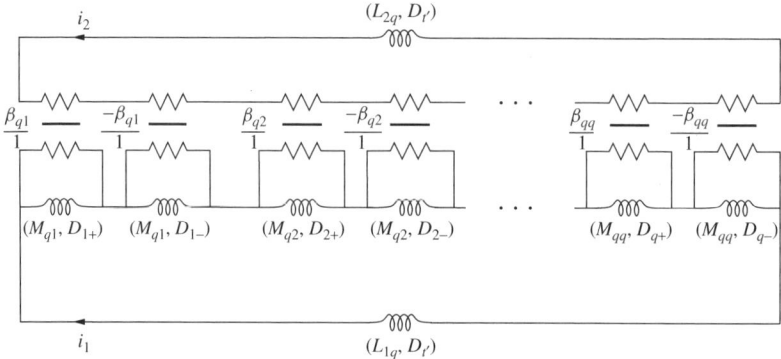

Figure 3.26: *MDKC for the lossless source-free transmission line equations, according to the decomposition given by (3.98).*

3.12. HIGHER-ORDER ACCURACY

positive, the resulting MDWD network will be a second-order stable accurate approximation to system (3.95). Suppose, however, that we apply a different set of discretization rules, namely

$$D_{j+} \to \frac{1}{\Delta} \left(1 + \delta_{j-}\delta_{j+}\right)^{-1} \left(1 - \delta_{j+}\right) \left(1 + \delta_{j-}\right) \tag{3.100a}$$

$$D_{j-} \to \frac{1}{\Delta} \left(1 + \delta_{j-}\delta_{j+}\right)^{-1} \left(1 - \delta_{j-}\right) \left(1 + \delta_{j+}\right) \tag{3.100b}$$

for $j = 1, \ldots, q$. Here, δ_{j+} and δ_{j-} are the shift operators in the directions t_{j+} and t_{j-} defined by

$$\delta_{j+} e(\mathbf{t}) = e(\mathbf{t} - \mathbf{T}_{j+}) \qquad \delta_{j-} e(\mathbf{t}) = e(\mathbf{t} - \mathbf{T}_{j-})$$

for a function $e(\mathbf{t})$, where $\mathbf{T}_{j+}$ and $\mathbf{T}_{j-}$ are vectors of length Δ in directions t_{j+} and t_{j-} respectively (see Figure 3.25 for a graphical representation of these shifts on the computational grid). These rules correspond, in the linear shift-invariant case, to pairs of spectral mappings of the type mentioned briefly in Section 3.4.4, with shift lengths equal to Δ; they are also MD-passivity preserving and are in general second-order accurate [89]. To the scaled time derivative, we apply the trapezoid rule with a doubled time step $T' = 2\Delta$, as defined by

$$D_{t'} \to \frac{2}{T'} \left(1 + \delta_{t'}^2\right)^{-1} \left(1 - \delta_{t'}^2\right) = \frac{1}{\Delta} \left(1 + \delta_{t'}^2\right)^{-1} \left(1 - \delta_{t'}^2\right)$$

Equation (3.98a) then becomes

$$L_{1q} \left(1 + \delta_{t'}^2\right)^{-1} \left(1 - \delta_{t'}^2\right) i_1 + \sum_{j=1}^{q} M_{qj} \left(1 + \delta_{j-}\delta_{j+}\right)^{-1} \left(1 - \delta_{j+}\right) \left(1 + \delta_{j-}\right) (i_1 + \beta_{qj} i_2)$$

$$+ \sum_{j=1}^{q} M_{qj} \left(1 + \delta_{j-}\delta_{j+}\right)^{-1} \left(1 - \delta_{j-}\right) \left(1 + \delta_{j+}\right) (i_1 - \beta_{qj} i_2)$$

$$= O(\Delta^2) \tag{3.101}$$

However, since $\delta_{j+}\delta_{j-} = \delta_{t'}^2$ and our system is time-invariant, the operator $1 + \delta_{j+}\delta_{j-}$ commutes with L_{1q} and M_{jq} and may be factored out of (3.101), giving

$$L_{1q} \left(1 - \delta_{t'}^2\right) i_1 + \sum_{j=1}^{q} M_{qj} \left(1 - \delta_{j+}\right) \left(1 + \delta_{j-}\right) (i_1 + \beta_{qj} i_2)$$

$$+ \sum_{j=1}^{q} M_{qj} \left(1 - \delta_{j-}\right) \left(1 + \delta_{j+}\right) (i_1 - \beta_{qj} i_2)$$

$$= O(\Delta^2)$$

which can be further simplified to

$$v_0 l \left(1 - \delta_{t'}^2\right) i_1 + r_0 \sum_{j=1}^{q} \frac{\alpha_{qj}}{j} \left(\delta_{j-} - \delta_{j+}\right) i_2 = O(\Delta^2)$$

or, writing $\delta_{j-} = \delta_{t'}\delta_x^{-j}$ and $\delta_{j+} = \delta_{t'}\delta_x^j$ where δ_x is a simple shift in the x direction of Δ, as

$$l\underbrace{\frac{\left(\delta_{t'}^{-1} - \delta_{t'}\right)}{2\Delta}}_{\approx \frac{\partial}{\partial t}} i_1 + r_0 \underbrace{\sum_{j=1}^{q} \alpha_{qj} \frac{\left(\delta_x^{-j} - \delta_x^{j}\right)}{2j\Delta}}_{\approx \frac{\partial}{\partial x}} i_2 = O(\Delta^2)$$

which is easily seen to be a simple difference approximation to (3.95a). The approximation is nominally second-order accurate in Δ, but we have not as yet made any special choice of the α_{qj}. This can be done via a conventional finite difference approach in such a way as to yield a higher-order accurate approximation to the spatial derivative.

We can write, expanding the shift operators in Taylor series,

$$\sum_{j=1}^{q} \alpha_{qj} \frac{\left(\delta_x^{-j} - \delta_x^{j}\right)}{2j\Delta} = \sum_{k>0,\ \text{odd}} \frac{\Delta^{k-1}}{k!} \sum_{j=1}^{q} \alpha_{qj} j^{k-1} \frac{\partial^k}{\partial x^k} \quad (3.102)$$

There are q degrees of freedom, corresponding to the parameters α_{qj}, $j = 1, \ldots, q$. We require, from (3.97) that the coefficient of the first derivative on the right-hand side of (3.102) equal one. We may then additionally require that the other coefficients, for $k = 3, \ldots, 2q + 1$ be zero; the resulting difference approximation will then be accurate to order $2q$. This yields the linear system

$$\mathbf{C}\boldsymbol{\alpha}_q = \mathbf{e}_q \quad (3.103)$$

where $\mathbf{C}$ is a $q \times q$ matrix with $[\mathbf{C}_{ij}] = j^{2(i-1)}$, $\boldsymbol{\alpha}_q = [\alpha_{q1}, \ldots, \alpha_{qq}]^T$, and $\mathbf{e}_q$ is a $q \times 1$ vector whose first entry is one, and whose others are zero. $\mathbf{C}$ is always full rank, so there is a unique solution for any q. The same $\boldsymbol{\alpha}_q$ will also give a higher-order approximation to (3.95b), and thus system (3.95) will be approximated to higher-order accuracy as a whole. For a fourth-order approximation, for example, we obtain $\boldsymbol{\alpha}_2 = [4/3, -1/3]^T$, and for a sixth-order approximation, we get $\boldsymbol{\alpha}_3 = [3/2, -3/5, 1/10]^T$. These values completely determine the MDKC pictured in Figure 3.26.

The passivity requirement is, as before, a condition on the positivity of L_{1q} and L_{2q}. Choosing $r_0 = \sqrt{l_{\min}/c_{\min}}$ gives

$$v_0 \geq \frac{3}{2\sqrt{l_{\min}c_{\min}}} \qquad \text{Fourth-order accurate scheme}$$

$$v_0 \geq \frac{11}{6\sqrt{l_{\min}c_{\min}}} \qquad \text{Sixth-order accurate scheme}$$

It is interesting to note that in the constant-coefficient case, this bound is distinct from the stability bound obtained from von Neumann analysis (see Appendix A) applied to the same difference method. For example, for the fourth-order accurate scheme defined by $\boldsymbol{\alpha}_2$, the stability bound is $v_0 \geq 1.37\gamma$, with $\gamma = 1/\sqrt{lc}$; there is thus a range of values of v_0 for which the scheme will be stable, but not MD-passive. We will comment extensively on the distinction between passive and stable methods in Appendix A.

3.12. HIGHER-ORDER ACCURACY

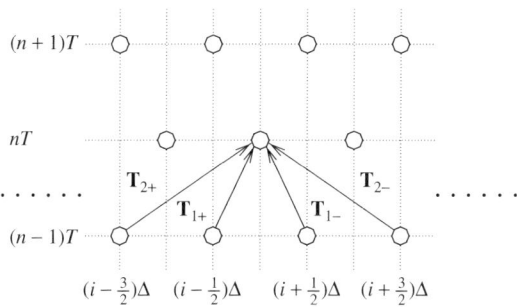

Figure 3.27: *Unit shifts in the coordinates defined by (3.104).*

It is also of interest to define a similar scheme with respect to the coordinate transformation defined by

$$\mathbf{H} = \begin{bmatrix} \frac{1}{2} & -\frac{1}{2} & \frac{3}{2} & -\frac{3}{2} & \cdots & \frac{2q-1}{2} & -\frac{2q-1}{2} \\ 1 & 1 & 1 & 1 & \cdots & 1 & 1 \end{bmatrix} \qquad (3.104)$$

Keeping the same notation for the new coordinates, the shifts are as shown in Figure 3.27; now we have a grid ideal for a staggered or interleaved algorithm, with alternating grid points at alternating time steps.

For this coordinate system, we follow through a development very similar to that in the previous pages. We again have an MD circuit representation as in Figure 3.26, where now we have

$$L_{1q} = v_0 l - r_0 \sum_{j=1}^{q} \frac{|\alpha_{qj}|}{j - \frac{1}{2}} \qquad L_{2q} = v_0 c r_0^2 - r_0 \sum_{j=1}^{q} \frac{|\alpha_{qj}|}{j - \frac{1}{2}} \qquad M_{qj} = \frac{r_0 |\alpha_{qj}|}{2(j - \frac{1}{2})}$$

for some set of α_{qj}, $j = 1, \ldots, q$ that sum to unity. The symbols D_{j+} and D_{j-} in the figure now refer to directional derivatives in the coordinate directions defined by (3.104). For higher-order accuracy, constraint equation (3.103) will apply, now with $[C_{ij}] = (j - 1/2)^{2(i-1)}$. For fourth-order accuracy, we obtain $\alpha_2 = [9/8, -1/8]^T$, and for a sixth-order approximation, we get $\alpha_3 \approx [1.179, -0.195, 0.0234]^T$.

Since we are using an alternative discretization rule here, the resulting MDWD networks are more appropriately discussed in the context of DWNs (which are the subject of the next chapter). We will return briefly to waveguide network representations of these higher-order accurate methods in Section 6.3.

Chapter 4

Digital Waveguide Networks

We now turn our attention to a different approach to numerical integration, which is, in many respects, very similar to the multidimensional wave digital network technique discussed in the last chapter. *Digital waveguide networks* (DWNs) [228], the invention of Julius Smith, are also based on the ideas of scattering and propagation of wave variables in multiple dimensions; indeed, the basic signal-processing block of the DWN, the scattering junction, is identical to the wave digital adaptor. As such, a waveguide network will possess the same discrete passivity properties as a wave digital network, and passivity in finite arithmetic also follows accordingly [227].

The process through which one arrives at a particular DWN intended to simulate the behavior of a distributed physical system has been, to date, quite different. Following the wave digital approach, one first writes a multidimensional circuit representation (MDKC) for a system of PDEs and then applies a set of coordinate transformations and spectral mappings in order to obtain a discrete time/space algorithm. As discussed in the previous chapter, all MD circuit elements (as well as Kirchhoff connections between elements) are to be interpreted as *distributed*, from the outset through to the final wave digital network. The integrity of each multidimensional circuit element (including its energetic properties) is preserved through the discretization step, as is network topology as a whole. As we mentioned in Section 1.1.3, however, the DWN is usually thought of as a collection of lumped elements, and as such lacks a convenient multidimensional representation. We will address this point in some detail in Chapter 6. A DWN always operates on a predefined grid, at the points of which are located scattering junctions. Even though the paired delay elements (*digital waveguides*) that connect the various scattering junctions behave like transmission lines, we will persist in calling them lumped elements, because they are typically connected between junctions at neighboring grid points, and their behavior is hence localized in a way that that of an MDWD element is not.

An MDWD network will behave consistently with the generating system of PDEs because the continuous-to-discrete spectral mapping applied approximates differential operators consistently; for a DWN, we must first show consistency with a particular physical system. For both approaches, convergence of simulation results to the true solution of the physical system follows from this consistency as well as stability implied by passivity [238].

It was shown by Van Duyne and Smith [267, 269] that the DWN structures designed to solve the *wave equation* in (2+1) and (3+1)D could be recast as finite difference approximations [238] (and in particular centered difference approximations) to these equations; we looked at the (2+1)D waveguide mesh briefly in Section 1.1.3. In infinite-precision arithmetic, these DWNs and centered differences yield identical results. A similar correspondence holds for the MDWD networks examined in the last chapter, though the equivalent difference methods are more involved (see Section 3.8). The distinction between a DWN and a finite difference approximation is in the types of signals used. Finite difference methods operate using grid variables that are approximations to the physical dependent variables of the problem at hand, but the DWN propagates wave variables; in this formulation, the solution to a system of PDEs is obtained as a by-product of the scattering of these waves. It is perhaps best to think of the difference between the finite difference scheme and DWN implementations as analogous to the distinction between direct form and lattice/ladder form digital filters [109]—both can be designed to implement the same transfer functions, but for the latter forms, stability is tightly controlled by the range of values that the filter multipliers ('reflection coefficients') can take. And indeed, as we saw in Section 1.1.2, a particular type of (1+1)D DWN can be shown to be directly related to these lattice/ladder forms [227]. One goal of the remainder of the book is to show how this correspondence between the DWN and centered differences may be extended to a wide variety of physical systems.

The immediate question that arises is then: If the DWN is equivalent to finite differences, then is there a compelling reason for using it? Finite differences, after all, are more straightforward to implement. The answer is twofold. First, although the approaches are equivalent in infinite-precision arithmetic, this is no longer true when we are forced, inevitably, to truncate both the signals and multipliers in a computer implementation; stability of a DWN can be simply maintained even in finite arithmetic. Second, the stability criterion for a DWN is, as for MDWD networks, a *positivity condition* on the values of the elements contained in the network (i.e., the immittances of the transmission line segments). It thus becomes very simple to check stability of a given DWN even in the presence of boundary conditions. Checking the stability of a finite difference scheme is considerably more involved, especially considering that a difference scheme that is stable over the interior of a domain may become unstable when boundary conditions are applied [112]. There is a theoretical machinery for performing such checks (known as *GKSO theory* [181, 113, 148, 112, 238, 253]), though it can be formidable even in the (1+1)D case. It is quite possible, of course, to *design* a convergent numerical method using a DWN, and then *apply* it as a finite difference scheme; as mentioned above, however, its stability in finite arithmetic is then no longer guaranteed.

Summary

In this chapter, we introduce DWNs and explore the relationship between DWN and finite difference methods of the finite difference time domain (FDTD) variety; we showed a very simple example of such a correspondence in Section 1.1.3. We reintroduce DWNs, now in the electrical context, so as to make easier an eventual comparison with the MDWD networks of Chapter 3—the acoustic tubes in the Kelly-Lochbaum model of Section 1.1.2 and the waveguide mesh of Section 1.1.3 are thus replaced by transmission line sections, and pressures and velocities become voltages and currents. After a brief review of the fundamentals, we reexamine the transmission line and parallel-plate test problems in Section 4.3 and Section 4.4. Here, DWNs structures that numerically integrate these systems are built 'the hard way' (i.e., by association with finite difference methods and FDTD, without the benefit of a multidimensional representation) and we uncover several distinct families of such networks, in Section 4.3.6 and Section 4.4.2, with different passivity properties. We also examine the initialization of these networks in Section 4.5, and the implementation of boundary conditions in Section 4.3.9 and Section 4.4.4. The chapter is concluded with a short overview of applications of digital waveguides in musical sound synthesis.

4.1 FDTD and TLM

Numerical integration methods for the transmission line equations and electromagnetic field problems have developed along two important directions. The first approach was pioneered by Yee [286] in the mid-1960s, and has since blossomed into what is now known as the *finite difference time domain* method or FDTD [246]. The idea behind the method is a straightforward application of centered differences to (in Yee's case) the defining equations of electromagnetics, namely Maxwell's equations. (2+1)D simplifications of Maxwell's equations that describe the evolution of transverse electric (TE) and transverse magnetic (TM) fields can also be treated as well, and are, with some trivial modifications, equivalent to the (2+1)D parallel-plate problem. The important advantage of Yee's method is that, because of the structure of the system of equations to be modeled, it is not necessary to calculate all the field components simultaneously—the field components are interleaved both temporally and spatially[1]. We will examine FDTD in the (2+1)D case explicitly in Section 4.4. The literature on FDTD is quite large; we refer to Chapters 2 and 3 of Taflove's excellent survey [246] for a succinct technical overview. As we mentioned in the Preface, the *transmission line matrix* method, or TLM [5, 44, 122, 121] appeared a bit later, in the early 1970s [131, 135]. It (like the wave digital filtering approach) is a descendant of the ground-breaking work of Kron [149], who developed circuit models of electromagnetic field problems before the widespread availability of electronic computers. TLM structures are very similar to DWNs in that they employ networks of discrete transmission lines connected at scattering junctions in order to simulate the behavior of distributed systems. The first formulation, known as the *expanded node* formulation, was derived from a lumped (RLC) model of the (2+1)D transmission line equations [122], and is identical to the type III DWN we will present in Section 4.3.6. The library of TLM structures has burgeoned since its inception; the most significant thrust has been toward formulations for which the various field components are not staggered but computed together at larger nodes. The *symmetric condensed node* [134] and its numerous offspring, such as the *hybrid symmetrical condensed node* [215], and the *symmetrical supercondensed node* [255] are the results of this work. A major distinction between TLM and the methods discussed in this book is that the TLM scattering node is often required to be orthogonal; here, as for WDFs, we do not require this (except in time-varying or nonlinear problems), as conservation of power holds even in the nonorthogonal case. Another distinction is that our adaptors or junctions are usually of the simplified Householder reflection form discussed in Section 2.2.5, which follows directly from Kirchhoff's Laws at a single loop or node; TLM scattering nodes are not necessarily or even usually of this form. This is of lesser importance, as it is of course possible to build more general adaptors by combining various nodes and loops into a single scattering equation.

TLM has developed in a multitude of ways away from pure electromagnetic problems; one obvious analogy is in linear acoustics [189, 207], which interestingly, was the starting point for the DWN. Another intriguing direction has been toward the modeling of diffusion processes [191, 132, 54, 50]; this is a significant point of contrast with methods based on the MD circuit representations of the last chapter, because a passive circuit representation only exists for hyperbolic systems; this difficulty can be simply circumvented (as it apparently has been in the case of the TLM models) by solving a numerically hyperbolic approximation to a parabolic problem (or perhaps by reverting to a more correct but also more complex

---
[1]This can, of course, be a disadvantage as well if one is forced to set mixed-type boundary conditions involving variables that are not present at the boundary in the staggered formulation.

hyperbolic model of diffusive processes). There do exist DWN structures that numerically integrate the Euler–Bernoulli beam equations [26], which are also not hyperbolic (in this case, borderline parabolic).

FDTD and TLM have been compared and linked in various ways [58, 133, 222, 271, 174], most significantly through the use of field expansions [150], and comparisons of numerical dispersion properties [170] and new variants of FDTD have been developed using TLM as a starting point [42]. We will take a different approach here. Beginning from the observations that have been made regarding the equivalence of certain DWNs to difference methods [95, 213, 267, 269], we will show that Yee's algorithm is equivalent to a family of scattering structures, some of which appear to be quite different from those that have been proposed in the TLM literature. The correspondence holds for media with spatially varying material parameters. We also note that the TLM community appears to be aware neither of the many valuable numerical properties that scattering-based numerical methods possess [68, 227], and in particular, their behavior in finite arithmetic, nor of other useful signal-processing manipulations (such as power-normalization of wave quantities and dynamic range minimization [229]), which have their roots in electrical network theory. Along these dimensions at least, the palette of possible scattering structures available in the DWN and MDWD frameworks is undeniably greater.

As a final word, we say that one apparent weakness of the TLM approach, in the opinion of this author, is that to date, with the exception of the field-theoretic techniques discussed by Krumpholz et al. [150, 151], there has not been a systematic way of *deriving* scattering structures from a model system. Very little, if anything, is written on this subject; existing structures may be analyzed and proved to be consistent with the model problem, but it seems as if the most well-known forms have been arrived at through guesswork, more often than not through the penetrating insights of P. B. Johns. This is quite a serious shortcoming, we feel, compared with the MDWD approach, which definitively links a model equation to a scattering method, without any guesswork. And as we will see in Chapter 6, this guesswork is unnecessary in the case of DWNs as well, once they are viewed as MDWD structures themselves (which they are). We fully realize that many TLM experts may not agree with these statements.

4.2 Digital Waveguides

We surveyed the basics of digital waveguide networks in Section 1.1.3. In this section, we review the main principles of waveguide networks, now in the transmission line setting. For a full treatment, we refer the reader to Smith's comprehensive technical report [228].

4.2.1 The Bidirectional Delay Line

The basic element in a waveguide network, and the one which does the work of moving energy from one part of the network to another, is the bidirectional delay line, shown in Figure 4.1. It is no more than a pair of digital delay lines whose delays are of equal length (m samples of duration T in Figure 4.1). It should be understood that, for realizability, all delay lengths in a given network should be multiples of a common smallest ('unit') delay. We will use the terms *waveguide* and *bidirectional delay line* interchangeably in this work.

4.2. DIGITAL WAVEGUIDES

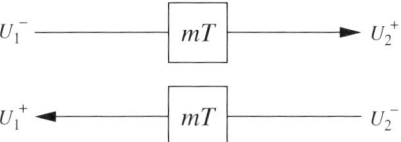

Figure 4.1: *Bidirectional delay line.*

Associated with the bidirectional delay line are two sets of signals, called *waves*: *voltage waves* U, and *current waves* I. Only voltage waves are shown in Figure 4.1. Waves of either type are indexed with respect to a particular end of the delay line; in Figure 4.1, waves at the left end of the delay line pair are subscripted with a '1,' and those at the right end with a '2.' In addition, one of the waves at either end *enters* the waveguide, and one *leaves*; the waves are superscripted with $-$ or $+$ respectively[2]. We can immediately read the relationship among the variables from Figure 4.1:

$$U_2^+(n) = U_1^-(n-m) \qquad U_1^+(n) = U_2^-(n-m) \tag{4.1}$$

The delay duration T is implicit, so that a wave variable indexed by n refers to the value of that quantity at time $t = nT$. In terms of z-transformed quantities [179] (which we will denote with a hat),

$$\hat{U}_2^+ = z^{-m}\hat{U}_1^- \qquad \hat{U}_1^+ = z^{-m}\hat{U}_2^- \tag{4.2}$$

We also define, at either end of the waveguide, the so-called physical voltage by

$$U_j = U_j^+ + U_j^- \qquad j = 1, 2 \tag{4.3}$$

The digital waveguide can be thought of as a trivial generalization of the WD unit element (see Section 2.2.4) to the case of multiple sample delay.

4.2.2 Impedance

From the point of view of a programmer, the above description of the operation of an isolated bidirectional delay line is complete. In order to connect one bidirectional delay line to others, however, we must introduce the *impedance* Z, a positive number associated with a particular waveguide. The impedance allows us to define the relationship between the voltage and current waves, which were mentioned in the last section, which is

$$U_j^+ = ZI_j^+ \tag{4.4a}$$

$$U_j^- = -ZI_j^- \tag{4.4b}$$

---

[2]We have chosen here to break with the notational tradition in [229], in which the superscripts are reversed. We choose the above notation so that signal nomenclature in a waveguide network is well defined. That is, a signal *leaving* a bidirectional delay line, and a signal *entering* a scattering junction (to which each end of the waveguide will ultimately be connected) are both superscripted with a $+$. This will simplify the derivation of difference schemes later in this chapter. The other indexing method is more useful from the point of view of a unified treatment of both scattering junctions and bidirectional delay lines as digital n-port devices (where we would want $+$ and $-$ to denote incoming and outgoing signals respectively, for any N-port).

where $j = 1, 2$ referring to Figure 4.1. This implies, from (4.1), that we have

$$I_2^+(n) = -I_1^-(n-m) \qquad I_1^+(n) = -I_2^-(n-m) \qquad (4.5)$$

Thus, current waves entering a bidirectional delay line are delayed by the same amount as their voltage wave counterparts, but with sign inversion. In view of (4.4), we need only propagate a particular type of wave (i.e., either voltage or current) in a particular waveguide. In a waveguide network, however, we are free to use different types of waves in different waveguides, converting between the different types with (4.4) where necessary.

The *admittance* Y of the waveguide is defined by

$$Y = \frac{1}{Z}$$

and we define the physical current at either end of the waveguide, like the voltage, to be the sum of the wave components. Thus, we have

$$I_j = I_j^+ + I_j^- \qquad j = 1, 2 \qquad (4.6)$$

4.2.3 Wave Equation Interpretation

The second-order PDE describing the voltage distribution $u(x, t)$ along an electrical transmission line with constant inductance and capacitance l and c per unit length and which runs parallel to the x-axis is the one-dimensional wave equation,

$$\frac{\partial^2 u}{\partial t^2} = \gamma^2 \frac{\partial^2 u}{\partial x^2} \qquad (4.7)$$

where the wave speed γ is $1/\sqrt{lc}$. As we saw in Section 1.1.2, the solution to this equation, if we set aside boundary conditions for the moment, can be written in terms of traveling waves:

$$u(x, t) = u^l(x + \gamma t) + u^r(x - \gamma t) \qquad (4.8)$$

That is, the solution at any time $t \geq 0$ is made up of a sum of two shifted copies of the initializing functions $u^l(x)$ and $u^r(x)$, which have traveled to the left and right respectively with velocity γ over a distance γt. For any Δ we have, for the leftward-traveling wave, the identity

$$u^l(x + \gamma t) = u^l\big((x + \Delta) + \gamma(t - \Delta/\gamma)\big) \qquad (4.9)$$

If we set $\gamma = \Delta/T$, then at time $t = nT$,

$$u^l(x + \gamma nT) = u^l\big((x + \Delta) + \gamma(n - 1)T)\big) \qquad (4.10)$$

Associate now with a particular waveguide a delay T and a physical length Δ, so that in Figure 4.1, U_1^+ represents an outgoing voltage wave quantity at position x, and U_2^-, an incoming wave at position $x + \Delta$. It is then clear that if we have $\Delta/T = \gamma$, then (4.10) is equivalent to the second equation of (4.1), with $m = 1$, and with $u^l(x + \gamma nT) = U_1^+(n)$ and $u^l(x + \Delta + \gamma(n-1)T) = U_2^-(n-1)$. A similar correspondence holds for the right-going traveling wave component u^r and the wave variables at either end of the rightward delay

4.2. DIGITAL WAVEGUIDES

line, U_1^- and U_2^+. A chain of bidirectional delay lines connected in cascade will then implement an exact traveling wave solution to the wave equation. The physical voltage u may be obtained (as should be clear from (4.8)) by summing the leftward and rightward-traveling components at any particular location in the cascade, as per equation (4.3). Note that because $\gamma = \Delta/T$, the delay period and the waveguide length cannot be chosen independently if the discrete wave quantities are to behave as traveling wave solutions to (4.7).

4.2.4 Note on the Different Definitions of Wave Quantities

Waveguides are sometimes defined in a slightly different way [227], as pictured in Figure 4.2. Now the superscripted $+$ and $-$ refer to a direction of propagation (to the right or left, respectively) rather than to outputs and inputs to the delay line pair. If we still assume (4.4a) and (4.4b) to hold for some positive impedance Z, then this definition of wave quantities implies that there is a *direction* associated with a particular waveguide—that is, a leftward ($-$) traveling current wave is sign-inverted with respect to the leftward-traveling voltage wave, but the same is not true for the rightward-traveling waves. It should be obvious that this definition of wave quantities also leads to a traveling wave solution of the wave equation (indeed, the bidirectional delay line of Figure 4.1 is identical to that of Figure 4.2 if we are using only voltage waves). The difference here is that we can now interpret the traveling wave pair (U, I) to be a solution to the transmission line or telegrapher's equations [43], a set of two first-order PDEs (from which the wave equation may be derived),

$$l\frac{\partial i}{\partial t} + \frac{\partial u}{\partial x} = 0 \tag{4.11a}$$

$$c\frac{\partial u}{\partial t} + \frac{\partial i}{\partial x} = 0 \tag{4.11b}$$

which, for constant l and c, has a solution

$$u(x,t) = u_l(x+\gamma t) + u_r(x-\gamma t)$$

$$i(x,t) = i_l(x+\gamma t) + i_r(x-\gamma t)$$

where

$$u_l = -\sqrt{\frac{l}{c}}i_l \qquad\qquad u_r = \sqrt{\frac{l}{c}}i_r$$

and γ is again given by $1/\sqrt{lc}$. It should be remarked that if we had chosen the relationship between the wave variables to be such that the $(+)$-superscripted current wave were to be

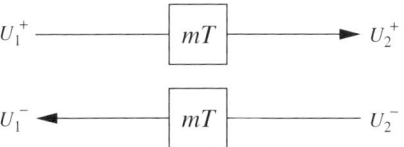

Figure 4.2: *Oriented bidirectional delay line.*

sign-inverted with respect to the (+)-superscripted voltage wave, then we would be solving the 'mirror-image' PDEs that one would get if one were to replace x by $-x$ in system (4.11). The definition of wave variables (which we might call the 'input–output' definition) given in Section 4.2.1 solves the wave equation that results from eliminating variables in system (4.11), or its mirror image, and hence does not have an orientation.

In practice, in order to develop numerical methods in systems of PDEs for which direction is important, we can either use the oriented wave variable definition given in this section, or we can use the input–output formulation, and reintroduce directionality into the network where appropriate. We have chosen the latter course, and we will indicate which changes must be made (usually through the use of transformers) explicitly on the signal flow graph. From a programmer's point of view, there is no substantial difference between using the different definitions. Both lead to the same scattering equations (to be discussed in the next section) and are identical from a power conservation point of view (that is to say, sign inversions of wave quantities do not affect the energy measure of the network).

An additional reason for choosing the input–output definition of wave variables is that it will require less notational juggling when we eventually link DWNs to MDWD networks in Chapter 6.

4.2.5 Scattering Junctions

Returning to the 'input–output' waveguide defined in Section 4.2.1 and Section 4.2.2, we now must deal with connecting bidirectional delay lines; this is done in the same way as in the wave digital filtering framework, namely through the use of Kirchhoff's Laws, which conserve instantaneous power at a connection. The resulting equations relating input to output waves at such a connection or *scattering junction* are identical to the adaptor equations for wave digital filters already mentioned in Section 2.2.5. For completeness sake, we will rederive the scattering equations for a series connection of M bidirectional delay lines of impedances Z_j, $j = 1, \ldots, M$.

At such a series connection, we must have

$$I_1 = I_2 = \ldots = I_M \triangleq I_J \tag{4.12}$$

$$U_1 + U_2 + \ldots + U_M = 0 \tag{4.13}$$

where I_J is defined to be the *junction current* common to all waveguides. Thus, we have

$$0 \stackrel{(4.13)}{=} \sum_{j=1}^{M} U_j \stackrel{(4.3)}{=} \sum_{j=1}^{M} \left(U_j^+ + U_j^-\right) \stackrel{(4.4a),(4.4b)}{=} \sum_{j=1}^{M} Z_j \left(I_j^+ - I_j^-\right) \stackrel{(4.6)}{=} \sum_{j=1}^{M} Z_j \left(2I_j^+ - I_J\right)$$

where the equation numbers appear over the equalities to which they pertain. Using (4.12), we can then write the equation used to calculate the junction current from the incoming current waves

$$I_J = \frac{2}{Z_J} \sum_{j=1}^{M} Z_j I_j^+$$

4.2. DIGITAL WAVEGUIDES

as well as the *scattering equation*

$$I_k^- = -I_k^+ + \frac{2}{Z_J} \sum_{j=1}^{M} Z_j I_j^+, \qquad k = 1, \ldots, M$$

where we have defined the *junction impedance* Z_J by

$$Z_J \triangleq \sum_{j=1}^{M} Z_j$$

In terms of voltage waves, using (4.4a) and (4.4b), the scattering equations can be written as

$$U_k^- = U_k^+ - \frac{2Z_k}{Z_J} \sum_{j=1}^{M} U_j^+, \qquad k = 1, \ldots, M$$

This is identical to the definition of a wave digital series adaptor (2.31) for voltage waves, where we replace U_k^- by b_k, U_k^+ by a_k and Z_k by R_k, for $k = 1, \ldots, M$.

The scattering equations for a dual parallel connection are similar under the replacement of U_k^+ and U_k^- by I_k^+ and I_k^-, Z_k by Y_k, and Z_J by the *junction admittance*, defined by

$$Y_J \triangleq \sum_{j=1}^{M} Y_j$$

so that we have a junction voltage U_J defined by

$$U_J = \frac{2}{Y_J} \sum_{j=1}^{M} Y_j U_j^+ \qquad (4.14)$$

and

$$U_k^- = -U_k^+ + \frac{2}{Y_J} \sum_{j=1}^{M} Y_j U_j^+, \qquad k = 1, \ldots, M \qquad (4.15)$$

The representation we will use for scattering junctions in the waveguide networks in this and the subsequent chapter will usually be as shown in Figure 4.3 (in the case of a connection of four waveguides). A waveguide's immittance is placed at the port at which it is connected to the junction, and the junction quantity to be calculated from incoming waves appears at the center of the junction. Sometimes, if there is no room in the figure, we will indicate the immittance of a waveguide by an overbrace (see, e.g., Figure 4.6). In the case of electrical variables, a junction current I_J is calculated at a series junction, and a junction voltage U_J at a parallel junction, but when we move to mechanical systems in Chapter 7, we will of course use different variable names. A small 's' or 'p' is placed in a corner of the junction in order to indicate that the junction is series or parallel respectively. In addition, because it is only necessary to propagate one type of wave in a bidirectional delay line, *a graphical representation of a waveguide network will always imply the use of voltage waves everywhere*. This is the same convention that is used in wave digital signal

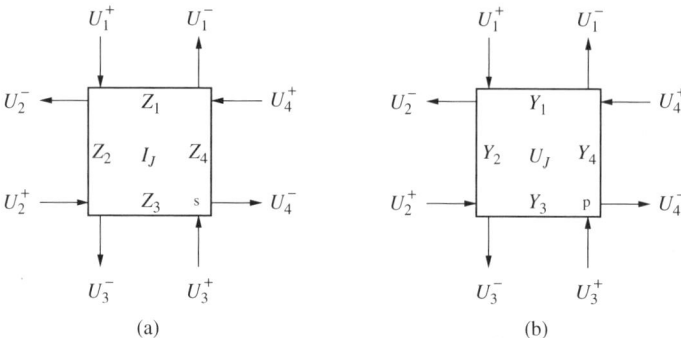

Figure 4.3: *Graphical representations of scattering 4-port junctions—(a) series and (b) parallel.*

flow graphs. This is important, because it will be recalled from Section 4.2.2 that current waves require an additional sign inversion that is not shown in the network diagrams.

Instantaneous power is preserved at the scattering junction (here again, as in the WDF case, the scattering junction is no more than a wave variable implementation of Kirchhoff's Laws, which preserve power by definition). The power-normalization strategy employed in the wave digital filter setting can also be used here as well, and gives rise to the same orthogonality property of the scattering junction in either the series or parallel case (see Section 2.2.5). Power-normalized waves can be used in order to construct time-varying passive waveguide networks [227], though for time-invariant problems, the use of such power-normalized quantities involves more arithmetic operations.

4.2.6 Vector Waveguides and Scattering Junctions

The DWN has been extended to the vector case by Smith and Rocchesso [229, 203]; this has been done in the WDF framework in [68, 176], as discussed in Section 2.2.7. We briefly introduce vector waveguides, because it will be necessary to make use of them when simulating the behavior of stiff systems and elastic solids; we will examine such problems in depth in Chapter 7.

A vector waveguide accepts two incoming signals $\hat{\mathbf{U}}_1^-$ and $\hat{\mathbf{U}}_2^-$ and outputs $\hat{\mathbf{U}}_1^+$ and $\hat{\mathbf{U}}_2^+$; all are assumed to be $q \times 1$ vectors (note that we have used z-transformed quantities here). The waveguide itself, like its scalar counterpart, is described by two parameters: its *impedance* $\mathbf{Z}$, a $q \times q$ matrix, which we will assume to be constant and symmetric positive definite (though it may be generalized to a para-Hermitian matrix function of the unit delay z^{-1} [203]) and its *generalized delay*, $\mathbf{H}(z^{-1})$, a $2q \times 2q$ matrix function of the unit delay, which we assume to be para-unitary (lossless) [261]. The input and output voltage waves are related by

$$\begin{bmatrix} \hat{\mathbf{U}}_1^+(z^{-1}) \\ \hat{\mathbf{U}}_2^+(z^{-1}) \end{bmatrix} = \mathbf{H}(z^{-1}) \begin{bmatrix} \hat{\mathbf{U}}_2^-(z^{-1}) \\ \hat{\mathbf{U}}_1^-(z^{-1}) \end{bmatrix}$$

(In applications in Chapter 7, in which bidirectional delay lines are extracted from an MDKC, we will always set $\mathbf{H}(z^{-1})$ to be a multiple of the $2q \times 2q$ identity matrix.) We

4.2. DIGITAL WAVEGUIDES

can define instantaneous current wave vectors $\mathbf{I}_j^+$ and $\mathbf{I}_j^-$, for $j = 1, 2$, and the relationship between voltage and current wave variables becomes

$$\mathbf{U}_j^+ = \mathbf{Z}\mathbf{I}_j^+ \qquad \mathbf{U}_j^- = -\mathbf{Z}\mathbf{I}_j^-$$

The scattering equations at a series or parallel junction of k waveguides generalize in a straightforward way to the vector case—we have

$$\mathbf{I}_k^- = -\mathbf{I}_k^+ + 2\mathbf{Z}_J^{-1} \sum_{j=1}^{M} \mathbf{Z}_j \mathbf{I}_j^+ \qquad \text{Series junction} \qquad (4.16a)$$

$$\mathbf{U}_k^- = -\mathbf{U}_k^+ + 2\mathbf{Y}_J^{-1} \sum_{i=1}^{M} \mathbf{Y}_j \mathbf{U}_j^+ \qquad \text{Parallel junction} \qquad (4.16b)$$

for $k = 1, \ldots, M$, where $\mathbf{Y}_j = \mathbf{Z}_j^{-1}$ is the admittance of the jth waveguide, which must exist because $\mathbf{Z}_j$ is assumed positive definite (for passivity) [229]. The matrix junction admittance and impedance are defined by

$$\mathbf{Z}_J \triangleq \sum_{j=1}^{M} \mathbf{Z}_j \qquad \text{Series junction}$$

$$\mathbf{Y}_J \triangleq \sum_{j=1}^{M} \mathbf{Y}_j \qquad \text{Parallel junction}$$

By virtue of the fact that they are sums of positive-definite matrices, they will also be positive definite, and thus their inverses, used in the scattering equations (4.16), must exist. Power normalization may also be applied by scaling the wave variables by a square root of the impedance (which is nonunique) [229]. Vector junction passivity has been shown to hold in the fixed word-length case in [229].

Vector/Scalar Waveguide Coupling

In a few cases, it is useful to have a means of connecting vector and scalar network elements. This comes up when designing networks to simulate mixed vector/scalar systems of PDEs; and in particular, it will be necessary to use vector methods when working in nonorthogonal coordinate systems (see Section 5.3) and for some of the mechanical systems of Chapter 7.

We will always assume that for any given scattering junction, the number of components of any approaching wave is the same at every port; it may not be, however, that every junction in the network accepts waves with some universal number of components. In particular, a vector waveguide, one of whose ends is connected to a vector junction, may be split into several scalar waveguides, or more generally into a number of vector waveguides, each with a smaller number of components. Similarly, it is possible to 'bundle' several waveguides into a single larger waveguide.

Let us assume that all splittings and bundlings are from vector to strictly scalar and scalar to vector respectively. An element that splits a single three-component waveguide into three

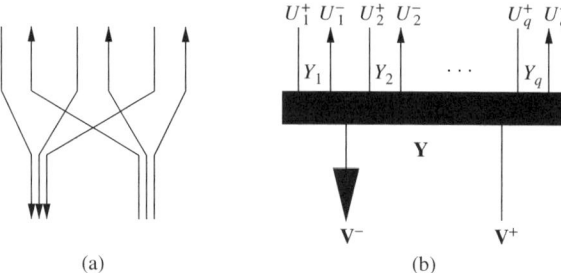

Figure 4.4: (a) *Element for splitting of a vector waveguide into three scalar waveguides and* (b) *a simplified graphical representation in the general case of q scalar waveguides.*

scalar waveguides is shown in Figure 4.4(a), and its simplified graphical representation for arbitrary length q vectors in (b). In (b) the admittance of the jth scalar waveguide, $j = 1, \ldots, q$ is Y_j, and the voltage waves entering and leaving the splitter at the connection with this waveguide are U_j^+ and U_j^-. On the other side of the connection, we have a single q-vector waveguide of matrix admittance $\mathbf{Y}$. The column vector voltage waves entering and leaving the connection are $\mathbf{V}^+$ and $\mathbf{V}^-$. In order for this connection to make sense, we must choose $\mathbf{Y} = \mathrm{diag}(Y_1, \ldots, Y_q)$, and order the splitting such that the jth components of $\mathbf{V}_-$ and $\mathbf{V}_+$ are equal to U_j^+ and U_j^- respectively; such an ordering for a 1:3 splitting is shown in (a). In this case, it is easy to see that energy is conserved across such a connection (indeed, if the scalar waveguides are thought of as acoustic tubes, then the black bar in (b) is merely equivalent to a 'rubber band' joining them). The power entering the connection from the vector side can be written as

$$\left(\mathbf{V}^+ + \mathbf{V}^-\right)^T \mathbf{Y} \left(\mathbf{V}^+ - \mathbf{V}^-\right) = -\sum_{j=1}^{q} \left(U_j^+ + U_j^-\right) Y_j \left(U_j^+ - U_j^-\right)$$

which is the total power leaving the connection through the scalar waveguides. It is important to note that such a connection cannot be viewed as a multiport element. In addition, passivity is contingent upon this choice of $\mathbf{Y}$. It is easy to generalize this picture to a connection that splits a vector waveguide into various smaller vector waves (instead of scalars). In that case, $\mathbf{Y}$ should be chosen to be the block diagonal 'direct sum' of the various matrix admittances on the other side of the connection.

4.2.7 Transitional Note

We have now finished reviewing the fundamentals of digital waveguide networks. On the more practical level of the implementation of DWNs, there are many more topics that deserve elaboration, including strategies for reducing the numbers of delays, and also various normalization techniques that can be used to vary the number of arithmetic operations required. In this last respect, we note that it is possible to establish formal links between chains of bidirectional delay lines and other similar filter designs such as the normalized ladder form [110], and so on. We refer the reader to Smith's technical report [228] for an in-depth treatment.

4.3 The (1+1)D Transmission Line

We return now to the transmission line system which served as a useful model problem for MDWD network methods in Section 3.6 and show that its numerical solution can also be approached using waveguide networks. We pay special attention to the relationship with finite difference schemes [25].

4.3.1 First-order System and the Wave Equation

We recall that the set of PDEs that describes the evolution of the voltage and current distributions along a lossless, source-free transmission line in (1+1)D is

$$l\frac{\partial i}{\partial t} + \frac{\partial u}{\partial x} = 0 \tag{4.17a}$$

$$c\frac{\partial u}{\partial t} + \frac{\partial i}{\partial x} = 0 \tag{4.17b}$$

where $i(x,t)$ and $u(x,t)$ are, respectively, the current in and voltage across the lines, and $l(x)$ and $c(x)$, both assumed strictly positive everywhere, are the inductance and capacitance per unit length. For the moment, we will leave aside the discussion of boundary conditions, and deal only with the Cauchy problem (i.e., we assume the spatial domain of the problem to be the entire x axis). Note also that this system includes the vocal tract model (1.20) as a special case, under an appropriate set of variable and parameter replacements.

As discussed in Section 4.2.3, if we assume that l and c are constant, then the set of equations can be reduced to a single second-order equation in the voltage alone[3]:

$$\frac{\partial^2 u}{\partial t^2} = \gamma^2 \frac{\partial^2 u}{\partial x^2} \tag{4.18}$$

where the wave speed γ is given by

$$\gamma = \frac{1}{\sqrt{lc}}$$

This equation and its analogues in higher dimensions (see Appendix A) are collectively known as the wave equation. The solution, as mentioned in Section 4.2.3, can be written in terms of traveling waves. In the (1+1)D case, we can write an identical wave equation in the current alone, but this does not hold in higher dimensions.

4.3.2 Centered Difference Schemes and Grid Decimation

Suppose we are interested in developing a finite difference scheme to calculate the solution to (4.17) numerically. We first define grid functions $I_i(n)$, and $U_i(n)$ which, for convenience, will run over half-integer values of i and n, that is,

$$i, n = \ldots -1, -\tfrac{1}{2}, 0, \tfrac{1}{2}, 1 \ldots$$

---

[3] Even if l and c are functions of x, it is still possible to reduce system (4.17) to a second-order equation in the voltage alone, but it does not have the simple form of (4.18).

They are intended to approximate i and u at the points $(i\Delta, nT)$, where Δ is the spatial grid spacing, and T the time step. We note that we have used the same variable, i, to stand for both the continuous-time current that solves (4.17), as well as the discrete-valued variable representing the spatial coordinate on the grid.

Centered difference approximations to the time and space derivatives can be written as

$$\left.\frac{\partial w}{\partial t}\right|_{i\Delta, nT} = \frac{w(i\Delta, (n+\frac{1}{2})T) - w(i\Delta, (n-\frac{1}{2})T)}{T} + O(T^2) \quad (4.19a)$$

$$\left.\frac{\partial w}{\partial x}\right|_{i\Delta, nT} = \frac{w((i+\frac{1}{2})\Delta, nT) - w((i-\frac{1}{2})\Delta, nT)}{\Delta} + O(\Delta^2) \quad (4.19b)$$

where w stands for either of i or u.

Employing these differences in (4.17), and replacing the continuous time/space variables i and u by their respective grid functions yields the difference scheme

$$I_i(n+\tfrac{1}{2}) - I_i(n-\tfrac{1}{2}) + \frac{1}{v_0 \bar{l}_i}\left(U_{i+\frac{1}{2}}(n) - U_{i-\frac{1}{2}}(n)\right) = 0 \quad (4.20a)$$

$$U_i(n+\tfrac{1}{2}) - U_i(n-\tfrac{1}{2}) + \frac{1}{v_0 \bar{c}_i}\left(I_{i+\frac{1}{2}}(n) - I_{i-\frac{1}{2}}(n)\right) = 0 \quad (4.20b)$$

Here, we have chosen

$$\bar{l}_i \triangleq l(i\Delta) + O(\Delta^2) \quad (4.21a)$$

$$\bar{c}_i \triangleq c(i\Delta) + O(\Delta^2) \quad (4.21b)$$

for half-integer i. Because the centered difference approximations (4.19) are second-order accurate, l and c may be approximated to the same order without any decrease in accuracy. We leave the exact form of these approximations, $\bar{l}$ and $\bar{c}$, unspecified for the moment, but will return to various settings in Section 4.3.6. Also, in order to remain consistent with the notation in the MDWD schemes of the last chapter, we have set

$$v_0 \triangleq \frac{\Delta}{T}$$

Thus, difference equations (4.20) are consistent with (4.17), and accurate to $O(\Delta^2, T^2)$.

In a difference scheme for a general system of PDEs, it would be necessary to update all the grid functions at every time step, and at every grid point—that is to say, at every increment in n and i of one-half, new values of the grid functions would have to be calculated, and indeed, we can proceed in this manner with the scheme (4.20) as well. In this case, however, it is easy to see that updating $U_k(m)$ for $2k$ and $2m$ even requires access only to $I_k(m)$ at the previous time step, and at *neighboring* grid locations (thus for $2m$ odd and $2k$ odd), as well as U at the same location, two time steps previously ($2m$ and $2k$ again even) [176, 246]. Similarly, updating $I_k(m)$ for $2m$ odd and $2k$ odd involves only values of U for $2m$ even and $2k$ even, and I for $2m$ odd and $2k$ odd. It is then obvious that only values of $U_k(m)$ for which $2m$ is even and $2k$ even (and values of $I_k(m)$ with $2m$ odd and $2k$ odd) need enter into our scheme. We can thus *decimate* the grid in the manner shown in Figure 4.5.

4.3. THE (1+1)D TRANSMISSION LINE

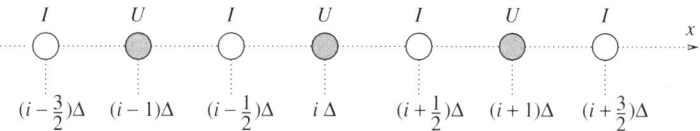

Figure 4.5: *Interleaved sampling grid for the (1 + 1)D transmission line.*

We calculate the values of $U_i(n)$ at the gray dots in Figure 4.5, and $I_{i+\frac{1}{2}}(n+\frac{1}{2})$ at the white dots. The difference scheme on the decimated grid can be written as

$$I_{i+\frac{1}{2}}(n+\tfrac{1}{2}) - I_{i+\frac{1}{2}}(n-\tfrac{1}{2}) + \frac{1}{v_0 \bar{l}_{i+\frac{1}{2}}}(U_{i+1}(n) - U_i(n)) = 0 \qquad (4.22a)$$

$$U_i(n) - U_i(n-1) + \frac{1}{v_0 \bar{c}_i}\left(I_{i+\frac{1}{2}}(n-\tfrac{1}{2}) - I_{i-\frac{1}{2}}(n-\tfrac{1}{2})\right) = 0 \qquad (4.22b)$$

for i, n integer. We perform the calculation on the decimated grid with no decrease in accuracy, although we are of course approximating the solution at fewer grid points. In analogy with the continuous case, when l and c are constant, it is possible to combine the difference equations (4.22) into a single equation for the voltage grid function U, which is

$$U_i(n+1) - 2U_i(n) + U_i(n-1) = \frac{\gamma^2}{v_0^2}(U_{i+1}(n) - 2U_i(n) + U_{i-1}(n)) \qquad (4.23)$$

and which solves the (1+1)D wave equation (4.18). For the so-called *magic* time step [246],

$$v_0 = \gamma \triangleq \frac{1}{\sqrt{lc}}$$

the difference scheme (4.23) reduces to

$$U_i(n+1) + U_i(n-1) = U_{i+1}(n) + U_{i-1}(n) \qquad (4.24)$$

a form that has great relevance to the discussion to follow on the waveguide implementation. It is interesting that in this case, the grid may be further decimated; we need only calculate $U_i(n)$ for $i + n$ even (or odd), for i, n integer. We will examine this point in more detail in higher dimensions in Appendix A.

4.3.3 A (1+1)D Waveguide Network

Consider the waveguide network pictured in Figure 4.6. Each scattering junction (in this case parallel) is connected to its two neighbors by unit-sample bidirectional delay lines. The spacing of the junctions is Δ and the waveguide delays are of duration T. The voltage at a junction with coordinate $i\Delta$ and at time nT is written[4] as $U_{J,i}(n)$ for integer i and n.

---

[4]In any case in which the time index n is omitted, we mean for the statement to hold at any time step.

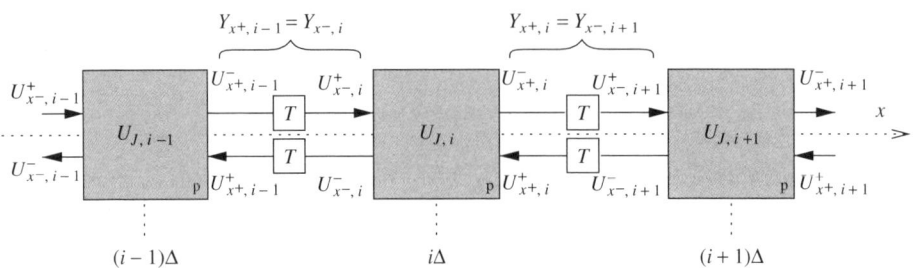

Figure 4.6: (1+1)D *waveguide network*.

We can name the voltages and current flows in individual waveguides in the following way. At junction i, the line voltages are

$$U_{x+,i} = \text{voltage in waveguide leading east}$$
$$U_{x-,i} = \text{voltage in waveguide leading west}$$

and the flows are

$$I_{x+,i} = \text{current flow in waveguide leading east}$$
$$I_{x-,i} = \text{current flow in waveguide leading west}$$

The constraints, imposed by Kirchhoff's Laws at a parallel junction, are as follows:

$$U_{J,i} = U_{x+,i} = U_{x-,i} \qquad I_{x+,i} + I_{x-,i} = 0 \qquad (4.25)$$

As discussed in Section 4.2, the voltages and current flows in the individual waveguides can be further broken up into incoming and outgoing waves. That is, we have, at a junction at grid location i,

$$U_{q,i} = U_{q,i}^+ + U_{q,i}^- \qquad I_{q,i} = I_{q,i}^+ + I_{q,i}^-$$

where q is either of x^+ or x^-. In a particular waveguide section, the current and voltage waves are related by

$$I_{q,i}^+ = Y_{q,i} U_{q,i}^+ \qquad I_{q,i}^- = -Y_{q,i} U_{q,i}^- \qquad (4.26)$$

where $Y_{q,i}$ is the admittance of the waveguide connected to junction i in direction q. In addition, because the junctions at i and $i+1$ are connected to opposite ends of the same waveguide, we have

$$Y_{x-,i+1} = Y_{x+,i}$$

As before, we will also define the *impedance* of any waveguide to be

$$Z_{q,i} = \frac{1}{Y_{q,i}}$$

4.3. THE (1 + 1)D TRANSMISSION LINE

At a particular parallel junction, the junction admittance will thus be

$$Y_{J,i} \triangleq Y_{x^+,i} + Y_{x^-,i}$$

In this case, from (4.14), the junction voltage can be written in terms of incoming wave variables as

$$U_{J,i} = \frac{2}{Y_{J,i}} \left(Y_{x^-,i} U^+_{x^-,i} + Y_{x^+,i} U^+_{x^+,i} \right) \tag{4.27}$$

and the outgoing voltage waves from any junction are related to the incoming waves by

$$U^-_{r,i} = -U^+_{r,i} + U_{J,i} \tag{4.28}$$

where r refers to either of the directions x^+ or x^-.

The incoming voltage wave entering each junction from a particular waveguide at time step n is simply the outgoing voltage wave leaving a neighboring junction, one time step before. Reading directly from Figure 4.6, we have

$$U^+_{x^+,i}(n) = U^-_{x^-,i+1}(n-1) \tag{4.29a}$$

$$U^+_{x^-,i}(n) = U^-_{x^+,i-1}(n-1) \tag{4.29b}$$

The case of flow waves is similar except for a sign inversion—that is, we have

$$I^+_{x^+,i}(n) = -I^-_{x^-,i+1}(n-1) \tag{4.30a}$$

$$I^+_{x^-,i}(n) = -I^-_{x^+,i-1}(n-1) \tag{4.30b}$$

As discussed in Section 4.2, we can perform all calculations using voltage waves; in the waveguide networks pictured in this chapter, we will always assume, without loss of generality, that we are dealing with voltage waves.

4.3.4 Waveguide Network and the Wave Equation

Now consider the case in which $Y_{q,i} = Y$ is invariant over i (and thus q, where again, q stands for either x^+ or x^-). At all junctions, then, we have $Y_{J,i} = 2Y$. From (4.28) and (4.29), it is possible to obtain a finite difference scheme purely in terms of the junction voltages $U_{J,i}$. Beginning from (4.27), we have

$$U_{J,i}(n+1) = \frac{2}{Y_{J,i}} \left(Y_{x^-,i} U^+_{x^-,i}(n+1) + Y_{x^+,i} U^+_{x^+,i}(n+1) \right)$$

$$\overset{(4.29)}{=} U^-_{x^+,i-1}(n) + U^-_{x^-,i+1}(n)$$

$$\overset{(4.28)}{=} U_{J,i-1}(n) + U_{J,i+1}(n) - U^+_{x^+,i-1}(n) - U^+_{x^-,i+1}(n)$$

$$\overset{(4.29)}{=} U_{J,i-1}(n) + U_{J,i+1}(n) - U^-_{x^-,i}(n-1) - U^-_{x^+,i}(n-1)$$

$$\overset{(4.28)}{=} U_{J,i-1}(n) + U_{J,i+1}(n) - U_{J,i}(n-1) \tag{4.31}$$

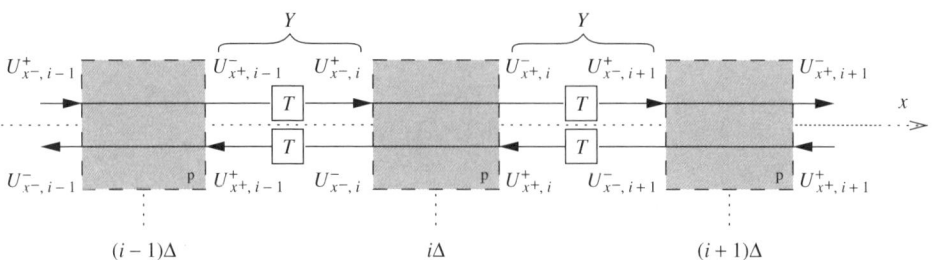

Figure 4.7: *Simplified (1+1)D waveguide network.*

This is identical to (4.24) if we replace U_J by U. In this case of identical impedances in all the waveguides there is no scattering, so the parallel junctions in Figure 4.6 reduce to simple 'throughs,' and Figure 4.6 becomes Figure 4.7.

We thus have a discrete equivalent to the traveling wave solution to the wave equation—this is to be expected when the impedance does not vary spatially along the line. This particular case, which is trivial to implement (as a *single* many-sample bidirectional delay line), has enormous applications to (1+1)D problems in homogeneous media, as will be mentioned in Section 4.6. We also note that if the impedances do vary from one waveguide to the next, as in Figure 4.6, then we have a useful model of a system such as a tube with varying cross-sectional area or horn [94]—in other words, a system whose impedance varies along its length, but whose *wave speed* remains constant. (In order to deal with local changes in the wave speed, we will have to introduce self-loops, which we will do shortly in Section 4.3.6.) This waveguide network is essentially equivalent to the Kelly–Lochbaum model used in speech synthesis [142], which we discussed in Section 1.1.2. It is interesting that linear predictive coding (LPC) [165], which is used to design filters to fit the spectrum of an analysis signal, essentially synthesizes a waveguide network like the one shown in Figure 4.6 (in effect it produces, as a by-product of the main calculation of direct-form filter coefficients, the *reflection coefficients* at the scattering junctions, from which impedances can then be deduced).

Comment on Numerical Instability

We have just shown, in the derivation ending with (4.31), that scattering in a particular waveguide network can be rewritten as a finite difference scheme purely in terms of the *junction* quantities. Thus, all numerical solutions obtained using the waveguide network implementation could also be obtained (at least in infinite-precision arithmetic) using such a scheme. It is interesting to note that certain solutions to the finite difference equation (4.31) cannot be obtained using the DWN if we require that the wave variables in the network be *bounded in magnitude*. As a very simple example, consider initializing scheme (4.24) with $U_i(0) = -1$ and $U_i(1) = 1$, for all i. Then we will have $U_i(2) = 3$, $U_i(3) = 5$, and in general, $U_i(n) = 2n - 1$ for all i. Similar linear growth will result from setting $U_i(0) = U_i(1) = (-1)^i$. We will then have, at any future time step n, $U_i(n) = (2n-1)(-1)^{i+n-1}$.

Though these solutions would appear to be completely unphysical, it is worth mentioning that the (1+1)D wave equation (to which (4.24) is an approximation) admits linear growth as well; $u = t$, for example, is a solution to (4.18). It is possible to view this solution as the sum of two traveling wave solutions $(x/\gamma + t)/2$ and $(-x/\gamma + t)/2$; these, however

4.3. THE (1+1)D TRANSMISSION LINE

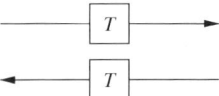

Figure 4.8: *Bidirectional delay line.*

are unbounded in magnitude, and thus the wave variables used to initialize the DWN will be as well; the finite difference scheme, on the other hand, produces this behavior for the bounded initial conditions mentioned above. It is important to note that this linear growth occurs at the spatial DC and Nyquist frequencies; it is simple to show that these are in fact the only spatial frequencies for which scheme (4.24) will admit such behavior. We will return to this point in some detail in Appendix A, because the analysis is somewhat easier in the frequency domain.

4.3.5 An Interleaved Waveguide Network

The simplified waveguide network described above solves the wave equation for voltage at the magic time step, $T = \Delta/\gamma$. That is, the junction voltages U_J solve the difference equation (4.24), and hence approximate u. We would, however, like to be able to have direct access to a discrete equivalent of the other variable as well, the current i.

Bearing in mind the discussion in Section 4.3.2 on interleaved grids, examine the identity pictured in Figures 4.8 and 4.9. We have merely split the unit-sample bidirectional delay line into two half-sample delay lines of equal impedance, and placed a *series* junction (in cascade with sign inverters) in between. In this case, since there is no scattering, the net behavior of the junction and sign inversion is that of a simple 'through,' with sign inversions exactly canceling those that appear in the signal path (these can be added formally using transformers). Later we will add additional ports to this new junction. We introduce these series junctions so as to be able to associate a junction current with them, which we will identify with the physical current in the transmission line.

If we now replace all the bidirectional delay lines in Figure 4.6 by the split pair of lines, then we get the arrangement in Figure 4.10. As at the parallel junctions, we can define wave voltages and currents at the series junctions, which we will index by $i + \frac{1}{2}$ for i integer. Furthermore, we name the impedances at the left- and right-hand ports of the series junctions $Z_{x^-, i-\frac{1}{2}}$ and $Z_{x^+, i-\frac{1}{2}}$ respectively. As indicated in Figure 4.10, we must also have

$$Z_{x^+, i-\frac{1}{2}} = \frac{1}{Y_{x^-, i}} \qquad Z_{x^-, i+\frac{1}{2}} = \frac{1}{Y_{x^+, i}}$$

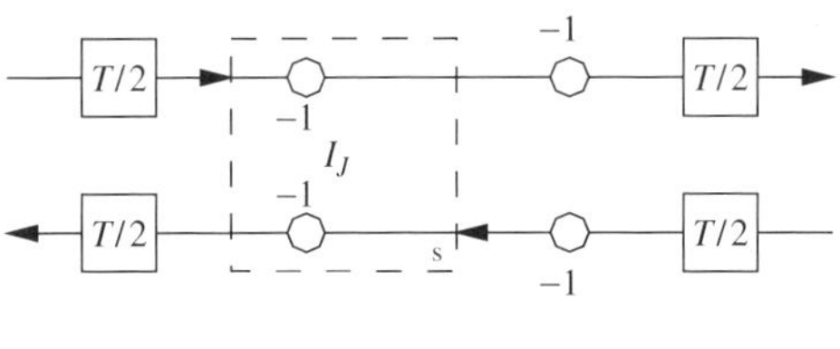

Figure 4.9: *Split equivalent to the bidirectional delay line.*

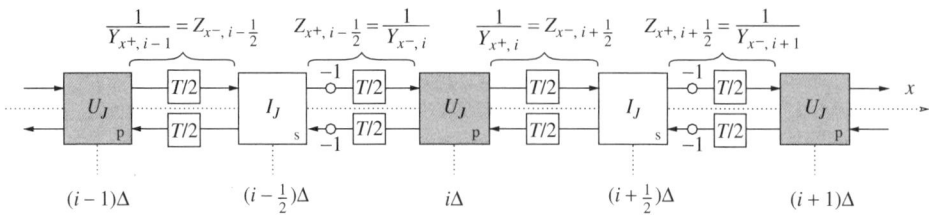

Figure 4.10: (1+1)D *interleaved waveguide network*.

The junction impedance at the series junctions will be

$$Z_{J,i+\frac{1}{2}} \triangleq Z_{x-,i+\frac{1}{2}} + Z_{x+,i+\frac{1}{2}} = \frac{1}{Y_{x+,i}} + \frac{1}{Y_{x-,i+1}}$$

See Figure 4.11 for a complete picture of the various wave quantities at the interleaved junctions.

Assuming that the impedances in all the delay lines are identical and equal to Z (and so $Z_{J,i-\frac{1}{2}} = 2Z$), we can now define

$$I_{J,i-\frac{1}{2}} = \frac{2}{Z_{J,i-\frac{1}{2}}} \left(Z_{x+,i-\frac{1}{2}} I^+_{x+,i-\frac{1}{2}} + Z_{x-,i-\frac{1}{2}} I^+_{x-,i-\frac{1}{2}} \right)$$

$$= I^+_{x+,i-\frac{1}{2}} + I^+_{x-,i-\frac{1}{2}}$$

We also have that

$$U^+_{x-,i}(n) = -U^-_{x+,i-\frac{1}{2}}(n - \tfrac{1}{2})$$

$$U^+_{x+,i-\frac{1}{2}}(n + \tfrac{1}{2}) = -U^-_{x-,i}(n)$$

and (4.26) holds as before.

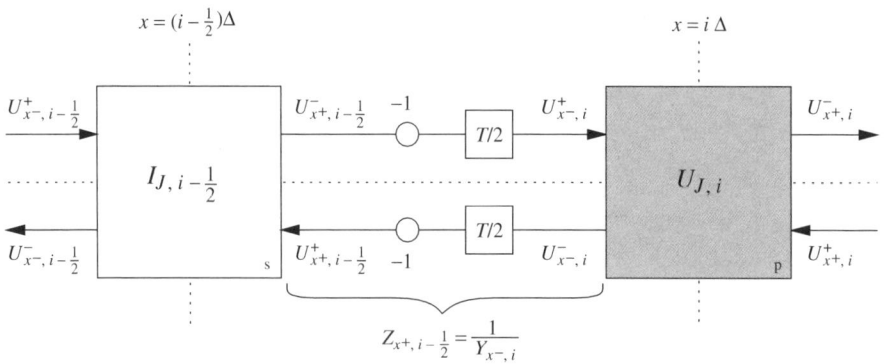

Figure 4.11: *Wave quantities in the interleaved network of Figure 4.10.*

4.3. THE (1 + 1)D TRANSMISSION LINE

We now show that this waveguide network performs a calculation identical to that which we would get for centered differences on a decimated grid, exactly as in Figure 4.5. For integer i and n, we have

$$\begin{aligned}
U_{J,i}(n) &= U^+_{x^+,i}(n) + U^+_{x^-,i}(n) \\
&= Z\left(I^+_{x^+,i}(n) + I^+_{x^-,i}(n)\right) \\
&= -Z\left(I^-_{x^-,i+\frac{1}{2}}(n-\tfrac{1}{2}) - I^-_{x^+,i-\frac{1}{2}}(n-\tfrac{1}{2})\right) \\
&= -Z\left(I_{J,i+\frac{1}{2}}(n-\tfrac{1}{2}) - I_{J,i-\frac{1}{2}}(n-\tfrac{1}{2})\right) \\
&\quad + Z\left(I^+_{x^-,i+\frac{1}{2}}(n-\tfrac{1}{2}) - I^+_{x^+,i-\frac{1}{2}}(n-\tfrac{1}{2})\right) \\
&= -Z\left(I_{J,i+\frac{1}{2}}(n-\tfrac{1}{2}) - I_{J,i-\frac{1}{2}}(n-\tfrac{1}{2})\right) \\
&\quad + U^-_{x^+,i}(n-1) + U^-_{x^-,i}(n-1) \\
&= -Z\left(I_{J,i+\frac{1}{2}}(n-\tfrac{1}{2}) - I_{J,i-\frac{1}{2}}(n-\tfrac{1}{2})\right) + U_{J,i}(n-1) \quad (4.32)
\end{aligned}$$

If we now identify I_J with I and U_J with U, we get (4.22b) (in the constant-coefficient case), with $Z = 1/(v_0 c)$. A similar derivation beginning from the series junctions yields (4.22a), with $Z = v_0 l$, for constant l. Together, these constraints imply that

$$v_0 = \frac{1}{\sqrt{lc}} \qquad Z = \sqrt{\frac{l}{c}}$$

so that we are again at the magic time step. Furthermore, the impedance of any waveguide in the network must be set equal to the characteristic impedance of the continuous-time/space transmission line described by (4.17), whereas in the network of Figure 4.6, the constant impedance value could be set arbitrarily, since it is *not used* in the simulation. It is important to realize that, at least in this constant-coefficient case, no scattering occurs at any of the junctions. We can still perform all operations at the original sampling rate, and on the original grid (i.e., with grid spacing Δ and time step T). It is, however, possible to see more clearly how initial (and boundary) conditions must be set, and also to extend the network to handle more complex problems. We will deal with one such generalization in the next section.

4.3.6 Varying Coefficients

We now return to the more general case in which the material parameters l and c have spatial dependence. The staggered, or interleaved network of delay lines and scattering junctions presented in the last section gives rise to a centered difference method that approximates the solution to system (4.17).

Consider the waveguide network in Figure 4.12. The picture is the same as in the constant-coefficient case, except that we have added an extra port to each scattering junction, which is connected to a delay line of impedance Z_c (marked as Y_c at the parallel junctions).

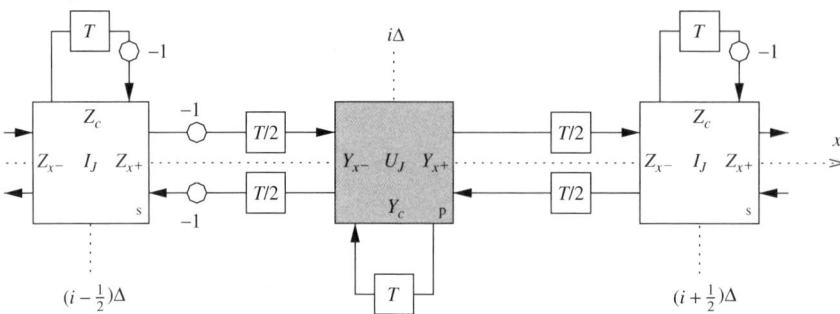

Figure 4.12: *Waveguide network for the (1+1)D transmission line equations with spatially varying coefficients.*

These *self-loops* [229] are bidirectional delay lines in their own right—each can be viewed as a pair of lines of delay $T/2$, terminated on an open or short-circuit. Note that the loop connected to the series junction contains a sign inversion (like a lumped wave digital inductor) and the loop at the parallel junction does not (like a capacitor). The delay in this line is thus a full sample, so that we are able to operate on the interleaved grid. The reason for this choice should be clear from Figure 4.5. Now, the immittances of the delay lines are in general no longer identical, so we expect nontrivial scattering to occur because of the spatially varying material parameters. The admittances of the lines connected to the parallel junction at grid point i, an integer, are denoted by $Y_{x+,i}$, $Y_{x-,i}$ and $Y_{c,i}$. The impedances of the self-loops at grid points $i + \frac{1}{2}$ for i integer will be called $Z_{c,i+\frac{1}{2}}$. (We have marked these new immittances in Figure 4.12.) The junction immittances are now

$$Y_{J,i} \triangleq Y_{x+,i} + Y_{x-,i} + Y_{c,i} \qquad \text{Parallel junction}$$

$$Z_{J,i+\frac{1}{2}} \triangleq Z_{x+,i+\frac{1}{2}} + Z_{x-,i+\frac{1}{2}} + Z_{c,i+\frac{1}{2}} \qquad \text{Series junction}$$

$$= \frac{1}{Y_{x-,i+1}} + \frac{1}{Y_{x+,i}} + Z_{c,i+\frac{1}{2}}$$

at a parallel junction at grid point i and at a series junctions at grid point $i + \frac{1}{2}$, for integer i.

We will call the new voltage wave variables entering and leaving the parallel junction from the new port $U^+_{c,i}(n)$ and $U^-_{c,i}(n)$ respectively, and those entering and leaving the series junction from its new port $U^+_{c,i+\frac{1}{2}}(n + \frac{1}{2})$ and $U^-_{c,i+\frac{1}{2}}(n + \frac{1}{2})$, for integer values of i and n. In addition, we will adopt the notation[5]

$$l_k \triangleq l(k\Delta) \qquad\qquad c_k \triangleq c(k\Delta)$$

for half-integer values of k. One physical interpretation of the need for these self-loops is that if l and c vary over the domain, the effective local wave speed does as well. Thus, if we choose a regular grid spacing, there are necessarily grid points at which the space step/time step ratio is *not* the magic ratio (it must be greater, by the Courant–Friedrichs–Lewy (CFL)

---

[5] We take care to distinguish these quantities, which are simply values of the continuous functions l and c, from the indeterminate quantities $\bar{l}_i$ and $\bar{c}_i$ defined in (4.21).

4.3. THE (1 + 1)D TRANSMISSION LINE

condition). Thus if we were to try to use a structure such as that pictured in Figure 4.10, regardless of how the waveguide impedances are set, energy would be moving 'too fast' through the network. The extra delays at the junctions serve to slow down the propagation of energy, by storing a portion of it for a time step. The amount of slow down is locally determined by the values of self-loop admittances and impedances. Self-loops are used for the same purpose in TLM [44, 122]—in this context they are called inductive or capacitive stubs, depending on whether they do or do not invert sign.

Beginning from a parallel junction in Figure 4.12, we can proceed through a derivation similar to that leading to (4.32)

$$\begin{aligned}
U_{J,i}(n) &= \frac{2}{Y_{J,i}} \left(Y_{x^+,i} U^+_{x^+,i}(n) + Y_{x^-,i} U^+_{x^-,i}(n) + Y_{c,i} U^+_{c,i}(n) \right) \\
&= \frac{2}{Y_{J,i}} \left(I^+_{x^+,i}(n) + I^+_{x^-,i}(n) + Y_{c,i} U^+_{c,i}(n) \right) \\
&= \frac{2}{Y_{J,i}} \left(-I^-_{x^-,i+\frac{1}{2}}(n - \tfrac{1}{2}) + I^-_{x^+,i-\frac{1}{2}}(n - \tfrac{1}{2}) + I^+_{c,i}(n) \right) \\
&= \frac{2}{Y_{J,i}} \left(-I_{J,i+\frac{1}{2}}(n - \tfrac{1}{2}) + I_{J,i-\frac{1}{2}}(n - \tfrac{1}{2}) \right) \\
&\quad + \frac{2}{Y_{J,i}} \left(I^+_{x^-,i+\frac{1}{2}}(n - \tfrac{1}{2}) - I^+_{x^+,i-\frac{1}{2}}(n - \tfrac{1}{2}) - I^-_{c,i}(n - 1) \right) \\
&= \frac{2}{Y_{J,i}} \left(-I_{J,i+\frac{1}{2}}(n - \tfrac{1}{2}) + I_{J,i-\frac{1}{2}}(n - \tfrac{1}{2}) \right) \\
&\quad - \frac{2}{Y_{J,i}} \left(I^-_{x^+,i}(n - 1) + I^-_{x^-,i}(n - 1) + I^-_{c,i}(n - 1) \right) \\
&= \frac{2}{Y_{J,i}} \left(-I_{J,i+\frac{1}{2}}(n - \tfrac{1}{2}) + I_{J,i-\frac{1}{2}}(n - \tfrac{1}{2}) \right) + U_{J,i}(n - 1) \quad (4.33)
\end{aligned}$$

In order to equate this difference relation with (4.22b), we must have, recalling (4.21b),

$$Y_{J,i} = 2v_0 \bar{c}_i \quad (4.34)$$

Beginning from the series junctions at $x = (i + \frac{1}{2})\Delta$, for i integer, we obtain, similarly, an interleaved difference approximation to (4.17a), under the constraint

$$Z_{J,i+\frac{1}{2}} = 2v_0 \bar{l}_{i+\frac{1}{2}} \quad (4.35)$$

Under the further condition that all impedances in the network be positive, these constraints give rise to a family of stable difference approximations to (4.17). We can distinguish three special types:

Type I: Voltage-centered Network

We set the admittances of the waveguides leading away from a parallel junction to be identical, that is,

$$Y_{x^-,i} = Y_{x^+,i} = v_0 c_i \quad (4.36)$$

and set
$$Y_{c,i,j} = 0$$
which satisfies (4.63) with $\bar{c}_i = c_i$. From (4.36), we have that
$$Z_{x^+,i+\frac{1}{2}} = \frac{1}{v_0 c_{i+1}} \qquad Z_{x^-,i+\frac{1}{2}} = \frac{1}{v_0 c_i}$$

Thus the series junction impedance at location $i + \frac{1}{2}$ will be
$$Z_{J,i+\frac{1}{2}} = \frac{1}{v_0 c_i} + \frac{1}{v_0 c_{i+1}} + Z_{c,i+\frac{1}{2}}$$

We can then set
$$Z_{c,i+\frac{1}{2}} = v_0(l_i + l_{i+1}) - \frac{1}{v_0}\left(\frac{1}{c_i} + \frac{1}{c_{i+1}}\right)$$
which satisfies (4.35) with $\bar{l}_{i+\frac{1}{2}} = \frac{1}{2}(l_i + l_{i+1})$.

Only the series self-loop impedances are possibly negative, so the network will be passive if $Z_{c,i+\frac{1}{2}} \geq 0$. This will certainly be true if we choose
$$v_0 \geq \max_i \left(\sqrt{\frac{1}{l_i c_i}}\right) \qquad (4.37)$$

Recall that in our earlier discussion of group velocities for symmetric hyperbolic systems in Section 3.1 and for the transmission line in particular in Section 3.6, the maximum group velocity for the transmission line is
$$v^g_{TL,\max} = \frac{1}{\sqrt{(lc)_{\min}}} \qquad (4.38)$$

The optimal space step/time step ratio from (4.37) is exactly the maximum of the local group velocity of the transmission line, at least over the range of values of l and c sampled at the parallel junction locations; it thus approaches the maximum group velocity for the continuous system in the limit as the grid spacing Δ becomes small.

Type II: Current-centered Network

This arrangement is the dual to the previous case. We now set
$$Z_{x^+,i+\frac{1}{2}} = Z_{x^-,i+\frac{1}{2}} = v_0 l_{i+\frac{1}{2}} \qquad Z_{c,i+\frac{1}{2}} = 0$$
and
$$Y_{c,i} = v_0\left(c_{i+\frac{1}{2}} + c_{i-\frac{1}{2}}\right) - \frac{1}{v_0}\left(\frac{1}{l_{i+\frac{1}{2}}} + \frac{1}{l_{i-\frac{1}{2}}}\right) \qquad (4.39)$$

We have thus chosen $\bar{c}_i = \frac{1}{2}(c_{i+\frac{1}{2}} + c_{i-\frac{1}{2}})$ and $\bar{l}_{i+\frac{1}{2}} = l_{i+\frac{1}{2}}$. Now, we must have
$$v_0 \geq \max_i \left(\sqrt{\frac{1}{l_{i+\frac{1}{2}} c_{i+\frac{1}{2}}}}\right)$$

4.3. THE (1 + 1)D TRANSMISSION LINE

which is very similar to condition (4.37), except that the maximum is now taken over the series junction locations.

Here, scattering at the series junctions is trivial, since the impedances at the two connecting ports are identical, and the self-loop impedance is zero (and we thus drop entirely any calculation of the value in the self-loop at the series nodes). We can operate at the down-sampled rate, with scattering occurring only at the parallel junctions. In this case, we are directly computing only junction voltages, and are in fact solving the second-order reduction of system (4.17), namely,

$$\frac{\partial^2 u}{\partial t^2} = \frac{1}{c}\frac{\partial}{\partial x}\left(\frac{1}{l}\frac{\partial u}{\partial x}\right) \qquad (4.40)$$

We could have made a similar statement about scattering at the parallel junctions in the previous case. This efficient configuration, unlike type I, however, generalizes to the (2+1)D case, as we will see in Section 4.4.2.

Type III: Mixed Network

Suppose we set all the impedances that connect one grid point to another to be equal to some constant Z_{const} that is independent of position. Thus,

$$Y_{x-,i} = Y_{x+,i} = 1/Z_{\text{const}}$$

We then choose, to satisfy (4.34) and (4.35),

$$Y_{c,i} = 2v_0 c_i - \frac{2}{Z_{\text{const}}}$$

$$Z_{c,i+\frac{1}{2}} = 2v_0 l_{i+\frac{1}{2}} - 2Z_{\text{const}}$$

and this leads to the conditions

$$v_0 \geq \max_i \frac{1}{c_i Z_{\text{const}}} = \frac{1}{\min_i(c_i) Z_{\text{const}}} \qquad v_0 \geq \max_i \frac{Z_{\text{const}}}{l_{i+\frac{1}{2}}} = \frac{Z_{\text{const}}}{\min_i(l_{i+\frac{1}{2}})}$$

The lower bounds on v_0 coincide (in the limit as the grid spacing becomes small) when

$$Z_{\text{const}} = \sqrt{\frac{\min_i(l_{i+\frac{1}{2}})}{\min_i(c_i)}}$$

in which case we have

$$v_0 \geq \sqrt{\frac{1}{\min_{i+\frac{1}{2}} l_i \min_i c_i}}$$

Thus, for $2i$ either even or odd, we are no longer at the optimal bound, and are forced to use a smaller time step than in the previous two cases if we wish the network to remain concretely passive. This arrangement bears a strong resemblance to the MDWD network discussed in Section 3.6. We will explore this similarity in more detail in Chapter 6. Many other choices of the waveguide immittances satisfying (4.34) and (4.35) are of course possible.

Comment: Passivity and Stability

At this point, we would like to mention an interesting property of the interleaved waveguide networks discussed in the earlier part of this section. We showed, in the last few pages, that three different types of immittance settings for the waveguide network could be used to solve the (1+1)D transmission line equations, and could, in fact, be interpreted as centered difference approximations. The three types of network integrate system (4.17) using slightly different *effective* inductances $\bar{l}$ and capacitances $\bar{c}$, which converge to l and c in the limit as the grid spacing becomes small. We had, for integer i,

$$\bar{l}_{i+\frac{1}{2}} = \frac{1}{2}(l_i + l_{i+1}) \qquad \bar{c}_i = c_i \qquad \text{Type I}$$

$$\bar{l}_{i+\frac{1}{2}} = l_{i+\frac{1}{2}} \qquad \bar{c}_i = \frac{1}{2}\left(c_{i+\frac{1}{2}} + c_{i-\frac{1}{2}}\right) \qquad \text{Type II}$$

$$\bar{l}_{i+\frac{1}{2}} = l_{i+\frac{1}{2}} \qquad \bar{c}_i = c_i \qquad \text{Type III}$$

The three types, however, yield different requirements for passivity on the space step/time step ratio v_0,

$$v_0 \geq \max_i \left(\sqrt{\frac{1}{l_i c_i}}\right) \qquad \text{Type I} \qquad (4.41)$$

$$v_0 \geq \max_{i+\frac{1}{2}} \left(\sqrt{\frac{1}{l_{i+\frac{1}{2}} c_{i+\frac{1}{2}}}}\right) \qquad \text{Type II} \qquad (4.42)$$

$$v_0 \geq \sqrt{\frac{1}{\min_{i+\frac{1}{2}} l_i \min_i c_i}} \qquad \text{Type III} \qquad (4.43)$$

The first two bounds are roughly the same, and are close to optimal, in the sense that v_0 is bounded by (in the limit as Δ approaches 0) the maximum of the local group velocity over the transmission line. The type III bound, however, may be substantially poorer, and is similar to that which arises in MDWD networks (see Section 3.6).

If we choose $l(x)$ and $c(x)$ to be positive affine functions (proportional to x with a constant offset), then $\bar{l}$ and $\bar{c}$ are the same in all the three cases, so the three networks will, in infinite-precision arithmetic, calculate identical solutions. But there will be a range of values of v_0 (namely, the range of v_0 greater than the bounds given in (4.41) and (4.42), but less than that of (4.43)) for which the type I and II networks are *concretely* passive [14], but for which the type III network is not. Over this range, some immittances in the type III network will necessarily be negative.

We can conclude that there is a large middle ground between passivity and global stability of networks. One important difference would seem to be that wave quantities in a concretely passive network are *power-normalizable*, whereas if a network is only *abstractly* passive—that is, if its global behavior is passive, even though it contains elements that are themselves not—they may not be. We do not investigate this further here, but add that it would be of great interest to make clearer the distinction between passive and stable numerical methods for solving PDEs. This subject has been broached in some detail for ODEs [49, 106], and we will see some other interesting examples of this distinction in Appendix A.

4.3. THE (1 + 1)D TRANSMISSION LINE

4.3.7 Incorporating Losses and Sources

We now reconsider the full (1+1)D transmission line equations, including the effects of losses and sources; this system was presented earlier in Section 3.6, and we repeat its definition here:

$$l\frac{\partial i}{\partial t} + \frac{\partial u}{\partial x} + ri + e = 0 \qquad (4.44a)$$

$$c\frac{\partial u}{\partial t} + \frac{\partial i}{\partial x} + gu + h = 0 \qquad (4.44b)$$

Here $r(x) \geq 0$ and $g(x) \geq 0$ represent resistance and shunt conductivity at any point in the domain, and e and h are driving terms and can be functions of x and t.

In order to add these terms to the difference approximation in such a way that we may still use an interleaved scheme, we can use *semi-implicit* [246] approximations to i and u given by

$$i(x,t) = \frac{1}{2}(i(x, t - T/2) + i(x, t + T/2)) + O(T^2)$$

$$u(x,t) = \frac{1}{2}(u(x, t - T/2) + u(x, t + T/2)) + O(T^2)$$

We also define

$$r_{i+\frac{1}{2}} \triangleq r\left((i + \tfrac{1}{2})\Delta\right) \qquad\qquad g_i \triangleq g(i\Delta)$$

$$e_{i+\frac{1}{2}}(n + \tfrac{1}{2}) \triangleq e\left((i + \tfrac{1}{2})\Delta, (n + \tfrac{1}{2})T\right) \qquad h_i(n) \triangleq h(i\Delta, nT)$$

and use the second-order approximations

$$\bar{e}_{i+\frac{1}{2}}(n) = \frac{1}{2}\left(e_{i+\frac{1}{2}}(n + \tfrac{1}{2}) + e_{i+\frac{1}{2}}(n - \tfrac{1}{2})\right)$$

$$\bar{h}_i(n - \tfrac{1}{2}) = \frac{1}{2}(h_i(n) + h_i(n - 1))$$

We then get, as an approximation to (4.44),

$$I_{i+\frac{1}{2}}(n + \tfrac{1}{2}) - \rho_{I,i+\frac{1}{2}} I_{i+\frac{1}{2}}(n - \tfrac{1}{2})$$
$$+ \sigma_{I,i+\frac{1}{2}}(U_{i+1}(n) - U_i(n)) + \Delta\sigma_{I,i+\frac{1}{2}}\bar{e}_{i+\frac{1}{2}}(n) = 0 \qquad (4.45a)$$

$$U_i(n) - \rho_{U,i} U_i(n-1)$$
$$+ \sigma_{U,i}\left(I_{i+\frac{1}{2}}(n - \tfrac{1}{2}) - I_{i-\frac{1}{2}}(n - \tfrac{1}{2})\right) + \Delta\sigma_{U,i}\bar{h}_i(n - \tfrac{1}{2}) = 0 \qquad (4.45b)$$

with

$$\rho_{I,i+\frac{1}{2}} \triangleq \frac{2\bar{l}_{i+\frac{1}{2}} - r_{i+\frac{1}{2}}T}{2\bar{l}_{i+\frac{1}{2}} + r_{i+\frac{1}{2}}T} \qquad\qquad \rho_{U,i} \triangleq \frac{2\bar{c}_i - g_i T}{2\bar{c}_i + g_i T}$$

$$\sigma_{I,i+\frac{1}{2}} \triangleq \frac{2}{2v_0\bar{l}_{i+\frac{1}{2}} + r_{i+\frac{1}{2}}\Delta} \qquad\qquad \sigma_{U,i} \triangleq \frac{2}{2v_0\bar{c}_i + g_i\Delta}$$

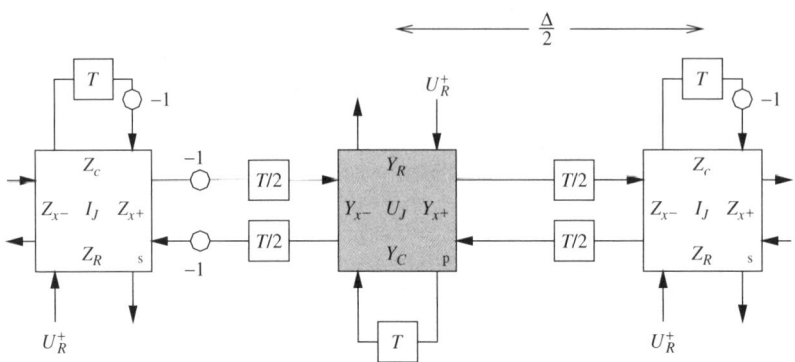

Figure 4.13: *Waveguide network for system* (4.44).

Losses and sources can be added to the waveguide network scheme rather easily, by introducing new ports at each series or parallel junction. In fact, as per wave digital filters, each pair of terms $ri + e$, and $gu + h$ can be interpreted as a *resistive source* [68], and only requires the addition of a single new port at each junction. (The resistive voltage source was discussed in Section 2.2.4.) For any parallel port we will call the new port admittance $Y_{R,i}$, and the voltage wave variable entering the port $U_{R,i}^+$. For a series port, we call the new impedance $Z_{R,i+\frac{1}{2}}$, and the incoming voltage wave variable $U_{R,i+\frac{1}{2}}^+$. The generalized network is shown in Figure 4.13, with the new loss/source port immittances marked.

As a result of the addition of this port, the junction admittances and impedances become

$$Y_{J,i} \triangleq Y_{x-,i} + Y_{x+,i} + Y_{c,i} + Y_{R,i}$$

$$Z_{J,i+\frac{1}{2}} \triangleq Z_{x+,i+\frac{1}{2}} + Z_{x-,i+\frac{1}{2}} + Z_{c,i+\frac{1}{2}} + Z_{R,i+\frac{1}{2}}$$

Beginning again from a parallel junction, and proceeding through a derivation similar to that which leads to (4.33), we obtain a difference relation among the junction voltages and currents:

$$U_{J,i}(n) - \frac{Y_{J,i} - 2Y_{R,i}}{Y_{J,i}} U_{J,i}(n-1) + \frac{2}{Y_{J,i}} \left(I_{J,i+\frac{1}{2}}(n-\tfrac{1}{2}) - I_{J,i-\frac{1}{2}}(n-\tfrac{1}{2}) \right)$$

$$- \frac{2Y_{R,i}}{Y_{J,i}} \left(U_{R,i}^+(n) + U_{R,i}^+(n-1) \right) = 0$$

In order to equate this relation with (4.45b), we can set

$$Y_{R,i} = g_i \Delta$$

$$U_{R,i}^+(n) = -\frac{h_i(n)}{2g_i}$$

Beginning from a series junction, we obtain an analogous relation, which becomes (4.45a) under the choices

$$Z_{R,i+\frac{1}{2}} = r_{i+\frac{1}{2}} \Delta$$

$$U_{R,i+\frac{1}{2}}^+(n+\tfrac{1}{2}) = -\frac{\Delta e_{i+\frac{1}{2}}(n+\tfrac{1}{2})}{2}$$

4.3. THE (1+1)D TRANSMISSION LINE

Note that in the case in which the loss parameter g is zero, or close to zero, $U_{R,i}^+(n)$ will become infinite, or very large. For this reason, it will be necessary in this case to use the dual type of wave; that is, if g_i is small, set $I_{R,i}^+ = Y_{R,i} U_{R,i}^+ = -\frac{\Delta}{2} h_i$, and use current waves at the series junctions. The other impedances in the network remain unchanged under the addition of losses and sources; thus all the stability criteria mentioned in Section 4.3.6 remain the same. It is rather interesting to note, however, that in the case of the current-centered network, for example (type II), scattering at the series junctions is no longer trivial if we have nonzero sources e or loss r, and the series junctions cannot be treated as simple throughs. A similar statement holds for the dual case of the voltage-centered network (type I) in (1+1)D, but will not be true when we generalize to the (2+1)D mesh (see Section 4.4).

4.3.8 Numerical Phase Velocity and Dispersion

We now make a few comments regarding the spectral properties of these difference methods; a detailed summary of spectral analysis is provided in Appendix A.

Consider again the type II DWN for the (1+1)D transmission line equations, as discussed in Section 4.3.6. In the lossless, source-free case, the difference scheme can be written purely in terms of the junction voltages, for integer time steps n, as

$$U_{J,i}(n+1) + U_{J,i}(n-1) = \frac{2}{Y_{J,i}} \left(Y_{x-,i} U_{J,i-1}(n) + Y_{x+,i} U_{J,i+1}(n) + Y_{c,i} U_{J,i}(n) \right)$$

(4.46)

where $Y_{x-,i} = 1/(v_0 l_{i-\frac{1}{2}})$ and $Y_{x+,i} = 1/(v_0 l_{i+\frac{1}{2}})$, and the self-loop admittance is given by (4.39). In effect, we are numerically solving the reduced form (4.40) of (4.17) obtained by elimination of the current i.

If the material parameters are constant, then (4.46) can be rewritten as

$$U_{J,i}(n+1) + U_{J,i}(n-1) = \lambda^2 \left(U_{J,i-1}(n) + U_{J,i+1}(n) \right) + 2(1-\lambda^2) U_{J,i}(n)$$

where $\lambda = \gamma/v_0$ and $\gamma = 1/\sqrt{lc}$ is the wave speed. It is possible to examine this scheme in terms of discrete spatial frequencies β, as per the methods discussed in [238]; the range of spatial frequencies that can be represented on this grid of spacing Δ are $-\pi/\Delta \leq \beta \leq \pi/\Delta$. The *spectral amplification factors* (defined in Section A.1) for this scheme are given by

$$G_{\beta,\pm} = \frac{1}{2} \left(-B_\beta \pm \sqrt{B_\beta^2 - 4} \right)$$

where

$$B_\beta = -2 \left(1 + \lambda^2 (1 - \cos(\beta \Delta)) \right)$$

These spectral amplification factors define the numerical phase velocities $v_{\beta,\text{phase}}$ [238] (see Section A.1.4) and thus the numerical dispersion of the scheme (in general, the numerical phase velocity is different from the physical velocity γ). It is of interest to plot the numerical phase velocity of this scheme versus that of the MDWD network for the same system; the spatial frequency dependence of the various modal frequencies of the MDWD network were discussed in Section 3.8.2.

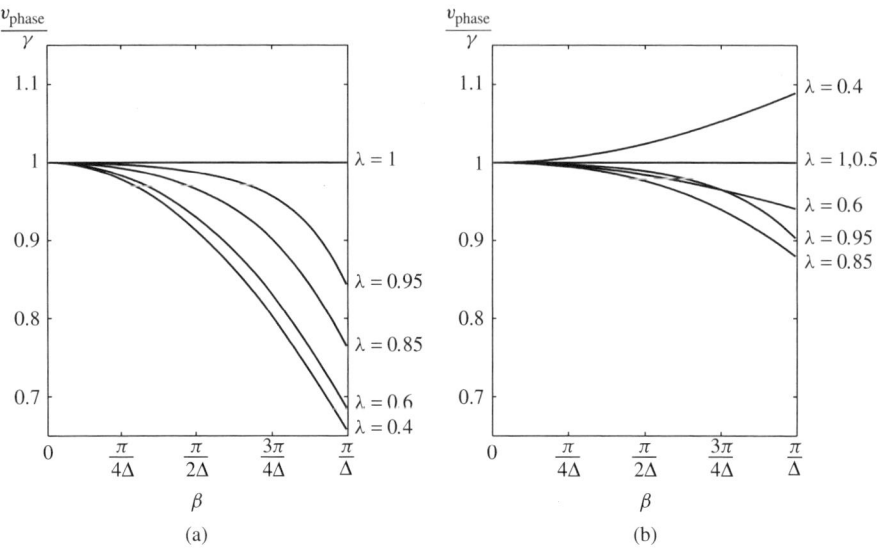

Figure 4.14: *Numerical dispersion curves for various values of* λ—(a) *for the DWN and* (b) *for the MDWD network.*

In Figure 4.14, the quantity v_{phase}/γ is plotted for various values of the parameter λ. At the stability bound, for $\lambda = 1$ (i.e., $v_0 = \gamma = 1/\sqrt{lc}$), both schemes are dispersionless. For the DWN, all spatial frequencies are slowed increasingly as λ is decreased, but for the MDWD network, wave speeds decrease for $\lambda > 1/2$, then are exact again at $\lambda = 1/2$, and are finally faster than the physical speed if $\lambda < 1/2$. This curious behavior of the phase velocities in the MDWD network was also mentioned in Section 3.8.2. In general, the phase velocities of the MDWD network are closer to the correct wave speed over the entire spatial frequency spectrum for a wide range of λ—this is mitigated, however, by the fact that this MDWD network corresponds to a three-step difference method (compared to two-step for the DWN), and is thus more computationally intensive.

4.3.9 Boundary Conditions

The (1+1)D transmission line equations, if they are to be solved on a domain of finite extent, require one supplementary boundary condition at any end point [112]. Suppose that $x = 0$ is such an end point, and furthermore assume that our grid has been constructed such that the point $x = 0$ coincides with one of the parallel (gray) scattering junctions of the interleaved waveguide network shown in Figure 4.13. We now are faced with *terminating* the waveguide network, by replacing the left-hand port of the parallel junction at $x = 0$ with a lumped network. If the lumped network is passive (that is, if its reflectance is bounded), then it should be clear that the network as a whole will be passive as well.

The two most important types of boundary condition are

$$u(0, t) = 0 \qquad \text{Short-circuit at } x = 0$$

$$i(0, t) = 0 \qquad \text{Open-circuit at } x = 0$$

4.3. THE (1 + 1)D TRANSMISSION LINE

Both are lossless, and have the form of (3.8). The first of these conditions is easy to deal with by short circuiting the left-hand port. In this case, we may also remove the self-loop, as well as the combined loss/source port from the junction at $x = 0$. $U_{J,0}$ is forced to zero by the short-circuit condition; this boundary termination is shown in Figure 4.15(a). We have assumed that $Y_{x+,0} = Y_{x-,0}$, so that the termination degenerates to a simple sign inversion of the wave entering from the right-hand side.

The other boundary termination requires a bit more analysis, because we do not have access to I_J at a parallel junction. It is sufficient to drop the left-hand port entirely in this case (or, equivalently, to terminate the junction with an open-circuit, which, when connected in parallel, may be ignored). In this case, though, we must retain the self-loop and the loss/source port. We now show that such a termination does indeed approximate the boundary condition $i(0, t) = 0$. The resulting termination is shown in Figure 4.15(b).

Beginning from system (4.44), where we assume no source (to avoid conflicting conditions at the boundary), we can apply centered differences in time, and use a *one-sided* [238] approximation for the spatial derivative,

$$\frac{\partial i}{\partial x}\bigg|_{x=0, t=(n+\frac{1}{2})T} = \frac{2}{\Delta}\left(i(\tfrac{1}{2}, (n+\tfrac{1}{2})T) - i(0, (n+\tfrac{1}{2})T)\right) + O(\Delta)$$

$$= \frac{2}{\Delta} i(\tfrac{1}{2}, (n+\tfrac{1}{2})T) + O(\Delta)$$

where we have set $i(0, (n + \frac{1}{2})T) = 0$ in accordance with the boundary condition. This yields the difference approximation to (4.44b) given by

$$U_0(n) - \frac{2\bar{c}_0 - g_0 T}{2\bar{c}_0 + g_0 T} U_0(n-1) + \frac{4}{2v_0\bar{c}_0 + g_0 \Delta} I_{\frac{1}{2}}(n - \tfrac{1}{2}) = 0 \qquad (4.47)$$

The difference formula obtained from the network termination of Figure 4.15(b) is

$$U_{0,J}(n) - \frac{Y_{J,0} - 2Y_{R,0}}{Y_{J,0}} U_{0,J}(n-1) + \frac{2}{Y_{J,0}} I_{\frac{1}{2}, J} = 0$$

where we now have

$$Y_{J,0} \triangleq Y_{x+,0} + Y_{R,0} + Y_{c,0} \qquad (4.48)$$

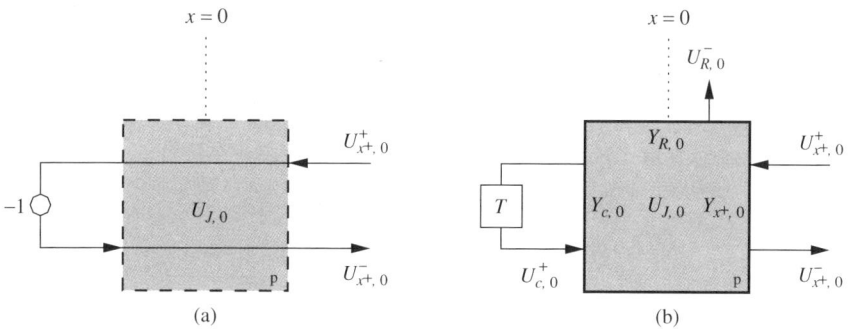

Figure 4.15: *Boundary terminations of the waveguide network for the (1+1)D transmission line equations at $x = 0$ when (a) $u(0, t) = 0$ and (b) $i(0, t) = 0$.*

It is easy to see that if we set

$$Y_{R,0} = \frac{1}{2}g_0\Delta \qquad Y_J = \frac{1}{2}v_0(2\bar{c}_0 + g_0 T) \qquad (4.49)$$

then the waveguide network is indeed performing a calculation equivalent to (4.47). $Y_{c,0}$ and $Y_{x+,0}$ may be set according to the type of network we are using (i.e., I, II or III in Section 4.3.6) as long as (4.48) and (4.49) are satisfied; the stability bounds are unchanged. It is important to note that this realization of the boundary condition $i(0,t) = 0$ will be first-order accurate in the grid spacing Δ, due to our use of a one-sided difference approximation. It is recommended, then, to align the waveguide network such that a series junction lies at the left-most grid point, where it can be simply terminated by an open-circuit. Such a termination will be second-order accurate.

Because *any* bounded reflectance may be used to terminate the waveguide network, it has been proposed [130] (in the context of the Karplus-Strong algorithm [141], which can be shown to be equivalent to a particular digital waveguide configuration) that one can model wave propagation in a dispersive medium (like a beam [108]) by using a nondispersive waveguide for the interior with an all-pass terminating reflectance. Such an all-pass filter will introduce a frequency-dependent phase delay that can be chosen so as to match the dispersion of the medium itself, without dissipating energy. It should be kept in mind, however, that one means of synthesizing such an all-pass filter would be by adding an additional chain of junctions on the opposing side of the boundary junction, and terminating it with an open or short-circuit. Synthesis (for a given dispersion profile) presumably proceeds along the lines of methods used in related filter-design areas [287].

4.4 The (2+1)D Parallel-plate System

We return now to the parallel-plate system that was discussed in Section 3.7. As we mentioned before, this is a useful model problem in that it can be simplified to model the transverse motion of a membrane, and (2+1)D linear acoustics. It can be modified in a trivial way to model TE or TM fields in a substance of spatially varying dielectric constant, magnetic permeability, and loss characteristics.

4.4.1 Defining Equations and Centered Differences

The set of PDEs describing a lossless, source-free parallel-plate transmission line in (2+1)D is a direct generalization of system (4.17):

$$l\frac{\partial i_x}{\partial t} + \frac{\partial u}{\partial x} = 0 \qquad (4.50a)$$

$$l\frac{\partial i_y}{\partial t} + \frac{\partial u}{\partial y} = 0 \qquad (4.50b)$$

$$c\frac{\partial u}{\partial t} + \frac{\partial i_x}{\partial x} + \frac{\partial i_y}{\partial y} = 0 \qquad (4.50c)$$

4.4. THE (2+1)D PARALLEL-PLATE SYSTEM

Now $i_x(x, y, t)$ and $i_y(x, y, t)$ are the components of the current density vector in the x and y directions, respectively, and $u(x, y, t)$ is the voltage between the plates. $l(x, y)$ and $c(x, y)$, both assumed positive everywhere, are the inductance and capacitance per unit length.

If we assume that l and c are constant, then as in the (1+1)D case, the set of equations can be reduced to a single second-order equation in the voltage alone as

$$\frac{\partial^2 u}{\partial t^2} = \gamma^2 \left(\frac{\partial^2 u}{\partial x^2} + \frac{\partial^2 u}{\partial y^2} \right) \tag{4.51}$$

and again, the *wave speed* γ is given by

$$\gamma = \frac{1}{\sqrt{lc}}$$

The centered difference scheme for system (4.50) also generalizes simply. Define grid functions $I_{x,i,j}(n)$, $I_{y,i,j}(n)$, and $U_{i,j}(n)$, which run over half-integer values of i, j, and n, that is,

$$i, j, n = \ldots -1, -\tfrac{1}{2}, 0, \tfrac{1}{2}, 1 \ldots$$

We will furthermore assume that the spatial step in the x direction and the y direction are the same and equal to $\Delta/2$. As before, the time step will be $T/2$. We can use the approximations (4.19), as well as an approximation to the derivative in the y direction,

$$\left. \frac{\partial w}{\partial y} \right|_{i\Delta, j\Delta, nT} = \frac{w(i, j+\tfrac{1}{2}, n) - w(i, j-\tfrac{1}{2}, n)}{\Delta} + O(\Delta^2)$$

where w stands for either of i_y or u. We obtain the difference scheme

$$I_{x,i,j}(n+\tfrac{1}{2}) - I_{x,i,j}(n-\tfrac{1}{2}) + \frac{1}{v_0 \bar{l}_{i,j}} \left(U_{i+\tfrac{1}{2},j}(n) - U_{i-\tfrac{1}{2},j}(n) \right) = 0 \tag{4.52a}$$

$$I_{y,i,j}(n+\tfrac{1}{2}) - I_{y,i,j}(n-\tfrac{1}{2}) + \frac{1}{v_0 \bar{l}_{i,j}} \left(U_{i,j+\tfrac{1}{2}}(n) - U_{i,j-\tfrac{1}{2}}(n) \right) = 0 \tag{4.52b}$$

$$U_{i,j}(n+\tfrac{1}{2}) - U_{i,j}(n-\tfrac{1}{2}) + \frac{1}{v_0 \bar{c}_{i,j}} \left(I_{x,i+\tfrac{1}{2},j}(n) - I_{x,i-\tfrac{1}{2},j}(n) \right)$$

$$+ \frac{1}{v_0 \bar{c}_{i,j}} \left(I_{y,i,j+\tfrac{1}{2}}(n) - I_{y,i,j-\tfrac{1}{2}}(n) \right) = 0 \tag{4.52c}$$

where we have written

$$\bar{l}_{i,j} \triangleq l(i\Delta, j\Delta) + O(\Delta^2)$$

$$\bar{c}_{i,j} \triangleq l(i\Delta, j\Delta) + O(\Delta^2)$$

and

$$v_0 \triangleq \frac{\Delta}{T}$$

As in the (1+1)D case, it is possible to subdivide the calculation scheme (4.52) into smaller, mutually exclusive subschemes. Using a decimated grid for the variable coefficient difference scheme amounts to rewriting scheme (4.52) as

$$I_{x,i+\frac{1}{2},j}(n+\tfrac{1}{2}) - I_{x,i+\frac{1}{2},j}(n-\tfrac{1}{2}) + \frac{1}{\overline{vol}_{i+\frac{1}{2},j}}\left(U_{i+1,j}(n) - U_{i,j}(n)\right) = 0 \qquad (4.53a)$$

$$I_{y,i,j+\frac{1}{2}}(n+\tfrac{1}{2}) - I_{y,i,j+\frac{1}{2}}(n-\tfrac{1}{2}) + \frac{1}{\overline{vol}_{i,j+\frac{1}{2}}}\left(U_{i,j+1}(n) - U_{i,j}(n)\right) = 0 \qquad (4.53b)$$

$$U_{i,j}(n) - U_{i,j}(n-1) + \frac{1}{v_0\overline{c}_{i,j}}\left(I_{x,i+\frac{1}{2},j}(n-\tfrac{1}{2}) - I_{x,i-\frac{1}{2},j}(n-\tfrac{1}{2})\right)$$
$$+ \frac{1}{v_0\overline{c}_{i,j}}\left(I_{y,i,j+\frac{1}{2}}(n-\tfrac{1}{2}) - I_{y,i,j-\frac{1}{2}}(n-\tfrac{1}{2})\right) = 0 \qquad (4.53c)$$

where we now compute solutions for i, j, and n integer. The interleaved grid is shown in Figure 4.16; a gray (white) dot at a grid location indicates that the adjacent named variable is to be calculated at times that are even (odd) multiples of $T/2$. This interleaved form was originally put forth by Yee [286] in the context of electromagnetic field problems, and forms the basis of the widely used *finite difference time domain* (FDTD) family of difference methods [246], which were discussed briefly in Section 4.1. If system (4.53) is rewritten as a TE or TM system, the interleaved arrangement of the field components also has an interesting physical interpretation as a discrete counterpart to the integral form of Ampere's and Faraday's Laws [246]. This result also extends easily to the discretization of *Maxwell's equations* in (3+1)D [286] (see Section 6.4 for more details).

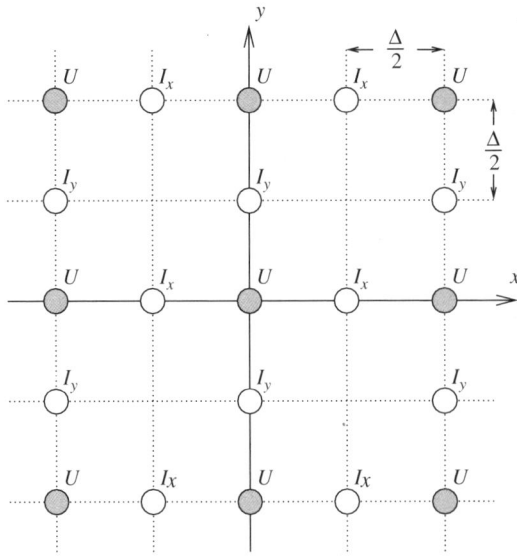

Figure 4.16: *Interleaved computational grid for the (2+1)D parallel-plate system.*

4.4. THE (2 + 1)D PARALLEL-PLATE SYSTEM

The (2+1)D analogue of (4.23), which holds in the case in which l and c are constant, is

$$\frac{v_0^2}{\gamma^2}\left(U_{i,j}(n+1) + U_{i,j}(n-1)\right) = U_{i+1,j}(n) + U_{i-1,j}(n) + U_{i,j+1}(n) + U_{i,j-1}(n)$$
$$+ 2\left(\frac{v_0^2}{\gamma^2} - 2\right)U_{i,j}(n) \quad (4.54)$$

The *magic* time step will now be

$$v_0 = \sqrt{2}\gamma$$

and (4.54) simplifies to

$$U_{i,j}(n+1) + U_{i,j}(n-1) = \frac{1}{2}\left(U_{i+1,j}(n) + U_{i-1,j}(n) + U_{i,j+1}(n) + U_{i,j-1}(n)\right) \quad (4.55)$$

As in (1+1)D, when we are solving the wave equation by centered differences at the magic time step (or at CFL), the calculation further decomposes into two independent calculations; we need only update $U_{i,j}(n)$ for $i + j + n$ even (or odd). We will examine this interesting decomposition property in detail in Appendix A.

4.4.2 The Waveguide Mesh

Consider the original form (2+1)D waveguide network, or *mesh* [267], operating on a rectilinear grid. (This was presented earlier in Section 1.1.3, and is also the original form of TLM [135].) Each scattering junction (parallel) is connected to four neighbors by unit-sample bidirectional delay lines. The spacing of the junctions is Δ (in either the x or y direction) and the time delay is T in the delay lines (see Figure 4.17). We now index a junction (and all its associated voltages and currents and wave quantities) at coordinates $(i\Delta, j\Delta)$ by the pair (i, j).

As in the (1+1)D case, at each parallel junction at location (i, j), we have voltages at every port, given by

$$U_{x+,i,j} = \text{voltage in waveguide leading east}$$
$$U_{x-,i,j} = \text{voltage in waveguide leading west}$$
$$U_{y+,i,j} = \text{voltage in waveguide leading north}$$
$$U_{y-,i,j} = \text{voltage in waveguide leading south}$$

and current flows

$$I_{x+,i,j} = \text{current flow in waveguide leading east}$$
$$I_{x-,i,j} = \text{current flow in waveguide leading west}$$
$$I_{y+,i,j} = \text{current flow in waveguide leading north}$$
$$I_{y-,i,j} = \text{current flow in waveguide leading south}$$

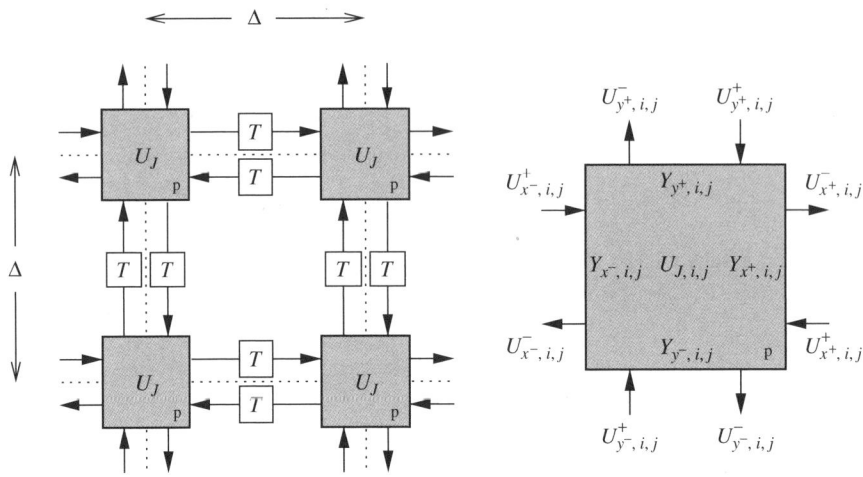

Figure 4.17: (2+1)D *waveguide mesh and a representative scattering junction.*

as well as wave quantities

$$U_{q,i,j} = U^+_{q,i,j} + U^-_{q,i,j}$$

$$I_{q,i,j} = I^+_{q,i,j} + I^-_{q,i,j}$$

where q is any of x^+, x^-, y^+ or y^-. The voltage and current waves are related by

$$I^+_{q,i,j} = Y_{q,i,j} U^+_{q,i,j} \qquad I^-_{q,i,j} = -Y_{q,i,j} U^-_{q,i,j} \qquad (4.56)$$

where $Y_{q,i,j}$ is the admittance of the waveguide connected to the junction with coordinates $(i\Delta, j\Delta)$ in direction q. The junction admittance is then

$$Y_{J,i,j} \triangleq Y_{x^-,i,j} + Y_{x^+,i,j} + Y_{y^-,i,j} + Y_{y^+,i,j} \qquad (4.57)$$

and the scattering equation, for voltage waves, will be, from (4.15),

$$U^-_{r,i,j} = -U^+_{r,i,j} + \frac{2}{Y_{J,i,j}}$$

$$\times \left(Y_{x^-,i,j} U^+_{x^-,i,j} + Y_{x^+,i,j} U^+_{x^+,i,j} + Y_{y^+,i,j} U^+_{y^+,i,j} + Y_{y^-,i,j} U^+_{y^-,i,j} \right) \qquad (4.58)$$

where r is any of x^+, x^-, y^+, or y^-. Voltage waves are propagated by

$$U^+_{x^+,i,j}(n) = U^-_{x^-,i+1,j}(n-1) \qquad U^+_{x^-,i,j}(n) = U^-_{x^+,i-1,j}(n-1)$$

$$U^+_{y^+,i,j}(n) = U^-_{y^-,i,j+1}(n-1) \qquad U^+_{y^-,i,j}(n) = U^-_{y^+,i,j-1}(n-1)$$

The case of flow waves is similar except for a sign inversion.

4.4. THE (2+1)D PARALLEL-PLATE SYSTEM

Similar to the (1+1)D case, it is possible to obtain a finite difference scheme purely in terms of the junction voltages $U_{J,i,j}$, under the assumption that the admittances of all the waveguides in the network are identical, and equal to some positive constant Y. Thus, from (4.57), $Y_{J,i,j} = 4Y$. We have, for the junction at location $x = i\Delta$, $y = j\Delta$,

$$\begin{aligned}
U_{J,i,j}(n+1) &= \frac{2}{Y_{J,i,j}} \sum_q Y_{q,i,j} U^+_{q,i,j}(n+1), \qquad q = \{x^-, x^+, y^-, y^+\} \\
&= \frac{1}{2}\left(U^-_{x^-,i+1,j}(n) + U^-_{x^+,i-1,j}(n) + U^-_{y^-,i,j+1}(n) + U^-_{y^+,i,j-1}(n)\right) \\
&= \frac{1}{2}\left(U_{J,i+1,j}(n) + U_{J,i-1,j}(n) + U_{J,i,j+1}(n) + U_{J,i,j-1}(n)\right) \\
&\quad - \frac{1}{2}\left(U^+_{x^-,i+1,j}(n) + U^+_{x^+,i-1,j}(n) + U^+_{y^-,i,j+1}(n) + U^+_{y^+,i,j-1}(n)\right) \\
&= \frac{1}{2}\left(U_{J,i+1,j}(n) + U_{J,i-1,j}(n) + U_{J,i,j+1}(n) + U_{J,i,j-1}(n)\right) \\
&\quad - \frac{2}{Y_{J,i,j}} \sum_q Y_{q,i,j} U^-_{q,i,j}(n-1) \\
&= \frac{1}{2}\left(U_{J,i+1,j}(n) + U_{J,i-1,j}(n) + U_{J,i,j+1}(n) + U_{J,i,j-1}(n)\right) \\
&\quad - U_{J,i,j}(n-1)
\end{aligned}$$

This is identical to (4.55) if we replace U_J by U. If we now replace all the bidirectional delay lines in Figure 4.17 by the same split pair of lines shown in Figure 4.9, then we get the arrangement in Figure 4.18.

We have placed the split lines such that the branches containing sign inversions are adjacent to the western and southern ports of the parallel junctions. We also introduce new junction variables I_{xJ} at the series junctions between two horizontal half-sample waveguides, and I_{yJ} at the series junctions between two vertical delay line pairs, as well as all the associated wave quantities at the ports of the new series junctions. It is straightforward to show that upon identifying I_{xJ}, I_{yJ}, and U_J with I_x, I_y, and U_J, the mesh will be calculated according to scheme (4.53) with constant coefficients, if we choose

$$v_0 = \sqrt{\frac{2}{lc}} \qquad Z = \sqrt{\frac{2l}{c}}$$

where Z is the impedance in all the delay lines. We are again at the magic time step, but the impedance has been set to be *larger* than the characteristic impedance of the medium. Also, notice that the speed of propagation along the delay lines is not the wave speed of the medium, which is $\gamma = 1/\sqrt{lc}$. Such a mesh is called a *slow-wave* structure [122] in the TLM literature.

Losses, Sources, and Spatially Varying Coefficients

We can deal with spatially varying material parameters as well as losses and sources in a manner similar to the (1+1)D case. The full parallel-plate system, as originally presented

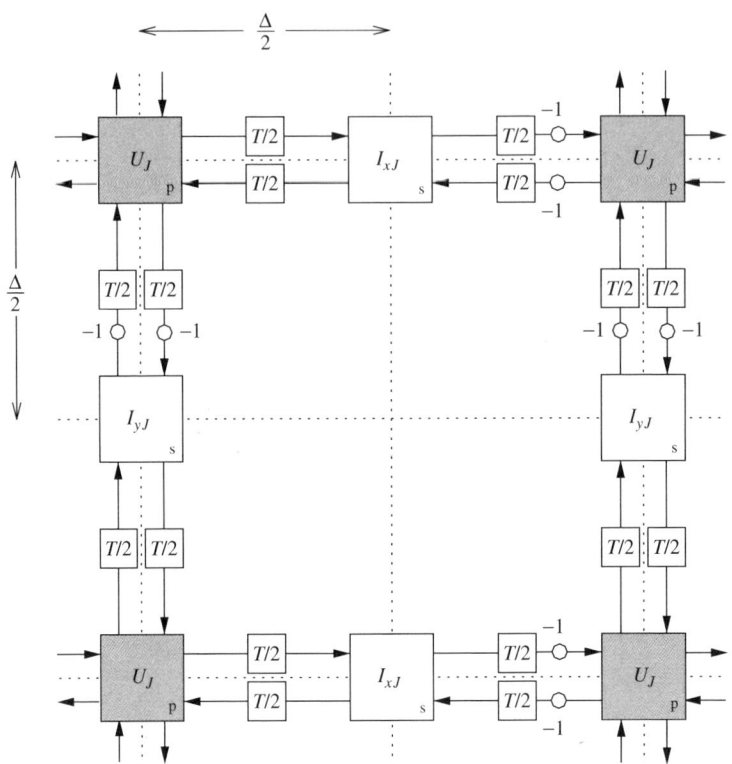

Figure 4.18: (2+1)D *interleaved waveguide mesh*.

in Section 3.7, is

$$l\frac{\partial i_x}{\partial t} + \frac{\partial u}{\partial x} + ri_x + e = 0 \tag{4.59a}$$

$$l\frac{\partial i_y}{\partial t} + \frac{\partial u}{\partial y} + ri_y + f = 0 \tag{4.59b}$$

$$c\frac{\partial u}{\partial t} + \frac{\partial i_x}{\partial x} + \frac{\partial i_y}{\partial y} + gu + h = 0 \tag{4.59c}$$

where we have $r(x, y) \geq 0$ and $g(x, y) \geq 0$, and e, f, and h are driving functions of x, y, and t.

The centered difference approximation to (4.59) is

$$I_{x,i+\frac{1}{2},j}(n+\tfrac{1}{2}) - \rho_{I,i+\frac{1}{2},j} I_{x,i+\frac{1}{2},j}(n-\tfrac{1}{2}) + \sigma_{I,i+\frac{1}{2},j}\left(U_{i+1,j}(n) - U_{i,j}(n)\right)$$
$$+ \Delta\sigma_{I,i+\frac{1}{2},j}\overline{e}_{i+\frac{1}{2},j}(n) = 0 \tag{4.60a}$$

$$I_{y,i,j+\frac{1}{2}}(n+\tfrac{1}{2}) - \rho_{I,i,j+\frac{1}{2}} I_{y,i,j+\frac{1}{2}}(n-\tfrac{1}{2}) + \sigma_{I,i,j+\frac{1}{2}}\left(U_{i,j+1}(n) - U_{i,j}(n)\right)$$
$$+ \Delta\sigma_{I,i,j+\frac{1}{2}}\overline{f}_{i,j+\frac{1}{2}}(n) = 0 \tag{4.60b}$$

4.4. THE (2 + 1)D PARALLEL-PLATE SYSTEM

$$U_{i,j}(n) - \rho_{U,i,j} U_{i,j}(n-1) + \sigma_{U,i,j} \left(I_{x,i+\frac{1}{2},j}(n-\tfrac{1}{2}) - I_{x,i-\frac{1}{2},j}(n-\tfrac{1}{2}) \right)$$
$$+ \sigma_{U,i,j} \left(I_{y,i,j+\frac{1}{2}}(n-\tfrac{1}{2}) - I_{y,i,j-\frac{1}{2}}(n-\tfrac{1}{2}) \right)$$
$$+ \Delta \sigma_{U,i,j} \bar{h}_{i,j}(n-\tfrac{1}{2}) = 0 \qquad (4.60c)$$

where

$$\rho_{I,k,p} = \frac{2\bar{l}_{k,p} - r_{k,p} T}{2\bar{l}_{k,p} + r_{k,p} T} \qquad \sigma_{I,k,p} = \frac{2}{2v_0 \bar{l}_{k,p} + r_{k,p} \Delta} \qquad (4.61)$$

$$r_{k,p} = r(k\Delta, p\Delta) \qquad \bar{l}_{k,p} = l(k\Delta, p\Delta) + O(\Delta^2) \qquad (4.62)$$

for k, p half-integer such that $2(k+p)$ is odd, and

$$\rho_{U,i,j} = \frac{2\bar{c}_{i,j} - g_{i,j} T}{2\bar{c}_{i,j} + g_{i,j} T} \qquad \sigma_{U,i,j} = \frac{2}{2v_0 \bar{c}_{i,j} + g_{i,j} \Delta}$$

$$g_{i,j} = g(i\Delta, j\Delta) \qquad \bar{c}_{i,j} = c(i\Delta, j\Delta) + O(\Delta^2)$$

for i, j integer. For the sources, we have used

$$\bar{e}_{i+\frac{1}{2},j}(n) = \frac{1}{2} \left(e_{i+\frac{1}{2},j}(n+\tfrac{1}{2}) + e_{i+\frac{1}{2},j}(n-\tfrac{1}{2}) \right)$$

$$\bar{f}_{i,j+\frac{1}{2}}(n) = \frac{1}{2} \left(f_{i,j+\frac{1}{2}}(n+\tfrac{1}{2}) + f_{i,j+\frac{1}{2}}(n-\tfrac{1}{2}) \right)$$

$$\bar{h}_{i,j}(n-\tfrac{1}{2}) = \frac{1}{2} \left(h_{i,j}(n) + h_{i,j}(n-1) \right)$$

where

$$e_{i+\frac{1}{2},j}(n+\tfrac{1}{2}) = e\left((i+\tfrac{1}{2})\Delta, j\Delta, (n+\tfrac{1}{2})T \right)$$

$$f_{i,j+\frac{1}{2}}(n+\tfrac{1}{2}) = f\left(i\Delta, (j+\tfrac{1}{2})\Delta, (n+\tfrac{1}{2})T \right)$$

$$h_i(n) = h(i\Delta, j\Delta, nT)$$

Again, we have applied a semi-implicit approximation to the constant-proportional terms of (4.59).

The waveguide network shown in Figure 4.19 is a direct generalization to (2+1)D of Figure 4.13. To the structure of Figure 4.18 we have added an extra port to each scattering junction, series or parallel, which is connected to a self-loop of impedance Z_c and doubled delay length, as well as a port with impedance Z_R to introduce losses and sources. All immittances are indexed by the coordinates of their associated junctions. As before, we set the admittance $Y = 1/Z$ for any impedance Z in the network. In Figure 4.19, the linking admittances of the bidirectional delay lines are indicated only at the parallel junction, since we must have

$$Y_{x^+,i,j} = \frac{1}{Z_{x^-,i+\frac{1}{2},j}} \qquad Y_{x^-,i,j} = \frac{1}{Z_{x^+,i-\frac{1}{2},j}}$$

$$Y_{y^+,i,j} = \frac{1}{Z_{y^-,i,j+\frac{1}{2}}} \qquad Y_{y^-,i,j} = \frac{1}{Z_{y^+,i,j-\frac{1}{2}}}$$

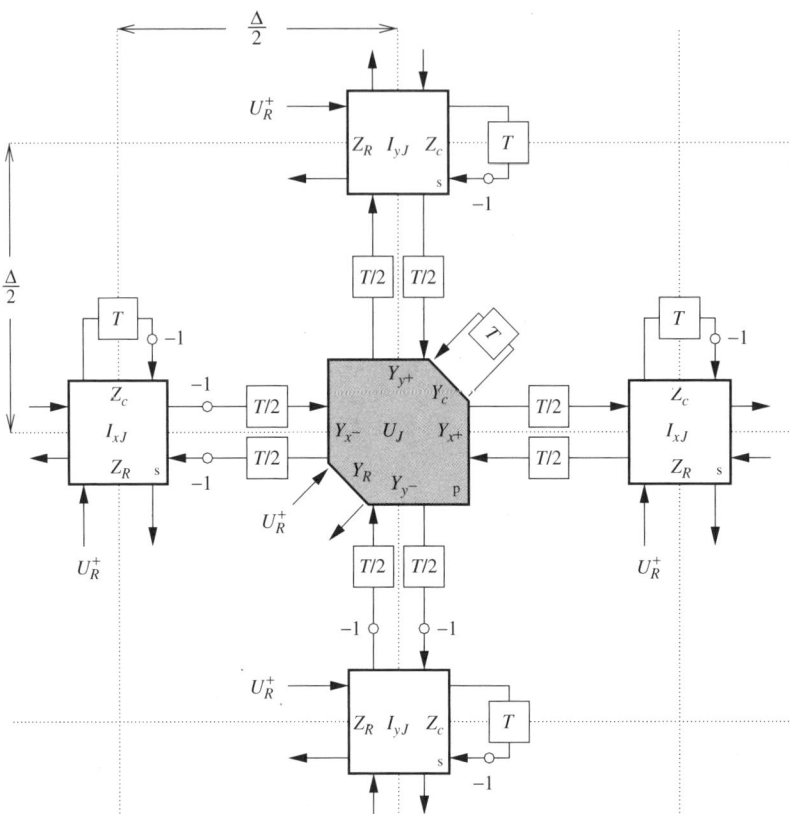

Figure 4.19: (2+1)D *waveguide mesh for the varying-coefficient system* (4.59), *with losses and sources.*

The junction admittances and impedances are thus

$$Y_{J,i,j} = Y_{x^-,i,j} + Y_{x^+,i,j} + Y_{y^-,i,j} + Y_{y^+,i,j} + Y_{c,i,j} + Y_{R,i,j}$$

$$Z_{J,i+\frac{1}{2},j} = Z_{x^-,i+\frac{1}{2},j} + Z_{x^+,i+\frac{1}{2},j} + Z_{c,i+\frac{1}{2},j} + Z_{R,i+\frac{1}{2},j}$$

$$Z_{J,i,j+\frac{1}{2}} = Z_{y^-,i,j+\frac{1}{2}} + Z_{y^+,i,j+\frac{1}{2}} + Z_{c,i,j+\frac{1}{2}} + Z_{R,i,j+\frac{1}{2}}$$

Beginning from series and parallel junctions, and proceeding through derivations similar to (4.33) yields the difference scheme (4.60) in the junction variables U_J, I_{xJ}, and I_{yJ}, provided we set

$$Y_{J,i,j} = 2v_0 \bar{c}_{i,j} + \Delta g_{i,j} \qquad Y_{R,i,j} = \Delta g_{i,j} \qquad U^+_{R,i,j} = -h_{i,j}/2g_{i,j}$$

$$Z_{J,i+\frac{1}{2},j} = 2v_0 \bar{l}_{i+\frac{1}{2},j} + \Delta r_{i+\frac{1}{2},j} \qquad Z_{R,i+\frac{1}{2},j} = \Delta r_{i+\frac{1}{2},j} \qquad I^+_{R,i+\frac{1}{2},j} = -e_{i+\frac{1}{2},j}/2r_{i+\frac{1}{2},j}$$

$$Z_{J,i,j+\frac{1}{2}} = 2v_0 \bar{l}_{i,j+\frac{1}{2}} + \Delta r_{i,j+\frac{1}{2}} \qquad Z_{R,i,j+\frac{1}{2}} = \Delta r_{i,j+\frac{1}{2}} \qquad I^+_{R,i,j+\frac{1}{2}} = -e_{i,j+\frac{1}{2}}/2r_{i+\frac{1}{2},j}$$

(4.63)

We can again identify three useful ways of setting the immittances:

4.4. THE (2+1)D PARALLEL-PLATE SYSTEM

Type I: Voltage-centered Mesh

At a parallel junction, we set the self-loop admittance $Y_{c,i,j}$ to zero, and the admittances of the branches leading away from a parallel junction at grid point (i, j) to be identical, thus

$$Y_{c,i,j} = 0 \qquad Y_{x-,i,j} = Y_{x+,i,j} = Y_{y-,i,j} = Y_{y+,i,j} = \frac{v_0}{2} c_{i,j}$$

and set

$$Z_{c,i+\frac{1}{2},j} = v_0(l_{i,j} + l_{i+1,j}) - \frac{2}{v_0}\left(\frac{1}{c_{i,j}} + \frac{1}{c_{i+1,j}}\right)$$

$$Z_{c,i,j+\frac{1}{2}} = v_0(l_{i,j} + l_{i,j+1}) - \frac{2}{v_0}\left(\frac{1}{c_{i,j}} + \frac{1}{c_{i,j+1}}\right)$$

The positivity requirements on $Z_{c,i+\frac{1}{2},j}$ and $Z_{c,i,j+\frac{1}{2}}$ force us to choose

$$v_0 \geq \max_{i,j} \left(\sqrt{\frac{2}{l_{i,j} c_{i,j}}}\right) \tag{4.64}$$

Thus we have a simple CFL-type bound on the space step/time step ratio, which must be chosen greater than $\sqrt{2}$ times the maximum value of the local group velocity $1/\sqrt{lc}$ over parallel junction locations $(i\Delta, j\Delta)$. The bound (4.64) converges to $\sqrt{2}\gamma^g_{PP,max}$, as defined by (3.66), in the limit as the grid spacing Δ becomes small.

Type II: Current-centered Mesh

This arrangement is the dual to the previous case. We now set

$$Z_{x+,i+\frac{1}{2},j} = Z_{x-,i+\frac{1}{2},j} = v_0 l_{i+\frac{1}{2},j} \tag{4.65}$$

$$Z_{y+,i,j+\frac{1}{2}} = Z_{y-,i,j+\frac{1}{2}} = v_0 l_{i,j+\frac{1}{2}} \tag{4.66}$$

$$Z_{c,i+\frac{1}{2},j} = Z_{c,i,j+\frac{1}{2}} = 0 \tag{4.67}$$

and

$$Y_{c,i,j} = -\frac{1}{v_0}\left(\frac{1}{l_{i+\frac{1}{2},j}} + \frac{1}{l_{i-\frac{1}{2},j}} + \frac{1}{l_{i,j+\frac{1}{2}}} + \frac{1}{l_{i,j-\frac{1}{2}}}\right)$$
$$+ \frac{v_0}{2}\left(c_{i+\frac{1}{2},j} + c_{i-\frac{1}{2},j} + c_{i,j+\frac{1}{2}} + c_{i,j-\frac{1}{2}}\right)$$

We then have

$$v_0 \geq \max_{2(k+p) \text{ odd}} \left(\sqrt{\frac{2}{l_{k,p} c_{k,p}}}\right) \tag{4.68}$$

for half-integer k and p. It is rather interesting that in (2+1)D, if we have $r = 0$ and $e = f = 0$, this arrangement (and not that of type I) allows the series junctions to be treated as throughs (with sign inversion). We may thus operate at a reduced sample rate in this case. This particular choice of immittances, in the constant-coefficient, lossless, and source-free case with $v_0 = \sqrt{2/(lc)}$, yields the original form of the waveguide mesh proposed in [267], and mentioned in Section 4.6. We also note that networks such as this and type I, for which the connecting immittances may vary spatially have also been explored in TLM [44, 215].

Type III: Mixed Mesh

We set

$$Z_{x^-,i+\frac{1}{2},j} = Z_{x^+,i+\frac{1}{2},j} = Z_{y^-,i,j+\frac{1}{2}} = Z_{y^+,i,j+\frac{1}{2}} = Z_{\text{const}}$$

where Z_{const} is some positive constant, and then choose

$$Y_{c,i,j} = 2v_0 c_{i,j} - \frac{4}{Z_{\text{const}}}$$

$$Z_{c,i+\frac{1}{2},j} = 2v_0 l_{i+\frac{1}{2},j} - 2Z_{\text{const}}$$

$$Z_{c,i,j+\frac{1}{2}} = 2v_0 l_{i,j+\frac{1}{2}} - 2Z_{\text{const}}$$

The optimal value of Z_{const} is easily shown to be

$$Z_{\text{const}} = \sqrt{\frac{2\min_{2(k+p)\ \text{odd}} l_{k,p}}{\min_{i,j} c_{i,j}}}$$

and this leads to the constraint

$$v_0 \geq \sqrt{\frac{2}{\min_{2(k+p)\ \text{odd}}(l_{k,p})\min_{i,j}(c_{i,j})}}$$

for i, j integer and k and p half-integer. As in (1+1)D, this bound is inferior to those obtained using the type I and type II meshes.

4.4.3 Reduced Computational Complexity and Memory Requirements in the Standard Form of the Waveguide Mesh

The standard form of the waveguide mesh [267] was proposed as a means of solving the (2+1)D wave equation (4.51). We would like to note that it is possible, in this special case, to reduce both computational complexity and memory requirements by taking advantage of the fact that the mesh calculation can be subdivided into two mutually exclusive schemes. We represent this grid decimation graphically in Figure 4.20 by coloring the two subgrids white and gray.

Voltages are calculated at all the junctions and at every time step. However, it should be clear from the figure that the calculation of U_J at a gray-colored junction will only depend on wave variables scattered from the white junctions at the previous time step. We then only need calculate half the junction voltages at any given time step: at the gray junctions at even multiples (say) of the time step, and at white junctions for odd multiples.

4.4. THE (2 + 1)D PARALLEL-PLATE SYSTEM

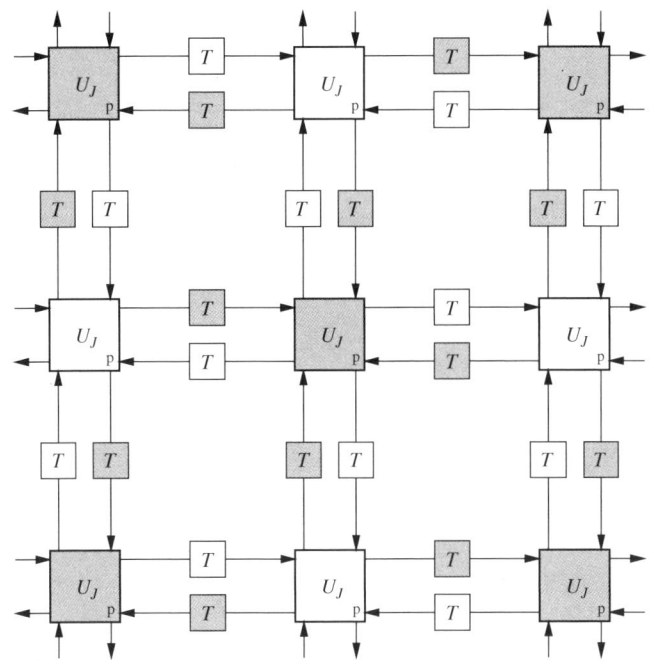

Figure 4.20: *Grid decomposition for the standard waveguide mesh.*

We can go further as well—if we have dropped half the junction calculations at any given time step, then we are in fact only using one of the two delay registers in any bidirectional delay line (gray or white in Figure 4.20). Thus only one delay register will be required for any waveguide. In addition, because in calculating the junction voltage at a gray junction (say) we make use only of incoming wave variables and not the junction voltages at the white junctions, calculated one time step previously, one set of registers may be used at alternating time steps to store both the gray and white voltages.

This amounts to a factor-of-two savings in terms of both memory requirements and the operation count. We are, of course, approximating the solution at only half the mesh points at any time step, but there is no reduction in accuracy, because the separate subcalculation that has been removed operated completely independently. (It is true, however, that we have sacrificed the higher end of the range of spatial frequencies which can be approximated without aliasing.) One additional advantage is that under this simplification, there will be no need for temporary registers (normally necessary during the scattering operation); at a given time step there is no danger of overwriting an incoming wave variable at a particular junction, because if we have just scattered there, we will not scatter there again for two time steps. The savings in terms of memory is of a factor of 14/5 over the standard mesh implementation. We note that in this implementation, each pass through the main loop in the computer program will contain two time steps worth of scattering operations. Once one has programmed a mesh a few times, the possibility of performing such a computational trick becomes obvious, and for this reason it is rather surprising that it has apparently not been explicitly mentioned in the TLM literature, even in papers specific to parallel computation [233]. We will look at grid decomposition issues in much more detail in Appendix A.

Such memory-sharing ideas also appeared early on in wave digital filter designs that use unit elements [62] and in WDF-based numerical integration schemes [176]. We note that this idea can be applied here only for the wave equation at the CFL bound; otherwise, if we are using, for example, a type II mesh, we necessarily have unit-delay self-loops at the junctions, so that the mesh calculation cannot be similarly decomposed.

4.4.4 Boundary Conditions

We now examine the termination of the waveguide mesh that simulates the behavior of the (2+1)D parallel-plate system. The two most important types of boundary conditions are

$$u = 0 \qquad \text{Short-circuit termination} \qquad (4.69)$$

$$i_n = 0 \qquad \text{Open-circuit termination} \qquad (4.70)$$

where i_n refers to the component of (i_x, i_y) that is normal to the boundary. Condition (4.69) corresponds to a transmission line plate pair that are connected (and thus short-circuited) at the boundary; the same condition holds for a clamped membrane for which u is interpreted as a transverse velocity, and (i_x, i_y) as in-plane forces. Condition (4.70) is an open-circuited termination; current cannot leave the plate at the edges. This second condition is analogous to the rigid termination of a (2+1)D acoustic medium, in which (i_x, i_y) are interpreted as flow velocities, and u as a pressure. Both conditions are of the form of (3.8), and are lossless. We will examine only the termination of the mesh on a rectangular domain (though the result extends easily to the radial mesh to be discussed in Section 5.1.2).

In the case of the (1+1)D transmission line, we could treat a staggered mesh terminated by a parallel junction, and through the duality of i and u extend the result to include termination by a series junction (see Section 4.3.9). This is no longer the case in (2+1)D, and we must treat the two types of termination separately. Consider a bottom (southern) boundary at $y = 0$ of an interleaved mesh of the type shown in Figure 4.19. The two possible types of termination are shown in Figure 4.21.

Grid Arrangement Requiring Voltage and Tangential Current Density Component on Boundary

Consider the termination arrangement of Figure 4.21(a). In the source-free case, if for $y = 0$ we have $u(x, 0, t) = 0$, then from (4.59a),

$$l\frac{\partial i_x}{\partial t} + r i_x = 0 \qquad \text{for} \qquad y = 0$$

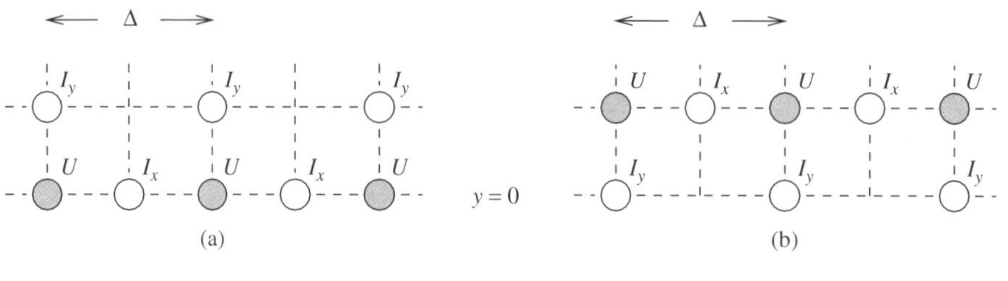

Figure 4.21: *Grid terminations at a southern boundary.*

4.4. THE (2+1)D PARALLEL-PLATE SYSTEM

Thus the current density component tangential to the boundary is uncoupled from the other dependent variables. It is convenient to assume, then, that $i_x(x, 0, t)$ is initially zero, so that it will remain so permanently. In this case, we can drop the series junctions corresponding to i_x on the southern boundary from the network. Otherwise, we may allow the junctions to remain, as lumped, damped (by a factor r/l) elements, still uncoupled from the rest of the network. In either case, the parallel junctions at the gray points in Figure 4.21(a) may be short circuited as in the (1+1)D case in order to realize boundary condition (4.69). The waveguide mesh termination corresponding to (4.69) is shown in Figure 4.22(a).

To deal with the boundary condition $i_y(x, 0, t) = 0$, we may proceed as in the (1+1)D case, and write down a difference approximation to (4.59c), where we use the one-sided difference approximation

$$\left.\frac{\partial i_y}{\partial y}\right|_{y=0} = \frac{2}{\Delta}\left(i_y(x, \frac{\Delta}{2}, t) - i_y(x, 0, t)\right) + O(\Delta)$$

$$= \frac{2}{\Delta} i_y(x, \frac{\Delta}{2}, t) + O(\Delta)$$

and centered differences in time and the x direction,

$$U_{i,0}(n) - \rho_{U,i,0} U_{i,0}(n-1) + \sigma_{U,i,0}\left(I_{x,i+\frac{1}{2},0}(n-\frac{1}{2}) - I_{x,i-\frac{1}{2},0}(n-\frac{1}{2})\right)$$
$$+ 2\sigma_{U,i,0} I_{y,i,\frac{1}{2}}(n-\frac{1}{2}) = 0$$

where $\sigma_{U,i,0}$ and $\rho_{U,i,0}$ are as given in (4.61).

Here, the voltages on the boundary are related to the tangential currents, which, from Figure 4.21(a), are also calculated on the boundary. This implies that the corresponding junctions will be connected to one another by waveguides that lie directly on the boundary. Also notice the doubled weighting of the I_y grid function at the boundary; this requires special care in the DWN implementation, though it also follows from a structurally passive termination, provided we make use of *transformers* along the boundary waveguides. Though we have not discussed transformers in the DWN context [228], they are identical to wave digital transformers, which were mentioned in Section 2.2.4. In effect, we may introduce

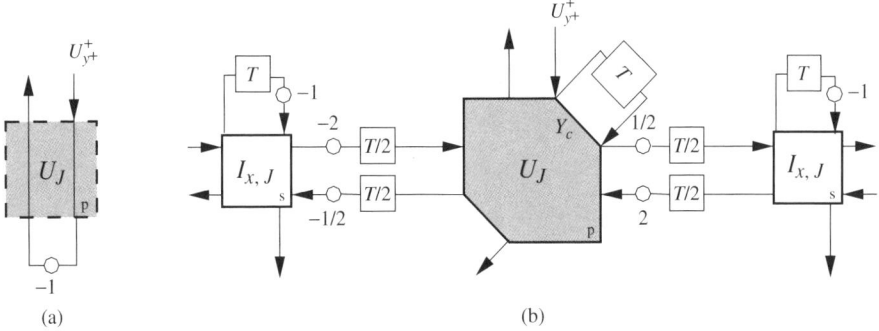

Figure 4.22: (2+1)D *waveguide mesh terminations at a southern boundary, for the grid arrangement of Figure 4.21(a)*—(a) *for* $u(x, 0, t) = 0$ *and* (b) $i(x, 0, t) = 0$.

multiplies by κ and $1/\kappa$, for any real κ in the two signal paths of any waveguide without affecting losslessness, provide we scale impedances at both ends of the waveguide accordingly. The DWN termination corresponding to $i_y = 0$ at a southern boundary is shown in Figure 4.22(b). We have used transformers of turns ratio 2, implying that the immittances on the boundary satisfy

$$Y_{x^-,i,0} = \frac{1}{4Z_{x^+,i-\frac{1}{2},0}} \qquad Y_{x^+,i,0} = \frac{1}{4Z_{x^-,i+\frac{1}{2},0}} \qquad (4.71)$$

The corresponding difference equation at a parallel boundary junction is then

$$U_{J,i,0}(n) - \frac{Y_{J,i,0} - 2Y_{R,i,0}}{Y_{J,i,0}} U_{J,i,0}(n-1) + \frac{1}{Y_{J,i,0}} \left(I_{xJ,i+\frac{1}{2},0}(n-\tfrac{1}{2}) - I_{xJ,i-\frac{1}{2},0}(n-\tfrac{1}{2}) \right)$$

$$+ \frac{2}{Y_{J,i,0}} I_{y,i,\frac{1}{2}}(n-\tfrac{1}{2}) = 0$$

where

$$Y_{J,i,0} \triangleq Y_{x^-,i,0} + Y_{x^+,i,0} + Y_{y^+,i,0} + Y_{c,i,0} + Y_{R,i,0} \qquad (4.72)$$

The junction updating will be equivalent to the centered difference scheme if we choose

$$Y_{J,i,0} = v_0 \bar{c}_{i,0} + \frac{g_{i,0}\Delta}{2} \qquad Y_{R,i,0} = \frac{g_{i,0}\Delta}{2} \qquad (4.73)$$

Given that the northward immittances at the boundary junctions must be set as interior values, the settings for the remaining immittances for the type I and II meshes discussed in Section 4.4.2 will be

Type I:

$$Y_{x^-,i,0} = Y_{x^+,i,0} = \frac{v_0 c_{i,0}}{4} \qquad Z_{x^+,i+\frac{1}{2},0} = \frac{1}{v_0 c_{i+1,0}} \qquad Z_{x^-,i+\frac{1}{2},0} = \frac{1}{v_0 c_{i-1,0}}$$

$$Y_{c,i,0} = 0 \qquad Z_{c,i+\frac{1}{2},0} = v_0 l_{i+1,0} + v_0 l_{i,0} - \frac{1}{v_0 c_{i+1,0}} - \frac{1}{v_0 c_{i,0}}$$

Type II:

$$Z_{x^-,i+\frac{1}{2},0} = Z_{x^-,i+\frac{1}{2},0} = v_0 l_{i+\frac{1}{2},0} \qquad Y_{x^+,i,0} = \frac{1}{4v_0 l_{i+\frac{1}{2},0}}$$

$$Y_{x^-,i,0} = \frac{1}{4v_0 l_{i-\frac{1}{2},0}} \qquad Z_{c,i+\frac{1}{2},0} = 0$$

$$Y_{c,i,0} = \frac{v_0 c_{i,\frac{1}{2}}}{2} + \frac{v_0 c_{i+\frac{1}{2},0} + v_0 c_{i-\frac{1}{2},0}}{4} - \frac{1}{v_0 l_{i,\frac{1}{2}}} - \frac{1}{4v_0 l_{i+\frac{1}{2},0}} - \frac{1}{4v_0 l_{i-\frac{1}{2},0}}$$

The passivity conditions that follow from the positivity of the boundary self-loop immittances will be

4.4. THE (2+1)D PARALLEL-PLATE SYSTEM

$$v_0 \geq \max_i \sqrt{\frac{1}{l_{i,0}c_{i,0}}} \qquad \text{Type I}$$

$$v_0 \geq \max\left(\max_i \sqrt{\frac{2}{l_{i,\frac{1}{2}}c_{i,\frac{1}{2}}}}, \max_{i+\frac{1}{2}} \sqrt{\frac{1}{l_{i+\frac{1}{2},0}c_{i+\frac{1}{2},0}}}\right) \qquad \text{Type II}$$

and clearly are *less restrictive* than the conditions (4.64) and (4.68) over the mesh interior, and hence do not affect the overall stability bound on v_0.

Grid Arrangement with Normal Current Density Component Required on Boundary

Termination of the other type of mesh, as shown in Figure 4.21(b), is comparatively simple because the series junctions are isolated from one another along the boundary itself. Terminations for both types of boundary conditions are shown in Figure 4.23.

The boundary condition $i_y = 0$ can be simply implemented by terminating the series boundary junctions in open circuits. The condition $u = 0$ can be ensured through proper adjustment of the self-loop impedance, depending on the type of waveguide mesh. The settings will be

$$Z_{c,i+\frac{1}{2},0} = v_0 l_{i+\frac{1}{2},\frac{1}{2}} - \frac{2}{v_0 c_{i+\frac{1}{2},\frac{1}{2}}} \qquad \text{Type I}$$

$$Z_{c,i+\frac{1}{2},0} = 0 \qquad \text{Type II}$$

The positivity condition on the boundary self-loop impedances for the type I mesh again does not degrade the stability bound over the mesh interior. As for interior series junctions, we will have $Z_R = r\Delta/2$ at the boundary junctions.

This type of mesh possesses an additional advantage—if we are working on a rectangular domain, then the 'holes' in the staggered grid (that is, those points at which neither U nor I is calculated, as per Figure 4.16) may be placed at the corners of the domain. The extra programming task of specializing the waveguide mesh at the corners can then be safely ignored.

All results can of course be extended, by symmetry, to any edge of a rectangular domain.

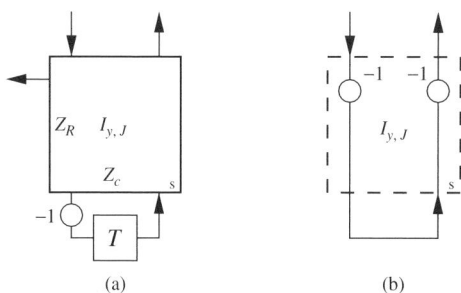

Figure 4.23: *(2 + 1)D waveguide mesh terminations at a southern boundary, for the grid of Figure 4.21(b), for (a) $u(x, 0, t) = 0$ and (b) $i_y(x, 0, t) = 0$.*

4.5 Initial Conditions

We have dealt so far with a method of integrating the transmission line and parallel-plate systems, but have not examined the necessary initialization of the algorithm. We will deal, here, with the lossless source-free cases.

In (1+1)D, the hyperbolic system (4.17) requires two initial conditions. That is, we require the knowledge of initial current and voltage distributions along the line. We would like to enter the discrete equivalent of this data into the delay registers somehow at the first time step, $n = 0$. From Figure 4.12, it should be clear that four sets of data are required: $U^+_{x^-,i}(0)$, $U^+_{x^+,i}(0)$, $U^+_{c,i}(0)$, which are the initial incoming waves at the parallel junctions, and $I^+_{c,i+1/2}(1/2)$, the values initially stored in the self-loops at the series junctions.

The first problem we encounter is that on our decimated grid we calculate values of U_J and I_J, the grid functions corresponding to voltage and current, at alternating time steps. We have chosen our grid such that for k and m half-integer, $I_k(m)$ is calculated for odd values of $2m$, and $U_k(m)$ for even values of $2m$; at time zero, then, U_J is accessible (as a combination of wave variables). How then do we enter the current initial data into the algorithm? It turns out that this problem is rather simply addressed. We can do one of two things: set the value of I_J at time step $\frac{1}{2}$ to be equal to a sampled version of $i(x, 0)$, and accept the error that this introduces, which will be $O(T)$, or we can use any available numerical method (i.e., one that does not operate on a staggered grid) to propagate the initial data $i_0(x)$ forward by $T/2$. Such a method should be at least $O(T^2)$ accurate, but it is allowed to even be *unstable*, since we will only be using it to update once [238].

Assume then that our initial data are $U_i^*(0) = u(i\Delta, 0)$, and $I^*_{i+1/2}(1/2)$, some approximation to $i((i + 1/2)\Delta, T/2)$ obtained by either of the methods mentioned above. At time step $n = 0$, we can write the junction voltages $U_{J,i}(0)$ as

$$U_{J,i}(0) = \frac{2}{Y_{J,i}} \left(Y_{x^+,i} U^+_{x^+,i}(0) + Y_{x^-,i} U^+_{x^-,i}(0) + Y_{c,i} U^+_{c,i}(0) \right) \quad (4.74)$$

and $I_{J,i+\frac{1}{2}}(\frac{1}{2})$ as

$$I_{J,i+\frac{1}{2}}(\tfrac{1}{2}) = \frac{2}{Z_{J,i+\frac{1}{2}}} \left(Z_{x^+,i+\frac{1}{2}} I^+_{x^+,i+\frac{1}{2}}(\tfrac{1}{2}) + Z_{x^-,i+\frac{1}{2}} I^+_{x^-,i+\frac{1}{2}}(\tfrac{1}{2}) + Z_{c,i+\frac{1}{2}} I^+_{c,i+\frac{1}{2}}(\tfrac{1}{2}) \right)$$

$$= \frac{2}{Z_{J,i+\frac{1}{2}}} \left(U^+_{x^+,i+\frac{1}{2}}(\tfrac{1}{2}) + U^+_{x^-,i+\frac{1}{2}}(\tfrac{1}{2}) + Z_{c,i+\frac{1}{2}} I^+_{c,i+\frac{1}{2}}(\tfrac{1}{2}) \right)$$

$$= \frac{2}{Z_{J,i+\frac{1}{2}}} \left(-U^-_{x^-,i+1}(0) + U^-_{x^+,i}(0) + Z_{c,i+\frac{1}{2}} I^+_{c,i+\frac{1}{2}}(\tfrac{1}{2}) \right)$$

$$= \frac{2}{Z_{J,i+\frac{1}{2}}} \left(-U_{J,i+1}(0) + U_{J,i}(0) + U^+_{x^-,i+1}(0) - U^+_{x^+,i}(0) + Z_{c,i+\frac{1}{2}} I^+_{c,i+\frac{1}{2}}(\tfrac{1}{2}) \right)$$

$$(4.75)$$

The safest and most general way of proceeding, given that the immittances $Z_{c,i+\frac{1}{2}}$ and $Y_{c,i}$ may be zero, depending on the type of network we are using, is to set the initial values in

4.5. INITIAL CONDITIONS

the self-loops to zero. In this case, we can set

$$U^+_{c,i}(0) = I^+_{c,i+\frac{1}{2}}(\tfrac{1}{2}) = 0 \tag{4.76a}$$

$$U^+_{x-,i}(0) = \frac{1}{4}\left(\frac{Y_{J,i}U^*_i(0)}{Y_{x-,i}} + Z_{J,i-\frac{1}{2}}I^*_{i-\frac{1}{2}}(\tfrac{1}{2})\right) \tag{4.76b}$$

$$U^+_{x+,i}(0) = \frac{1}{4}\left(\frac{Y_{J,i}U^*_i(0)}{Y_{x+,i}} - Z_{J,i+\frac{1}{2}}I^*_{i+\frac{1}{2}}(\tfrac{1}{2})\right) \tag{4.76c}$$

It can be easily verified that with these initial values for the wave variables, the junction voltages $U_{J,i}(0)$ and currents $I_{J,i+\frac{1}{2}}(\tfrac{1}{2})$, calculated from the DWN by (4.74) and (4.75) respectively, will be consistent with the initial values of the continuous problem to first order in Δ. These settings may be used with any of the three types of network mentioned in Section 4.3.6.

We may ask, however, whether there is a way of setting the initial values such that we achieve better initial accuracy. For a network of type I, say, we have $Y_{c,i} = 0$. Then, if $Z_{c,i+\frac{1}{2}}$ is nonzero everywhere (this can always be arranged by operating slightly away from the CFL bound), we may use

$$U^+_{x-,i}(0) = U^+_{x+,i}(0) = \frac{1}{2}U^*_i(0)$$

$$I^+_{c,i+\frac{1}{2}}(\tfrac{1}{2}) = \frac{1}{2Z_{c,i+\frac{1}{2}}}\left(Z_{J,i+\frac{1}{2}}I^*_{i+\frac{1}{2}}(\tfrac{1}{2}) - U^*_i(0) + U^*_{i+1}(0)\right)$$

in which case the waveguide network reproduces the initial currents and voltages with no error. Similarly, for a type II network, we may set

$$U^+_{x+,i}(0) = U^*_i(0) - \frac{1}{2Y_{x+,i}}I^*_{i+\frac{1}{2}}(\tfrac{1}{2})$$

$$U^+_{x-,i}(0) = U^*_i(0) + \frac{1}{2Y_{x-,i}}I^*_{i-\frac{1}{2}}(\tfrac{1}{2})$$

$$U^+_{c,i}(0) = \frac{1}{2Y_{c,i}}\left((Y_{c,i} - Y_{x-,i} - Y_{x+,i})U^*_i(0) + I^*_{i+\frac{1}{2}}(\tfrac{1}{2}) - I^*_{i-\frac{1}{2}}(\tfrac{1}{2})\right)$$

which also yields an exact calculation. Either of these two means of initializing wave variables may also be used in the type III DWN.

These initialization procedures generalize simply to (2+1)D, and the parallel-plate system requires three initial conditions: $u(x,y,0)$, $i_x(x,y,0)$, and $i_y(x,y,0)$. In general, we now have seven wave variables to set: the waves approaching any parallel junction with coordinates (i,j) at $n=0$, namely $U^+_{x-,i,j}(0)$, $U^+_{x+,i,j}(0)$, $U^+_{y-,i,j}(0)$, $U^+_{y+,i,j}(0)$, and $U^+_{c,i,j}(0)$, as well as the values stored in the self-loop registers at the series junctions, $I^+_{x,c,i+1/2,j}(1/2)$ and $I^+_{y,c,i,j+1/2}(1/2)$. For the sake of brevity, we provide only the settings for the general case, analogous to (4.76),

$$U^+_{x-,i,j}(0) = \frac{1}{4}\left(\frac{Y_{J,i,j}U^*_{i,j}(0)}{2Y_{x-,i,j}} + Z_{J,i-\frac{1}{2},j}I^*_{x,i-\frac{1}{2},j}(\tfrac{1}{2})\right)$$

$$U^+_{x^+,i,j}(0) = \frac{1}{4}\left(\frac{Y_{J,i,j}U^*_{i,j}(0)}{2Y_{x^+,i,j}} - Z_{J,i+\frac{1}{2},j}I^*_{x,i+\frac{1}{2},j}(\tfrac{1}{2})\right)$$

$$U^+_{y^-,i,j}(0) = \frac{1}{4}\left(\frac{Y_{J,i,j}U^*_{i,j}(0)}{2Y_{y^-,i,j}} + Z_{J,i,j-\frac{1}{2}}I^*_{y,i,j-\frac{1}{2}}(\tfrac{1}{2})\right)$$

$$U^+_{y^+,i,j}(0) = \frac{1}{4}\left(\frac{Y_{J,i,j}U^*_{i,j}(0)}{2Y_{y^+,i,j}} - Z_{J,i,j+\frac{1}{2}}I^*_{y,i,j+\frac{1}{2}}(\tfrac{1}{2})\right)$$

with, in addition,

$$U^+_{c,i,j}(0) = I^+_{x,i+\frac{1}{2},j}(\tfrac{1}{2}) = I^+_{y,i,j+\frac{1}{2}}(\tfrac{1}{2}) = 0$$

where

$$U^*_{i,j}(0) = u(i\Delta, j\Delta, 0)$$
$$I^*_{x,i+\frac{1}{2},j}(\tfrac{1}{2}) = i_x((i+\tfrac{1}{2})\Delta, j\Delta, \tfrac{1}{2})$$
$$I^*_{y,i,j+\frac{1}{2}}(\tfrac{1}{2}) = i_y(i\Delta, (j+\tfrac{1}{2})\Delta, \tfrac{1}{2})$$

Because DWNs of the forms discussed in the previous sections are equivalent to two-step finite difference methods, problems with parasitic modes do not arise as they do in wave digital networks that simulate the same systems. This problem was discussed in detail in Section 3.8.

4.6 Music and Audio Applications of Digital Waveguides

Digital waveguide networks, pioneered by Julius Smith at Stanford University, have been widely applied toward the synthesis of musical sound [231]. Significant portions of many musical instruments can be simply modeled as nearly lossless uniform transmission lines: strings support transverse wave motion, and stiff strings and bars allow longitudinal and torsional motion as well; acoustic waves travel in the tubes that make up brass and wind instruments, organ pipes, as well as the human vocal tract, as we saw in Section 1.1.2. As such, there is a *traveling-wave decomposition* of the motion in these systems.

As we already mentioned in Section 4.2.3, a bidirectional delay line can be thought of as a discrete-time description of traveling-wave propagation in a uniform transmission line. Thus, a single waveguide, which is in itself no more than a pair of delay lines, can be used to model an uninterrupted stretch of a tube or string, without requiring any machine arithmetic. Scattering occurs only at the ends of the waveguide, and in fact, it is possible to use bidirectional delay lines to model wave propagation even in lossy [216, 3] or dispersive [268] media by consolidating ('commuting') these effects at the terminations [230]. It is also possible to derive the characteristics of these terminating filters through comparison with a model PDE [17]. If the length of the string or tube does not correspond to an integer number of delays at a given sample rate, then one may employ *fractional delay lines* [155, 263] that approximate noninteger delay lengths using all-pass (lossless) filters[6].

---

[6]This is often essential because working at the audio sampling rate often forces a large grid spacing. In an acoustic tube, for example, the wave speed is approximately $\gamma = 330$ m/s. A sampling rate of 44.1 kHz will imply

4.6. MUSIC AND AUDIO APPLICATIONS OF DIGITAL WAVEGUIDES

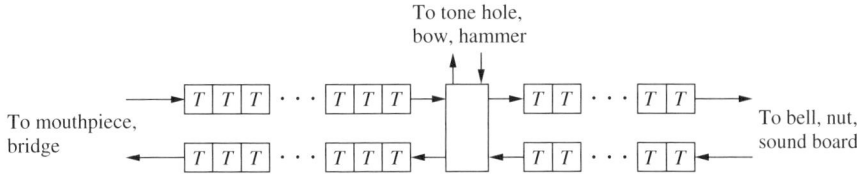

Figure 4.24: *Typical digital waveguide configuration for musical sound synthesis.*

It is possible to go further along these lines, in the particular case of strings, in order to model the nonlinearity due to large amplitude oscillations, in which case string tension is 'modulated.' A good deal of work has gone into the design problem for time-varying filter termination of digital waveguides in order to model this effect [250, 57, 265].

A typical situation is shown in Figure 4.24. The string (or tube) is modeled as two bidirectional delay lines; at the extreme left and right, digital filters may be employed that model bridge terminations [226], horn bells and acoustic radiation [20, 216], coupling with an instrument body or resonator such as a stringed instrument body [139], and, conceivably, coupling between different strings, and for a stiff string, even coupling between different types of motion (i.e., transverse, longitudinal, and torsional) [8]. Excitation mechanisms (such as mouth pressure for a woodwind instrument [28] and lip pressure in brasses [20]) may be modeled as sources and are also used to terminate the waveguide; these may be linearly or nonlinearly coupled to the instrument body. If wave propagation is disturbed along the length of the tube or string, either by an excitation (such as a piano hammer [266, 16, 15] or bow [226, 218]), or by an impedance change (due perhaps to a woodwind tone hole [216, 262, 264, 273], or a change in the cross-sectional area of the vocal tract [45]), then these effects may be modeled at the junction between the two waveguides. It is possible to model many of these effects by making use of lumped wave digital elements, usually of nonlinear type [28, 15] in an energy-conserving formulation; we will return to a discussion of the hammer-string nonlinearity and the single reed in Section 8.3. In some situations, it may be necessary to employ a larger network of interconnected waveguides, as when the vocal tract is to be coupled with the nasal passageways. A full articulatory model of the human vocal tract has been built in this way, by Perry Cook, to simulate the singing voice [45]. Another related example is the simulation of birdsong through a digital waveguide model of the syrinx [232]. Digital waveguides have been recently applied to increasingly exotic instruments such as the panpipes [48], rubbed bowls [220], whirled corrugated tubes [219], and bells [140].

Digital waveguide networks have also been used to simulate wave motion in higher dimensions, in which case they are called waveguide meshes [267, 269]; cases of particular interest have been (2+1)D meshes (see Section 1.1.3) used to simulate the vibration of a uniform membrane [95] and linear acoustics [171], and (3+1)D meshes used to model acoustic spaces [212, 211], as well as instrument bodies such as the violin [96, 127]. Many different types of meshes have been proposed; they differ chiefly in their numerical dispersion properties [213, 30, 26], and we will analyze these forms in detail in Appendix A. A good deal of recent work has gone into the problem of correcting numerical dispersion by

---

a waveguide length of $\Delta = 330/44100$ m $= 0.75$ cm. For a woodwind instrument, this distance is on the order of the tone hole separation distance, and will thus be far too crude for good physical modeling.

introducing terminating filters at the boundaries, and by using interpolation and frequency-warping techniques [213]. The standard (2+1)D rectilinear mesh is shown in Figure 4.25(a). Unit-sample bidirectional delay lines (here represented by two-headed arrows) are connected to scattering junctions (white circles) located at the nodes of a rectangular lattice. Such a mesh has been used to model drum heads as well as gongs (where a nonlinear mesh termination has been applied) [266]. We mentioned in Section 1.1.3 that this mesh indeed solves the (2+1)D wave equation numerically [267] and have elaborated extensively on this idea throughout the rest of the current chapter.

An interesting recent development has been work on modular automated environments for musical sound synthesis. Synthbuilder [188] is based mainly on digital waveguide models; others make use of wave digital methods exclusively [184]. Another, called Block-Compiler [137, 138], is especially relevant to the subject of this book in that it makes use of not only digital waveguides, but also wave digital elements and finite differences, with conversion between the physical and wave variables occurring at specialized module interfaces.

Waveguide networks have also been used in a quasi-physical manner in order to effect artificial reverberation [225]. In this case, a network of waveguides of possibly time-varying impedance is used; such a network is shown in Figure 4.25(b), in which the number of samples of delay in each waveguide (integers a through h) may be different. Such networks are passive, so that signal energy injected into the network from a dry source signal will produce an output whose amplitude will gradually attenuate, with frequency-dependent decay times dependent on the delays and immittances of the various waveguides—some of the delay lengths can be interpreted as implementing 'early reflections'[225]. Such networks provide a cheap and stable way of generating rich impulse responses. Rocchesso and Smith have explored generalizations of waveguide networks to feedback delay networks (FDNs) [201] and circulant delay networks [202] with an eye toward applications in digital reverberation. We will call these DWNs used for reverberation *unstructured*; by this we mean that the waveguides and scattering junctions are not necessarily arranged according to a regular grid in any coordinate system. Yet, such a network is, by construction, passive.

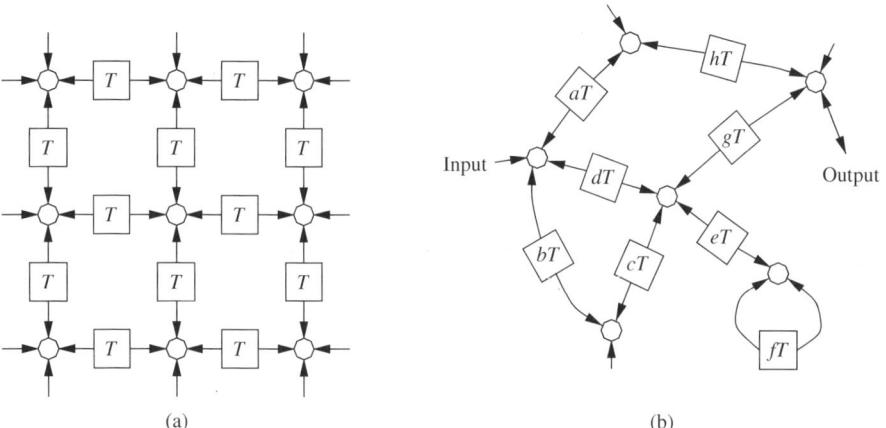

Figure 4.25: *Other waveguide network configurations*—(a) *a (2+1)D waveguide mesh, and* (b) *an unstructured network suitable for implementing artificial reverberation.*

4.6. MUSIC AND AUDIO APPLICATIONS OF DIGITAL WAVEGUIDES

This contrasts sharply with the MDWD networks discussed in the previous chapter. In that case, discretization is performed through the use of a spectral mapping or integration rule; implicit in such an approach is that the algorithm operates on a regular grid in some system of coordinates (and the same will be true of the DWNs that are derived through an MDWD-like discretization procedure, as will be discussed in Chapter 6). The reason for this is that the DWN, as we have described it in this section, is essentially a large network of lumped elements, whereas the MDWD network is a multidimensional object. In certain cases (see Section 5.4), unstructured DWNs may come in handy.

Chapter 5

Extensions of Digital Waveguide Networks

This chapter is of an applied nature—here we discuss several variations on the DWN form presented in the last chapter, specifically for the transmission line and parallel-plate problems. These structures are often referred to as *waveguide meshes*.

Summary

We first take a cursory look at some other two-dimensional DWN structures in Section 5.1, and in particular those operating over hexagonal or triangular grids, and in radial coordinates, extending them to the variable-coefficient case where necessary; three-dimensional DWNs are then examined in Section 5.2. We then introduce some generalized DWNs that may be useful in 'real-world' problems, and in particular, those involving irregular boundary configurations and sharp variations in material parameters, beginning with an extension of DWNs to general curvilinear coordinate systems in Section 5.3. Another means of tackling such irregularities, with an emphasis on computational efficiency considerations, involves the use of multi-grid DWNs; a 'fine' DWN can be used over any part of the problem domain where greater detail is required, and may be interfaced to a 'coarse' DWN operating over the remainder of the domain. The interface between such DWNs can be designed so as to maintain perfect losslessness, while introducing minimal numerical reflection. We look at several types of such layers, in two and three spatial dimensions, as well as at a way of interfacing grids in different coordinate systems, in Section 5.4. Several simulations are presented, in Section 5.1.2, Section 5.4.1, and Section 5.4.4.

5.1 Alternative Grids in (2 + 1)D

We have looked, so far, at waveguide mesh solutions to the $(2 + 1)$D parallel-plate system on a rectilinear grid. We now mention some other possible grid arrangements, first hexagonal and triangular regular grids, then a radial grid. Because Appendix A is devoted to an in-depth exploration of the various types of mesh that have appeared in the literature, we will only take a brief look at these meshes here, with an emphasis on spatially varying media.

5.1.1 Hexagonal and Triangular Grids

Two other ways of regularly sampling the (x, y) plane are shown in Figure 5.1. It was shown by Van Duyne and Smith [213, 269] that a waveguide mesh can be constructed that solves the $(2+1)$D wave equation on a grid of either type. We will extend this result here to include the lossless, source-free but varying-coefficient parallel-plate problem (4.50).

For both types of grids, lines between the grid points (marked by black dots in Figure 5.1) indicate that the points are to be connected by bidirectional delay lines. Connected points are separated by a distance Δ. We will look at meshes of the type II form—that is, meshes for which we will need to calculate using only parallel junctions. (Recall that if we did not have losses or sources in (4.59a) and (4.59b) for the type II mesh of Section 4.4.2, the impedances of the bidirectional delay lines were set so that the series junctions degenerated to simple throughs.) We still allow l and c to vary smoothly over the domain. Additionally, we could allow sources and losses in (4.59c), but as mentioned above, we consider only the fully lossless source-free case here. We remark that we now have a mesh for which all delays are of equal duration (i.e., in both connecting waveguides and self-loops), unlike the interleaved mesh of type I or III.

The construction of the waveguide mesh equivalent difference schemes on a hexagonal or triangular grid is very similar to the rectilinear case and we will omit most steps. Referring to Figure 5.1(a), for the hexagonal mesh we have a four-port parallel junction at each grid point. At grid point 0, for example, we have unit-sample bidirectional delay lines connecting the junction to those at locations 1, 2, and 3, and we will name the admittances of the three connecting lines Y_{01}, Y_{02}, and Y_{03} respectively. In order to allow variation in local wave speed, we also add a self-loop of admittance $Y_{c,0}$. The junction voltages at points 0, 1, 2, and 3 will be named $U_{J,0}$, $U_{J,1}$, $U_{J,2}$, and $U_{J,3}$, and we have, for the junction admittance at point 0,

$$Y_{J,0} \triangleq Y_{01} + Y_{02} + Y_{03} + Y_{c,0}$$

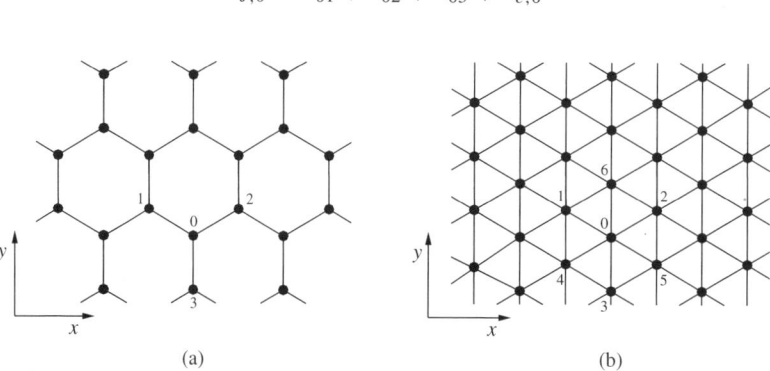

Figure 5.1: *Alternative sampling grids—(a) hexagonal and (b) triangular.*

5.1. ALTERNATIVE GRIDS IN (2 + 1)D

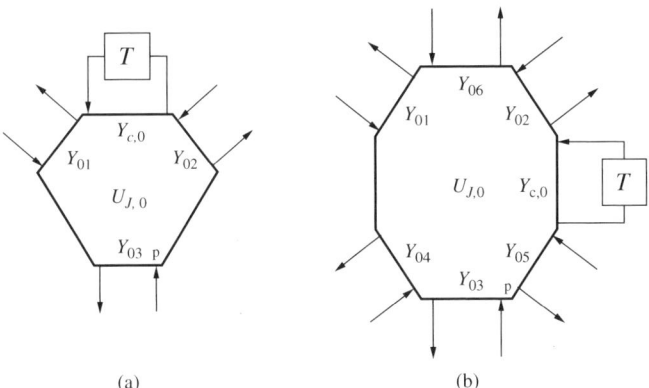

Figure 5.2: *Scattering junctions—(a) on a hexagonal grid and (b) on a triangular grid.*

The scattering junction at grid location 0 is shown in Figure 5.2(a). It should be clear from Figure 5.1(a) that the scattering junctions will be upside-down with respect to that of grid point 0 at half the grid points in the domain, but the waveguide mesh we will develop holds, by symmetry, at such points as well[1].

Beginning at junction 0 and performing manipulations similar to those on a rectilinear grid (i.e., breaking the junction voltage into wave components, and tracing their movement through the network), we get a difference relation among the voltages at point 0 and its neighbors:

$$U_{J,0}(n+1) + U_{J,0}(n-1) = \frac{2}{Y_{J,0}} \left(Y_{01} U_{J,1}(n) + Y_{02} U_{J,2}(n) + Y_{03} U_{J,3}(n) + Y_{c,0} U_{J,0}(n) \right)$$
(5.1)

which we would like to identify with

$$\frac{\partial^2 u}{\partial t^2} = \frac{1}{c} \left(\frac{\partial}{\partial x} \left(\frac{1}{l} \frac{\partial u}{\partial x} \right) + \frac{\partial}{\partial y} \left(\frac{1}{l} \frac{\partial u}{\partial y} \right) \right)$$
(5.2)

which is simply the reduced form of system (4.50) where we have eliminated i_x and i_y. If we were to identify this mesh with the type II mesh of Section 4.4, we might guess that we should set the impedances of the bidirectional delay lines to be equal to v_0 times the value of l at the midpoint of the line, where we again have $v_0 = \Delta/T$. That is, if grid point 0 is situated at coordinates (x, y), then we should set

$$Y_{01} = \frac{1}{v_0 l_{01}} \qquad Y_{02} = \frac{1}{v_0 l_{02}} \qquad Y_{03} = \frac{1}{v_0 l_{03}}$$

with

$$l_{01} = l\left(x - \frac{\sqrt{3}\Delta}{4}, y + \frac{\Delta}{4}\right) \qquad l_{02} = l\left(x + \frac{\sqrt{3}\Delta}{4}, y + \frac{\Delta}{4}\right) \qquad l_{03} = l\left(x, y - \frac{\Delta}{2}\right)$$

---

[1] In fact, we are being somewhat cavalier here (as we will be again when we look at the tetrahedral mesh in Section 5.2), because the updating is not the same at every junction. We will deal with this aspect in full detail in Appendix A.

With this choice of the waveguide admittances, we must have, in order to force consistency of (5.1) with (5.2),

$$Y_{J,0} = \frac{3}{2} v_0 \bar{c}_0 \tag{5.3}$$

where $\bar{c}_0$ is some second-order approximation to c at the location of junction 0. We can now choose, in order to satisfy (5.3),

$$Y_{c,0} = \frac{v_0}{2}(c_{01} + c_{02} + c_{03}) - \frac{1}{v_0 l_{01}} - \frac{1}{v_0 l_{02}} - \frac{1}{v_0 l_{03}}$$

where c_{01}, c_{02}, and c_{03} are values of $c(x, y)$ at the midpoints of the waveguides connecting the junction at point 0 to its neighbors.

The stability bound for this mesh, resulting from a positivity condition on the self-loop admittance $Y_{c,X}$ over all grid points X is

$$v_0 \geq \max_{\text{waveguide midpoints}} \sqrt{\frac{2}{lc}} \tag{5.4}$$

The triangular mesh is very similar. We now have, at each grid point, a seven-port junction to accommodate incoming waves from six directions and a self-loop (see Figure 5.2(b)). We now make the choices, at a grid point 0, of

$$Y_{0j} = \frac{1}{v_0 l_{0j}}, \qquad j = 1, \ldots, 6$$

where l_{0j}, $j = 1, \ldots, 6$ are the values of the inductance at the midpoints of the attached waveguides in directions 1 through 6. The relevant condition for consistency with (5.2) can be shown to be

$$Y_{J,0} = 3 v_0 \bar{c}$$

and we choose

$$Y_{c,0} = \sum_{j=1}^{6} \left(\frac{1}{2} v_0 c_{0j} - \frac{1}{v_0 l_{0j}} \right)$$

where c_{0j}, $j = 1, \ldots, 6$ are the values of the capacitance at the adjacent waveguide midpoints. The stability bound is again

$$v_0 \geq \max_{\text{waveguide midpoints}} \sqrt{\frac{2}{lc}} \tag{5.5}$$

If l and c are constant, and we are operating at the CFL bound (i.e., $v_0 = \sqrt{2/lc}$), then in both the hexagonal and the triangular meshes $Y_{c,0}$ vanishes and we have, respectively, three-port and six-port scattering junctions, all of whose port admittances are identical. These meshes, although they are slightly more difficult to program than the rectilinear mesh, possess better numerical dispersion properties [213]; a comparison of the directional dispersion of various types of $(2+1)$D meshes in the constant-coefficient case

5.1. ALTERNATIVE GRIDS IN (2 + 1)D

is given in Section A.2. We also note that in this same constant-coefficient case, when we are at CFL, the hexagonal mesh, like the rectilinear mesh, can be decomposed into two meshes that operate on half the grid points, and thus a cut in computational and memory costs of a factor of two is possible (see Section 4.4.3). This is not true for the triangular mesh (see Appendix A).

We do not include any material about boundary termination of hexagonal or triangular grids, although we do conjecture that it should be comparatively less tricky than the termination of multidimensional MDWD networks in the same coordinate systems (we mentioned a hexagonal coordinate system in passing in Section 3.2.3), where the available results are extremely unsatisfactory [284, 283].

5.1.2 The Waveguide Mesh in Radial Coordinates

We will look at waveguide meshes in general curvilinear coordinates in Section 5.3, but radial coordinates are an important special case.

In terms of radial coordinates (ρ, θ), where

$$x = \rho \cos \theta \qquad\qquad y = \rho \sin \theta \qquad (5.6)$$

the parallel-plate system (4.59) becomes

$$l_\rho \frac{\partial i_\rho}{\partial t} + \frac{\partial u}{\partial \rho} + r_\rho i_\rho + e_\rho = 0 \qquad (5.7a)$$

$$l_\theta \frac{\partial i_\theta}{\partial t} + \frac{\partial u}{\partial \theta} + r_\theta i_\theta + e_\theta = 0 \qquad (5.7b)$$

$$c_u \frac{\partial u}{\partial t} + \frac{\partial i_\rho}{\partial \rho} + \lambda \frac{\partial i_\theta}{\partial \theta} + g_u u + e_u = 0 \qquad (5.7c)$$

where we define radial and angular current densities by

$$i_\rho = \rho \left(i_x \cos \theta + i_y \sin \theta \right) \qquad i_\theta = \frac{1}{\lambda} \left(-i_x \sin \theta + i_y \cos \theta \right) \qquad (5.8)$$

and the effective radial material parameters and sources by

$$l_\rho = \frac{l}{\rho} \qquad\qquad l_\theta = \rho l \lambda \qquad\qquad c_u = \rho c$$

$$r_\rho = \frac{r}{\rho} \qquad\qquad r_\theta = \rho r \lambda \qquad\qquad g_u = \rho g \qquad (5.9)$$

$$e_\rho = e \cos \theta + f \sin \theta \qquad e_\theta = \rho(-e \sin \theta + f \cos \theta) \qquad e_u = \rho h$$

λ is a scaling coefficient, which we will set, in anticipation of discretization, equal to $\Delta_\theta / \Delta_\rho$, the ratio of the grid spacings in the θ and ρ directions. We will allow these spacings to be, in general, different.

It is evident that system (5.7) has a form similar to its counterpart in rectilinear coordinates, apart from the extra factor of λ in (5.7c). The chief difference is that we now have different effective inductances l_ρ and l_θ in the two coordinate directions, but, as

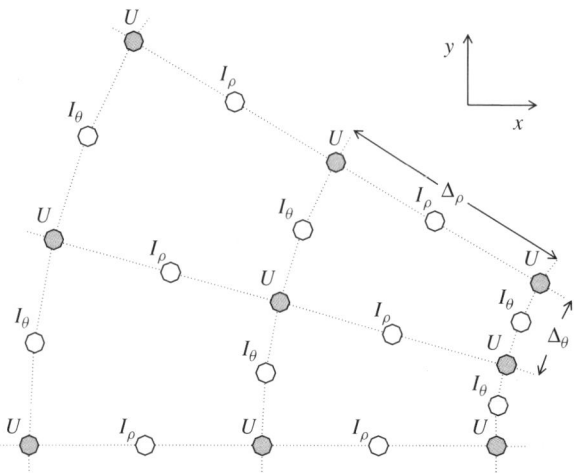

Figure 5.3: *Interleaved grid in radial coordinates.*

we shall see, this anisotropy is easily taken care of (indeed, we could have defined anisotropic inductances l_x and l_y in the rectilinear case without greatly complicating matters). We can see immediately that when centered differences are applied to system (5.7), we will be able to operate on an interleaved grid in the (ρ, θ) coordinates. A version of Yee's algorithm in arbitrary curvilinear coordinates has existed for some time [123, 281]. The interleaved grid, viewed in rectilinear coordinates, is shown in Figure 5.3, where, as before, the dependent variable to be calculated at a particular grid point is indicated next to the point. Gray and white coloring of points indicates operation at alternating time steps.

Centered differencing yields a scheme nearly identical to (4.60), with, again, the difference that the inductance has a directional character. We can thus proceed directly to the waveguide mesh, and, furthermore, can use the same indexing as in the rectilinear case; now, the grid indices (i, j, n) will refer to points $(\rho, \theta, t) = (i\Delta_r, j\Delta_\theta, nT)$. Because of the interleaved nature of the resulting difference approximations, we will have series junctions at locations $(i + \frac{1}{2}, j, n + \frac{1}{2})$ and $(i, j + \frac{1}{2}, n + \frac{1}{2})$ for $i > 0$, j and n integer (with associated junction currents $I_{\rho J, i+\frac{1}{2}, j}(n + \frac{1}{2})$ and $I_{\theta J, i, j+\frac{1}{2}}(n + \frac{1}{2})$) and parallel junctions at locations (i, j, n), where we will calculate junction voltages $U_{J, i, j}(n)$, for $i > 0$, j and n integer (we return to the central grid point at $i = 0$ later in this section). The computational molecule of the mesh is shown in Figure 5.4.

Referring to Figure 5.5, which gives the immittance nomenclature in the waveguide network, and where, in addition, we have the junction immittances defined by

$$Y_{J,i,j} \triangleq Y_{\rho-,i,j} + Y_{\rho+,i,j} + Y_{\theta+,i,j} + Y_{\theta-,i,j} + Y_{c,i,j} + Y_{R,i,j}$$

$$Z_{J,i+\frac{1}{2},j} \triangleq Z_{\theta+,i+\frac{1}{2},j} + Z_{\theta-,i+\frac{1}{2},j} + Z_{c,i+\frac{1}{2},j} + Z_{R,i+\frac{1}{2},j}$$

$$Z_{J,i,j+\frac{1}{2}} \triangleq Z_{\rho+,i,j+\frac{1}{2}} + Z_{\rho-,i,j+\frac{1}{2}} + Z_{c,i,j+\frac{1}{2}} + Z_{R,i,j+\frac{1}{2}}$$

5.1. ALTERNATIVE GRIDS IN (2 + 1)D

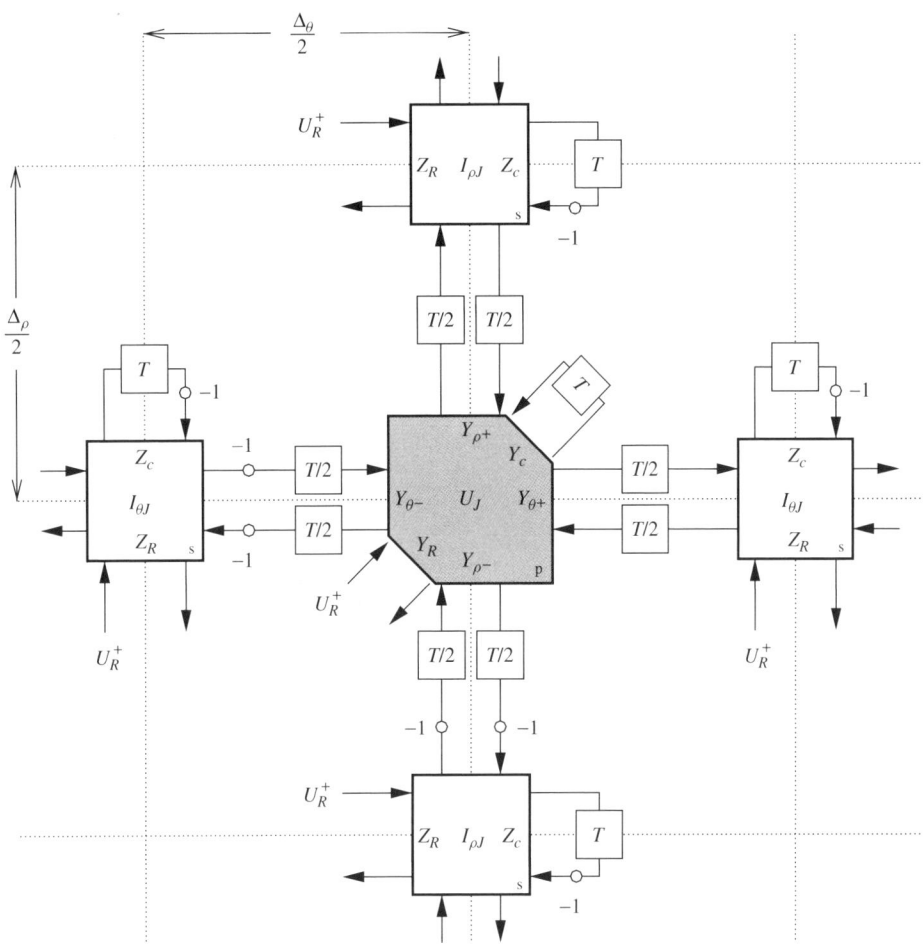

Figure 5.4: *Waveguide mesh for the (2 + 1)D parallel-plate system, in radial coordinates.*

for $i > 0$ and j integer, we can perform an analysis similar to the rectilinear case in order to determine that we must have

$$Y_{J,i,j} = 2\Delta_\rho \left(\frac{\bar{c}_{u,i,j}}{T} + \frac{1}{2} g_{u,i,j} \right) \qquad Y_{R,i,j} = \Delta_\rho g_{u,i,j}$$

$$Z_{J,i+\frac{1}{2},j} = 2\Delta_\rho \left(\frac{\bar{l}_{\rho,i+\frac{1}{2},j}}{T} + \frac{1}{2} r_{\rho,i+\frac{1}{2},j} \right) \qquad Z_{R,i+\frac{1}{2},j} = \Delta_\rho r_{\rho,i+\frac{1}{2},j}$$

$$Z_{J,i,j+\frac{1}{2}} = 2\Delta_\theta \left(\frac{\bar{l}_{\theta,i,j+\frac{1}{2}}}{T} + \frac{1}{2} r_{\theta,i,j+\frac{1}{2}} \right) \qquad Z_{R,i,j+\frac{1}{2}} = \Delta_\theta r_{\theta,i,j+\frac{1}{2}}$$

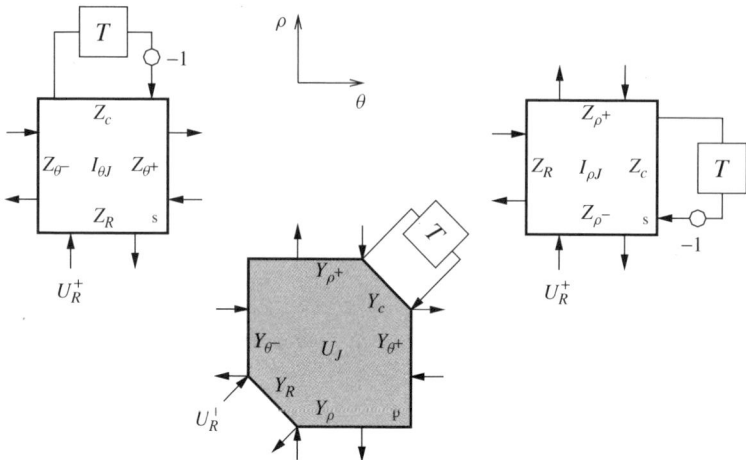

Figure 5.5: *Representative scattering junctions for the waveguide mesh for the $(2+1)$D parallel-plate system, in radial coordinates.*

(The junction admittance $Y_{J,0,0}$ will be dealt with shortly.) The source waves should be chosen as

$$U^+_{R,i,j}(n) = -\frac{e_{u,i,j}(n)}{2g_{u,i,j}}$$

$$I^+_{R,i+\frac{1}{2},j}(n+\tfrac{1}{2}) = -\frac{e_{\rho,i+\frac{1}{2},j}(n+\tfrac{1}{2})}{2r_{\rho,i+\frac{1}{2},j}}$$

$$I^+_{R,i,j+\frac{1}{2}}(n+\tfrac{1}{2}) = -\frac{e_{\theta,i,j+\frac{1}{2}}(n+\tfrac{1}{2})}{2r_{\theta,i,j+\frac{1}{2}}}$$

where we may of course use the dual type of wave in regions where the loss parameters become small, as discussed in Section 4.3.7. Just as in the rectilinear case, these conditions define a family of waveguide networks that solve the radial transmission line equations. We provide the impedance settings as well as stability bounds for voltage- and current-centered meshes.

Type I: Voltage-centered Mesh

$$Y_{\rho^+,i,j} = Y_{\rho^-,i,j} = Y_{\theta^+,i,j} = Y_{\theta^-,i,j} = \frac{\Delta_\rho c_{u,i,j}}{2T} \qquad Y_{c,i,j} = 0 \qquad (5.10)$$

$$Z_{c,i+\frac{1}{2},j} = \frac{\Delta_\rho l_{\rho,i,j}}{T} - \frac{2T}{\Delta_\rho c_{u,i,j}} + \frac{\Delta_\rho l_{\rho,i+1,j}}{T} - \frac{2T}{\Delta_\rho c_{u,i+1,j}} \qquad (5.11)$$

$$Z_{c,i,j+\frac{1}{2}} = \frac{\Delta_\theta l_{\theta,i,j}}{T} - \frac{2T}{\Delta_\rho c_{u,i,j}} + \frac{\Delta_\theta l_{\theta,i,j+1}}{T} - \frac{2T}{\Delta_\rho c_{u,i,j+1}} \qquad (5.12)$$

5.1. ALTERNATIVE GRIDS IN (2 + 1)D

The stability constraints (which follow from the requirement of positivity of Z_c everywhere) are

$$\frac{\Delta\rho}{T} \geq \max_{i,j} \sqrt{\frac{2}{l_{i,j}c_{i,j}}} \qquad \frac{\Delta\theta}{T} \geq \max_{i,j}\left(\frac{1}{\rho_i}\sqrt{\frac{2}{l_{i,j}c_{i,j}}}\right) \qquad (5.13)$$

There is thus a dependence on ρ in the second condition (relating the angular spacing $\Delta\theta$ to T), which we expect, since the spacing between the junctions at a given radius now varies linearly with the radius. Stability bounds are, for a radial mesh, necessarily more severe than in the rectilinear case, due to this variation in grid spacing.

Type II: Current-centered Mesh

$$Z_{\rho+,i+\frac{1}{2},j} = Z_{\rho-,i+\frac{1}{2},j} = \frac{\Delta\rho l_{\rho,i+\frac{1}{2},j}}{T} \qquad (5.14)$$

$$Z_{\theta+,i,j+\frac{1}{2}} = Z_{\theta-,i,j+\frac{1}{2}} = \frac{\Delta\theta l_{\theta,i,j+\frac{1}{2}}}{T} \qquad (5.15)$$

$$Z_{c,i+\frac{1}{2},j} = 0 \qquad (5.16)$$

$$Z_{c,i,j+\frac{1}{2}} = 0 \qquad (5.17)$$

$$Y_{c,i,j} = \frac{2\Delta\rho c_{u,i+\frac{1}{2},j}}{T} - \frac{T}{\Delta\rho l_{\rho,i+\frac{1}{2},j}} + \frac{2\Delta\rho c_{u,i-\frac{1}{2},j}}{T} - \frac{T}{\Delta\rho l_{\rho,i-\frac{1}{2},j}}$$
$$+ \frac{2\Delta\rho c_{u,i,j+\frac{1}{2}}}{T} - \frac{T}{l_{\theta,i,j+\frac{1}{2}}} + \frac{2\Delta\rho c_{u,i,j-\frac{1}{2}}}{T} - \frac{T}{l_{\theta,i,j-\frac{1}{2}}}$$

The stability bound is the same as given by (5.13), except that we take maxima over the series junction locations.

We leave out a discussion of the type III mesh because it was already shown in Section 4.4.2 to be relatively inefficient in terms of the maximum allowable time step for a given grid spacing (when compared to types I and II).

Central Gridpoint in a Radial Mesh

We have so far restricted our attention to interior points of the grid (for which $i > 0$). If the center of the (ρ,θ) coordinate system is to be contained in the grid, a special treatment is required. We have indexed the grid variables such that, in our interleaved mesh, a single parallel junction lies at the origin (for $i = 0$). If the problem domain includes a full circle, we also must assume that $\Delta\theta$ divides 2π evenly so that we have a positive integer N such that

$$N = \frac{2\pi}{\Delta\theta}$$

178   CHAPTER 5. EXTENSIONS OF DIGITAL WAVEGUIDE NETWORKS

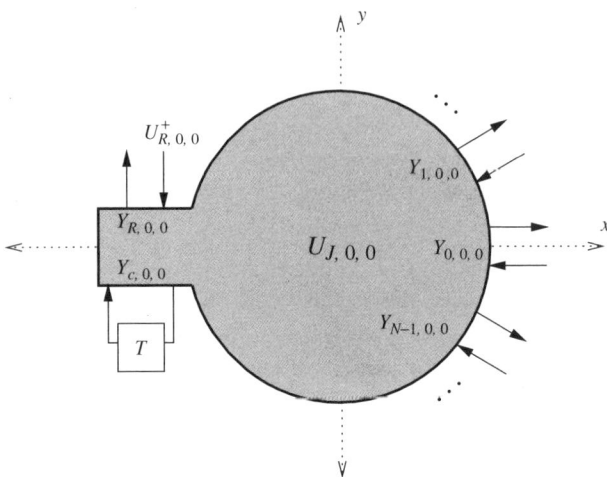

Figure 5.6: *Central scattering junction for the waveguide mesh in radial coordinates.*

Thus, the central parallel junction will be connected to N series junctions at locations $(\Delta_\rho/2, j\Delta_\theta)$, $j = 0, \ldots, N-1$. We will name the admittances of the N waveguides radiating from the central hub $Y_{j,0,0}$, $j = 0, \ldots, N-1$, and the admittance of the self-loop and loss/source ports will be $Y_{c,0,0}$ and $Y_{R,0,0}$ respectively (see Figure 5.6).

The difference scheme relating the junction voltage at the central grid point to the radial currents at the surrounding junctions will then be

$$U_{J,0,0}(n) - \frac{Y_{J,0,0} - 2Y_{R,0,0}}{Y_{J,0,0}} U_{J,0,0}(n-1) + \frac{2}{Y_{J,0,0}} \sum_{j=0}^{N-1} I_{\rho J, \frac{1}{2}, j}\left(n - \frac{1}{2}\right)$$

$$- \frac{2Y_{R,0,0}}{Y_{J,0,0}} \left(U^+_{R,0,0}(n) + U^+_{R,0,0}(n-1)\right) = 0 \quad (5.18)$$

The sum over the junction currents, which we now look at in the continuous time/space domain so as to develop a series approximation, may be rewritten using (5.8) in terms of the rectilinear current variables as

$$\sum_{j=0}^{N-1} i_\rho\left(\frac{\Delta_\rho}{2}, j\Delta_\theta\right) = \sum_{j=0}^{N-1} \frac{\Delta_\rho}{2} \left(i_x\left(\frac{\Delta_\rho}{2}, j\Delta_\theta\right) \cos(j\Delta_\theta) + i_y\left(\frac{\Delta_\rho}{2}, j\Delta_\theta\right) \sin(j\Delta_\theta)\right)$$

$$= \frac{\Delta_\rho}{2} i_x(0,0) \sum_{j=0}^{N-1} \cos(j\Delta_\theta) + \frac{\Delta_\rho}{2} i_y(0,0) \sum_{j=0}^{N-1} \sin(j\Delta_\theta)$$

$$+ \frac{\Delta_\rho^2}{4} \sum_{j=0}^{N-1} \left(\left.\frac{\partial i_x}{\partial x}\right|_{(0,0)} \cos^2(j\Delta_\rho) + \left.\frac{\partial i_y}{\partial y}\right|_{(0,0)} \sin^2(j\Delta_\rho)\right)$$

$$= \frac{N\Delta_\rho^2}{8} \left(\left.\frac{\partial i_x}{\partial x}\right|_{(0,0)} + \left.\frac{\partial i_y}{\partial y}\right|_{(0,0)}\right) \quad (5.19)$$

5.1. ALTERNATIVE GRIDS IN (2 + 1)D

where we have neglected higher-order terms in Δ_ρ and used, in the last line, the identities

$$\sum_{j=0}^{N-1} \cos\left(\frac{2\pi j}{N}\right) = \sum_{j=0}^{N-1} \sin\left(\frac{2\pi j}{N}\right) = 0 \qquad \sum_{j=0}^{N-1} \cos^2\left(\frac{2\pi j}{N}\right) = \sum_{j=0}^{N-1} \sin^2\left(\frac{2\pi j}{N}\right) = \frac{N}{2}$$

which hold for $N > 2$. Using approximation (5.19) and by comparing (5.18) with (5.7c), we must choose the junction and loss/source port admittance to be

$$Y_{J,0,0} = \frac{1}{4} N \Delta_\rho^2 \left(\frac{\bar{c}_{0,0}}{T} + \frac{g_{0,0}}{2} \right) \qquad Y_{R,0,0} = \frac{1}{8} N \Delta_\rho^2 g_{0,0}$$

and the source wave variable $U^+_{R,0,0}(n)$ to be

$$U^+_{R,0,0}(n) = -\frac{h_{0,0}(n)}{g_{0,0}}$$

A mesh of type I is infeasible because it would require access to $l_{\rho,0,0}$ in order to set $Z_{c,\frac{1}{2},j}$ as prescribed in (5.11), but l_ρ as defined in (5.9) is singular at the origin (although if we are working with a radial geometry, which does not contain the origin, this problem does not arise). For a mesh of type II, we have, from (5.14),

$$Y_{j,0,j} = \frac{1}{Z_{\rho-,\frac{1}{2},j}} = \frac{T}{\Delta_\rho l_{\rho,\frac{1}{2},j}} = \frac{T}{2l_{\frac{1}{2},j}}$$

and so we may set, for the self-loop admittance at the central junction,

$$Y_{c,0,0} = \sum_{j=0}^{N-1} \left(\frac{\Delta_\rho^2 c_{\frac{1}{2},j}}{4T} - \frac{T}{2l_{\frac{1}{2},j}} \right)$$

which is positive when

$$\frac{\Delta_\rho}{T} \geq \max_{j=0,\ldots,N-1} \sqrt{\frac{2}{l_{\frac{1}{2},j} c_{\frac{1}{2},j}}}$$

Thus, the stability requirement at the central junction does not interfere with the requirements over the interior of the mesh given for the type II mesh.

We note that a different type of central node has been proposed for use in radial TLM simulations. The singularity at the origin has also been approached in the context of FDTD modeling of Maxwell's equations in rotationally symmetric geometries [41].

Simulation: Circular Region with Varying Inductance

Consider the parallel-plate problem to be solved over a circular region, radius 1, with short-circuited boundary conditions ($u = 0$ on the outer rim). The capacitance is 1 everywhere, as is the inductance, except over four circular regions of radius 0.2 with centers at radius 0.5 and that are equally spaced around the circle (circled in black in Figure 5.7). In these smaller regions, the inductance has the form of a 2D raised cosine distribution—l takes on

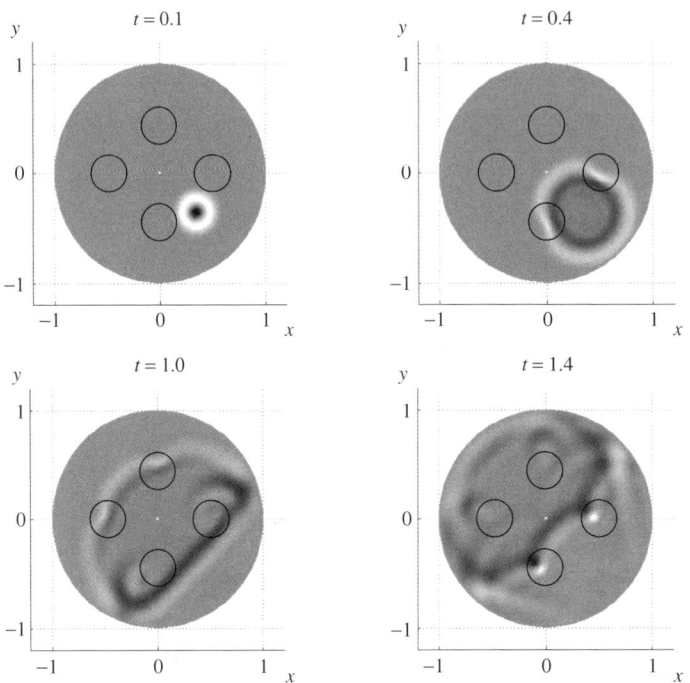

Figure 5.7: *Simulation of a circular parallel-plate system using a radial waveguide mesh.*

a maximum value of 3 at the centers and decreases to 1 at the edges. The initial voltage distribution is a raised 2D cosine of radius 0.15 and amplitude 1, centered at radius 0.5, directly between two of the circular regions of higher inductance. The grid spacings are $\Delta_\rho = \frac{1}{60}$ and $\Delta_\theta = \frac{2\pi}{100}$, and we use a radial waveguide mesh of type II.

The time evolution of the voltage distribution is shown in Figure 5.7, where light- and dark-colored areas indicate regions of positive and negative voltage respectively. The plot is normalized to show voltages between −0.3 (black) and 0.3 (white). We remark that the voltage distribution on this pair of plates will behave identically to the transverse velocity distribution on a clamped membrane that has regions of increased density. Interpolation has been performed for better plotting results.

5.2 The (3 + 1)D Wave Equation and Waveguide Meshes

In this brief section, we summarize, for completeness sake, (3 + 1)D waveguide meshes. We will return to these mesh formulations in Section 5.4.5, where we will discuss interfaces between meshes of different grid densities. We will also analyze the spectral characteristics of these methods in some detail in Appendix A.

The transmission line problem with spatially varying material parameters can not be generalized in a meaningful way to (3 + 1)D; there is no commonly known physical system that would behave according to such a set of equations (though linear acoustics in non-Cartesian

5.2. THE (3+1)D WAVE EQUATION AND WAVEGUIDE MESHES

coordinates might serve as one example). Physical systems of interest in (3 + 1)D generally have a more complex form than would be implied by such a straightforward generalization. We will have occasion to examine two such systems in detail, namely, the (3 + 1)D equations describing the vibration of a linear, isotropic elastic solid, in Section 7.4, and Maxwell's equations in Section 6.4.

The (3 + 1)D wave equation, however, is of interest in linear acoustics (and it is arrived at by linearizing a system of conservation laws, namely *Euler's equations* [153], to which we will return briefly in Appendix A). It is written as

$$\frac{1}{\gamma^2}\frac{\partial^2 p}{\partial t^2} = \frac{\partial^2 p}{\partial x^2} + \frac{\partial^2 p}{\partial y^2} + \frac{\partial^2 p}{\partial z^2} \qquad (5.20)$$

where $p(x, y, z, t)$ is pressure deviation from ambient pressure, and $\gamma = \sqrt{K/\rho}$ is the wave speed (K and ρ are the bulk modulus and density of the medium [20, 94]). In order to enforce notational consistency, we will assume that we can write $\gamma^2 = 1/lc$, for some positive constants l and c. Obviously, any such choices of l and c will be appropriate if we are only interested in solving for the pressure. (In particular, a reasonable choice would be $l = \rho$ and $c = 1/K$.)

Three regular grids are shown in Figure 5.8, and we have indicated waveguide couplings between neighboring points (where scattering junctions will be placed) by double-headed arrows. Delays in these bidirectional delay lines are assumed to be of length T and are identical over the entire network, in all the three cases. Junctions are separated by a distance Δ. The first, shown in (a), is the standard rectilinear mesh [212, 267], and the second, shown in (b), is a mesh obtained by superimposing one rectilinear grid on top of a shifted copy of itself and then connecting each point to its eight nearest neighbors; an appropriate name for such a configuration might be an 'octahedral mesh.' A third structure, the so-called tetrahedral mesh [269], is shown in (c). Self-loops, necessary when we are operating away from the CFL bound, are not shown, and the immittances of the connecting waveguides are assumed to be all identical. Other structures are also conceivable.

We remark here on a computational aspect of these junctions: as mentioned by Van Duyne and Smith [269], if we are at CFL (and so do not need self-loops), it is useful

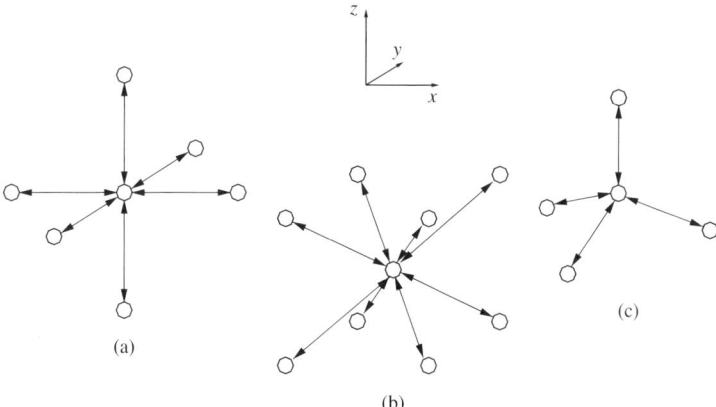

Figure 5.8: *Regular grids in (3 + 1)D— (a) rectilinear, (b) octahedral and (c) tetrahedral.*

to have the number of waveguides connected to a particular junction be a power of two; if this can be arranged, then all multiplies carried out during the scattering step may be implemented as simple bit-shifting operations in a fixed-point implementation. Because this is not true for the rectilinear mesh, (i.e., there are six waveguides connected to each junction), the tetrahedral mesh was proposed as a more efficient structure. We note, however, that the octahedral mesh, with eight waveguides connected to each junction, also can be implemented efficiently in fixed point. Furthermore, it may be easier to deal with from the programmer's point of view because unlike the tetrahedral mesh, it will not involve any special indexing strategy (for a tetrahedral mesh, half of the junctions will have an inverse orientation with respect to the other half).

If we are at the CFL bound—that is, if we have

$$v_0 = \sqrt{3}\gamma \tag{5.21}$$

where Δ is the physical length of a waveguide in any of the three types of mesh, then we may choose any constant to be the admittance of the connecting waveguide. In this case, all the three types of meshes can be decomposed into two meshes, to be used at alternating time steps; as in $(2+1)$D, this can be exploited to increase computational efficiency (see Section 4.4.3). If we are away from CFL (as we may be in a multi-grid setting—see Section 5.4), then we must set, for the self-loop admittances,

$$Y_c = 2v_0 c - \frac{6}{v_0 l} \qquad \text{Rectilinear mesh}$$

$$Y_c = \frac{4}{\sqrt{3}} v_0 c - \frac{4\sqrt{3}}{v_0 l} \qquad \text{Octahedral mesh}$$

$$Y_c = \frac{4}{3} v_0 c - \frac{4}{v_0 l} \qquad \text{Tetrahedral mesh}$$

where the connecting waveguide admittance has been chosen as $\frac{1}{v_0 l}$.

5.3 The Waveguide Mesh in General Curvilinear Coordinates

A generalization of the waveguide mesh to arbitrary curvilinear coordinates is useful in that it becomes possible to model boundary conditions that may not be simply aligned with a rectilinear grid. The resulting structure is quite similar to the interleaved forms discussed earlier, and for this reason we will give only a brief description of the coordinate transformation procedure. Consider the following system:

$$l_\mathbf{x} \frac{\partial \mathbf{i_x}}{\partial t} + \nabla_\mathbf{x} u + r_\mathbf{x} \mathbf{i_x} + \mathbf{e_x} = \mathbf{0} \tag{5.22a}$$

$$c_\mathbf{x} \frac{\partial u}{\partial t} + \nabla_\mathbf{x} \cdot \mathbf{i_x} + g_\mathbf{x} u + h_\mathbf{x} = 0 \tag{5.22b}$$

Here, we may assume any number k of physical spatial coordinates $\mathbf{x} = [x_1, \ldots, x_k]^T$, so that $\nabla_\mathbf{x} = \frac{\partial}{\partial \mathbf{x}} = [\frac{\partial}{\partial x_1}, \ldots, \frac{\partial}{\partial x_k}]^T$. $\mathbf{i_x}$ and $\mathbf{e_x}$ are both assumed to be k-dimensional column

5.3. THE WAVEGUIDE MESH IN GENERAL CURVILINEAR COORDINATES

vectors. $l_\mathbf{x}$, $c_\mathbf{x}$, $r_\mathbf{x}$, and $g_\mathbf{x}$ are all positive functions of $\mathbf{x}$ ($l_\mathbf{x}$ and $c_\mathbf{x}$ are strictly positive), and $\mathbf{e}_\mathbf{x}$ and $h_\mathbf{x}$ are the source terms. If k is 1 or 2, then we have the transmission line or parallel-plate transmission line system respectively, and if $k = 3$, we have the system describing linear acoustic phenomena (assuming that the material parameters are constant). Curvilinear coordinate systems have been touched upon in the multidimensional wave digital framework as well; we follow here the approach of Fries [99], and that of Nitsche [176] is similar.

Consider the mapping

$$\mathbf{x} = \boldsymbol{\zeta}(\mathbf{w})$$

where $\mathbf{w} = [w_1, \ldots, w_k]^T$ are the transformed coordinates. A rectilinear grid in the $\mathbf{w}$ coordinates can be mapped to a curvilinear grid in the physical $\mathbf{x}$ coordinates. We can then define the $k \times k$ matrix of partial derivatives $\mathbf{J}$ by

$$[\mathbf{J}_{\alpha,\beta}] = \frac{\partial \zeta_\alpha}{\partial w_\beta} \qquad \alpha, \beta = 1, \ldots, k$$

where ζ_α is the αth component of $\boldsymbol{\zeta}$. We assume $\mathbf{J}$ to be nonsingular everywhere in the problem domain (though this assumption may be relaxed as we will discuss later in this section). Defining the differential operator $\nabla_\mathbf{w}$ by $\nabla_\mathbf{w} = \frac{\partial}{\partial \mathbf{w}} = [\frac{\partial}{\partial w_1}, \ldots, \frac{\partial}{\partial w_k}]^T$, it then follows [99] that

$$\nabla_\mathbf{w} = \mathbf{J}^T \nabla_\mathbf{x} \qquad \nabla_\mathbf{x} \cdot \mathbf{a} = \frac{1}{|\mathbf{J}|} \nabla_\mathbf{w} \cdot \left(|\mathbf{J}| \mathbf{J}^{-1} \mathbf{a} \right)$$

for any $k \times 1$ column vector $\mathbf{a}$. Here, $|\mathbf{J}|$ is the so-called *Jacobian determinant* [235]. Using these relationships, the system (5.22) may be rewritten as

$$l_\mathbf{x} \mathbf{J}^T \frac{\partial \mathbf{i}_\mathbf{x}}{\partial t} + \nabla_\mathbf{w} u + r_\mathbf{x} \mathbf{J}^T \mathbf{i}_\mathbf{x} + \mathbf{J}^T \mathbf{e}_\mathbf{x} = 0$$

$$c_\mathbf{x} |\mathbf{J}| \frac{\partial u}{\partial t} + \nabla_\mathbf{w} \cdot \left(|\mathbf{J}| \mathbf{J}^{-1} \mathbf{i}_\mathbf{x} \right) + |\mathbf{J}| g_\mathbf{x} u + |\mathbf{J}| h_\mathbf{x} = 0$$

or

$$\mathbf{L}_\mathbf{w} \frac{\partial \mathbf{i}_\mathbf{w}}{\partial t} + \nabla_\mathbf{w} u + \mathbf{R}_\mathbf{w} \mathbf{i}_\mathbf{w} + \mathbf{e}_\mathbf{w} = 0 \qquad (5.23a)$$

$$c_\mathbf{w} \frac{\partial u}{\partial t} + \nabla_\mathbf{w} \cdot \mathbf{i}_\mathbf{w} + g_\mathbf{w} u + h_\mathbf{w} = 0 \qquad (5.23b)$$

where

$$\mathbf{i}_\mathbf{w} = |\mathbf{J}| \mathbf{J}^{-1} \mathbf{i}_\mathbf{x}$$

and

$$\mathbf{L}_\mathbf{w} = \frac{l_\mathbf{x}}{|\mathbf{J}|} \mathbf{J}^T \mathbf{J} \qquad \mathbf{R}_\mathbf{w} = \frac{r_\mathbf{x}}{|\mathbf{J}|} \mathbf{J}^T \mathbf{J} \qquad \mathbf{e}_\mathbf{w} = \mathbf{J}^T \mathbf{e}_\mathbf{x}$$

$$c_\mathbf{w} = |\mathbf{J}| c_\mathbf{x} \qquad g_\mathbf{w} = |\mathbf{J}| g_\mathbf{x} \qquad h_\mathbf{w} = |\mathbf{J}| h_\mathbf{x}$$

System (5.23) is similar to (5.22), except that we now have 'vector' inductance and resistance coefficients (note that both $\mathbf{L_w}$ and $\mathbf{R_w}$ are positive-definite matrices if $\mathbf{J}$ is nonsingular). In particular, it is still symmetric hyperbolic (see Section 3.1), so we may expect that it is possible to derive a waveguide structure.

Consider now the transformed system (5.23) in (2 + 1)D. If $\mathbf{J}^T\mathbf{J}$ is diagonal, then, $\mathbf{L_w}$ and $\mathbf{R_w}$ will be as well; in this case, system (5.23) is in the same form[2] as the parallel-plate system in radial coordinates (5.7), so we need not discuss this case further here. Indeed, the radial system is a special case of (5.23) with $\mathbf{w} = [\rho, \theta]^T$ and

$$\mathbf{J} = \begin{bmatrix} \cos\theta & -\rho\sin\theta \\ \sin\theta & \rho\cos\theta \end{bmatrix}$$

On the other hand, if $\mathbf{J}^T\mathbf{J}$ is not diagonal (so that we are working in nonorthogonal or oblique coordinates), then the situation is more complex. Because of the cross-coupling between the components of $\mathbf{i_w}$ through the matrices $\mathbf{L_w}$ and $\mathbf{R_w}$, it will no longer be possible to stagger all the components of the solution; and in particular, it will be necessary to use *vector* scattering junctions. Let us look at the case $k = 2$, so that (5.23) are the equations of the parallel-plate system in the curvilinear coordinates $\mathbf{w}$. Furthermore, we will set $w_1 = p$ and $w_2 = q$. A centered difference approximation to (5.23), over grid points with coordinates $p = i\Delta$ and $q = j\Delta$, and at times $t = nT$ for i, j, and n half-integer is

$$v_0 \overline{\mathbf{L}}_{\mathbf{w},i+\frac{1}{2},j}\left(\mathbf{I}_{i+\frac{1}{2},j}(n+\tfrac{1}{2}) - \mathbf{I}_{i+\frac{1}{2},j}(n-\tfrac{1}{2})\right) + \begin{bmatrix} U_{i+1,j}(n) - U_{i,j}(n) \\ U_{i,j+\frac{1}{2}}(n) - U_{i,j-\frac{1}{2}}(n) \end{bmatrix}$$

$$+ \frac{\Delta}{2}\mathbf{R}_{\mathbf{w},i+\frac{1}{2},j}\left(\mathbf{I}_{i+\frac{1}{2},j}(n+\tfrac{1}{2}) + \mathbf{I}_{i+\frac{1}{2},j}(n-\tfrac{1}{2})\right)$$

$$+ \frac{\Delta}{2}\left(\mathbf{e}_{\mathbf{w},i+\frac{1}{2},j}(n+\tfrac{1}{2}) + \mathbf{e}_{\mathbf{w},i+\frac{1}{2},j}(n-\tfrac{1}{2})\right) = \mathbf{0} \quad (5.24\text{a})$$

$$v_0 \overline{c}_{\mathbf{w},i,j}\left(U_{i,j}(n) - U_{i,j}(n-1)\right) + I_{p,i+\frac{1}{2},j}(n-\tfrac{1}{2}) - I_{p,i-\frac{1}{2},j}(n-\tfrac{1}{2})$$

$$+ I_{q,i,j+\frac{1}{2}}(n-\tfrac{1}{2}) - I_{q,i,j-\frac{1}{2}}(n-\tfrac{1}{2})$$

$$+ \frac{\Delta}{2}g_{\mathbf{w},i,j}\left(U_{i,j}(n) + U_{i,j}(n-1)\right)$$

$$+ \frac{\Delta}{2}\left(h_{\mathbf{w},i,j}(n) + h_{\mathbf{w},i,j}(n-1)\right) = 0 \quad (5.24\text{b})$$

Here, we have the vector grid function $\mathbf{I}_{i+1/2,j}(n + 1/2)$, which is a two-vector with components $I_{p,i+1/2,j}(n + 1/2)$ and $I_{q,i+1/2,j}(n + 1/2)$ as well as the scalar grid function $U_{i,j}(n)$. $\overline{\mathbf{L}}_{\mathbf{w},i+1/2,j+1/2}$ and $\overline{c}_{\mathbf{w},i,j}$ are second-order approximations to $\mathbf{L_w}$ and $c_\mathbf{w}$ at the indicated grid points. The scheme above has been written so that it is clear that it can operate for n integer, and for i and j such that $i + j$ is an integer; notice that U and $\mathbf{I}$ are calculated at alternating time instants and grid locations, but the components of $\mathbf{I}$ cannot,

---

[2] It is in the same form except for the scaling parameter λ, which was introduced so as to allow a different grid spacing in the two radial coordinate directions; such a scaling parameter may be used here to exactly the same effect.

5.3. THE WAVEGUIDE MESH IN GENERAL CURVILINEAR COORDINATES

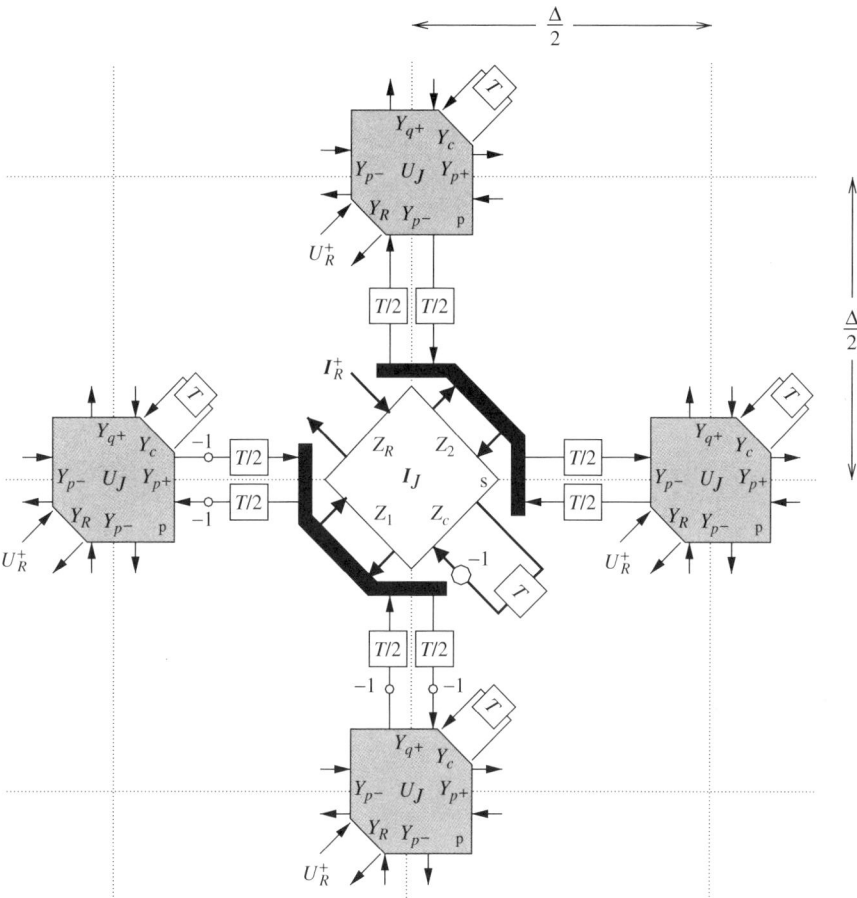

Figure 5.9: $(2+1)$D *DWN* for the parallel-plate system (5.23) in general nonorthogonal curvilinear coordinates.

in general, be calculated at separate locations. v_0, again, is equal to Δ/T, and (5.24) will be a second-order accurate approximation to (5.23).

We will skip the tedious procedure of deriving a waveguide mesh and simply present the resulting structure in Figure 5.9. Junction vector currents $\mathbf{I}_J$ are calculated at the series vector scattering junctions; the black bars surrounding this junction in the figure are the splitting elements that were discussed in Section 4.2.6. Although we have not drawn them in the figure, there will be similar vector junctions at the four grid points neighboring any of the parallel junctions where the voltages U_J are calculated. This vector junction has four 2×2 matrix impedances associated with it: $\mathbf{Z}_c$, the self-loop impedance, $\mathbf{Z}_R$, the loss/source impedance, and $\mathbf{Z}_1$ and $\mathbf{Z}_2$, which are constrained (see Section 4.2.6) to be

$$\mathbf{Z}_{1,i+\frac{1}{2},j} = \begin{bmatrix} \frac{1}{Y_{p^+,i,j}} & 0 \\ 0 & \frac{1}{Y_{q^+,i+\frac{1}{2},j-\frac{1}{2}}} \end{bmatrix} \qquad \mathbf{Z}_{2,i+\frac{1}{2},j} = \begin{bmatrix} \frac{1}{Y_{p^-,i+1,j}} & 0 \\ 0 & \frac{1}{Y_{q^-,i+\frac{1}{2},j+\frac{1}{2}}} \end{bmatrix} \qquad (5.25)$$

The junction impedance $\mathbf{Z}_J$ is defined to be the sum of these four matrices. The admittances at the parallel junctions are defined in a manner similar to those of the DWN in rectilinear coordinates. Also, we have the source voltage waves $U^+_{R,i,j}$ at the parallel junctions and vector source current waves $\mathbf{I}^+_{R,i+1/2,j}$ at the series junctions.

This DWN can be identified with the difference system (5.24) if we set

$$\mathbf{Z}_J = 2v_0\overline{\mathbf{L}}_\mathbf{w} + \Delta\mathbf{R}_\mathbf{w} \qquad \mathbf{Z}_R = \Delta\mathbf{R}_\mathbf{w} \qquad \mathbf{I}^+_R = -\Delta\mathbf{R}_\mathbf{w}^{-1}\mathbf{e}_\mathbf{w}/2$$

$$Y_J = 2v_0\overline{c}_\mathbf{w} + \Delta g_\mathbf{w} \qquad Y_R = \Delta g_\mathbf{w} \qquad U^+_R = -\Delta g_\mathbf{w} h_\mathbf{w}/2$$

at the grid points for which such quantities are defined.

There are, of course, various realizations, depending on how the self-loop and connecting immittances are chosen. First, note that because $\mathbf{Z}_J$ is not diagonal, it will not be possible to distribute it equally among the two connecting impedances $\mathbf{Z}_1$ and $\mathbf{Z}_2$, which are constrained to be diagonal from (5.25). Thus, a type II (current-centered) realization analogous to that which was discussed in the case of the rectilinear mesh will not be possible, even in the absence of losses and sources. A type I realization is certainly possible, but for brevity sake, we will only provide the settings for the type III DWN. Here, all the connecting impedances are all set to be some constant value Z_{const}. This then implies that

$$\mathbf{Z}_c = 2v_0\overline{\mathbf{L}}_\mathbf{w} - 2Z_{\text{const}}\mathbf{I}_2 \qquad Y_c = 2v_0\overline{c}_\mathbf{w} - 4/Z_{\text{const}}$$

where $\mathbf{I}_2$ is the 2×2 identity matrix. Requiring the positivity of Y_c and the positive definiteness of $\mathbf{Z}_c$ gives the constraint

$$v_0 \geq \sqrt{\frac{2}{\mathbf{L}_{\mathbf{w},\min} c_{\mathbf{w},\min}}}$$

for $c_{\mathbf{w},\min}$, the minimum of $c_\mathbf{w}$ over parallel junction locations and $\mathbf{L}_{\mathbf{w},\min}$, the minimum of the eigenvalues of $\mathbf{L}_\mathbf{w}$ over series junction locations, and where we have made the choice

$$Z_{\text{const}} = \sqrt{\frac{2\mathbf{L}_{\mathbf{w},\min}}{c_{\mathbf{w},\min}}}$$

In general, this bound will depend on the choice of coordinates.

FDTD in general curvilinear coordinates has developed in a similar way; most formulations are slightly different in that they are based on a tensor density formulation [123, 281] and employ a double set of variables (covariant and contravariant) in the nonorthogonal case; differencing involves interleaving these two sets of components at alternating time steps. They have also been used as a starting point for developing FDTD methods in 'local' coordinates defined with respect to an automatically generated grid [102, 103].

5.4 Interfaces between Grids

In $(2+1)$D, we have looked so far at numerically solving the parallel-plate system over *regular* grids—that is to say, grids whose points can be indexed with respect to some regular coordinate system. We now examine ways of connecting grids of different types and, in particular, grids of differing densities of points. The ability to decompose a domain into regions of different grid point densities is especially useful when dealing with boundaries and irregular features (i.e., variations in material parameters) throughout the problem domain; we may use a fine grid to calculate the solution to a problem in such regions, and then a coarse grid everywhere else. The problem, then, is in connecting the various subgrids so that consistency of a numerical method with the original set of equations to be solved can be maintained at the boundaries between the regions. The use of multi-grid techniques in numerical integration had developed into a very large field recently, and we cannot hope to summarize the many developments that have taken place, not even the basic theory. Multi-grid methods have been used in the TLM framework, but the structures employed there are somewhat different. In particular, proposed methods [119, 118, 44] do not, in general, enforce passivity at a coarse/fine grid interface, though they are capable of operating using different time steps in the coarse and fine meshes. One method [279] is perhaps closer in spirit to that presented here, in that the time step is everywhere the same, but in that case, certain *a priori* assumptions are made about the fields at the interface.

We will show that it is in fact possible to devise passive connections between waveguide networks (themselves generally passive, unless sources are present) that operate on different types of grids, so that passivity can be maintained in a global sense—there is, as before, an energy measure for the network that can be expressed as a weighted sum of the squares of the wave variables in the network. In this section, we will look in particular at the lossless, source-free parallel-plate equations in $(2+1)$D (system (4.59) with $r = g = e = f = h = 0$ everywhere). Furthermore, for simplicity, we will confine our attention to waveguide networks of type II as described in Section 4.4; it will be recalled that for type II networks, when $r = e = f = 0$, there is no scattering at the series junctions, and we need only calculate voltages at the parallel junctions. l and c will still be allowed to vary over the problem domain. (It may be possible to extend the multi-grid methods to be described here to the full lossy system including sources, but we will not pursue this direction here.)

5.4.1 Doubled Grid Density Across an Interface

In Figure 5.10(a) is shown an interface between a regular rectilinear grid with spacing Δ (region I) and a grid with spacing $\Delta/\sqrt{2}$ (region II) whose orientation is rotated by $45°$ with respect to that of region I. Clearly then, the density of grid points in region II is double that of region I. At any point in the interior of either regions I or II, if we are interested in solving the parallel-plate transmission line equations, we can use the rectilinear mesh described in Section 4.4. We indicate waveguide connections between junctions located at the gridpoints by black lines. At points lying on the interface between the two regions (labeled B), however, we need to develop special scattering junctions. The most straightforward arrangement requires a six-port junction at a boundary point (waves enter the junction from five irregularly spaced directions, as well as through a self-loop). Such a junction is shown in Figure 5.10(b). The problem, then, is in finding the correct admittance settings for the waveguides connected to such boundary points. If these admittances can be chosen positive and in such a way that the resulting scheme is consistent with the parallel-plate system, then we are assured convergence over the entire

188   CHAPTER 5. EXTENSIONS OF DIGITAL WAVEGUIDE NETWORKS

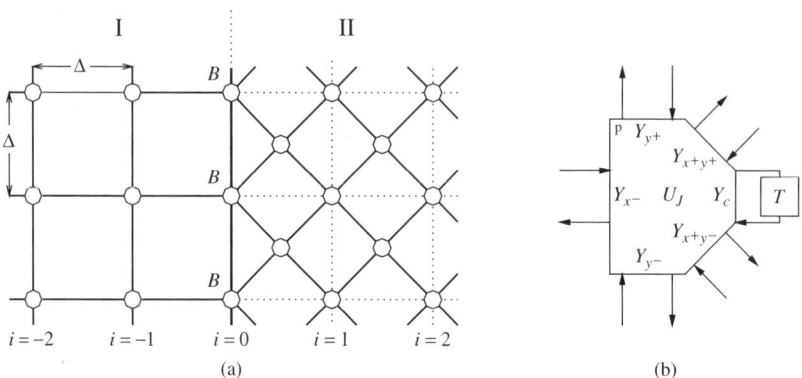

Figure 5.10: *Interface between mesh and mesh with doubled grid density*— (a) *grid arrangement, where boundary junctions are labeled B and* (b) *scattering junction at such a boundary junction.*

problem domain. We note that such interfaces bear a resemblance to the very early work of MacNeal [164], who developed asymmetric resistive networks as a means of solving elliptic problems via relaxation.

We assume that the boundary is aligned with the y-axis, so that the boundary junctions are located at coordinates $(0, j\Delta)$, for the j integer. We also assume, for the moment, that all delays in the network will be unit-sample delays (we will return to interfaces between grids with differing delay lengths in the next section). As before, we will set $v_0 = \Delta/T$. At a boundary junction at coordinates $(0, j\Delta)$, we will have six port admittances: $Y_{x^-,0,j}$, $Y_{y^-,0,j}$, and $Y_{y^+,0,j}$ corresponding to waveguide connections with junctions to the west, south, and north respectively, $Y_{x^+y^+,0,j}$ and $Y_{x^+y^-,0,j}$ for connections to junctions to the northeast and southeast respectively in region II and a self-loop admittance $Y_{c,0,j}$. The junction admittance at a boundary point B is then

$$Y_{J,0,j} \triangleq Y_{x^-,0,j} + Y_{y^-,0,j} + Y_{y^+,0,j} + Y_{x^+y^+,0,j} + Y_{x^+y^-,0,j} + Y_{c,0,j}$$

Because this waveguide mesh is an extension of the type II mesh described in Section 4.4.2, we might expect that the waveguide admittances will be related to values of the material parameters l and c at the midpoints of the waveguides. This is, in fact, true even at the boundary junctions, though because of the asymmetric nature of these junctions with respect to the coordinate axes, we must perform a judicious scaling of some of these admittances. In fact, we must only scale the admittances of the waveguides that lie along the boundary itself and that of the self-loop. The admittances of waveguides connected to interior points in region I or II should be treated as 'interior,' so that the scattering will be correct at junctions neighboring the boundary.

The difference scheme operating at a junction on the boundary will be

$$\frac{Y_{J,0,j}}{2}\left(U_{J,0,j}(n+1) + U_{J,0,j}(n-1)\right) = Y_{x^+y^+,0,j}U_{J,\frac{1}{2},j+\frac{1}{2}}(n) + Y_{x^+y^-,0,j}U_{J,\frac{1}{2},j-\frac{1}{2}}(n)$$
$$+ Y_{x^-,0,j}U_{J,-1,j}(n) + Y_{y^+,0,j}U_{J,0,j+1}(n)$$
$$+ Y_{y^-,0,j}U_{J,0,j-1}(n) + Y_{c,0,j}U_{J,0,j}(n)$$

5.4. INTERFACES BETWEEN GRIDS

If we now treat the junction voltages as samples of a continuous function u, then the difference scheme above can be expanded in a Taylor series about $(0, j\Delta, nT)$ to give

$$Y_{J,0,j} T^2 \frac{\partial^2 u}{\partial t^2} = 2\Delta \left(-Y_{x-,0,j} + \frac{1}{2}(Y_{x+y-,0,j} + Y_{x+y+,0,j}) \right) \frac{\partial u}{\partial x}$$

$$+ 2\Delta \left(Y_{y+,0,j} - Y_{y-,0,j} + \frac{1}{2}(Y_{x+y+,0,j} - Y_{x+y-,0,j}) \right) \frac{\partial u}{\partial y}$$

$$+ \Delta^2 \left(Y_{x-,0,j} + \frac{1}{4}(Y_{x+y-,0,j} + Y_{x+y+,0,j}) \right) \frac{\partial^2 u}{\partial x^2}$$

$$+ \Delta^2 \left(Y_{y-,0,j} + Y_{y+,0,j} + \frac{1}{4}(Y_{x+y-,0,j} + Y_{x+y+,0,j}) \right) \frac{\partial^2 u}{\partial y^2}$$

$$+ \frac{\Delta^2}{2} (Y_{x+y+,0,j} - Y_{x+y-,0,j}) \frac{\partial^2 u}{\partial x \partial y}$$

$$+ O(\Delta^3, T^4) \qquad (5.26)$$

In order to associate this expansion with the reduced form of the parallel-plate equations of (5.2), we may set

$$Y_{x-,0,j} = \frac{1}{vol_{-\frac{1}{2},j}} \qquad Y_{y+,0,j} = \frac{1}{2vol_{0,j+\frac{1}{2}}} \qquad Y_{y-,0,j} = \frac{1}{2vol_{0,j-\frac{1}{2}}}$$

$$Y_{x+y+,0,j} = \frac{1}{vol_{\frac{1}{4},j+\frac{1}{4}}} \qquad Y_{x+y-,0,j} = \frac{1}{vol_{\frac{1}{4},j-\frac{1}{4}}}$$

and for the self-loop admittance, we set

$$Y_{c,0,j} = \frac{3}{16} v_0 \left(2c_{-\frac{1}{2},j} + 2c_{\frac{1}{4},j+\frac{1}{4}} + 2c_{\frac{1}{4},j-\frac{1}{4}} + c_{0,j+\frac{1}{2}} + c_{0,j-\frac{1}{2}} \right)$$

$$- Y_{x-,0,j} - Y_{y-,0,j} - Y_{y+,0,j} - Y_{x+y+,0,j} - Y_{x+y-,0,j} \qquad (5.27)$$

These settings yield a difference scheme that is consistent with the parallel-plate equations, and which is first-order accurate in the grid spacing Δ. It is important to note that the admittances of the waveguides connecting two junctions on the boundary itself are set to be *half* of what they would be in the interior of region I. Also note that the mixed-derivative term in (5.26) becomes $O(\Delta^3)$ (because $Y_{x+y+,0,j}$ and $Y_{x+y-,0,j}$ are the same to zeroth order, and hence their difference, which is the coefficient of the mixed-derivative term, will be $O(\Delta)$).

The additional stability requirement, from (5.27) is

$$v_0 \geq \max_{\text{boundary waveguide midpoints}} \sqrt{\frac{8}{3lc}}$$

which is marginally more restrictive than the requirement on the interior of region I (by a factor of $\sqrt{4/3}$). This deterioration in the stability bound is offset, however, by the fact that in region II, the grid spacing is $\Delta/\sqrt{2}$ and we must have

$$\frac{\Delta}{\sqrt{2}T} \geq \max_{\text{waveguide midpoints in region II}} \sqrt{\frac{2}{lc}} \quad \Rightarrow \quad v_0 \geq \max_{\text{waveguide midpoints in region II}} 2\sqrt{\frac{1}{lc}}$$

because we are by necessity operating away from the CFL bound in this particular multi-grid setting, which incorporates different grid spacings and yet maintains the same time step throughout the mesh.

The choice of $Y_{x^-,0,j} = 1/(v_0 l_{-1/2,j})$ means that all other admittances in region I may be set as previously discussed in Section 4.4.2 for a type II mesh.

Corners

If we are interested in using a grid of doubled density over a particular region of the problem domain (in order to surround a particular feature or an irregular part of the boundary), then we are faced with specializing the scattering junctions at the corners of the region. An example of an irregular partitioning of the problem domain into two regions, I and II, is shown in Figure 5.11(a). Boundary points (labeled B) were treated previously, and it was found that waveguides connected to points B, which lie along the boundary must have their admittances set to one-half of what they would be in the interior of region I (i.e., to $\frac{1}{2v_0 l}$, where l is the inductance at the center of the particular boundary waveguide).

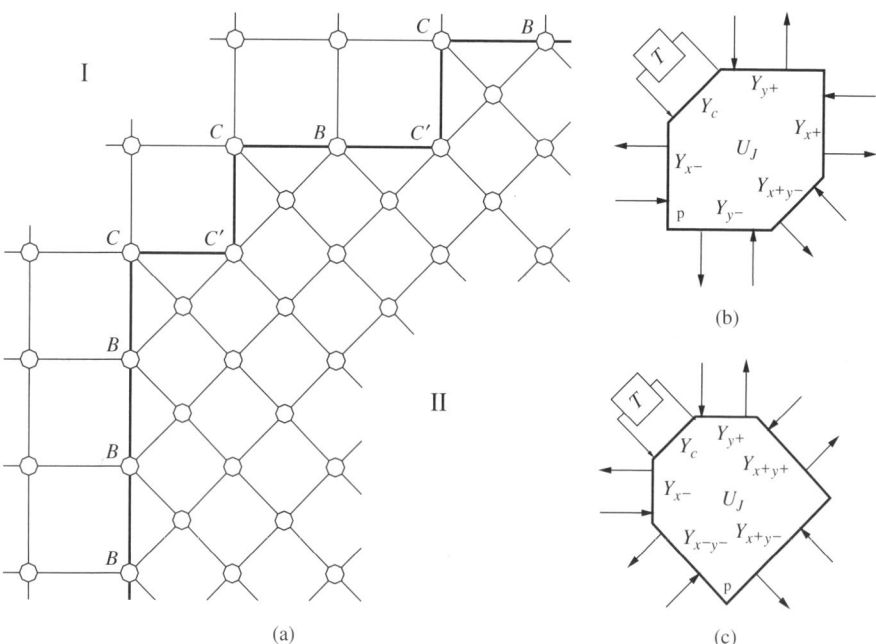

Figure 5.11: (a) *A particular grid arrangement between a rectilinear mesh I and one of doubled grid density II, (b) a scattering junction at a corner point (labeled C), and (c) a scattering junction at another type of corner point (labeled C').*

5.4. INTERFACES BETWEEN GRIDS

There are two types of corners that can occur in an irregular domain decomposition of this kind: those that are concave with respect to region I (the grid point at such a corner is labeled C in Figure 5.11(a)), and those that are concave with respect to region II (labeled C'). There are obviously four possible orientations for each type of corner, though, by symmetry, we need only treat the type shown. Because the admittances of the boundary waveguides (represented by thick lines connecting boundary points) are now prescribed, then for a corner junction, the linking admittances are fixed; only the self-loop admittance may be varied. Scattering junctions corresponding to corners of type C and C' are shown in Figure 5.11(b) and (c).

Suppose we have a corner point of type C located at coordinates $(0, 0)$. Then, from the results previously given in this section, the admittances of the five waveguides connecting this corner to its neighbors will be

$$Y_{y^+,0,0} = \frac{1}{v_0 l_{0,\frac{1}{2}}} \qquad Y_{x^-,0,0} = \frac{1}{v_0 l_{-\frac{1}{2},0}} \qquad Y_{x^+y^-,0,0} = \frac{1}{v_0 l_{\frac{1}{4},-\frac{1}{4}}}$$

$$Y_{x^+,0,0} = \frac{1}{2v_0 l_{\frac{1}{2},0}} \qquad Y_{y^-,0,0} = \frac{1}{2v_0 l_{0,-\frac{1}{2}}}$$

and the difference equation relating the corner junction voltage $U_{J,0,0}$ to those of its neighbors is

$$\frac{Y_{J,0,0}}{2}(U_{J,0,0}(n+1) + U_{J,0,0}(n-1)) = \frac{1}{v_0 l_{-\frac{1}{2},0}} U_{J,-1,0} + \frac{1}{v_0 l_{0,\frac{1}{2}}} U_{J,0,1}$$

$$+ \frac{1}{2v_0 l_{\frac{1}{2},0}} U_{J,1,0} + \frac{1}{2v_0 l_{0,-\frac{1}{2}}} U_{J,0,-1}$$

$$+ \frac{1}{v_0 l_{\frac{1}{4},-\frac{1}{4}}} U_{J,\frac{1}{2},-\frac{1}{2}} + Y_{c,0,0} U_{J,0,0}$$

where we have yet not specified $Y_{c,0,0}$ or $Y_{J,0,0}$. A Taylor expansion about $(0, 0)$ gives, in terms of the continuous variable u and neglecting higher-order terms,

$$\frac{4Y_{J,0,0}}{7v_0} \frac{\partial^2 u}{\partial t^2} = \left(\frac{\partial}{\partial x}\left(\frac{1}{l}\right) - \frac{1}{7}\frac{\partial}{\partial y}\left(\frac{1}{l}\right)\right)\frac{\partial u}{\partial x} + \left(\frac{\partial}{\partial y}\left(\frac{1}{l}\right) - \frac{1}{7}\frac{\partial}{\partial x}\left(\frac{1}{l}\right)\right)\frac{\partial u}{\partial y}$$

$$+ \frac{1}{l}\left(\frac{\partial^2 u}{\partial x^2} + \frac{\partial^2 u}{\partial y^2}\right) + \frac{2}{7l}\frac{\partial^2 u}{\partial x \partial y}$$

We can thus conclude that the differencing occurring at the corner junction is *not* consistent with the lossless source-free parallel-plate system, regardless of our choice of the self-loop admittance. The coefficients of the spatial first-derivative terms are incorrect, and there is an extra mixed-derivative term; neither vanishes in the limit as $\Delta \to 0$ (although if l is constant in a neighborhood surrounding the corner, then the first-derivative terms vanish).

The corner junction is, however, still *lossless*, as are all junctions in this waveguide network, and hence a simulation of the parallel-plate system using such a mesh will be stable, regardless of inconsistencies at the corners. It is possible to argue, loosely speaking, that if the number of corners in the interface does not grow as the grid spacing is decreased (an example of this would be an enclosed rectangular doubled density grid, for which the number of corners will be four, independently of the grid spacing), then the error at the corners will become negligible. Although we will make no attempt to prove this, the simulation that we will present later in this section concurs readily with this assertion; indeed, in all the tests we have run, any anomalous scattering at the corners is certainly far less important than the first-order scattering error (i.e., numerical reflection) along the interface itself. In the interest, however, of making any scattering error at the corner junctions as small as possible, we should set, at a corner with coordinates C (in order that the wave speed at the corner is correct, in a gross sense)

$$Y_{c,C} = \frac{7}{4}v_0 c_C - \frac{4}{v_0 l_C}$$

where l_C and c_C are the inductance and capacitance at a corner point C. The stability requirement at such a corner is then

$$v_0 \geq \sqrt{\frac{16}{7 l_C c_C}} \qquad \text{Corner } C$$

which is, like the stability condition at boundary points B, only marginally worse than the CFL bound, and mitigated by the somewhat worse bound to be found in region II (due to the decreased interjunction spacing). See the discussion earlier in this section.

A similar argument follows for points C', and we also ideally have l constant in the neighborhood of such points. The setting of the self-loop admittance $Y_{c,C'}$ should be

$$Y_{c,C'} = \frac{5}{4}v_0 c_{C'} - \frac{4}{v_0 l_{C'}}$$

The resulting stability bound is

$$v_0 \geq \sqrt{\frac{16}{5 l_{C'} c_{C'}}} \qquad \text{Corner } C'$$

Simulation

To demonstrate the behavior of the interface discussed above, we will simulate the parallel-plate system (4.59), assuming no losses or sources, over a square region with side-length 1, with short-circuited boundary conditions (i.e., $u = 0$). Over a central square region with side lengths of $\frac{1}{2}$ (bounded by the white square in Figure 5.12), we use a mesh of doubled density with respect to the outer region, where a simple rectilinear mesh is in place. The grid spacing in the outer region is $\Delta = 0.01$. We set the capacitance per unit length $c = 1$ everywhere, and the inductance per unit length l is equal to 1 in the outer region. In the inner region, it also takes the value 1, except in the interior of the black circle in Figure 5.12 (radius 0.1, center at $x = 0.6$, $y = 0.5$), where it rises, in a 2D raised cosine distribution, to a maximal value of $l = 11$.

The initial voltage distribution takes the form of a 2D raised cosine over a circular region of radius 0.07, centered at coordinates $x = 0.4$ and $y = 0.5$. The initial voltage takes a maximum of 1, and is 0 everywhere outside this circle.

5.4. INTERFACES BETWEEN GRIDS

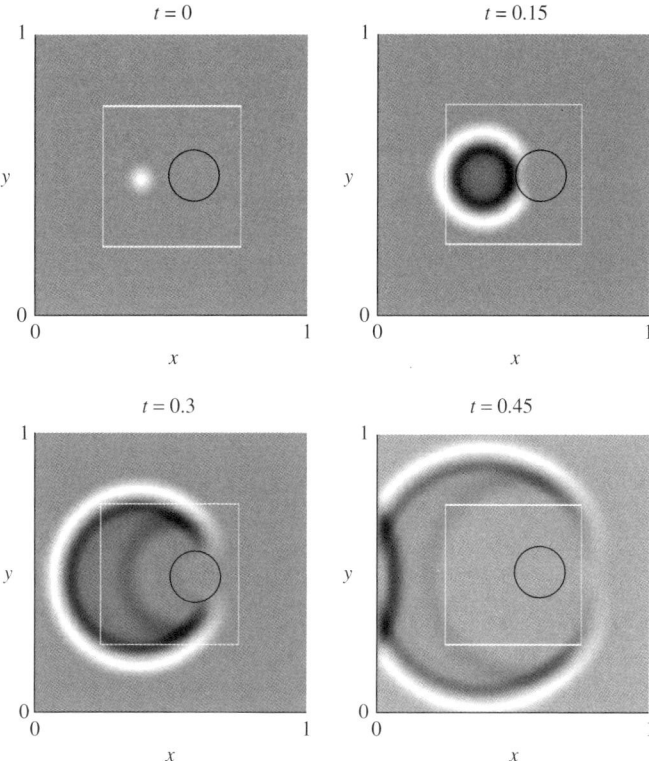

Figure 5.12: *Simulation of a parallel-plate system, with an inductive feature surrounded by a region of doubled grid density.*

The voltage distribution is shown at three successive time instants, with light and dark coloring indicating areas of positive and negative plate voltage respectively; the plots have been normalized and interpolated for better plotting results. At time $t = 0.15$, the voltage distribution has spread and begun to scatter from the inductive region inside the black circle. Note that the local wave speed decreases in the interior of the circle. By $t = 0.3$, it has progressed through the interface, with little visible numerical reflection, and by $t = 0.45$, the outward moving voltage distribution has reflected with inversion from the boundary at $x = 0$.

5.4.2 Progressive Grid Density Doubling

The delays in the bidirectional delay lines of the combined coarse/fine grid of the previous section were everywhere identical. In this section, we will look at networks for which this is not true—all delay line lengths, though, will be multiples of a common smallest unit delay, in order that the network remain synchronous [68]. It should be said, though, that even in a portion of the network where the delay line lengths are longer than a single sample delay, we will still be scattering at the rate of the smallest delay line length in the system as a whole. We will, however, be performing scattering operations at fewer points

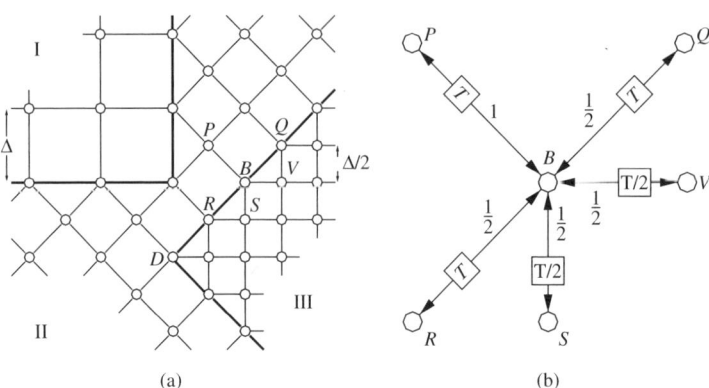

Figure 5.13: (a) *Mesh I, with grid spacing Δ, is adjoined to a doubled density mesh II, which is in turn adjoined to a quadrupled density mesh III, with grid spacing $\frac{\Delta}{2}$ and a halved waveguide delay.* (b) *A scattering junction at a point B on the boundary between layers II and III.*

in the coarse regions. We will call such structures (for lack of a better word) multirate, though it should be understood that such networks are not multirate in the signal processing sense [261] (see above comment). Specifically, we will focus on the use of such structures in order to extend the grid refinement technique introduced in the previous section.

Consider the grid arrangement of Figure 5.13(a). The interface between region I and region II was discussed in the previous section; in that case, we assumed the waveguide delays to be identical everywhere in regions I and II, including the boundary. We can, of course, apply the same idea again in order to introduce a grid of quadrupled point density, by adjoining it to region II. In this case, however, we would like to take advantage of the fact that waveguide lengths in region III are half those of region I; it is more natural, then, to use a delay of half that of region I throughout the interior of region III. Waveguides that run along the boundary will still operate at the rate of regions I and II, as will the self-loop (of admittance Y_c, not shown). A scattering junction at a typical point B on the boundary between regions II and III is shown in Figure 5.13(b). Here, we have abbreviated the depiction of a bidirectional delay line to a single double-headed arrow and have omitted the self-loop (which contains a full unit delay). Note in particular that for such a boundary junction, the delays of the connecting waveguides are now not all identical. As might be expected, the difference scheme relating the junction voltage at such a point to those of its neighbors is no longer a simple two-step difference method, but a four-step scheme, where each time step is now $\frac{T}{2}$; a full derivation of this difference scheme is very lengthy but rewarding, in the sense that it becomes clear why it takes four steps for the wave variables to fully 'recombine' into junction voltages. We will, however, only present the resulting difference equation for a boundary junction at location B as in Figure 5.13(b). Here, the junction voltage will be $U_{J,B}$ and the junction voltages at the neighboring points are, referring to Figure 5.13(b), $U_{J,P}, \ldots, U_{J,V}$. The admittances of the connecting waveguides will be called $Y_{PB}, \ldots, Y_{VB}$, the self-loop admittance $Y_{c,B}$ and the junction admittance at point B is defined as

$$Y_{J,B} = Y_{PB} + Y_{QB} + Y_{RB} + Y_{SB} + Y_{VB} + Y_{c,B}$$

5.4. INTERFACES BETWEEN GRIDS

We have

$$\frac{Y_{J,B}}{2} U_{J,B}(n+1) = Y_{VB} U_{J,V}(n+\tfrac{1}{2}) + Y_{SB} U_{J,S}(n+\tfrac{1}{2})$$
$$+ Y_{QB} U_{J,Q}(n) + Y_{PB} U_{J,P}(n) + Y_{RB} U_{J,R}(n)$$
$$+ \left(Y_{c,B} - Y_{SB} - Y_{VB}\right) U_{J,B}(n)$$
$$+ Y_{VB} U_{J,V}(n-\tfrac{1}{2}) + Y_{SB} U_{J,S}(n-\tfrac{1}{2})$$
$$- \frac{Y_{J,B}}{2} U_{J,B}(n-1)$$

A Taylor series expansion allows us to set the admittances of the waveguides to be

$$Y_{PB} = \frac{1}{v_0 l_{PB}} \quad Y_{QB} = \frac{1}{2v_0 l_{QB}} \quad Y_{RB} = \frac{1}{2v_0 l_{RB}}$$

$$Y_{SB} = \frac{1}{2v_0 l_{SB}} \quad Y_{VB} = \frac{1}{2v_0 l_{VB}}$$

where l_{XY} is the material inductance at the point midway between points X and Y. Note in particular that the setting of Y_{PB} coincides with an interior point setting of a connecting waveguide admittance in the interior of region II, from (4.65) and (4.66). These relative strengths of the connecting waveguide admittances are indicated by adjacent small numbers in Figure 5.13(b).

Because we now have a four-step scheme, the determination of $Y_{c,B}$ is no longer as simple as in the two-step case, but it can be found, nevertheless, to be

$$Y_{c,B} = \frac{v_0}{8} \left(2c_{PB} + c_{QB} + c_{RB} + c_{SB} + c_{VB}\right)$$
$$- \frac{5}{12v_0} \left(\frac{2}{l_{PB}} + \frac{1}{l_{QB}} + \frac{1}{l_{RB}} + \frac{1}{l_{SB}} + \frac{1}{l_{VB}}\right)$$

where c_{XY} signifies a waveguide midpoint evaluation of c between any points X and Y. The positivity requirement yields the bound

$$v_0 \geq \max_{\substack{\text{II/III boundary waveguide} \\ \text{midpoints}}} \sqrt{\frac{10}{3lc}}$$

Corners present essentially the same problems as before, and we will not discuss them further other than to repeat that given the settings derived above for the waveguide admittances at the II/III boundary, which determine completely the scattering behavior at the corners (an example of which is point D in Figure 5.13(a)), the mesh will not be consistent with the parallel-plate system at these points.

5.4.3 Grid Density Quadrupling

Instead of progressing in two steps from a grid to one of quadrupled point density, as we did in the last section, we might ask whether it is possible to design a direct passive interface in order to quadruple grid density in one step. Such an arrangement is shown in Figure 5.14(a): a quadrupled density region (II), for which the delay in all waveguides is a

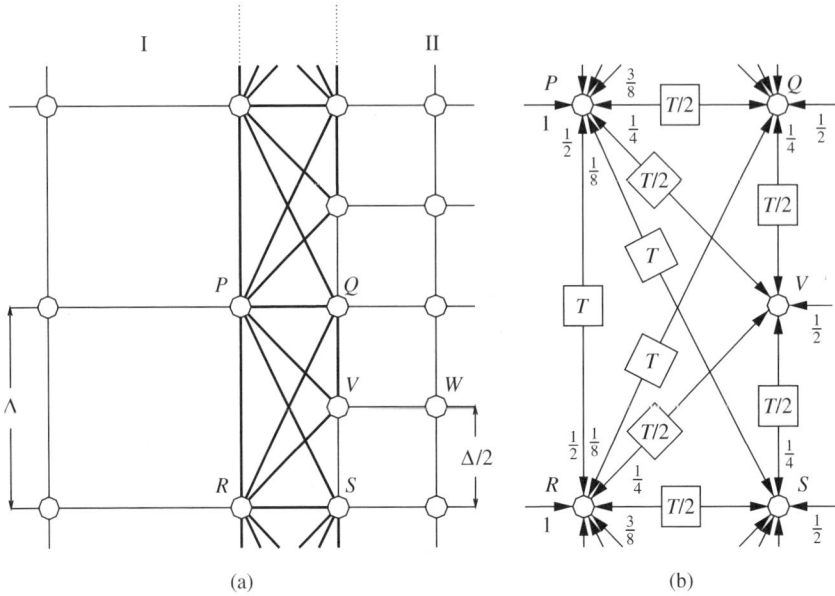

Figure 5.14: (a) *Grid I, with grid spacing* Δ, *is adjoined to a quadrupled density region II with grid spacing* $\frac{\Delta}{2}$. (b) *Detail of the interface.*

half-time step, or $\frac{T}{2}$ is adjoined to a rectilinear mesh (I) via a matching layer composed of various waveguides connecting each point on the boundary of region I with five neighboring points in region II. The admittances and delays of the waveguides in the matching layer must be set to particular values relative to the values in the interiors of regions I and II so as to satisfy the parallel-plate equations at all grid points (junctions). These delays and admittance settings are shown in Figure 5.14(b), where we have used the following notation for the admittance settings: first suppose that admittances in the interior of region I are set to the normal values for a type II mesh—namely, the admittance of a particular connecting waveguide is equal to $\frac{1}{v_0 l}$, where l is evaluated at the midpoint of the waveguide. All other connecting admittances in the network, including those in region II and in the matching later, will also be set to $\frac{1}{v_0 l}$, where l is evaluated at the center of the particular waveguide and multiplied by a scaling coefficient. In Figure 5.14(b), this scaling coefficient is shown next to the waveguide. For example, for the waveguide connecting point P to point S, the admittance should be set to

$$Y_{PS} = \frac{1}{8} \frac{1}{v_0 l_{PS}}$$

where l_{PS} is the inductance evaluated midway between points P and S. Note, in particular, that even though the space step/time step ratio is the same in region II (both quantities are halved), the admittances of the interior waveguides should be scaled by a factor of $\frac{1}{2}$. This, however, has no bearing on stability, since scaling all connecting waveguide admittances in a network by the same factor does not affect the *reflection coefficients* (though we must change the self-loop admittance settings in region II; we provide this setting shortly).

It is simple (after some tedious algebra) to set the self-loop admittances at the junctions in the matching layer such that the parallel-plate equations are approximated at these

5.4. INTERFACES BETWEEN GRIDS

points. We first define average capacitances and inverse inductances at a junction at point X by

$$\left(\overline{\frac{1}{l}}\right)_X = \frac{1}{\sum_j \beta_{Xj}} \sum_j \frac{\beta_{Xj}}{l_{Xj}} \qquad \bar{c}_X = \frac{1}{\sum_j \beta_{Xj}} \sum_j \beta_{Xj} c_{Xj}$$

where the index j runs over all the junctions to which the junction at point X is connected and where β_{Xj} is the scaling factor of the waveguide connecting point X to point j. l_{Xj} and c_{Xj} are the inductance and capacitance at the midpoint of the same waveguide. Obviously, we have $(\overline{1/l})_X = 1/l_X$ and $\bar{c}_X = c_X$ to first order in Δ. For example, we would have, from Figure 5.14,

$$\bar{c}_V = \left(\frac{1}{2} + \frac{1}{4} + \frac{1}{4} + \frac{1}{4} + \frac{1}{4}\right)^{-1} \left(\frac{1}{2} c_{VW} + \frac{1}{4} c_{VQ} + \frac{1}{4} c_{VS} + \frac{1}{4} c_{VP} + \frac{1}{4} c_{VR}\right)$$

Referring to Figure 5.14(a), it is easy to see that we need only examine self-loop admittances $Y_{c,P}$, $Y_{c,V}$, and $Y_{c,Q}$ at points P, V, and Q; all other junctions in the matching layer will behave similarly (because they map to these three points under translation in the vertical direction). We set

$$Y_{c,P} = \frac{3}{2} v_0 \bar{c}_P - \frac{43}{16 v_0}\left(\overline{\frac{1}{l}}\right)_P \qquad Y_{c,V} = v_0 \bar{c}_V - \frac{3}{2 v_0}\left(\overline{\frac{1}{l}}\right)_V$$

$$Y_{c,Q} = \frac{1}{2} v_0 \bar{c}_Q - \frac{15}{8 v_0}\left(\overline{\frac{1}{l}}\right)_Q$$

and, at an interior point in region II (such as point W),

$$Y_{c,W} = v_0 \bar{c}_W - 2\left(\overline{\frac{1}{l}}\right)_W$$

It is interesting that, in contrast to situation at the grid-doubling interface presented in the last section, the positivity conditions that arise from these self-loop admittances do not degrade the stability bound which arises from self-loop admittances in the interior of either region I or II. We note that it would appear (through trial and error) that this is the simplest density-quadrupling layer possible.

We also show some numerical results, comparing the numerical reflection error of the density-doubling and -quadrupling layers for normally incident waves. Waves (one period of a raised cosine) impinge on the layer from the side of smaller grid density (region I) in both cases. In Figure 5.15, we have plotted the log of the ratio of the reflected energy to the incident energy ($E_{\text{refl}}/E_{\text{inc}}$) versus the log of the number of grid spacings Δ per wavelength λ. E_{refl} and E_{inc} were computed by taking the sum of the squares of the junction quantities U_J over region I, before and after the passage of the wave through the interface.

The density-doubling layer (solid line in Figure 5.15) leads to a smaller reflected energy than the quadrupling layer (dashed line). This is to be expected—in general, the more abrupt a change in grid density, the more numerical reflection will result. As indicated

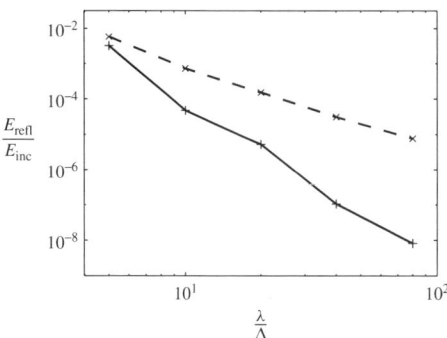

Figure 5.15: *Log plot of the ratio of reflected energy to incident energy versus number of grid points per wavelength, for a wave normally incident from a region containing a standard rectilinear mesh on two types of interface—density doubling (solid line) and density quadrupling (dashed line).*

in Figure 5.15, however, the reflected energy will always tend to zero as grid density is increased on both sides of the interface. The case of oblique incidence has not been examined; it would appear, also, to be possible to derive an analytic expression for the numerical reflectance, though we have not done so here.

It would be very interesting to know whether the coarse/fine mesh arrangements discussed in this section and the last could be made truly multirate—that is, could we have scattering operations that recur with period T in the coarse mesh and period $\frac{T}{2}$ in the fine mesh? As mentioned earlier, there are TLM structures that are capable of this, but they, in general, perform time-averaging of wave quantities, which may render the interface nonpassive [118, 119].

We also note that because interfaces between grids operating at different rates by necessity correspond to multistep methods, we may also see parasitic modes [238] appearing along the interface (though we are assured convergence in the limit as the grid spacing becomes small). A discussion of parasitic modes in MDWD networks appears in Section 3.8.2.

5.4.4 Connecting Rectilinear and Radial Grids

As another example of a passive interface between different types of waveguide meshes, we examine the means by which grids defined in different coordinate systems may be connected for the special (but quite practically important) case of the connection between a rectilinear and radial grid. Such a grid would be useful in cases where it is desired to solve the parallel-plate system or wave equation over some region that has boundaries that are straight in some places but circularly curved in others. One could, in general, proceed by attempting to find a global coordinate transformation that maps an irregular region to a regular one (like a rectangle) and then developing a waveguide mesh in the new coordinates, as per the methods discussed in Section 5.3. It is perhaps simpler, however, to use rectilinear and radial meshes at appropriate places in the domain, and then define a matching layer at the boundary between the regions, which should also be locally consistent with the equations to be solved. Consider the grid arrangement of Figure 5.16.

5.4. INTERFACES BETWEEN GRIDS

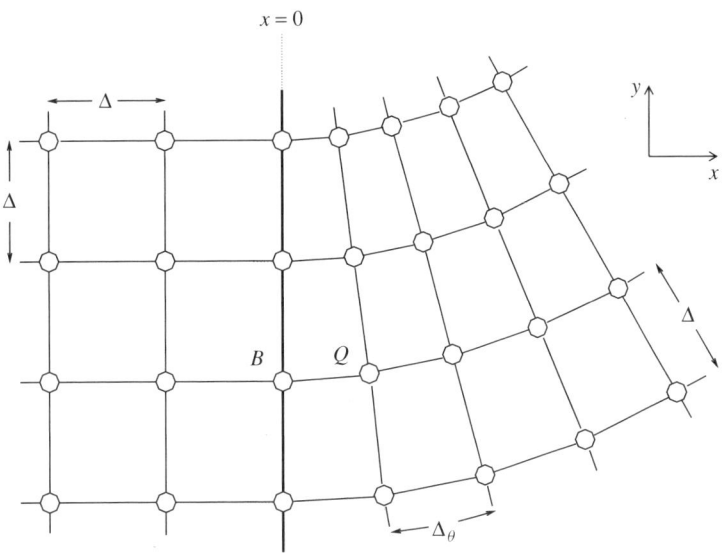

Figure 5.16: *Interface between radial and rectilinear meshes.*

We have a type II radial waveguide mesh in $x > 0$ and a type II rectilinear mesh in $x < 0$; parallel junctions are to be placed at all the grid points and waveguide connections (bidirectional delay lines of delay T) are indicated by connecting lines. Special boundary waveguides, which lie along the y axis, are drawn in bold. All interior waveguide admittances in either region (i.e., all admittances except for those of the boundary waveguides) are assumed to be set to the values that they must take in the interior in order to solve the lossless source-free parallel-plate equations, as given in (4.65)–(4.67) and (5.14)–(5.17). Self-loops are of course required in general at all junctions, though for simplicity they are not represented in Figure 5.16. The spacing Δ between junctions in the rectilinear mesh is assumed to be equal to the radial grid spacing in the radial mesh. The angular spacing in this same grid, Δ_θ, may be set independently. We will use $v_0 = \Delta/T$ here.

In order to derive the admittance and self-loop settings at the boundary, we may examine a junction at coordinates $(0, j\Delta)$, j integer (one such point is labeled B in Figure 5.16). In keeping with the notation for a rectilinear mesh (used in $x < 0$), we call the admittances of the four connecting waveguides at such a point $Y_{x-,0,j}$, $Y_{x+,0,j}$, $Y_{y-,0,j}$, and $Y_{y+,0,j}$, and the self-loop admittance $Y_{c,0,j}$. The junction admittance $Y_{J,0,j}$ is then the sum of these five admittances. The junction voltage at point B will be called $U_{J,0,j}$, and we will call the junction voltage at point Q directly to the right $U_{J,Q}$. The difference scheme in the junction voltages resulting from such a mesh is then

$$\frac{Y_{J,0,j}}{2}\left(U_{J,0,j}(n+1) + U_{J,0,j}(n-1)\right) = Y_{x-,0,j}U_{J,-1,j}(n) + Y_{x+,0,j}U_{J,Q}(n)$$
$$+ Y_{y-,0,j}U_{J,0,j-1}(n) + Y_{y+,0,j}U_{J,0,j+1}(n)$$
$$+ Y_{c,0,j}U_{J,0,j}(n)$$

Expansion in terms of a Taylor series about $(0, j\Delta, nT)$ gives, in terms of the continuous function $u(x, y, t)$,

$$\frac{Y_{J,0,j}T^2}{2}\frac{\partial^2 u}{\partial t^2} = \Delta\left(Y_{y+,0,j} - Y_{y-,0,j}\right)\frac{\partial u}{\partial y} + \Delta\left(\rho_j\lambda Y_{x+,0,j} - Y_{x-,0,j}\right)\frac{\partial u}{\partial x}$$

$$+ \frac{\Delta^2}{2}\left(Y_{y+,0,j} + Y_{y-,0,j}\right)\frac{\partial^2 u}{\partial y^2}$$

$$+ \frac{\Delta^2}{2}\left(Y_{x-,0,j} + \rho_j^2\lambda^2 Y_{x+,0,j}\right)\frac{\partial^2 u}{\partial x^2} \tag{5.28}$$

where we have discarded higher-order terms in Δ and used $\lambda = \Delta_\theta/\Delta$ and ρ_j as in Section 5.1.2.

First examine, in (5.28), the coefficient of $\frac{\partial u}{\partial x}$ on the right-hand side. Notice that in order for this term to behave as $O(\Delta^2)$, we must have $Y_{x-,0,j} = \rho_j\lambda Y_{x+,0,j} + O(\Delta)$. Since $Y_{x+,0,j}$ is assumed set as an interior admittance in the type II radial waveguide mesh, namely, from (5.15) as

$$Y_{x+,0,j} = \frac{1}{\lambda^2 v_0 l_{BQ}\rho_j}$$

where l_{BQ} is the value of the inductance at the midpoint of the waveguide connecting points B and Q, we may choose

$$Y_{x-,0,j} = \frac{1}{\lambda v_0 l_{-\frac{1}{2},j}}$$

Because $Y_{x-,0,j}$ is to be interpreted as an interior admittance in the rectilinear mesh, it is clear that all connecting admittances in this mesh should incorporate this same scaling factor of λ.

Since a boundary waveguide can be interpreted as belonging to *both* waveguide meshes, a good initial guess as to its admittance might be a simple linear average of the admittances of interior radial and rectilinear waveguides located at the same position. This would give

$$Y_{y+,0,j} = Y_{y-,0,j+1} = \frac{1}{2v_0\lambda}l_{0,j+\frac{1}{2}}\left(1 + \frac{1}{\rho_{j+\frac{1}{2}}\lambda}\right)$$

It is straightforward (but tedious) to show that these admittances do indeed yield a difference scheme that is consistent with the lossless source-free parallel-plate system, provided that we set

$$Y_{c,0,j} = \frac{c_{-\frac{1}{2},j}v_0}{2\lambda} + \frac{c_{0,j+\frac{1}{2}}v_0}{4\lambda}\left(\rho_{j+\frac{1}{2}} + \frac{1}{\lambda}\right) + \frac{c_{0,j-\frac{1}{2}}v_0}{4\lambda}\left(\rho_{j-\frac{1}{2}} + \frac{1}{\lambda}\right) + \frac{v_0 c_{BQ}\rho_j}{2}$$

$$- Y_{x-,0,j} - Y_{x+,0,j} - Y_{y-,0,j} - Y_{y+,0,j}$$

where c_{BQ} is the capacitance at the midpoint of the waveguide joining points B and Q, for any j. The stability bound is identical to that obtained in the interior of the radial mesh, and this type of matching layer requires very little extra programming effort in an implementation.

5.4. INTERFACES BETWEEN GRIDS

Simulation: Solving the Acoustic Wave Equation in a U-shaped Tube

As a simple application of an interface between radial and rectilinear meshes, we solve the $(2+1)$D acoustic wave equation (1.18) in a U-shaped tube—that is, a tube consisting of two straight segments connected by a semicircular radial tube. We note that the $(3+1)$D version of this problem was first solved using a waveguide mesh, in the context of physical modeling of brass instruments, in [269]. In that case, a rectilinear mesh was used over the entire domain, which forces a staircase-type approximation to the radial boundary. Here, however, because we are modeling the semicircular tube in radial coordinates, boundary conditions can be implemented in a well-defined manner; the trade-off will be in the added numerical reflection at the radial/rectilinear mesh interface.

The evolution of the voltage in the tube is shown in Figure 5.17. We assume a wave speed of 1 throughout the interior and an open-circuited boundary condition (i_n, the normal current density component is 0 everywhere on the boundary). The thickness of the tube is 0.2, the inner radius of the circular portion is 0.1, and the lengths of the two straight tube segments are each 0.3. The grid spacing is set to $\Delta = 0.005$, and white lines indicate the interfaces between the radial mesh and the adjacent rectilinear meshes. The input signal is a raised cosine voltage of period 0.2 at the upper left-hand entrance to the tube. Here, u can be interpreted as a pressure in a tube with hard boundaries; as mentioned above, this situation comes up in the modeling of brass instruments, and in particular in curved sections of tubes such as trombone slides or crooks (though we have used, for illustrative purposes, an unnaturally short wavelength; longer wavelengths in the musical range can be modeled more cheaply using a coarse grid, and numerical reflections will be even smaller as the number of grid points per wavelength increases).

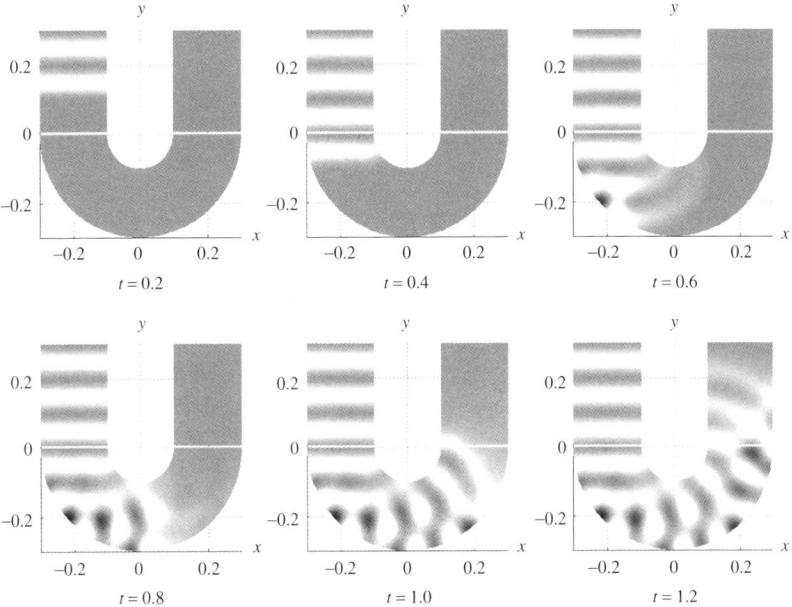

Figure 5.17: *Simulation of wave propagation in a U-shaped tube.*

5.4.5 Grid Density Doubling in (3 + 1)D

It is straightforward to extend the grid density-doubling technique presented in Section 5.4.1 to (3 + 1)D. As in Section 5.2, we will assume that our mesh is to simulate the (3 + 1)D wave equation (5.20).

If we have a rectilinear mesh (region I in Figure 5.18(a)) with interjunction spacing Δ to which we would like to adjoin a mesh of doubled grid density, we may introduce an octahedral mesh (region II) with a grid spacing of $\sqrt{(3/4)}\Delta$ to be connected to region I across an interface; special matching waveguide connections at this interface are shown in bold in Figure 5.18(a). The spacing of $\sqrt{(3/4)}\Delta$ in region II is chosen so that the octahedral grid may be decomposed into two offset rectilinear grids of spacing Δ, and may hence be aligned with region I at the interface.

The relative admittances at the special boundary 10-ports (nine connecting waveguides and a self-loop) are shown in Figure 5.18(b); only the waveguides in bold in (a) take on special values; all others may be set as interior admittances. For consistency with the wave equation, we must choose the admittances in region I to be double those in region II. In addition, the self-loop admittance at a boundary junction (labeled B in Figure 5.18) must be chosen to be

$$Y_{c,B} = \frac{3}{2}v_0 c - \frac{5}{v_0 l} \qquad \text{Boundary junction}$$

where we have chosen the connecting admittance in region I to be $\frac{1}{v_0 l}$, and where we must have $\gamma = \sqrt{1/lc}$ as discussed in Section 5.2. The stability requirement at such a boundary junction B is then

$$v_0 \geq \sqrt{\frac{10}{3}}\gamma \qquad \text{Boundary junction}$$

which is worse than the bound over the interior of region I, as given in (5.21). On the other hand, because the grid spacing is $\sqrt{3/4}\Delta$ in region II, we must set, at an interior point in

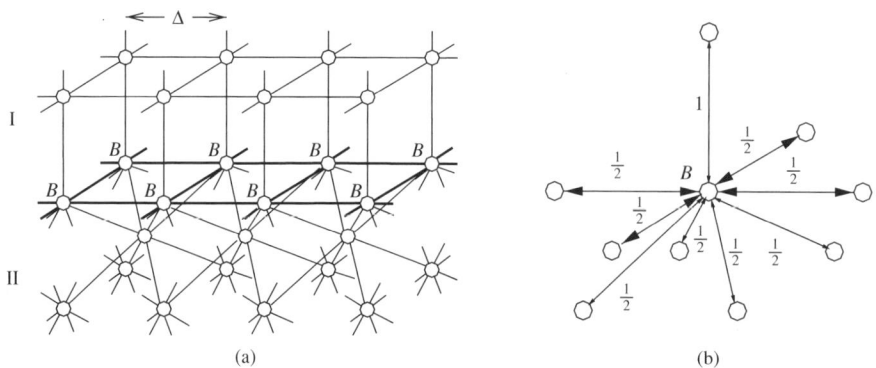

Figure 5.18: (a) *Grid I, with grid spacing* Δ, *is adjoined to a doubled density region II with grid spacing* $\sqrt{3}\Delta/2$. (b) *A junction at the interface.*

region II,

$$Y_c = cv_0 - \frac{4}{v_0 l} \quad \Rightarrow \quad v_0 \geq 2\gamma \quad \text{in region II}$$

This bound is, as expected, worse still than the boundary requirement and hence boundary scattering, as in the (2 + 1)D case, does not compromise the overall requirement on space step/time step ratio that is forced by the settings in region II.

We briefly mention that at edges and corners of a doubled density region, the waveguide mesh cannot be made consistent with the wave equation. We may invoke the same argument as in the (2 + 1)D case, namely, that numerical reflection should be minimal, and should vanish in the limit as the grid spacing becomes small. Special admittance values for the connecting waveguides along such edges and for the self-loops at both edges and corners may be chosen such that numerical reflection is made as small as possible. We do not provide here those values, due to the ease with which they can be calculated, and the relatively large number of cases that must be considered (two edge types and four corner types).

We have not investigated ways of redoubling (quadrupling) grid density as was done in the (2 + 1)D case, though we would conjecture that it should be possible, either via a direct quadrupling layer with a special interface (analogous to the scheme of Section 5.4.3) or via successive redoubling (as per Section 5.4.2).

5.4.6 Note

We would like to add that an interesting future direction of development of waveguide meshes might involve the use of fully unstructured grids—that is, grids whose points cannot be ordered according to some regular indexing system. An unstructured DWN is certainly conceivable (and has been used for artificial reverberation [225]), though it is difficult to design a structure that is locally consistent with the continuous parallel-plate problem (though it will be passive, regardless). Such a structure would, of course, be very useful for dealing with irregular geometries, which arise in many real-world problems. On the other hand, FDTD has evolved in this direction, specifically by making use of the *finite volume method* [120], long known in the fluid dynamics community. The FVTD (*finite volume time domain*) method is the result—the general idea is that instead of using finite differences to approximate derivative terms, the integral forms of the governing equations (Maxwell's, for the FDTD people, but mechanical systems can be treated equally easily) are discretized over cells of finite size, which may not have any particular ordering. For a look at some recent work in this area, we refer the reader to [34] and [200].

Chapter 6

Scattering Methods: A Unified Perspective

At this point, the reader may have noticed more than a few similarities between the treatment of PDE systems in Chapter 3 and in Chapters 4 and 5. It should be recalled that the multidimensional wave digital (MDWD) networks that numerically integrate these systems are derived from multidimensional Kirchhoff circuits (MDKCs) under a set of coordinate transformations and continuous-to-discrete spectral mappings. Up until now, we have treated digital waveguide networks (DWN) as large collections of scattering junctions connected by paired delay lines, just like the Kelly–Lochbaum model of Section 1.1.2 and the mesh of Section 1.1.3. While this is a useful vantage point, especially when it comes to constructing irregular networks such as those discussed in Section 5.4, and for finding proper passive boundary terminations, it is somewhat lacking in that it does not allow the algorithm designer any guidance in the construction of these methods for more complex systems. Indeed, when faced with a many-variable system (such as, for example, the 13-variable system of PDEs that models vibration in a stiff cylindrical shell, to be discussed in Section 7.3.2), it becomes difficult to proceed as was done for the comparatively simple transmission line test problems in Section 4.3 and Section 4.4. An MDWD-based method does not fall prey to these design difficulties because it follows directly from a multidimensional circuit representation of a given defining system of PDEs; in other words, a passive numerical method can be automatically generated from the model system, regardless of its complexity. It is true, however, that this circuit representation is highly nonunique—we have all of classical network theory at our disposal in order to manipulate it. What is more, while WDF discretization is based on the use of the trapezoid rule (or bilinear transform), in multiple dimensions, the family of passive integration rules is much more general.

In this chapter, we use the flexibilities mentioned above in order to show that a DWN also can be viewed as the discrete image of an MDKC. In this way, the DWN may be considered to be the product of a multidimensional wave digital network in its own right; the range of physical systems to which the DWN can be applied as a simulation method is thus considerably enlarged to include any system that has been dealt with using MDWD networks. MDWD networks and DWNs are now on an equal footing (and we will emphasize

Wave and Scattering Methods for Numerical Simulation   Stefan D. Bilbao
© 2004 John Wiley & Sons, Ltd   ISBNs: 0-470-87017-6

CHAPTER 6. SCATTERING METHODS: A UNIFIED PERSPECTIVE

the fraternal relationship between the two methods repeatedly in Chapter 7). The behavior of TLM and wave digital numerical integration methods have been previously compared [101, 176], and some of the material in this chapter has appeared previously [24].

Summary

We develop, in Section 6.1, an alternative MDKC representation suitable for DWN discretization for the $(1+1)$D transmission line system through the introduction of *multidimensional unit elements* and an alternative spectral mapping. We apply the same techniques to the $(2+1)$D parallel-plate system in Section 6.2, then continue our previously postponed treatment of higher-order spatially accurate DWNs in Section 6.3, and finally conclude with a DWN for Maxwell's equations in Section 6.4. The DWN for this last system, in that it is equivalent to Yee's original FDTD formulation, completes the circle of ideas begun in Chapter 4.

6.1 The $(1+1)$D Transmission Line Revisited

It is instructive to first reconsider the $(1+1)$D transmission line system in the lossless source-free case. In Figure 6.1 are presented both the type III DWN and the MDWD network for the same system using offset sampling, with spatial dependence expanded out. For the

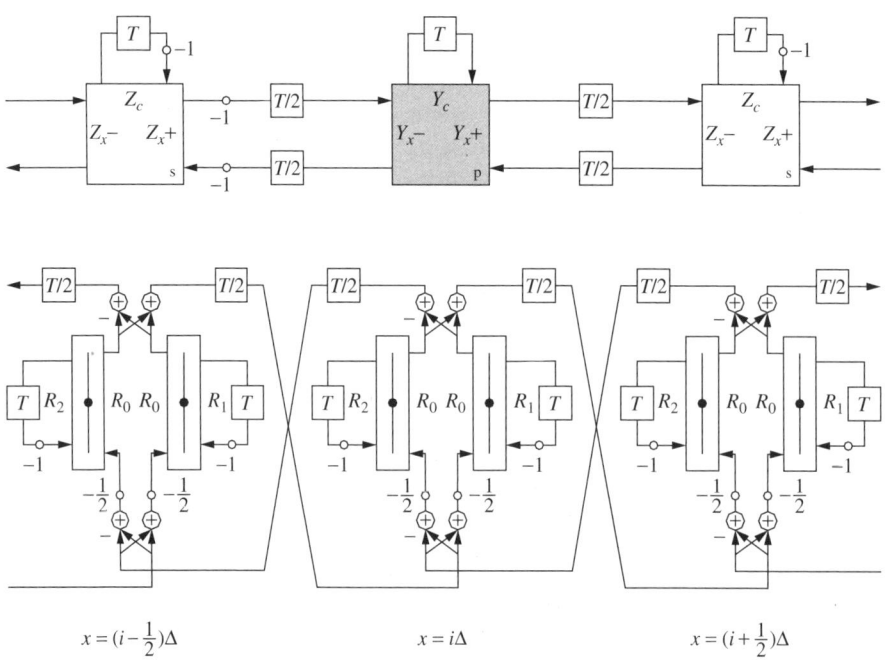

Figure 6.1: *Signal flow diagrams of the DWN and MDWD networks for the $(1+1)$D lossless source-free transmission line.*

6.1. THE (1 + 1)D TRANSMISSION LINE REVISITED

MDWD network, we have chosen a grid spacing of $\Delta/2$ and a time step of $T/2$ so as to align it with the DWN. (In other words, we have used $T_1 = T_2 = \Delta/\sqrt{2}$—see Section 3.6 for details.)

First, notice that for the DWN we have one three-port scattering junction at each grid point, and that parallel junctions are interleaved with series junctions. Approximations to i and u are calculated at alternating grid points, and at alternating multiples of the time step, $T/2$. For the MDWD network, we have two two-port series adaptors at each grid point; we are approximating both i and u *together* at the same locations, though due to offset sampling these locations alternate from one time step to the next. Both variables are treated as *currents*. For both the networks, all the material variation of the transmission line is expressed in the immittances of self-loops at every junction or adaptor, the delay in which is twice that of the linking delay between adjacent grid points. It is also useful to compare the waveguide immittances to the port resistances of the MDWD network. We have

$$R_1 = \frac{2}{\Delta}(v_0 l - r_0) \qquad R_2 = \frac{2}{\Delta}\left(v_0 r_0^2 c - r_0\right) \qquad R_0 = \frac{4r_0}{\Delta}$$

and

$$Z_c = 2(v_0 l - r_0) \qquad Y_c = 2\left(v_0 c - \frac{1}{r_0}\right) \qquad Z_{x-} = Z_{x+} = \frac{1}{Y_{x-}} = \frac{1}{Y_{x+}} = r_0$$

where we recall, from the discussion of the type III waveguide network in Section 4.3.6, that the connecting impedances were chosen to be some constant Z_{const}, in this case $Z_{\text{const}} = r_0$. Enforcing the positivity of R_1 and R_2 or Z_c and Y_c leads to identical stability conditions, and the self-loop immittances and inductances are simply related to one another by

$$Z_c = \Delta R_1 \qquad Y_c = \Delta R_2 / r_0^2$$

at locations where both quantities coexist in their respective discrete networks.

Most important, though, is the observation that whereas the DWN can be considered to be made up of an array of lumped two-port bidirectional delay lines, the signal flow diagram of the MDWD network in Figure 6.1 does not have such an interpretation—the port in this setting is defined only as a multidimensional object and instances of this port in the discrete domain are *not* connected portwise. It is crucial to recognize that passivity of such a discrete network is enforced by the power conservation of the scattering operation and not by where wave variables go in the network after they have been scattered; and in particular, they need not be paired as they are for waveguide networks, as long as the shifting operation that they undergo subsequently does not increase energy in the network. On the other hand, as we shall see in Chapter 7, boundary conditions are much easier to implement in a lumped network.

6.1.1 Multidimensional Unit Elements

As a first step toward reintroducing this port structure to a MDWD network (and hence toward relating the DWN to the MDWD network), we can extend the definition of the unit element (and recall from Section 2.2.4 that the unit element is a wave digital two-port, which is equivalent to a single-sample bidirectional delay line) to multiple dimensions in the following

208   CHAPTER 6. SCATTERING METHODS: A UNIFIED PERSPECTIVE

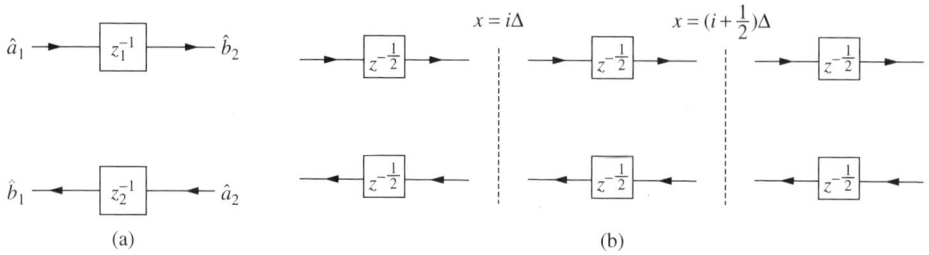

Figure 6.2: (a) *Multidimensional unit element at steady state employing shifts in directions t_1 and t_2 and* (b) *its steady state schematic when spatial dependence is expanded out.*

way[1]. Suppose we are dealing with a (1+1)D system, and new coordinates t_1 and t_2 are as defined by (3.19). We thus have two transform frequencies s_1 and s_2 as well as two frequency-domain shift operators z_1^{-1} and z_2^{-1} in the two directions t_1 and t_2. The multidimensional two-port defined, at steady state, by

$$\begin{bmatrix} \hat{b}_1 \\ \hat{b}_2 \end{bmatrix} = \begin{bmatrix} 0 & z_2^{-1} \\ z_1^{-1} & 0 \end{bmatrix} \begin{bmatrix} \hat{a}_1 \\ \hat{a}_2 \end{bmatrix} \quad (6.1)$$

and shown in Figure 6.2(a) bears some resemblance to the lumped unit element discussed initially in Section 2.2.4; it is clearly lossless (because it merely corresponds to a pair of shifts), but, unlike the unit element, it is no longer reciprocal. In this last respect, we remark that a multidimensional element so defined is perhaps closer in spirit to a generalization of the so-called quasi-reciprocal line (QUARL) proposed by Fettweis [68]. The two port resistances are assumed identical and equal to some positive constant R. When the spatial dependence is expanded out, it appears as an entire array of unit elements, as in Figure 6.2(b), where we have assumed

$$z_1^{-1} = z^{-\frac{1}{2}} w^{-\frac{1}{2}} \qquad\qquad z_2^{-1} = z^{-\frac{1}{2}} w^{\frac{1}{2}}$$

where $z^{-\frac{1}{2}}$ and $w^{-\frac{1}{2}}$ correspond, respectively, to unit shifts in time and space by $T/2$ and $\Delta/2$. We have thus chosen $T_1 = T_2 = \Delta/\sqrt{2}$.

This element, like the standard unit element, is defined in the discrete (time and space) domain using wave variables. Rewriting the scattering relation in terms of steady state discrete voltage and current amplitudes, (6.1) becomes the impedance relationship

$$\begin{bmatrix} \hat{v}_1 \\ \hat{v}_2 \end{bmatrix} = \frac{R}{1 - z_1^{-1} z_2^{-1}} \begin{bmatrix} 1 + z_1^{-1} z_2^{-1} & 2 z_2^{-1} \\ 2 z_1^{-1} & 1 + z_1^{-1} z_2^{-1} \end{bmatrix} \begin{bmatrix} \hat{i}_1 \\ \hat{i}_2 \end{bmatrix} \quad (6.2)$$

6.1.2 Hybrid Form of the Multidimensional Unit Element

We now have an impedance relationship describing the multidimensional unit element in terms of the discrete frequency variables z_1^{-1} and z_2^{-1}. It is of interest, however, to introduce

---

[1] Fettweis has already defined a multidimensional unit element [66], but in that case, shifts in a single direction were used in both delay paths; the multidimensional unit element defined here can be thought of as a simple generalization of this structure.

a particular type of *hybrid form* [14]. The reason for doing this ultimately has to do with the fact that in a DWN in interleaved form, such as that shown in Figure 4.12 or 4.19, a typical linking waveguide (or unit element) is connected in parallel at one port and in series at the other; it is somewhat easier to make the transition from wave digital filters to digital waveguide networks if we take account of this asymmetry.

Suppose we have a two-port that is defined, at steady state, by the relationship $\hat{\mathbf{v}} = \mathbf{Z}\hat{\mathbf{i}}$, or

$$\begin{bmatrix} \hat{v}_1 \\ \hat{v}_2 \end{bmatrix} = \begin{bmatrix} Z_{11} & Z_{12} \\ Z_{21} & Z_{22} \end{bmatrix} \begin{bmatrix} \hat{i}_1 \\ \hat{i}_2 \end{bmatrix}$$

This can be rewritten in a so-called hybrid form as

$$\begin{bmatrix} \hat{v}_1 \\ \hat{i}_2 \end{bmatrix} = \frac{1}{Z_{22}} \begin{bmatrix} Z_{11}Z_{22} - Z_{12}Z_{21} & Z_{12} \\ -Z_{21} & 1 \end{bmatrix} \begin{bmatrix} \hat{i}_1 \\ \hat{v}_2 \end{bmatrix}$$

or as

$$\hat{\mathbf{p}} = \mathbf{K} \hat{\mathbf{q}}$$

For the multidimensional unit element defined by (6.1), the hybrid matrix is, given the impedance relation (6.2),

$$\mathbf{K}_{ue}(z_1^{-1}, z_2^{-1}) = \frac{1}{1+z_1^{-1}z_2^{-1}} \begin{bmatrix} R(1 - z_1^{-1}z_2^{-1}) & 2z_2^{-1} \\ -2z_1^{-1} & \frac{1}{R}(1 - z_1^{-1}z_2^{-1}) \end{bmatrix}$$

It should be clear that the definition of the unit element holds regardless of which complex frequencies we choose. In particular, the delays z_1^{-1} and z_2^{-1} could be replaced by delays in higher-dimensional spaces (we will make use of these in Sections 6.2, 6.3 and 6.4). We will enforce the order of the arguments of $\mathbf{K}_{ue}$ so that, for example, $\mathbf{K}_{ue}(z_1^{-1}, z_2^{-1})$ and $\mathbf{K}_{ue}(z_2^{-1}, z_1^{-1})$ refer to unit elements of mirror-image orientation.

Suppose now that we have N two-ports defined by their impedance and hybrid relationships

$$\hat{\mathbf{v}}_k = \mathbf{Z}_k \hat{\mathbf{i}}_k \qquad \hat{\mathbf{p}}_k = \mathbf{K}_k \hat{\mathbf{q}}_k \qquad k = 1, \ldots, N$$

If we are interested in connecting the first ports of all N two-ports to each other in series, and the second ports in parallel as in Figure 6.3, then we will have, for the total voltages and currents

$$v_1 = \sum_{k=1}^{N} v_{1k} \qquad i_2 = \sum_{k=1}^{N} i_{2k}$$

and

$$i_{1k} = i_1 \qquad v_{2k} = v_2 \qquad k = 1, \ldots, N$$

which hold instantaneously. In order to describe the two-port resulting from the connection, we may write

$$\hat{\mathbf{p}} = \sum_{k=1}^{N} \hat{\mathbf{p}}_k = \sum_{k=1}^{N} \mathbf{K}_k \hat{\mathbf{q}}_k = \left(\sum_{k=1}^{N} \mathbf{K}_k \right) \hat{\mathbf{q}}$$

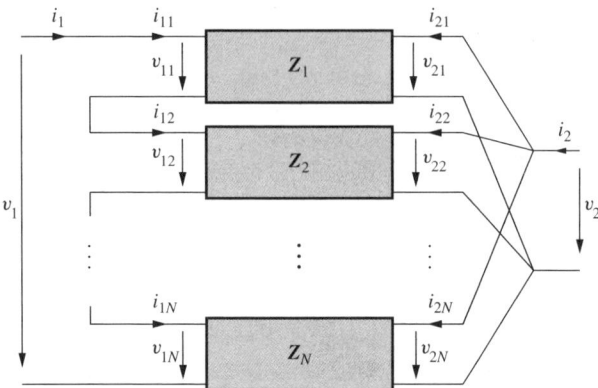

Figure 6.3: *Series/parallel connection of N two-ports.*

For such a series/parallel combination of two-ports, the hybrid matrix of the connection is simply the sum of the hybrid matrices of the individual two-ports:

$$\mathbf{K} = \sum_{k=1}^{N} \mathbf{K}_k \qquad \text{Series/parallel combination of } N \text{ two-ports}$$

6.1.3 Alternative MDKC for the (1+1)D Transmission Line

We now reexamine the lossless, source-free (1+1)D transmission line equations and show how, after a simple network manipulation and under the alternative spectral mappings mentioned in Section 3.4.4, we end up with an interleaved DWN identical to that discussed in Section 4.3.6.

We begin by first scaling the (1+1)D transmission line equations by a factor of Δ, and where, as before, we introduce a scaled time variable defined by $t' = v_0 t$, giving

$$\Delta v_0 l \frac{\partial i}{\partial t'} + \Delta \frac{\partial u}{\partial x} = 0 \qquad (6.3\text{a})$$

$$\Delta v_0 c \frac{\partial i}{\partial t'} + \Delta \frac{\partial u}{\partial x} = 0 \qquad (6.3\text{b})$$

It should be clear that the scaling by Δ will have no effect on the solution to equations (6.3), even in the limit as $\Delta \to 0$. The MDKC for this system is shown in Figure 6.4(a), where we have used the coordinate transformation defined by (3.19). Apart from the scaling of the element values by Δ, this is identical to the MDKC of Figure 3.14(a), where the resistors and voltage sources (corresponding to loss and source terms) have been omitted (they can be simply reintroduced at a later stage).

The element values are given by

$$L_1 = \Delta \left(v_0 l - r_0 \right) \qquad L_2 = \Delta \left(v_0 c r_0^2 - r_0 \right) \qquad L_0 = \Delta r_0 / \sqrt{2}$$

In this representation, the transmission line voltage u is considered, after a scaling by $1/r_0$, to be a *current*. This is somewhat unsatisfying from a physical point of view; it is easy,

6.1. THE (1+1)D TRANSMISSION LINE REVISITED

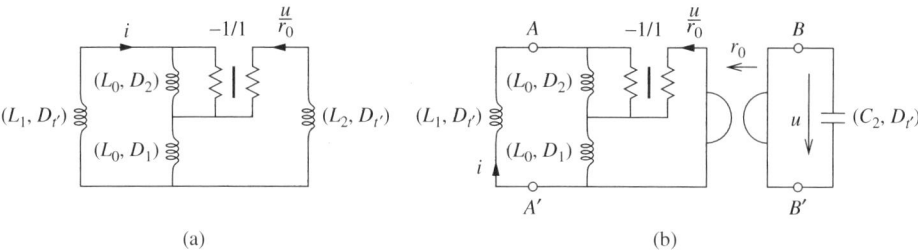

Figure 6.4: *MDKCs for the lossless source-free (1+1)D transmission line equations— (a) the standard representation and (b) a modified form.*

however, to rectify this: by a simple network transformation, the inductor in the right-hand loop (through which current u/r_0 flows) may be replaced by a *gyrator* of gyrator coefficient r_0 terminated on a *capacitor*. The voltage across this capacitor of capacitance $C_2 = \Delta(v_{0c} - 1/r_0)$ will then be exactly u. (See Figure 6.4(b)).

Consider now the two-port of Figure 6.4(b), with terminals A, A', B, and B'. The hybrid matrix for this linear shift-invariant two-port is, in terms of the frequencies s_1 and s_2,

$$\mathbf{K}(s_1, s_2) = \frac{\Delta}{\sqrt{2}} \begin{bmatrix} r_0(s_1 + s_2) & s_1 - s_2 \\ s_1 - s_2 & \frac{1}{r_0}(s_1 + s_2) \end{bmatrix}$$

We now apply the alternative spectral mapping introduced in Section 3.4.4, namely,

$$s_1 \to \frac{1}{T_1} \frac{(1-z_1^{-1})(1+z_2^{-1})}{1+z_1^{-1}z_2^{-1}} \qquad s_2 \to \frac{1}{T_2} \frac{(1-z_2^{-1})(1+z_1^{-1})}{1+z_1^{-1}z_2^{-1}} \qquad (6.4)$$

with $T_1 = T_2 = \Delta/\sqrt{2}$ (recall that for the interleaved scheme, we will have increments in the time step of $T/2$ and in the space step of $\Delta/2$, and thus we have chosen T_1 and T_2 to be half the values they previously took). It will be recalled that this mapping, like the trapezoidal rule, is passivity-preserving. The hybrid matrix becomes

$$\mathbf{K}(z_1^{-1}, z_2^{-1}) = \frac{2}{1+z_1^{-1}z_2^{-1}} \begin{bmatrix} r_0\left(1 - z_1^{-1}z_2^{-1}\right) & z_2^{-1} - z_1^{-1} \\ z_2^{-1} - z_1^{-1} & \frac{1}{r_0}\left(1 - z_1^{-1}z_2^{-1}\right) \end{bmatrix} \qquad (6.5)$$

which, upon inspection, can be written as the sum

$$\mathbf{K}(z_1^{-1}, z_2^{-1}) = \mathbf{K}_{ue}(z_1^{-1}, z_2^{-1}) + \mathbf{K}_{ue}(-z_2^{-1}, -z_1^{-1})$$

where it will be recalled that $\mathbf{K}_{ue}(z_1^{-1}, z_2^{-1})$ is the hybrid matrix for the multidimensional unit element defined in (6.1), with $R = r_0$. Because for a series/parallel connection of two-ports, hybrid matrices sum, we have thus decomposed our connecting two-port into two multidimensional unit elements of opposing directions, one of which incorporates a sign inversion in both of its signal paths. The MDWD network (if indeed it is fair to call it that) can then immediately be constructed as in Figure 6.5. The port resistances are

$$R_1 = \frac{2}{\Delta}L_1 = 2v_0l - 2r_0 \qquad R_2 = \frac{\Delta}{2C_2} = \frac{1}{2v_{0c} - 2/r_0} \qquad R_0 = r_0$$

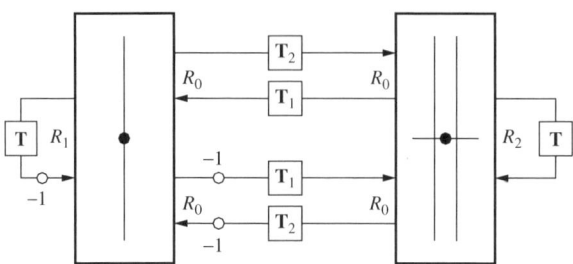

Figure 6.5: *MD network equivalent to type III DWN for the lossless source-free* $(1+1)$D *transmission line equations under an alternative spectral mapping.*

We have chosen here a doubled delay length (of $T' = \Delta$) in the self-loops, according to the offset scheme mentioned in Section 3.8 (for these one-ports, we use the trapezoid rule as for MDWD networks).

This network is, when spatial dependence is expanded out, identical to the DWN shown in Figure 6.1 under the replacement of the series and parallel adaptor symbols by series and parallel scattering junction symbols—recall that they perform identical operations. This MD network, then, operates on a decimated grid, unlike the standard form shown in Figure 6.1. Losses and sources can easily be added back into the alternative MDKC of Figure 6.4(b) and the resulting MD network will be identical to that of Figure 4.13.

We would conjecture that it is possible to find similar equivalences for the type I and II DWNs for the same system (which have better stability bounds); in these cases, however, recall from Section 4.3.6 that the immittances of the connecting waveguides did indeed vary from one grid location to the next. In order to design a MDKC corresponding to such a network, it would be necessary to extend the definition of the multidimensional unit element to include spatially varying port resistances (indeed, this extension is automatic, since the port resistances do not appear explicitly in the definition of this element in (6.1)). The problem, then, is that the two-port connecting the series and parallel adaptors will no longer be shift-invariant, so we must take special care with application of the discretization rule, which can no longer be treated as a spectral mapping.

6.2 Alternative MDKC for the $(2+1)$D Parallel-plate System

The same idea extends simply to the $(2+1)$D parallel-plate system. In this case, we again look at the lossless source-free system (4.50), scaled by a factor of Δ, and can develop an MDKC along the lines of Figure 6.4(b), with u now treated as a voltage instead of a current as in Figure 3.17. We remind the reader that we use the notation D_k, $k = 1, \ldots, 5$ in circuit diagrams to indicate directional derivatives in the directions t_k defined by the coordinate transformation (3.23). Also, we have written i_1 and i_2 for i_x and i_y, in keeping with the MDWD literature [90]. The alternative MDKC is shown in Figure 6.6, and the element values are given by

$$L_1 = L_2 = \Delta(v_0 l - r_0) \qquad C_3 = \Delta\left(v_0 c - \frac{2}{r_0}\right) \qquad L_0 = \Delta r_0 / 2$$

6.2. ALTERNATIVE MDKC FOR THE (2 + 1)D PARALLEL-PLATE SYSTEM

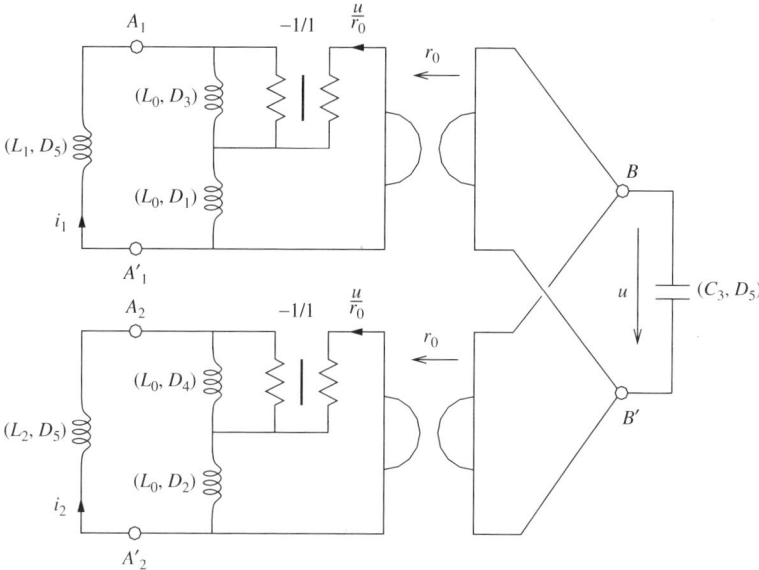

Figure 6.6: *Alternative MDKC for the (2 + 1)D parallel-plate system in rectangular coordinates.*

$r_0 > 0$ is, as before, a free parameter, which can be scaled to achieve an optimal space step/time step ratio.

The two two-ports with terminals A_1, A'_1, B, B' and A_2, A'_2, B, B' are both linear and shift-invariant, and their right-hand pairs of terminals are both connected in parallel with the capacitor. The hybrid matrices for these two-ports are

$$\mathbf{K}_1(s_1, s_3) = \frac{\Delta}{2} \begin{bmatrix} r_0(s_1 + s_3) & s_1 - s_3 \\ s_1 - s_3 & \frac{1}{r_0}(s_1 + s_3) \end{bmatrix} \qquad \text{Two-port } A_1 A'_1 B B'$$

$$\mathbf{K}_2(s_2, s_4) = \frac{\Delta}{2} \begin{bmatrix} r_0(s_2 + s_4) & s_2 - s_4 \\ s_2 - s_4 & \frac{1}{r_0}(s_2 + s_4) \end{bmatrix} \qquad \text{Two-port } A_2 A'_2 B B'$$

Under the spectral mappings

$$s_1 \to \frac{1}{T_1} \frac{(1 - z_1^{-1})(1 + z_3^{-1})}{1 + z_1^{-1} z_3^{-1}} \qquad s_3 \to \frac{1}{T_3} \frac{(1 - z_3^{-1})(1 + z_1^{-1})}{1 + z_1^{-1} z_3^{-1}} \qquad (6.6a)$$

$$s_2 \to \frac{1}{T_2} \frac{(1 - z_2^{-1})(1 + z_4^{-1})}{1 + z_2^{-1} z_4^{-1}} \qquad s_4 \to \frac{1}{T_4} \frac{(1 - z_4^{-1})(1 + z_2^{-1})}{1 + z_2^{-1} z_4^{-1}} \qquad (6.6b)$$

with $T_1 = T_2 = T_3 = T_4 = \Delta/2$, we again have a decomposition of the discrete hybrid two-ports into pairs of multidimensional unit elements (impedance r_0) connected in series/parallel, that is,

$$\mathbf{K}_1(z_1^{-1}, z_3^{-1}) = \mathbf{K}_{\text{ue}}(z_1^{-1}, z_3^{-1}) + \mathbf{K}_{\text{ue}}(-z_3^{-1}, -z_1^{-1})$$

$$\mathbf{K}_2(z_2^{-1}, z_4^{-1}) = \mathbf{K}_{\text{ue}}(z_2^{-1}, z_4^{-1}) + \mathbf{K}_{\text{ue}}(-z_4^{-1}, -z_2^{-1})$$

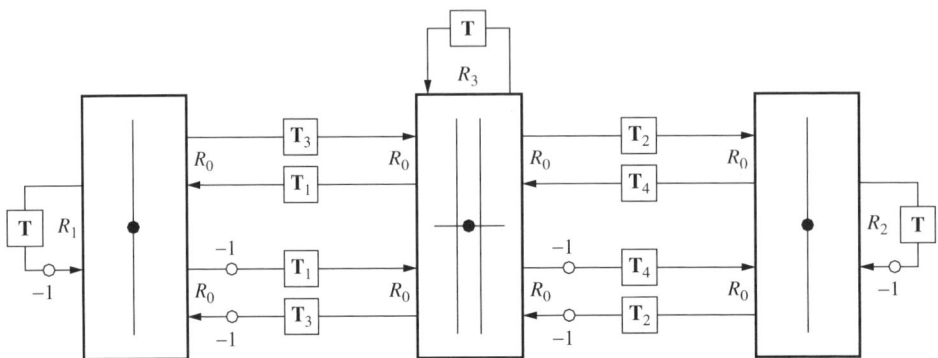

Figure 6.7: *MD network equivalent to type III DWN for the lossless source-free $(2+1)$D parallel-plate system under an alternative spectral mapping.*

The MD network, corresponding to 'checkerboard'-type sampling, is as shown in Figure 6.7. $\mathbf{T_1}, \ldots, \mathbf{T_4}$ are the shifts in the coordinate directions $t_1, \ldots, t_4$, and the port resistances are given by

$$R_1 = R_2 = 2(v_0 l - r_0) \qquad R_3 = \frac{1}{2(v_0 c - 2/r_0)} \qquad R_0 = r_0$$

For the one-port inductors and the capacitor, we have again applied the trapezoid rule, with a step size of $T_5 = \Delta$ (implying a delay of T, written as **T** in Figure 6.7). When spatial dependence is expanded out, the signal flow graph is identical to the interleaved type III DWN for the parallel-plate equations shown in Figure 4.19, without the loss/source ports. As in $(1+1)$D, loss and source elements can be reintroduced into the MDKC of Figure 6.6 without difficulty.

It should also be possible to derive DWNs from MDKCs for the same system on alternative grids, such as the hexagonal and triangular grids mentioned in Section 5.1.1, though we have not investigated this in any detail. In this case, one would presumably begin from the MDKC under the appropriate coordinate transformation. One of these coordinate transformations (which generates a hexagonal grid under uniform sampling in the new coordinates) was discussed briefly in Section 3.2.3. The full MDKC for the parallel-plate equations in these coordinates was given by Fettweis and Nitsche [90].

6.3 Higher-order Accuracy Revisited

Recall that in Section 3.12 we arrived at two MDKCs that were suitable for solving the transmission line equations to higher-order spatial accuracy; we were forced, however, to employ a set of alternative spectral mappings very similar to those that have appeared earlier in this section. For this reason, we postponed showing the full scattering network until now. This is a good opportunity to see the flexibility of having a multidimensional representation of a DWN.

The MDKC in Figure 3.26 represents the lossless source-free $(1+1)$D transmission line equations in a set of $2q$ coordinates defined by (3.22) using the transformation matrix of

6.3. HIGHER-ORDER ACCURACY REVISITED

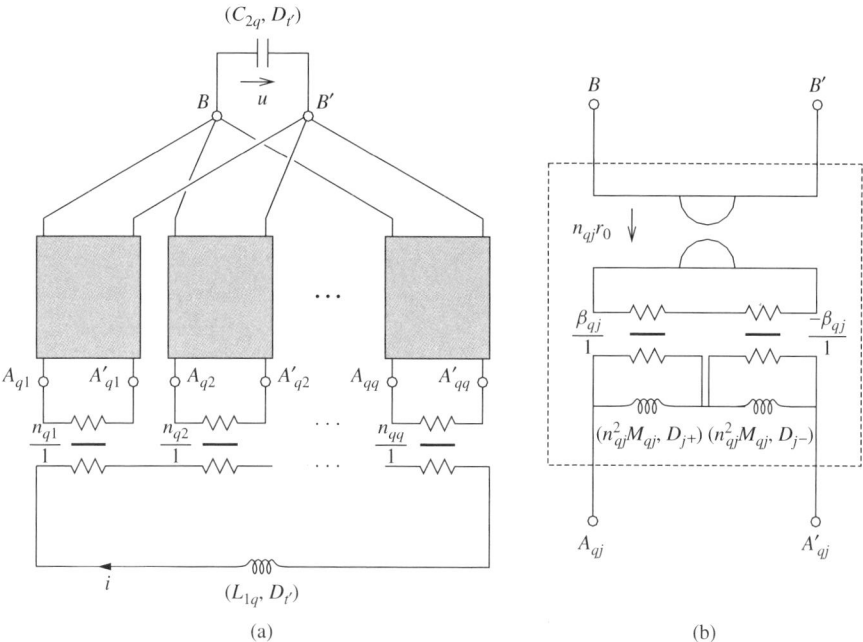

Figure 6.8: (a) *A modified MDKC for the lossless, source-free* (1 + 1)D *transmission line system and* (b) *a detail of connecting two-ports with terminals* A_{qj}, A'_{qj}, B, B' *for any* j, $j = 1, \ldots, q$ *from* (a).

(3.96). These new coordinates allow us to define directional shifts that refer to points other than the nearest neighbors. The general approach to deriving a DWN for this MDKC is the same as in the previous sections; first we perform some network manipulations on the MDKC, and then we apply alternative spectral mappings or integration rules to the connecting LSI two-ports, which then reduce to multidimensional unit elements, which are then interpreted as arrays of digital waveguides. The circuit manipulations in this case are slightly more involved; skipping several steps, we note that we can rewrite the MDKC as shown in Figure 6.8.

As before, we treat u as the voltage across a capacitor, and have introduced a gyrator in each of the connecting two-ports with terminals A_{qj}, A'_{qj}, B and B', $j = 1, \ldots, q$. In addition, we have extracted a transformer of turns ratio n_{qj} for each of these two-ports; the gyrator constant is similarly scaled in order to compensate. The effect of this extracted transformer will be to weight the port resistances of the various multidimensional unit elements that result (and hence the waveguide impedances in the DWN). In addition, we will also need to rescale the inductances from (3.99) by a factor of Δ, giving

$$L_{1q} = \Delta \left(v_0 l - r_0 \sum_{j=1}^{q} \frac{|\alpha_{qj}|}{j} \right) \qquad C_{2q} = \Delta \left(v_0 c - \sum_{j=1}^{q} \frac{|\alpha_{qj}|}{r_0 j} \right) \qquad M_{qj} = \frac{\Delta r_0 |\alpha_{qj}|}{2j}$$

The two-port shown in Figure 6.8(b) contains two inductors of inductances M_{qj} defined with respect to the directions t_{j+} and t_{j-}. Its hybrid matrix will be, in terms of the two associated complex frequencies s_{j+} and s_{j-},

$$\mathbf{K}_{qj}(s_{j+}, s_{j-}) = \begin{bmatrix} M_{qj}n_{qj}^2 \left(s_{j+} + s_{j-}\right) & \frac{M_{qj}n_{qj}\beta_{qj}}{r_0} \left(s_{j+} - s_{j-}\right) \\ \frac{M_{qj}n_{qj}\beta_{qj}}{r_0} \left(s_{j+} - s_{j-}\right) & \frac{M_{qj}}{r_0^2} \left(s_{j+} + s_{j-}\right) \end{bmatrix}$$

Under the spectral mappings from s_{j+} and s_{j-} to the frequency-domain unit shifts z_{j+}^{-1} and z_{j-}^{-1} given by

$$s_{j+} \to \frac{1}{\Delta} \frac{(1 - z_{j+}^{-1})(1 + z_{j-}^{-1})}{1 + z_{j+}^{-1} z_{j-}^{-1}} \qquad s_{j-} \to \frac{1}{\Delta} \frac{(1 - z_{j-}^{-1})(1 + z_{j+}^{-1})}{1 + z_{j+}^{-1} z_{j-}^{-1}}$$

(which are passive and correspond to the integration rules (3.100), with a shift length of Δ), the discrete hybrid matrix becomes

$$\mathbf{K}_{qj}(z_{j+}^{-1}, z_{j-}^{-1}) = \frac{2}{\Delta(1 + z_{j+}^{-1} z_{j-}^{-1})} \begin{bmatrix} M_{qj}n_{qj}^2 \left(1 - z_{j+}^{-1} z_{j-}^{-1}\right) & \frac{M_{qj}n_{qj}\beta_{qj}}{r_0} \left(z_{j-}^{-1} - z_{j+}^{-1}\right) \\ \frac{M_{qj}n_{qj}\beta_{qj}}{r_0} \left(z_{j-}^{-1} - z_{j+}^{-1}\right) & \frac{M_{qj}}{r_0^2} \left(1 - z_{j+}^{-1} z_{j-}^{-1}\right) \end{bmatrix}$$

It should be clear that, in general, this hybrid matrix does not reduce to a pair of series/parallel connected multidimensional unit elements (in which case it should have the form of (6.5), with z_{j-}^{-1} and z_{j+}^{-1} in place of z_2^{-1} and z_1^{-1}). Under the special choice of $n_{qj} = 2j/\alpha_{qj}$, however, it does reduce to such a connection, so we have

$$\mathbf{K}_{qj}(z_{j+}^{-1}, z_{j-}^{-1}) = \mathbf{K}_{\text{ue}}(z_{j+}^{-1}, z_{j-}^{-1}) + \mathbf{K}_{\text{ue}}(-z_{j-}^{-1}, -z_{j+}^{-1})$$

where the port resistances of the unit elements are

$$R_{qj} = \frac{2jr_0}{|\alpha_{qj}|}$$

The two-port multidimensional unit elements are connected to transformers of turns ratio n_{qj}, as per Figure 6.8(a). The port resistance at one end of each transformer can be set to that of the unit element to which it is connected, which is R_{qj}. It is possible to implement these transformers as multiplication by n_{qj} and $1/n_{qj}$ directly in the signal paths if we make the choice of the other port resistance (which we call R'_{qj}) according to the rule discussed in Section 2.2.4; we thus choose

$$R'_{qj} = R_{qj}/n_{qj}^2 = \frac{r_0 |\alpha_{qj}|}{2j}$$

For the one-port inductor and capacitor (of inductance L_{1q} and C_{2q} respectively), we use the trapezoid rule with a doubled time step $T' = 2\Delta$, and the wave digital one-ports (of port resistances $R_{1q} = L_{1q}/\Delta$ and $R_{2q} = \Delta/C_{2q}$) result. The multidimensional network shown in Figure 6.9 can be interpreted as a DWN, and if the parameters α_{qj} are chosen according to the method discussed in Section 3.12, then the DWN will give a qth-order spatially accurate solution to the (1+1)D lossless source-free transmission line system. In order not to belabor this point any further, we leave the explicit construction of the expanded signal

6.4. MAXWELL'S EQUATIONS

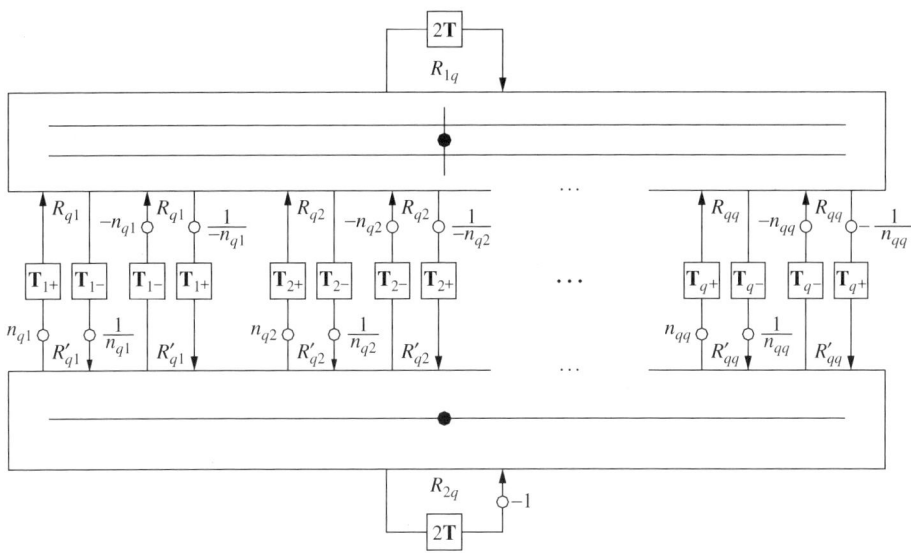

Figure 6.9: *Multidimensional DWN suitable for a qth-order spatially accurate solution to the* $(1+1)$D *transmission line equations.*

flow graph for this multidimensional DWN from Figure 6.9 (as well as the interleaved DWN that results from the use of coordinates defined by (3.104)) as an exercise to the reader.

6.4 Maxwell's Equations

We now take a brief look at Maxwell's equations, the (3+1)D system of PDEs that describes the time evolution of electromagnetic fields. Numerical simulation of this system was the original motivation behind the development of FDTD [246, 286] and TLM, and MDWD network methods for Maxwell's system were explored early on by Fettweis and Nitsche [72]. In the interest of solidifying the link between these two types of methods, we show how a passive circuit representation yields a DWN, which is no more than a scattering form of FDTD.

Maxwell's equations, for a linear isotropic (though not necessarily spatially homogeneous) medium, are usually written in vector form as

$$\epsilon \frac{\partial \mathbf{E}}{\partial t} = \nabla \times \mathbf{H} \qquad \mu \frac{\partial \mathbf{H}}{\partial t} = -\nabla \times \mathbf{E} \qquad (6.7)$$

where $\mathbf{E} = [E_x, E_y, E_z]^T$ and $\mathbf{H} = [H_x, H_y, H_z]^T$ are, respectively, the electric and magnetic field vectors, and $\epsilon(x, y, z)$ and $\mu(x, y, z)$ are, respectively, the dielectric constant and magnetic permeability of the medium, assumed positive and bounded away from 0. (We have left out losses and sources here.) This system has the form of (3.1), with $\mathbf{w} = [\mathbf{E}^T, \mathbf{H}^T]^T$, and

$$\mathbf{P} = \begin{bmatrix} \epsilon \mathbf{I}_3 & \cdot \\ \cdot & \mu \mathbf{I}_3 \end{bmatrix} \qquad \mathbf{A}_j = \begin{bmatrix} \cdot & \mathbf{A}_{j\times} \\ \mathbf{A}_{j\times}^T & \cdot \end{bmatrix} \qquad j = 1, 2, 3$$

where $\mathbf{I}_3$ is the 3×3 identity matrix, · stands for zero entries, and where we also have

$$\mathbf{A}_{1\times} = \begin{bmatrix} 0 & 0 & 0 \\ 0 & 0 & 1 \\ 0 & -1 & 0 \end{bmatrix} \quad \mathbf{A}_{2\times} = \begin{bmatrix} 0 & 0 & -1 \\ 0 & 0 & 0 \\ 1 & 0 & 0 \end{bmatrix} \quad \mathbf{A}_{3\times} = \begin{bmatrix} 0 & 1 & 0 \\ -1 & 0 & 0 \\ 0 & 0 & 0 \end{bmatrix}$$

Phase and Group Velocity

If ϵ and μ are constant, then from (3.10), the dispersion relation for Maxwell's equations has the form

$$\omega^2 \left(\epsilon\mu\omega^2 - \|\boldsymbol{\beta}\|_2^2\right)^2 = 0$$

in terms of frequencies ω and wavenumber magnitudes $\|\boldsymbol{\beta}\|_2$. This equation has solutions

$$\omega = 0 \qquad \omega = \pm\frac{\|\boldsymbol{\beta}\|_2}{\sqrt{\epsilon\mu}}$$

Leaving aside the nonpropagating mode with $\omega = 0$, the phase and group velocities will then be given by

$$v^p_{\text{Maxwell}} = v^g_{\text{Maxwell}} = \pm\frac{1}{\sqrt{\epsilon\mu}}$$

For spatially inhomogeneous problems, the maximum group velocity will be

$$v^g_{\text{Maxwell},\max} = \frac{1}{\sqrt{(\epsilon\mu)_{\min}}}$$

Scattering Networks for Maxwell's Equations

This system has been represented previously by an MDKC [72, 176], where the coordinate transformation defined by (3.25) has been employed, and the current variables are defined by

$$(i_1, i_2, i_3, i_4, i_5, i_6) = (E_x/r_0, E_y/r_0, E_z/r_0, H_x, H_y, H_z)$$

for some positive constant r_0. We have reproduced this MDKC in Figure 6.10. This network can be viewed as two coupled (2+1)D parallel-plate networks (see Section 3.7), and this is not surprising, given that the (2+1)D parallel-plate system is essentially equivalent to the *transverse electric* (TE) or *transverse magnetic* (TM) system alone. The resulting MDWD network is shown, at bottom, in Figure 6.10; here, we have assumed the directional shifts $\mathbf{T}_j$, $j = 1, \ldots, 7$ to be of length Δ (and thus $T_7 = T' = \Delta/v_0 \triangleq T$).

The passivity condition is again a condition on the positivity of the network inductances in the MDKC; these values are given in Figure 6.10, and the resulting conditions are

$$v_0 \geq \frac{2}{r_0 \epsilon_{\min}} \qquad v_0 \geq \frac{2r_0}{\mu_{\min}}$$

Under the choice of $r_0 = \sqrt{\frac{\mu_{\min}}{\epsilon_{\min}}}$, the passivity condition becomes

$$v_0 \geq \frac{2}{\sqrt{\mu_{\min}\epsilon_{\min}}} \geq 2v^g_{\text{Maxwell},\max} \tag{6.8}$$

and the numerical scheme is passive and hence stable over this range of v_0.

6.4. MAXWELL'S EQUATIONS

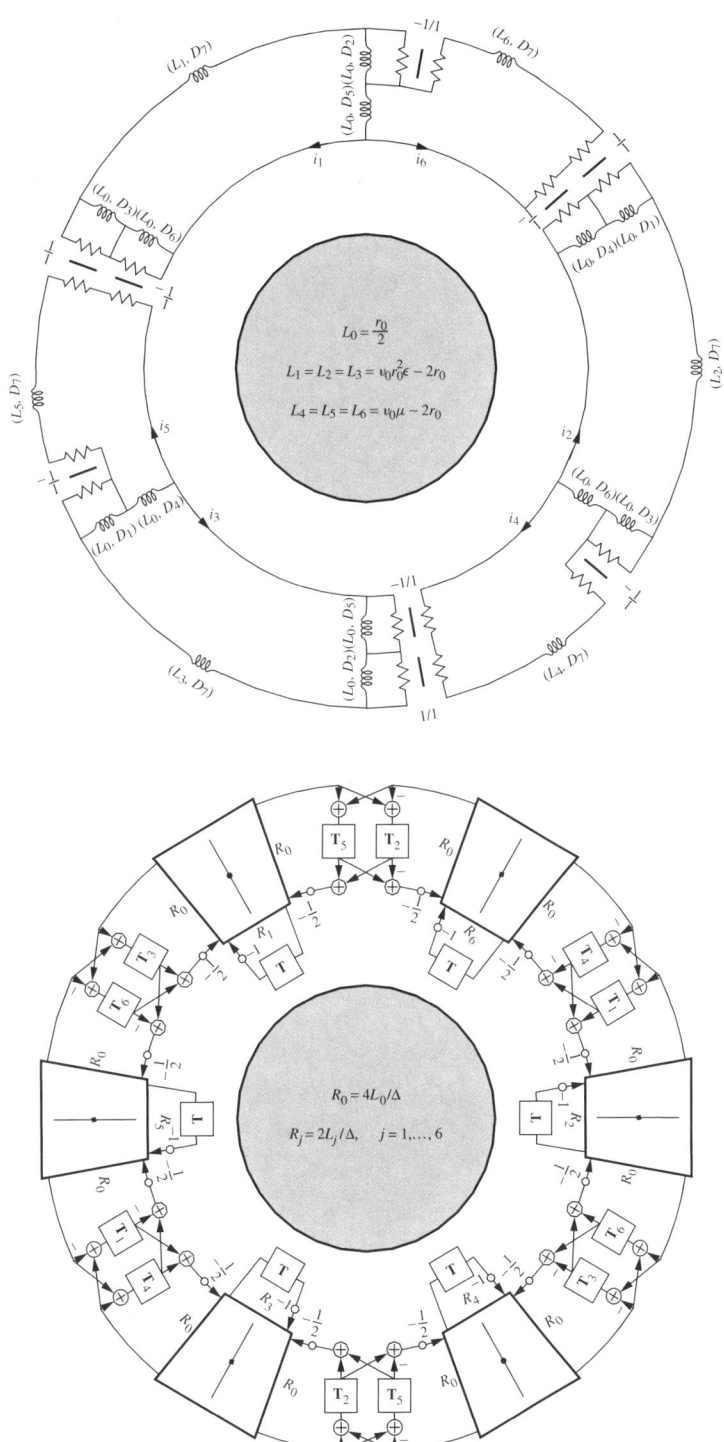

Figure 6.10: *MDKC and MDWD network for Maxwell's equations* (6.7).

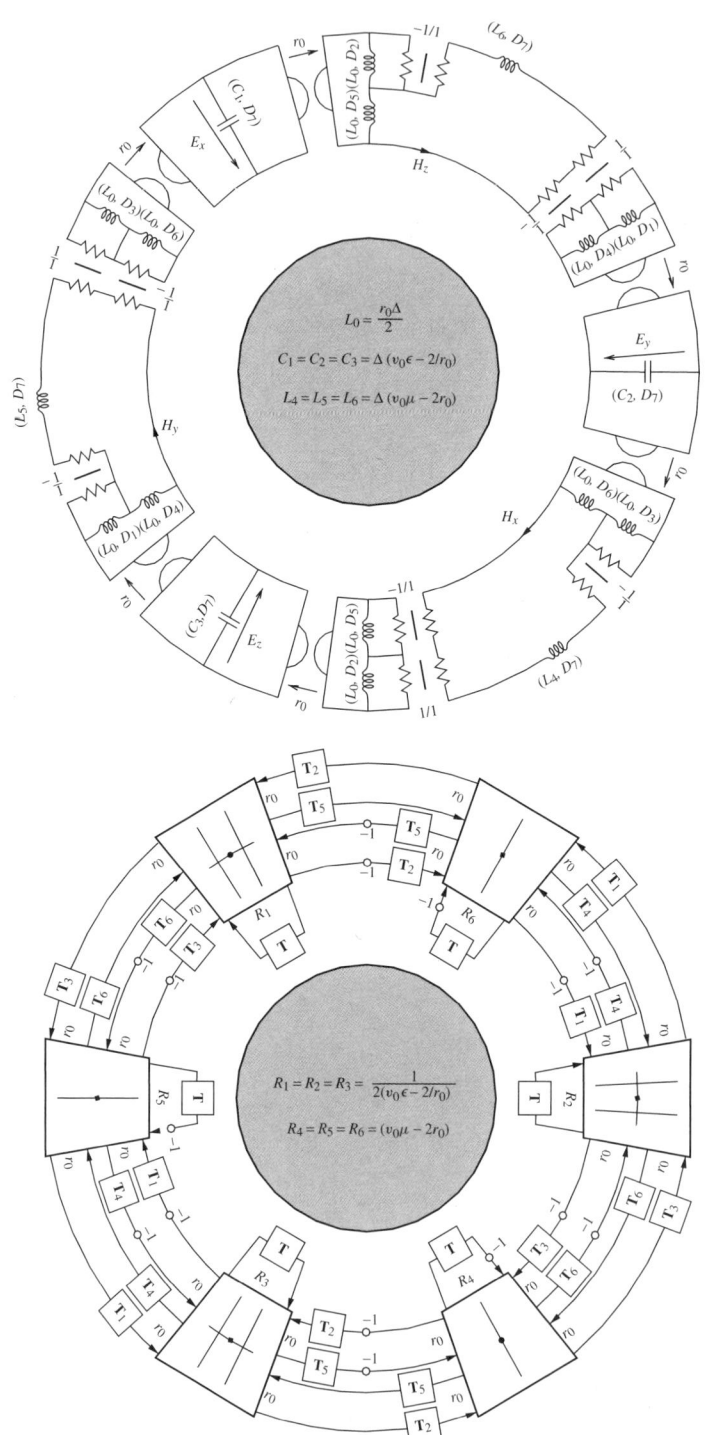

Figure 6.11: *Modified MDKC and multidimensional DWN for Maxwell's equations* (6.7).

6.4. MAXWELL'S EQUATIONS

In order to generate a digital waveguide network, we may proceed as for the (1+1)D transmission line and (2+1)D parallel-plate systems discussed in Section 6.1.3 and Section 6.2, and apply the by now familiar network transformations to yield the modified circuit shown in Figure 6.11. Electric field quantities are now treated as voltages across capacitors, and the six LSI connecting two-ports will all become multidimensional unit elements under the application of alternative spectral mappings.

The coordinate transformation from coordinates $\mathbf{u} = [x, y, z, t]^T$ to coordinates $\mathbf{t} = [t_1, \ldots, t_7]^T$ defined by (3.22), using the transformation matrix of (3.25) gives us, in the LSI case, seven frequencies $s_1, \ldots, s_7$. For the connecting two-ports, we may use pairwise spectral mappings defined by

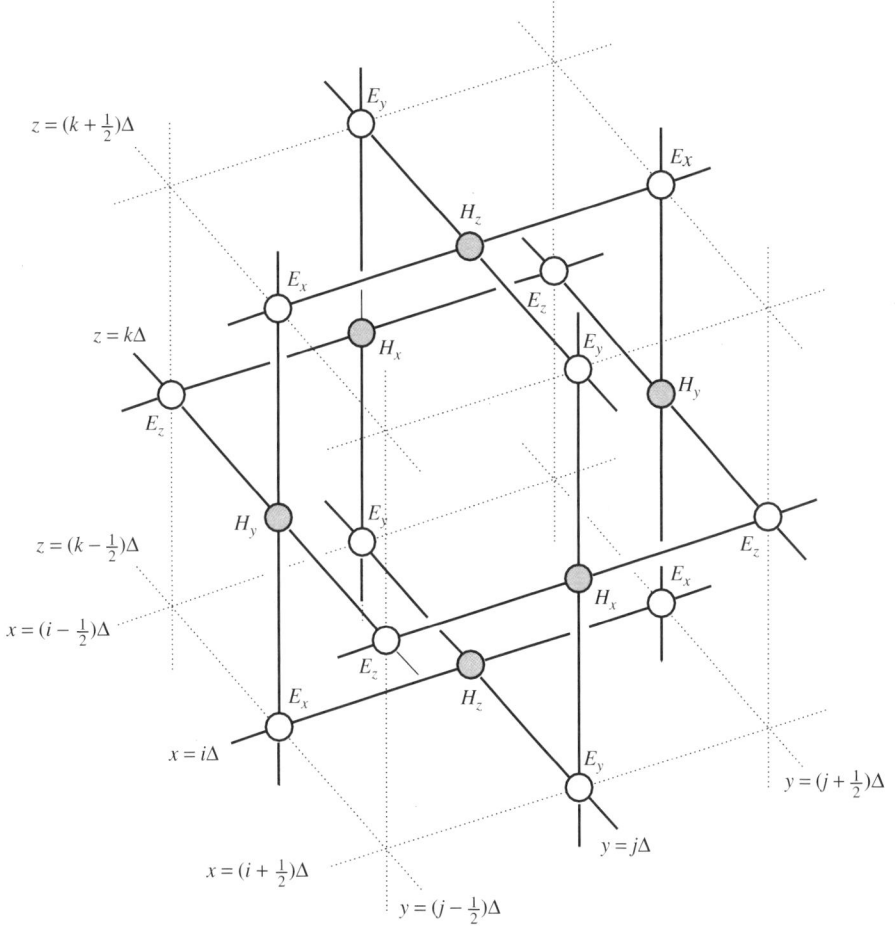

Figure 6.12: *Computational grid for FDTD applied to Maxwell's equations (6.7); electric and magnetic field quantities are calculated at alternating multiples of $T/2$, and at alternating grid locations. In the DWN implementation, waveguide connections (of delay length $T/2$) between series junctions (gray) and parallel junctions (white) are shown as dark lines; waveguide sign inversions and self-loops are not shown here.*

$$s_j \to \frac{1}{T_j}\frac{(1-z_j^{-1})(1+z_{j+3}^{-1})}{1+z_j^{-1}z_{j+3}^{-1}} \qquad s_{j+3} \to \frac{1}{T_{j+3}}\frac{(1-z_{j+3}^{-1})(1+z_j^{-1})}{1+z_{j+3}^{-1}z_j^{-1}} \qquad j=1,2,3$$

(6.9)

where z_j^{-1}, $j = 1, \ldots, 6$ corresponds to a unit shift in direction t_j, and the step-sizes T_j, $j = 1, \ldots, 6$ are all chosen equal to $\Delta/2$. For the one-port time inductors and capacitors, we use the trapezoid rule with a step size of $T_7 = T' = \Delta$. The resulting multidimensional DWN is shown at the bottom in Figure 6.11, and the stability bound is unchanged from (6.8).

When the spatial dependence is expanded out, we have a DWN operating on an interleaved numerical grid as shown in Figure 6.12, with the electric and magnetic field components calculated at parallel (white) and series (gray) junctions respectively. The connecting waveguide impedances (of delay $T/2$, shown as solid lines) are all equal to r_0, and the self-loops (of delay T, not shown) have impedances $2v_0\mu \quad 2r_0$ and admittances $2v_0\epsilon - 2/r_0$ at the series and parallel junctions respectively, where these expressions are evaluated at the junction location. It is also possible to derive DWNs of the type I and II forms (see Section 4.3.6), for which the stability bound is improved to CFL.

It is easy to verify that this scheme is indeed a scattering form of FDTD. Referring to Figure 6.12, we will have six sets of junction quantities: at the parallel junctions, we will have $E_{xJ,i,j+\frac{1}{2},k+\frac{1}{2}}(n+\frac{1}{2})$, $E_{yJ,i,j+\frac{1}{2},k}(n+\frac{1}{2})$, and $E_{zJ,i+\frac{1}{2},j+\frac{1}{2},k}(n+\frac{1}{2})$, and at the series junctions, we will have $H_{xJ,i+\frac{1}{2},j,k}(n)$, $H_{yJ,i,j+\frac{1}{2},k}(n)$, and $H_{zJ,i,j,k+\frac{1}{2}}(n)$. The indices i, j, k, and n take on integer values. Examine the DWN at a parallel junction with 'voltage' $E_{xJ,i,j+\frac{1}{2},k+\frac{1}{2}}(n+\frac{1}{2})$. The DWN updates this grid function according to

$$E_{xJ,i,j+\frac{1}{2},k+\frac{1}{2}}(n+\tfrac{1}{2}) = E_{xJ,i,j+\frac{1}{2},k+\frac{1}{2}}(n-\tfrac{1}{2})$$

$$+\frac{1}{v_0\epsilon_{i,j+\frac{1}{2},k+\frac{1}{2}}}\Big(H_{zJ,i,j+1,k+\frac{1}{2}}(n) - H_{zJ,i,j,k+\frac{1}{2}}(n)$$

$$-H_{yJ,i,j+\frac{1}{2},k+1}(n) + H_{yJ,i,j+\frac{1}{2},k}(n)\Big)$$

with $\epsilon_{i,j+\frac{1}{2},k+\frac{1}{2}} \triangleq \epsilon(i\Delta, (j+\frac{1}{2})\Delta, (k+\frac{1}{2})\Delta)$. This is exactly centered differences applied to the equation in E_x, H_y, and H_z according to the Yee algorithm [246]. It is also worth comparing this DWN to the original form of TLM [5].

Chapter 7

Applications to Vibrating Systems

In the previous chapters, we dealt with the numerical integration of systems of PDEs that were all, in some sense, generalizations of the *wave equation*. In the case where the material parameters (l and c in the case of the transmission line) have no spatial variation, this amounts to saying that wave propagation in these media is *dispersionless*; a plane wave travels at a fixed speed, regardless of its wavelength. We now turn to sets of equations that fundamentally engender some degree of dispersion, namely those describing the motion of stiff systems such as beams, plates, and shells. As a result, we move toward the use of mechanical quantities, as opposed to electrical, but the analogy should be clear. We will show how waveguide and wave digital filter principles can be used in order to obtain numerical solutions of such equations.

It will be necessary to introduce several new techniques in order to develop numerical methods for such systems, which are considerably more complex than the transmission line test problems that we examined in the previous chapters. First, although these systems are all symmetric hyperbolic, we will need to make use of non-reciprocal circuit elements in order to model some asymmetric couplings that occur. Second, we may have to perform some additional initial work on these same systems in order to symmetrize them, as they are not always symmetric hyperbolic in their commonly encountered forms. Third, in some cases we will be forced to make use of vector-valued wave variables and scattering junctions [176] (see Section 2.2.7).

Summary

In this chapter, we look at such systems in order of increasing dimensionality, loosely following the organization of the text by Graff [108]. Liberal use is made of the unifying result of Chapter 6 in order to develop wave digital and digital waveguide simulation networks in a parallel fashion for these systems. We first examine the Timoshenko theory of beams, which was treated by Nitsche [176], and present scattering methods in Section 7.1, while indicating the relevant differences, especially with respect to stability. We also apply the system balancing approach (introduced in Section 3.11) to the Timoshenko beam in Section 7.1.5 in order to show that it is possible to drastically reduce the computational requirements in certain cases and take an extended look at boundary conditions in Section 7.1.3. Then follows a look at stiff plate theory, and, in particular, the two-dimensional analogue of the Timoshenko beam, called the Mindlin plate, in Section 7.2. Here, due to the couplings between the variables, we are forced to make use of vector scattering elements, which were introduced in Section 2.2.7 for this very purpose. Boundary conditions for waveguide networks for the Mindlin plate are dealt with in detail in Section 7.2.2. We next spend some time examining network representations for two cylindrical shell models, first the membrane shell in Section 7.3.1 and then the more modern model of Naghdi and Cooper in Section 7.3.2. Finally, for completeness sake, we revisit in Section 7.4 Nitsche's MDKC for the full three-dimensional elastic solid dynamic system [176]; as for all the systems in this chapter, we show the alternative network form suitable for DWN discretization in Section 7.4.1. In keeping with the more applied flavor of this chapter, we also present simulation results for Timoshenko's beam system and the Mindlin plate in Section 7.1.4 and Section 7.2.3 respectively, under both uniform and spatially varying material parameter conditions.

7.1 Beam Dynamics

Consider a thin beam, or rod, aligned parallel to the x axis. We will be interested in the transverse motion of the beam, which we assume to be restricted to one perpendicular direction; we will call the deflection of the beam from the x axis $w(x, t)$. The relevant material parameters of the beam are the mass density ρ, the cross-sectional area A, Young's modulus E, and I, the moment of inertia of the beam about the perpendicular axis. The material parameters are, in general, allowed to be slowly varying functions of x. Under the further assumptions that the beam deflection $w(x, t)$ is small and that the beam cross section remains perpendicular to the so-called 'neutral axis,' it is possible to arrive at the *Euler–Bernoulli* equation [108]:

$$\rho A \frac{\partial^2 w}{\partial t^2} = -\frac{\partial^2}{\partial x^2}\left(EI \frac{\partial^2 w}{\partial x^2}\right) \quad (7.1)$$

Notice that this equation contains a fourth-order spatial derivative, resulting from the fact that the beam provides its own restoring stiffness, proportional to its curvature, in marked contrast to the equation for a string that requires externally applied tension in order to support wave motion. If the material properties of the beam do not vary spatially, then (7.1) reduces to the more familiar form

$$\frac{\partial^2 w}{\partial t^2} = -b^2 \frac{\partial^4 w}{\partial x^4} \quad (7.2)$$

where $b = \sqrt{EI/\rho A}$.

It is easy to show that phase and group velocities for the ideal beam are wavenumber dependent and thus wave propagation is dispersive. In addition, the group velocities are not bounded; this is a direct consequence of the fact that the Euler–Bernoulli equation does not result from the elimination of variables in a hyperbolic system (see Section 3.1). Though it is nevertheless possible to develop scattering methods for this equation [26],

7.1. BEAM DYNAMICS

approaches based on passive MD circuits are not directly applicable. For this reason, we turn immediately to the more modern theory of Timoshenko [108].

Timoshenko's theory of beams constitutes an improvement over the Euler–Bernoulli theory, in that it incorporates shear and rotational inertia effects [108]. This is one of the few cases in which a more refined modeling approach allows more tractable numerical simulation; the reason for this is that Timoshenko's theory gives rise to a *hyperbolic* system, unlike the Euler–Bernoulli system, for which propagation velocity is unbounded. It is this partially parabolic character of the Euler–Bernoulli system that engenders severe restrictions on the maximum allowable time step (at least in the case of explicit methods, of which type are all the scattering-based methods included in this book). For a physical derivation of Timoshenko's system, we refer the reader to the literature [108, 197, 205, 249, 251], and simply present it here

$$\rho A \frac{\partial^2 w}{\partial t^2} = \frac{\partial}{\partial x} \left(A\kappa G \left(\frac{\partial w}{\partial x} - \psi \right) \right) \tag{7.3a}$$

$$\rho I \frac{\partial^2 \psi}{\partial t^2} = \frac{\partial}{\partial x} \left(EI \frac{\partial \psi}{\partial x} \right) + A\kappa G \left(\frac{\partial w}{\partial x} - \psi \right) \tag{7.3b}$$

As before, $w(x, t)$ represents the transverse displacement of the beam from an equilibrium state, and the new dependent variable $\psi(x, t)$ is the angle of deflection of the cross section of the beam with respect to the vertical direction. Here, the quantities ρ, A, E, and I are as for the Euler–Bernoulli equation (7.1). G is the shear modulus (sometimes called μ in other contexts) and κ is a constant that depends on the geometry of the beam. For generality, we assume that all these material parameters are functions of x. Losses or sources are not modeled.

Nitsche [176], in his MDWD network-based approach, preferred to use the more fundamental set of four first-order PDEs from which system (7.3) is condensed

$$\rho A \frac{\partial v}{\partial t} = \frac{\partial q}{\partial x} \tag{7.4a}$$

$$\frac{1}{A\kappa G} \frac{\partial q}{\partial t} = \frac{\partial v}{\partial x} - \omega \tag{7.4b}$$

$$\rho I \frac{\partial \omega}{\partial t} = \frac{\partial m}{\partial x} + q \tag{7.5a}$$

$$\frac{1}{EI} \frac{\partial m}{\partial t} = \frac{\partial \omega}{\partial x} \tag{7.5b}$$

We have introduced here the quantities

$$v \triangleq \frac{\partial w}{\partial t} \qquad \omega \triangleq \frac{\partial \psi}{\partial t} \qquad m \triangleq EI \frac{\partial \psi}{\partial x} \qquad q \triangleq A\kappa G \left(\frac{\partial w}{\partial x} - \psi \right)$$

v is interpreted as transverse velocity, ω as an angular velocity, m as the bending moment, and q as the shear force on the cross section. Each of the subsystems (7.4) and (7.5) has the form of a lossless $(1 + 1)$D transmission line system; they are coupled by constant-proportional terms, and it is this coupling that gives the Timoshenko system its dispersive character. The Euler–Bernoulli equation is recovered in the limit as $A\kappa G \to \infty$ and $\rho I \to 0$ [176].

This is a symmetric hyperbolic system of the form given in (3.1), with $\mathbf{w} = [v, q, \omega, m]^T$, $\mathbf{f} = \mathbf{0}$ and

$$\mathbf{P} = \begin{bmatrix} \rho A & 0 & 0 & 0 \\ 0 & \frac{1}{A\kappa G} & 0 & 0 \\ 0 & 0 & \rho I & 0 \\ 0 & 0 & 0 & \frac{1}{EI} \end{bmatrix} \quad \mathbf{A}_1 = \begin{bmatrix} 0 & -1 & 0 & 0 \\ -1 & 0 & 0 & 0 \\ 0 & 0 & 0 & -1 \\ 0 & 0 & -1 & 0 \end{bmatrix} \quad \mathbf{B} = \begin{bmatrix} 0 & 0 & 0 & 0 \\ 0 & 0 & 1 & 0 \\ 0 & -1 & 0 & 0 \\ 0 & 0 & 0 & 0 \end{bmatrix}$$

Dispersion

The characteristic polynomial equation from (3.10) with the system matrices given above, in the case of constant coefficients, is

$$\omega^4 - \frac{\omega^2}{\rho}\left(\frac{A\kappa G}{I} + \beta^2(E + G\kappa)\right) + \frac{E\kappa G}{\rho^2}\beta^4 = 0 \tag{7.6}$$

where ω and β are frequency and spatial wavenumber respectively. There are two pairs of solutions to this equation, which can be written as

$$\omega_{1\pm} = \pm\sqrt{\frac{1}{2\rho}\left(\frac{A\kappa G}{I} + \beta^2(E + G\kappa) + \sqrt{\left(\frac{A\kappa G}{I} + \beta^2(E + G\kappa)\right)^2 - 4E\kappa G\beta^4}\right)}$$

$$\omega_{2\pm} = \pm\sqrt{\frac{1}{2\rho}\left(\frac{A\kappa G}{I} + \beta^2(E + G\kappa) - \sqrt{\left(\frac{A\kappa G}{I} + \beta^2(E + G\kappa)\right)^2 - 4E\kappa G\beta^4}\right)}$$

and it is simple to show that in contrast with the Euler–Bernoulli beam, the group velocities will be bounded. Indeed, we have, in particular, that

$$\lim_{\beta\to\infty} \omega_{1\pm} = \pm\beta\sqrt{\frac{E}{\rho}} \qquad \lim_{\beta\to\infty} \omega_{2\pm} = \pm\beta\sqrt{\frac{G\kappa}{\rho}},$$

where the first of these relations is similar to that which describes longitudinal wave propagation in a bar and the second corresponds to shear vibration [108]. For the full varying-coefficient problem, the maximum group velocity, as defined in (3.13), will be

$$\gamma^g_{T,\max} = \sqrt{\left(\frac{E}{\rho}\right)_{\max}} \tag{7.7}$$

7.1.1 MDKC and MDWD network for Timoshenko's System

Nitsche [176] showed how to write a MDKC and MDWD network corresponding to Timoshenko's system. In order to deal with the asymmetric coupling of the system of equations, he constructed a network using both MD capacitors and inductors, but here we will take the more conventional approach, and use a gyrator. (For computational purposes, there is no essential difference between this representation and his.)

7.1. BEAM DYNAMICS

Consider again the Timoshenko system of (7.4) and (7.5). We can scale the variables just as for the transmission line (see Section 3.6). That is, we can write

$$v = r_1 i_1 \qquad q = i_2 \qquad \omega = i_3 \qquad m = r_2 i_4 \qquad (7.8)$$

where the constants r_1 and r_2 are strictly positive. We introduce, as before, the scaled time variable $t' = v_0 t$ where v_0 is the space step/time step ratio. Then the Timoshenko system can be rewritten as

$$v_0 r_1^2 \rho A \frac{\partial i_1}{\partial t'} = r_1 \frac{\partial i_2}{\partial x} \qquad (7.9a)$$

$$\frac{v_0}{A\kappa G} \frac{\partial i_2}{\partial t'} = r_1 \frac{\partial i_1}{\partial x} - i_3 \qquad (7.9b)$$

$$v_0 \rho I \frac{\partial i_3}{\partial t'} = r_2 \frac{\partial i_4}{\partial x} + i_2 \qquad (7.10a)$$

$$\frac{v_0 r_2^2}{EI} \frac{\partial i_4}{\partial t'} = r_2 \frac{\partial i_3}{\partial x} \qquad (7.10b)$$

The constant-proportional terms on the right-hand side appear antisymmetrically, and can be interpreted as a lossless gyrator coupling. We can now write down an MDKC for the scaled system of equations; it is shown, along with element values in Figure 7.1. Its MDWD counterpart is pictured in Figure 7.2. Here, we have used the coordinate transformation defined in (3.19) with step sizes $T_1 = T_2 = \sqrt{2}\Delta$. We have used $T' = \Delta$ for the one-port time inductors. An MDWD network can obviously also be designed to operate on alternating grids, just as in the case of the (1 + 1)D transmission line.

A comment is necessary regarding the gyrator in Figure 7.2. In order to deal with the delay-free loop, which arises from the placement of a gyrator between two series junctions, we have set the corresponding ports of the series junctions on either side of the gyrator to be reflection-free. This, however, means that the two port resistances of the gyrator are not, in general, equal to the gyrator constant, which, in this case, will be 1. In terms of wave variables, the signal flow diagram of the gyrator will not be of the simple form of (2.26), but takes the more general form of (2.25) mentioned in Section 2.2.4. It is of course also possible to set only one of the ports connected to the gyrator to be reflection-free, (say $R_5 = R_7 + R_2$), and then the other port resistance to be $R_6 = R_5$, in which case the general gyrator form degenerates to a pair of scalings.

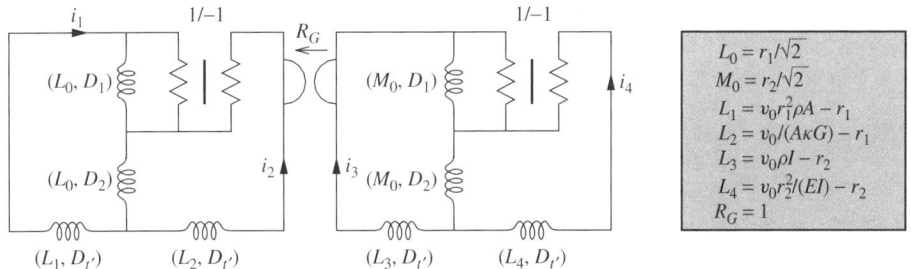

Figure 7.1: *MDKC for Timoshenko's system.*

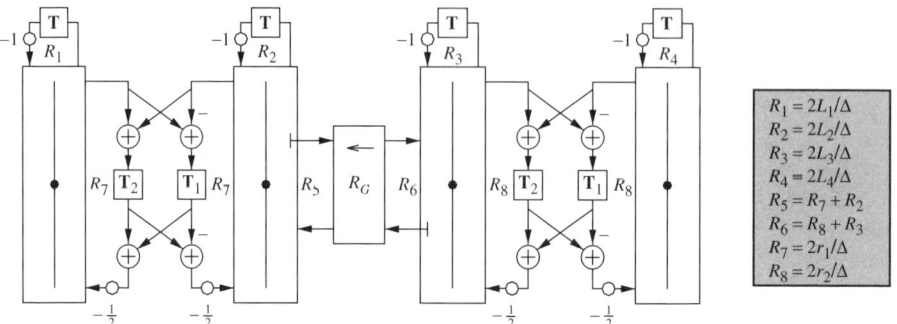

Figure 7.2: MDWD network for Timoshenko's system.

As for the parameters r_1 and r_2, an optimal choice can be shown to be

$$r_1 = ((AG\kappa)_{\max}(\rho A)_{\min})^{-\frac{1}{2}}$$
$$r_2 = ((EI)_{\max}(\rho I)_{\min})^{\frac{1}{2}}$$

which yields the bound

$$v_0 \geq \max\left(\sqrt{\frac{(AG\kappa)_{\max}}{(\rho A)_{\min}}}, \sqrt{\frac{(EI)_{\max}}{(\rho I)_{\min}}}\right) \geq v_{T,\max}^g \quad (7.11)$$

which is the same as that which is derived in [176]. We will show how to improve upon this bound in Section 7.1.5.

7.1.2 Waveguide Network for Timoshenko's System

Recall from Chapter 6 that for the $(1+1)$D transmission line problem, it is possible to obtain a DWN from an MDKC after a few network manipulations and under the application of an alternative spectral mapping, or integration rule. We may proceed in the same way for Timoshenko's system and we will skip most of the steps that were detailed in the earlier treatment. We do recall, however, that the original system of equations should be scaled by a factor of Δ, the grid spacing, before making the switch to DWNs.

We first transform the MDKC of Figure 7.1 such that the quantities v and m represent voltages across capacitors, instead of currents through inductors. The transformed MDKC is shown in Figure 7.3. The inductances are all as in Figure 7.1, except that they are scaled by Δ, the new capacitance values will be

$$C_1 = \Delta\left(v_0 \rho A - \frac{1}{r_1}\right) \qquad C_4 = \Delta\left(\frac{v_0}{EI} - \frac{1}{r_2}\right)$$

and the gyrator coefficient R_G will be equal to Δ. As before, it is possible to interpret the two two-ports $AA'BB'$ and $PP'QQ'$ as MD representations of digital waveguide pairs if we apply the alternative spectral mapping or integration rule as in Chapter 6. The MD waveguide network is shown in Figure 7.4. Here, we have chosen the step sizes such that

7.1. BEAM DYNAMICS

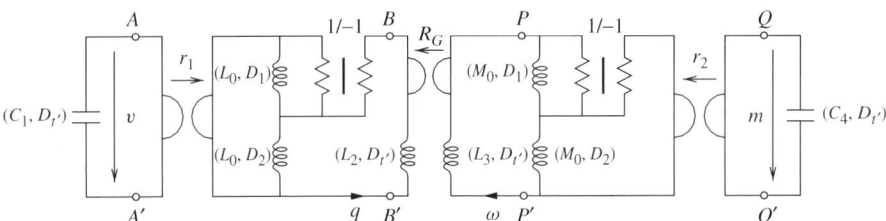

Figure 7.3: *Transformed MDKC for Timoshenko's system.*

an interleaved algorithm results, just as for the transmission line problem, as discussed in Chapter 6. Thus, we have $T_1 = T_2 = \Delta/\sqrt{2}$ and $T' = \Delta$ so that in a computer implementation, approximations to v and m are alternated with those of q and ω. The signal flow diagram corresponding to Figure 7.4 is shown, as a DWN, in Figure 7.5.

The junction quantities V_J, Q_J, Ω_J, and M_J approximate v, q, ω, and m respectively, and for consistency with the DWN notation of Chapter 4, we have replaced the port resistances by waveguide admittances (at the parallel junctions) and impedances (at the series junctions). Because there are now two sets of immittances at any grid point (corresponding to the upper and lower rails in Figure 7.5), we have indexed half of them with a tilde. Referring to Figure 7.5, the self-loop immittances will be given by

$$Y_{c,i} = 2v_0(\rho A)_i - \frac{2}{r_1} \qquad \tilde{Y}_{c,i} = \frac{2v_0}{(EI)_i} - \frac{2}{r_2}$$

$$Z_{c,i+\frac{1}{2}} = \frac{2v_0}{(AG\kappa)_{i+\frac{1}{2}}} - 2r_1 \qquad \tilde{Z}_{c,i+\frac{1}{2}} = 2v_0(\rho I)_{i+\frac{1}{2}} - 2r_2$$

at alternating grid locations indexed by integer i. The connecting immittances will be, referring to the series junctions,

$$Z_{x^-,i+\frac{1}{2}} = Z_{x^+,i+\frac{1}{2}} = r_1 \qquad \tilde{Z}_{x^-,i+\frac{1}{2}} = \tilde{Z}_{x^+,i+\frac{1}{2}} = r_2$$

The DWN incorporates a gyrator between the two series junctions, and as such we must employ reflection-free ports at at least one of the two connected junction ports. Though

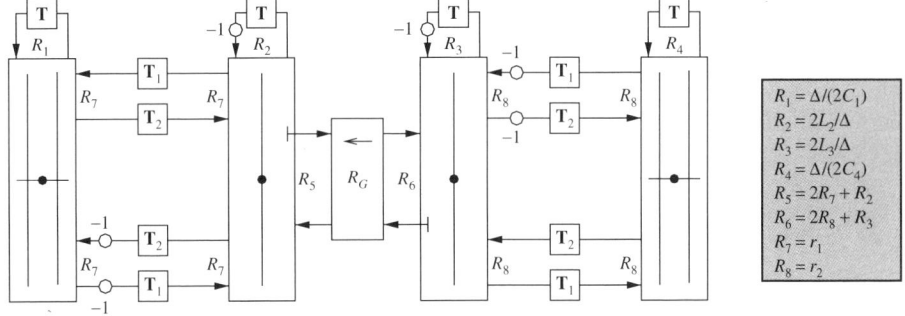

Figure 7.4: *Waveguide network for Timoshenko's system, in a multidimensional form.*

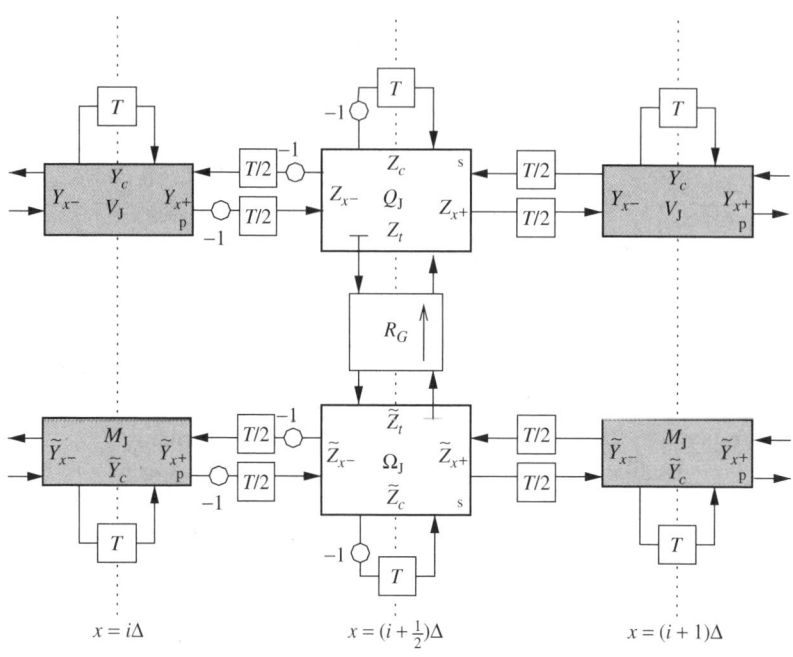

Figure 7.5: *DWN for Timoshenko's system.*

reflection-free ports and gyrators have not as yet appeared in the DWN context, it is straightforward (indeed immediate, if we are considering DWNs derived from MDKCs), to transfer them from wave digital filters. These port impedances (subscripted with a t) at the series junctions can be chosen to be

$$Z_{t,i+\frac{1}{2}} = Z_{c,i+\frac{1}{2}} + Z_{x-,i+\frac{1}{2}} + Z_{x+,i+\frac{1}{2}} \qquad \tilde{Z}_{t,i+\frac{1}{2}} = \tilde{Z}_{c,i+\frac{1}{2}} + \tilde{Z}_{x-,i+\frac{1}{2}} + \tilde{Z}_{x+,i+\frac{1}{2}}$$

This is a structure of the type III form (i.e., the connecting immittances are spatially invariant; see Section 4.3.6), and the bound on v_0 is suboptimal (and the same as that for the MDWD network structure discussed in the last section). The possible interleaving of the calculated junction quantities in this structure is indicated, as in Chapter 4, by gray/white coloring of scattering junctions.

7.1.3 Boundary Conditions in the DWN

The three common types of boundary conditions used to terminate a Timoshenko beam are

$w = m = 0$	Fixed end, allowed to pivot	(7.12a)
$w = \psi = 0$	Fixed clamped end	(7.12b)
$q = m = 0$	Free end	(7.12c)

All of these conditions are of the form of (3.8) and are lossless.

There are several possibilities for the implementation of these boundary conditions (7.12) at a boundary grid point in the MDWD network or any of the DWN structures

7.1. BEAM DYNAMICS

mentioned. Through simulation we have determined that the use of reflection-canceling waves at the boundary, as per the method of [146], does *not* lead to a passive termination. This statement also holds for the termination of the plate and shell models that we will discuss shortly; violent instabilities may appear in these systems, even though the termination of the simpler transmission line and parallel-plate networks by this method is not problematic. At present, the termination of a MDWD network is very poorly understood, even by experts[1]. Indeed, as we mentioned in Section 3.10, there is not, as yet, a general theory of boundary termination of MDWD networks. We refer to Section 3.10.1 for a discussion of a possible avenue of approach.

It is more straightforward to work with the termination of the DWN in its conventional lumped form as shown in Figure 7.5. When a network is viewed in this way, it is much simpler to see how boundary conditions may be set such that passivity may be maintained. The difficulty in working with the expanded signal flow diagram for a MDWD network is that unlike the DWN, there is no port structure in this case; in the DWN, applying lumped terminations to junctions on the boundary is straightforward.

Because we looked at the termination of the $(1+1)$D transmission line and $(2+1)$D parallel-plate systems in this way in Section 4.3.9 and Section 4.4.4 respectively, we simply present the terminations corresponding to boundary conditions (7.12). We align the parallel junctions (at which V_J and M_J are calculated) with the left end point (say) of the beam, located at $x = 0$. In this way, we avoid the slight additional complication of the coupling that occurs if the series junctions are placed at the boundary (though we will be forced to face this issue when we set boundary conditions in the Mindlin plate network in Section 7.2.2). The three terminations are shown in Figure 7.6.

A condition $w = 0$ (implying $v = 0$) or $m = 0$ is easily implemented by short-circuiting the appropriate junction. For the condition $\psi = 0$, it is easy to show that we should choose

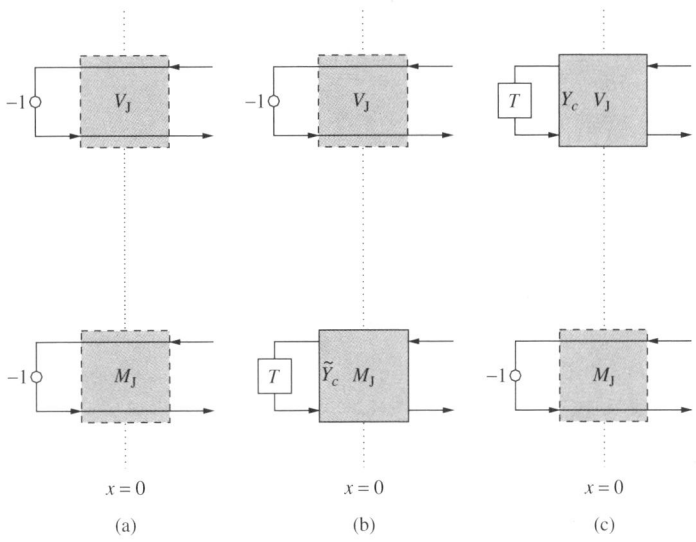

Figure 7.6: *Boundary terminations for the DWN for Timoshenko's system—*(a) *fixed end, allowed to pivot,* (b) *fixed clamped end, and* (c) *free end.*

---

[1] Stuart Lawson, private communication, 2000.

the self-loop admittance to be $\tilde{Y}_c = v_0/(EI) - 1/r_2$ at the terminating junction at which M_J is calculated. Similarly, for the condition $q = 0$, we should choose $Y_c = v_0 \rho A - 1/r_1$ at the junction at which V_J is calculated.

7.1.4 Simulation: Timoshenko's System for Beams of Uniform and Varying Cross-sectional Areas

We present here two simple DWN simulations of the Timoshenko beam equations. We simulate the behavior of a square prismatic steel beam, of length $L = 1$m, under the application of an initial transverse velocity distribution at the beam center, which has the form of one period of a raised cosine, of wavelength 5 cm and amplitude 0.0005 m/s. In the first simulation, the beam is assumed to have a uniform thickness of 2 cm, and the boundary conditions are of the free type at either end; the evolution of the velocity distribution is shown in Figure 7.7. In the second simulation, the beam is assumed to be of linearly varying thickness, from 1 cm at the left end to 3 cm at the right end. The area A thus varies quadratically and the moment of inertia quartically. In this second case (shown in Figure 7.8), the boundary conditions are assumed clamped. The material parameters for steel are taken to be $\rho = 5.38 \times 10^4$kg/m^3, $E = 1.4 \times 10^{12}$N/m^2, $G = 5.39 \times 10^{11}$N/m^2, and Timoshenko's coefficient for a beam of square cross section is $\kappa = 5/6$ [114]. In both simulations, we operate using a grid spacing of $1/400$ m and the time step is chosen to be at the passivity limit. From (7.11), and given the above values of the material parameters of the beam, v_0 is chosen to be 5.1×10^3 m/s for the uniform beam, and 4.59×10^4 m/s for the beam of varying thickness.

In both simulations, it is easy to see that due to dispersion the coherence of the initial velocity distribution is lost; short wavelengths tend to move faster (and hence reflect first from the boundary), as can be seen in the plot at $t = 8.65 \times 10^{-5}$s. In the beam with linearly varying thickness, velocities are amplified in the thin region of the beam and attenuated in the thick region; the propagation velocities themselves, however, are not significantly altered.

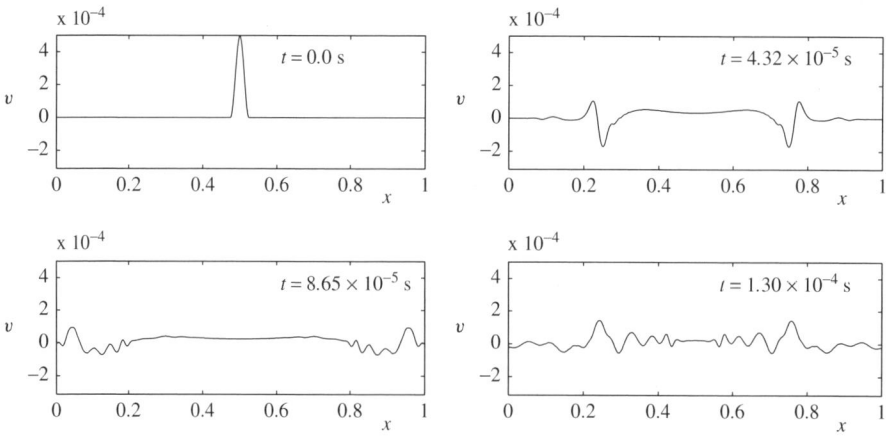

Figure 7.7: *Simulation: evolution of the transverse velocity distribution along a Timoshenko beam of uniform thickness, with free ends.*

7.1. BEAM DYNAMICS

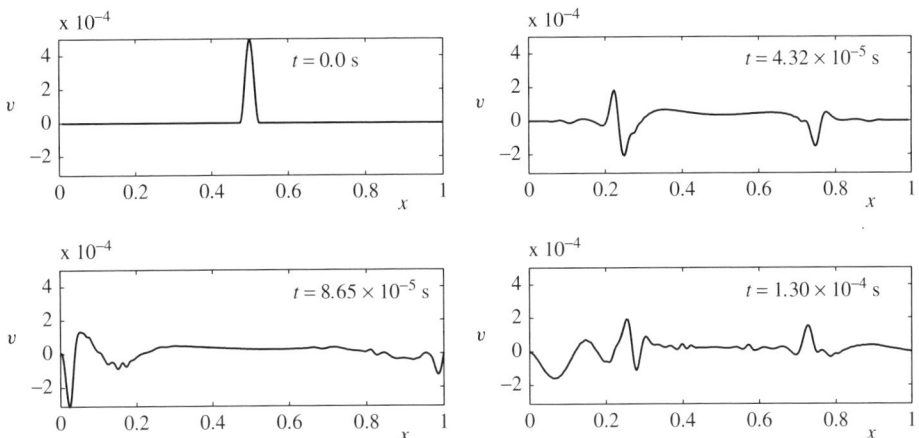

Figure 7.8: *Simulation: evolution of the transverse velocity distribution along a Timoshenko beam of linearly varying thickness, with clamped ends.*

7.1.5 Improved MDKC for Timoshenko's System via Balancing

In the preceding simulation of the steel beam of rectangular cross section and linearly varying thickness, we have, from (7.11), $v_0 = 4.59 \times 10^4$; the time step is thus restricted to be quite small. We will now show how the balancing or preconditioning approach applied to the $(1+1)$D transmission line problem in Section 3.11 can be used to drastically increase the maximum allowable time step for a given grid spacing.

Suppose we scale the dependent variables according to

$$v = r_1 i_1 \quad q = r_2 i_2 \quad \omega = r_3 i_3 \quad m = r_4 i_4$$

and allow the scaling parameters $r_1, \ldots, r_4$ to be arbitrary smooth functions of x. Timoshenko's system (7.4)–(7.5) can then be rewritten as

$$v_0 \left(\frac{r_1}{r_2}\right) \rho A \frac{\partial i_1}{\partial t'} = \frac{\partial i_2}{\partial x} + \frac{r_2'}{r_2} i_2 \tag{7.13a}$$

$$v_0 \left(\frac{r_2}{r_1}\right) \frac{1}{A\kappa G} \frac{\partial i_2}{\partial t'} = \frac{\partial i_1}{\partial x} + \frac{r_1'}{r_1} i_1 - \frac{r_3}{r_1} i_3 \tag{7.13b}$$

$$v_0 \left(\frac{r_3}{r_4}\right) \rho I \frac{\partial i_3}{\partial t'} = \frac{\partial i_4}{\partial x} + \frac{r_4'}{r_4} i_4 + \frac{r_2}{r_4} i_2 \tag{7.13c}$$

$$v_0 \left(\frac{r_4}{r_3}\right) \frac{1}{EI} \frac{\partial i_4}{\partial t'} = \frac{\partial i_3}{\partial x} + \frac{r_3'}{r_3} i_3 \tag{7.13d}$$

where primes above the r_i, $i = 1, \ldots, 4$ indicate x-differentiation. If we choose

$$r_1 = \left(\rho A^2 G \kappa\right)^{-\frac{1}{4}} \quad r_2 = \left(\rho A^2 G \kappa\right)^{\frac{1}{4}} \quad r_3 = \left(\rho E I^2\right)^{-\frac{1}{4}} \quad r_4 = \left(\rho E I^2\right)^{\frac{1}{4}} \tag{7.14}$$

then system (7.13) becomes

$$v_0 \sqrt{\frac{\rho}{G\kappa}} \frac{\partial i_1}{\partial t'} = \frac{\partial i_2}{\partial x} + \frac{r'_2}{r_2} i_2 \tag{7.15a}$$

$$v_0 \sqrt{\frac{\rho}{G\kappa}} \frac{\partial i_2}{\partial t'} = \frac{\partial i_1}{\partial x} - \frac{r'_2}{r_2} i_1 - \frac{r_3}{r_1} i_3 \tag{7.15b}$$

$$v_0 \sqrt{\frac{\rho}{E}} \frac{\partial i_3}{\partial t'} = \frac{\partial i_4}{\partial x} + \frac{r'_4}{r_4} i_4 + \frac{r_2}{r_4} i_2 \tag{7.15c}$$

$$v_0 \sqrt{\frac{\rho}{E}} \frac{\partial i_4}{\partial t'} = \frac{\partial i_3}{\partial x} - \frac{r'_4}{r_4} i_3 \tag{7.15d}$$

and the constant-proportional terms appear antisymmetrically (note, from (7.14), that $r_2/r_4 = r_3/r_1$). In the MDKC shown in Figure 7.9, these terms are all interpreted as gyrator couplings, where the gyrator coefficients are spatially varying. It is easy to check that this system is still symmetric hyperbolic according to definition (3.1).

The MDWD network (not shown) implied by the MDKC will be slightly more difficult to program because of the additional reflection-free ports, which will necessarily be introduced, but it has the same memory requirements, and the operation count is slightly larger (due chiefly to the post-scaling of the MDKC currents, which must now be performed in order to obtain the physical dependent variables). We now have, however, that

$$v_0 \geq \max\left(\sqrt{\left(\frac{E}{\rho}\right)_{\max}}, \sqrt{\left(\frac{G\kappa}{\rho}\right)_{\max}}\right) = \gamma^g_{T,\max}$$

where $\gamma^g_{T,\max}$ is the maximum group velocity given in (7.7). v_0 is now optimal (in the CFL sense) for a constant grid spacing. Referring to the simulation of Section 7.1.4, it is easy to see that due to the quartic dependence of the moment of inertia I on x (for a beam of linearly varying thickness), the maximum time step allowed by the previous approach will be severely constrained. Using a balanced formulation and MDKC, we now have $v_0 = 5.10 \times 10^3$. Thus for a given grid spacing, the maximum time step is now nine times larger. From a practical standpoint, this is a huge computational advantage.

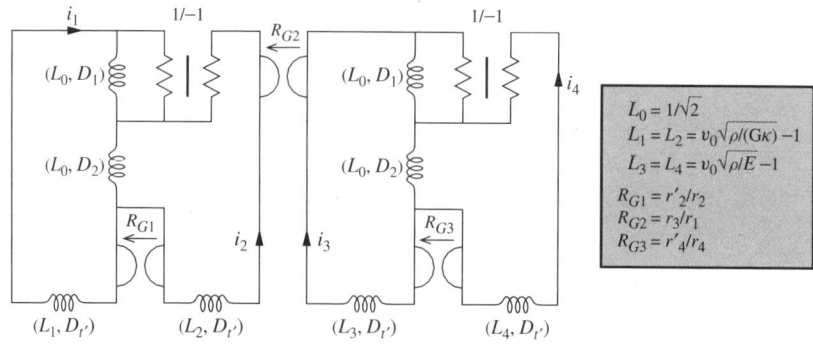

Figure 7.9: *Balanced MDKC for Timoshenko's system.*

7.2 Plates

The equations of motion of a stiff plate are the $(2+1)$D generalization of those of a beam. According to the classical theory, the plate is assumed to lie, when at rest, in the (x, y) plane, and to be of thickness $h(x, y)$; the deflection $w(x, y)$ of the plate from its equilibrium state is taken to be perpendicular to the (x, y) plane. The plate material has density ρ, as well as Young's modulus E and Poisson's ratio v, all of which are assumed, for the sake of generality, to be smooth positive functions of x and y. In particular, v must be less than one-half. The classical development depends on neglecting rotational inertia effects and makes various assumptions analogous to the 'plane sections remain plane and perpendicular to the neutral axis' hypothesis that was used as the basis for the Euler–Bernoulli beam model [108]. The resulting equation of motion [7, 154] can be written as

$$\rho h \frac{\partial^2 w}{\partial t^2} = -\nabla^2 \left(D \nabla^2 w \right) + (1-v) \left(\frac{\partial^2 D}{\partial x^2} \frac{\partial^2 w}{\partial y^2} - 2 \frac{\partial^2 D}{\partial x \partial y} \frac{\partial^2 w}{\partial x \partial y} + \frac{\partial^2 D}{\partial y^2} \frac{\partial^2 w}{\partial x^2} \right)$$

where

$$D = \frac{Eh^3}{12(1-v^2)}$$

and $\nabla^2 = \partial^2/\partial x^2 + \partial^2/\partial y^2$. If the material parameters and the thickness are constant, then

$$\frac{\partial^2 w}{\partial t^2} = -\frac{D}{\rho h} \nabla^2 \nabla^2 w \qquad (7.16)$$

which is easily seen to be a direct generalization of (7.2). As such, we expect to find the same anomalous behavior of the resulting propagation velocities, which can become infinitely large in the high-frequency limit.

Because the development is so similar to the $(1+1)$D case, we will proceed directly to the more refined model of plate motion, which is a direct generalization of the Timoshenko theory for beams. First proposed by Mindlin, the model [108, 161] can be written as a system of eight PDEs:

$$\rho h \frac{\partial v}{\partial t} = \frac{\partial q_x}{\partial x} + \frac{\partial q_y}{\partial y} \qquad (7.17\text{a})$$

$$\frac{1}{\kappa^2 G h} \frac{\partial q_x}{\partial t} = \frac{\partial v}{\partial x} + \omega_x \qquad (7.17\text{b})$$

$$\frac{1}{\kappa^2 G h} \frac{\partial q_y}{\partial t} = \frac{\partial v}{\partial y} + \omega_y \qquad (7.17\text{c})$$

$$\frac{\rho h^3}{12} \frac{\partial \omega_x}{\partial t} = \frac{\partial m_x}{\partial x} + \frac{\partial m_{xy}}{\partial y} - q_x \qquad (7.18a)$$

$$\frac{\rho h^3}{12} \frac{\partial \omega_y}{\partial t} = \frac{\partial m_{xy}}{\partial x} + \frac{\partial m_y}{\partial y} - q_y \qquad (7.18b)$$

$$\frac{1}{D} \frac{\partial m_x}{\partial t} = \frac{\partial \omega_x}{\partial x} + \nu \frac{\partial \omega_y}{\partial y} \qquad (7.18c)$$

$$\frac{1}{D} \frac{\partial m_y}{\partial t} = \frac{\partial \omega_y}{\partial y} + \nu \frac{\partial \omega_x}{\partial x} \qquad (7.18d)$$

$$\frac{2}{D(1-\nu)} \frac{\partial m_{xy}}{\partial t} = \frac{\partial \omega_y}{\partial x} + \frac{\partial \omega_x}{\partial y} \qquad (7.18e)$$

Here, we have written

$$\omega_x \triangleq \frac{\partial \psi_x}{\partial t} \qquad \omega_y \triangleq \frac{\partial \psi_y}{\partial t} \qquad v \triangleq \frac{\partial w}{\partial t}$$

where (ψ_x, ψ_y) is the pair of angles giving the orientation of the sides of a deformed differential element of the plate with respect to the perpendicular. (In the classical theory, for which cross sections of the plate are assumed to remain parallel to the plate normal, we have $(\psi_x, \psi_y) = (-\partial w/\partial x, -\partial w/\partial y)$.) In addition, we have the shear forces (q_x, q_y) and moments (m_x, m_y, m_{xy}), which are the $(2+1)$D generalizations of q and m from Timoshenko's system. The system (7.17)–(7.18) as a whole is known as *Mindlin's system*, although it is commonly reduced to a system of three second-order equations in the variables w, ψ_x and ψ_y [108]. We have written Mindlin's system so that it is easy to see the decomposition into two separate subsystems, one in (v, q_x, q_y) and the other in $(\omega_x, \omega_y, m_x, m_y, m_{xy})$, with the coupling occurring via constant-proportional terms in ω_x, ω_y, q_x, and q_y. In particular, subsystem (7.17) is identical, under reflection symmetries, to the lossless parallel-plate system (see Section 4.4), except for the coupling terms.

It is easy to see that this system is not, as written, symmetric hyperbolic. It is easy to symmetrize it by taking linear combinations of (7.18c) and (7.18d), in which case we get, in terms of the variable $\mathbf{w} = [v, q_x, q_y, \omega_x, \omega_y, m_x, m_y, m_{xy}]^T$,

$$\mathbf{P} = \mathbf{P}_M = \begin{bmatrix} \mathbf{P}_M^+ & \cdot \\ \cdot & \mathbf{P}_M^- \end{bmatrix} \quad \mathbf{A}_1 = \mathbf{A}_{M1} = \begin{bmatrix} \mathbf{A}_{M1}^+ & \cdot \\ \cdot & \mathbf{A}_{M1}^- \end{bmatrix} \quad \mathbf{A}_2 = \mathbf{A}_{M2} = \begin{bmatrix} \mathbf{A}_{M2}^+ & \cdot \\ \cdot & \mathbf{A}_{M2}^- \end{bmatrix}$$

$$\mathbf{B} = \mathbf{B}_M = \begin{bmatrix} \cdot & \mathbf{B}_{M\times} \\ -\mathbf{B}_{M\times}^T & \cdot \end{bmatrix} \qquad (7.19)$$

where the $\cdot$ stands for zero entries, and

$$\mathbf{P}_M^+ = \begin{bmatrix} \rho h & 0 & 0 \\ 0 & \frac{1}{\kappa^2 G h} & 0 \\ 0 & 0 & \frac{1}{\kappa^2 G h} \end{bmatrix} \qquad \mathbf{P}_M^- = \begin{bmatrix} \frac{\rho h^3}{12} & 0 & 0 & 0 & 0 \\ 0 & \frac{\rho h^3}{12} & 0 & 0 & 0 \\ 0 & 0 & \frac{12}{Eh^3} & \frac{-12\nu}{Eh^3} & 0 \\ 0 & 0 & \frac{-12\nu}{Eh^3} & \frac{12}{Eh^3} & 0 \\ 0 & 0 & 0 & 0 & \frac{2}{D(1-\nu)} \end{bmatrix}$$

7.2. PLATES

$$\mathbf{A}_{M1}^+ = \begin{bmatrix} 0 & -1 & 0 \\ -1 & 0 & 0 \\ 0 & 0 & 0 \end{bmatrix} \qquad \mathbf{A}_{M1}^- = \begin{bmatrix} 0 & 0 & -1 & 0 & 0 \\ 0 & 0 & 0 & 0 & -1 \\ -1 & 0 & 0 & 0 & 0 \\ 0 & 0 & 0 & 0 & 0 \\ 0 & -1 & 0 & 0 & 0 \end{bmatrix} \qquad (7.20)$$

$$\mathbf{A}_{M2}^+ = \begin{bmatrix} 0 & 0 & -1 \\ 0 & 0 & 0 \\ -1 & 0 & 0 \end{bmatrix} \qquad \mathbf{A}_{M2}^- = \begin{bmatrix} 0 & 0 & 0 & 0 & -1 \\ 0 & 0 & 0 & -1 & 0 \\ 0 & 0 & 0 & 0 & 0 \\ 0 & -1 & 0 & 0 & 0 \\ -1 & 0 & 0 & 0 & 0 \end{bmatrix} \qquad (7.21)$$

$$\mathbf{B}_{M\times} = \begin{bmatrix} 0 & 0 & 0 & 0 & 0 \\ -1 & 0 & 0 & 0 & 0 \\ 0 & -1 & 0 & 0 & 0 \end{bmatrix}$$

The system defined by (7.19) is lossless, due to the antisymmetry of $\mathbf{B}_M$. Also, note that $\mathbf{P}_M$ is positive definite (recall that ν is positive and less than one-half) but not diagonal[2]; this has not come up in any of the systems we have looked at previously and will have interesting consequences in the circuit representations in the next section.

Maximum Group Velocity

For the constant-coefficient problem, the characteristic polynomial relating frequencies ω to spatial wavenumber magnitude $\|\boldsymbol{\beta}\|_2 = \sqrt{\beta_x^2 + \beta_y^2}$, from (3.10), will be

$$\left(\omega^4 - \frac{\omega^2}{\rho}\left(\frac{12\kappa^2 G}{h^2} + \|\boldsymbol{\beta}\|_2^2\left(\frac{E}{1-\nu^2} + G\kappa^2\right)\right) + \frac{E\kappa^2 G}{\rho^2(1-\nu^2)}\|\boldsymbol{\beta}\|_2^4\right)$$

$$\times \left(\omega^2 - \frac{G}{\rho}\|\boldsymbol{\beta}\|_2^2 - \frac{12G\kappa^2}{\rho h^2}\right)\omega^2 = 0$$

The first factor, which is similar in form to that which defines the Timoshenko system, from (7.6), has four roots, $\omega_{1\pm}$, $\omega_{2\pm}$ which behave as

$$\lim_{\|\boldsymbol{\beta}\|_2 \to \infty} \omega_{1\pm} = \pm\|\boldsymbol{\beta}\|_2 \sqrt{\frac{E}{\rho(1-\nu^2)}} \qquad \lim_{\|\boldsymbol{\beta}\|_2 \to \infty} \omega_{2\pm} = \pm\|\boldsymbol{\beta}\|_2 \sqrt{\frac{G\kappa^2}{\rho}}$$

and the second factor has a pair of roots $\omega_{3\pm}$, which have the limiting behavior

$$\lim_{\|\boldsymbol{\beta}\|_2 \to \infty} \omega_{3\pm} = \pm\|\boldsymbol{\beta}\|_2 \sqrt{\frac{G}{\rho}}$$

---

[2] It is possible to write a symmetric hyperbolic form of Mindlin's system for which the $\mathbf{P}$ matrix is diagonal by introducing the variables $m_1 = (m_x + m_y)/2$ and $m_2 = (m_x - m_y)/2$. Though the resulting MDKC will be simpler, boundary conditions become overdetermined and difficult to set properly because $\mathbf{A}_1$ and $\mathbf{A}_2$ are consequently less sparse (implying greater network connectivity). The form described by the matrices in (7.19) is more fundamental in this respect.

All phase and group velocities are thus bounded. For the varying-coefficient problem, the maximum global group velocity will be

$$v^g_{M,\max} = \sqrt{\left(\frac{E}{\rho(1-v^2)}\right)_{\max}}$$

7.2.1 MDKCs and Scattering Networks for Mindlin's System

We now introduce scaled dependent variables

$$(i_1, i_2, i_3, i_4, i_5, i_6, i_7, i_8) = (r_1 v, q_x, q_y, \omega_x, \omega_y, r_2 m_x, r_2 m_y, r_3 m_{xy})$$

where again, r_1, r_2, and r_3 are positive constants, as well as the scaled time variable $t' = v_0 t$. The MD-passive circuit representation of Mindlin's system is shown at top in Figure 7.10, where we have used the coordinates defined by (3.22) with the transformation matrix (3.23). Here, the port with terminals A and A' is assumed to be short-circuited (we will return to this port in Section 7.3.2). It is best to view this as a three-loop network (at left, in Figure 7.10) corresponding to the subsystem (7.17), coupled to a five-loop network (at right in Figure 7.10) corresponding to subsystem (7.18). Because $\mathbf{P}_M$ from (7.19) is not diagonal, the coupling between the loops with currents i_6 and i_7 (corresponding to the moments m_x and m_y) is of a type not previously encountered in the systems examined in this book. It can be interpreted in terms of a coupled inductance between the loops (see Section 2.2.7); in Figure 7.10, self-inductances are indicated by directed arrows and mutual inductance by bidirectional arrows. The element values are as indicated in the figure.

Optimal choices of r_1, r_2, and r_3, and the passivity bound on v_0 are a little more difficult to determine in this case. As before, however, they follow from a positivity requirement on the inductance values defined in Figure 7.10. This requirement is simply applied to L_1, L_2, L_3, L_4, L_5, and L_8, but L_6 and L_7 define a coupled inductance between the loops with currents i_6 and i_7. The coupling matrix will be

$$\begin{bmatrix} L_6 & L_7 \\ L_7 & L_6 \end{bmatrix}$$

and is required to be positive semidefinite for passivity. This is true if

$$L_6 \geq |L_7| \tag{7.22}$$

In complete analogy with the case of the parallel-plate system, an optimal choice for r_1 is easily shown to be

$$r_1 = \sqrt{\frac{2}{(\rho h)_{\min}(\kappa^2 G h)_{\max}}} \tag{7.23}$$

and gives a first bound on v_0, which is

$$v_0 \geq v_{M+} \triangleq \sqrt{\frac{2(\kappa^2 G h)_{\max}}{(\rho h)_{\min}}} \tag{7.24}$$

7.2. PLATES

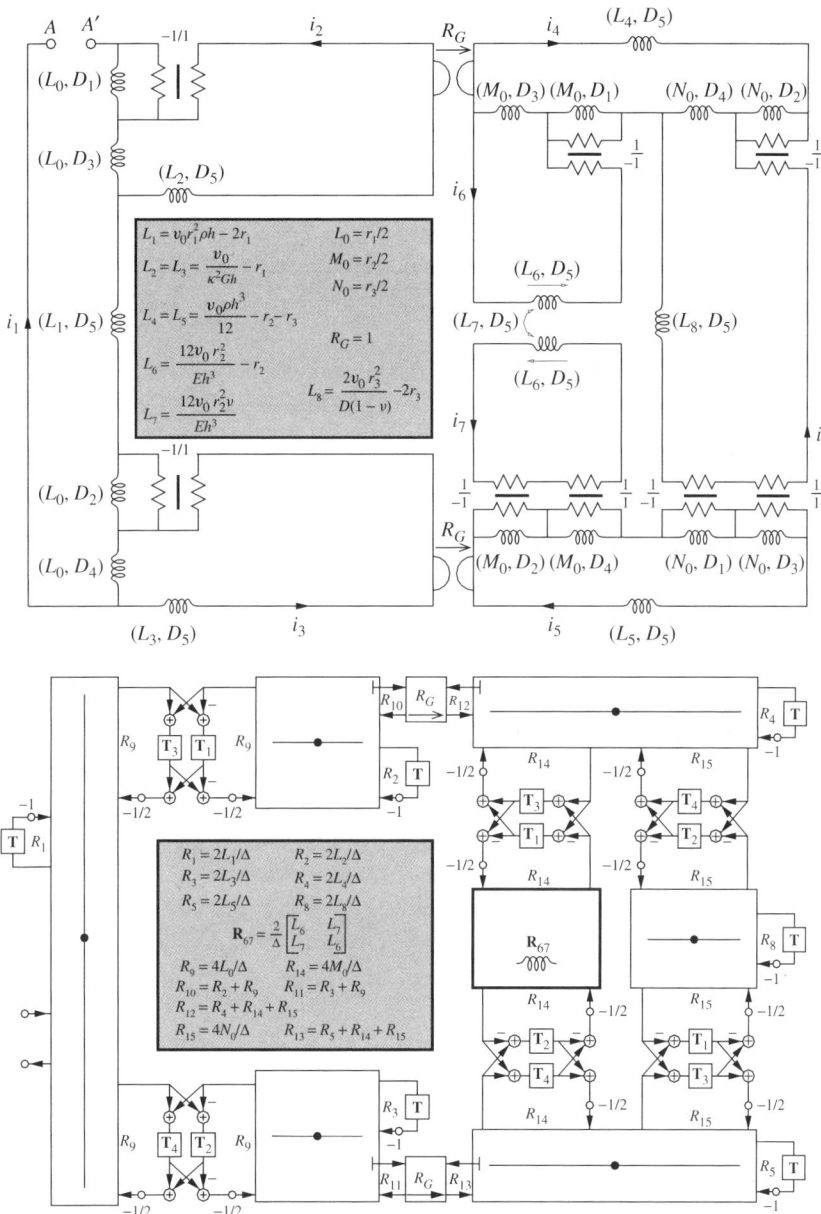

Figure 7.10: *MDKC and MDWD network for Mindlin's system.*

From the positivity requirement on L_4, L_5, L_8, as well as condition (7.22), we have a second bound on v_0,

$$v_0 \geq \max_{r_2, r_3 > 0} \left(\frac{12(r_2 + r_3)}{(\rho h^3)_{\min}}, \frac{\left(\frac{Eh^3}{1-\nu}\right)_{\max}}{12 r_2}, \frac{\left(\frac{Eh^3}{1+\nu}\right)_{\max}}{12 r_3} \right) \quad (7.25)$$

This is a simple minimax-type problem—we would like to minimize the bound on v_0, which is the maximum of three quantities as per (7.25), with respect to the parameters r_2 and r_3. The solution is

$$r_2 = \frac{1}{12}\left(\frac{Eh^3}{1-\nu}\right)_{\max}\sqrt{\frac{(\rho h^3)_{\min}}{\left(\frac{Eh^3}{1-\nu}\right)_{\max} + \left(\frac{Eh^3}{1+\nu}\right)_{\max}}} \qquad (7.26)$$

$$r_3 = \frac{1}{12}\left(\frac{Eh^3}{1+\nu}\right)_{\max}\sqrt{\frac{(\rho h^3)_{\min}}{\left(\frac{Eh^3}{1-\nu}\right)_{\max} + \left(\frac{Eh^3}{1+\nu}\right)_{\max}}} \qquad (7.27)$$

which gives the second bound

$$v_0 \geq v_{M-} \triangleq \sqrt{\frac{\left(\frac{Eh^3}{1-\nu}\right)_{\max} + \left(\frac{Eh^3}{1-\nu}\right)_{\max}}{(\rho h^3)_{\min}}} \qquad (7.28)$$

and the overall stability bound for the combined network will be

$$v_0 \geq v_M \triangleq \max(v_{M-}, v_{M+}) \qquad (7.29)$$

When the material parameters and the thickness are constant, this bound reduces to

$$v_0 \geq \sqrt{2}\max\left(\sqrt{\frac{G\kappa^2}{\rho}}, \sqrt{\frac{E}{\rho(1-\nu^2)}}\right) = \sqrt{2}\gamma^g_{M,\max}$$

The MDWD network, shown at the bottom in Figure 7.10, follows immediately from the MDKC; here, as for the parallel-plate problem discussed in Section 3.7.1, we have used step sizes $T_j = \Delta$, $j = 1, \ldots, 5$. Recall that because coordinate $t_5 = t' = v_0 t$, a step size of $T_5 = \Delta$ implies a time step of $\Delta/v_0 = T$, and we have indicated pure time delays of duration T by **T**. As for the Timoshenko network of Figure 7.2, reflection-free ports will be necessary due to the memoryless gyrator couplings between the loops with currents i_2 and i_4, and i_3 and i_5 in the MDKC. The coupled inductance has been treated as a vector scattering junction terminated on a vector inductor, as discussed in Section 2.2.7. We also note in passing that this network may be balanced in the same way as the Timoshenko system (see Section 7.1.5) in order to obtain a much better bound on v_0 (at the expense of increased network complexity).

It is also, of course, possible to put the MDKC into a form that yields, upon discretization, a DWN. This new form is shown in Figure 7.11; now the transverse velocity v and bending moments m_x, m_y, and m_{xy} are treated as voltages, and inductors in these loops are replaced by gyrators terminated on capacitances. In particular, the coupled inductance in Figure 7.10 is replaced by a coupled capacitance. In order to discretize this MDKC, we apply the trapezoid rule to all the inductances and capacitances with direction t_5 (using a step size of $T_5 = \Delta$), and for the Jaumann two-ports, we make use of the alternative spectral mappings defined by (6.6), with step sizes $T_j = \Delta/2$, $j = 1, \ldots, 4$. We have chosen these step sizes such that an interleaved algorithm results; the computational grid is shown in Figure 7.12. Grid quantities (capitalized) are shown next to the points at which they are

7.2. PLATES

to be calculated. The grid on the right, which operates on grid functions V, Q_x, and Q_y is identical, again under reflection symmetry, to the grid for the DWN for the $(2+1)$D parallel-plate problem (see Figure 4.16), which is to be expected, since the related subnetwork of the MDKC shown in Figure 7.11 is the same as that for the parallel-plate problem (see Figure 6.6). It is coupled via gyrators (these couplings are indicated by curved arrows) to a second grid, over which grid functions Ω_x, Ω_y, M_x, M_y, and M_{xy} are calculated. In particular, M_x and M_y are calculated together as a vector quantity at vector parallel

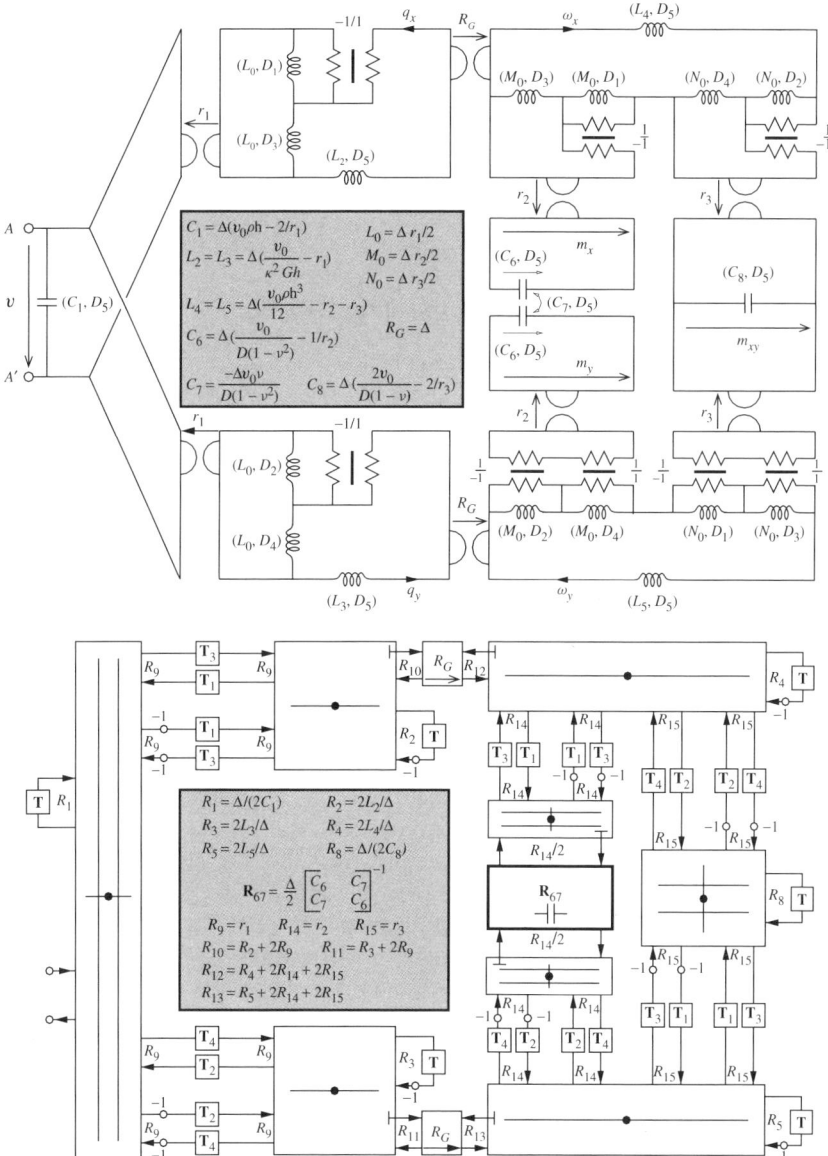

Figure 7.11: *Modified MDKC and multidimensional DWN for Mindlin's system.*

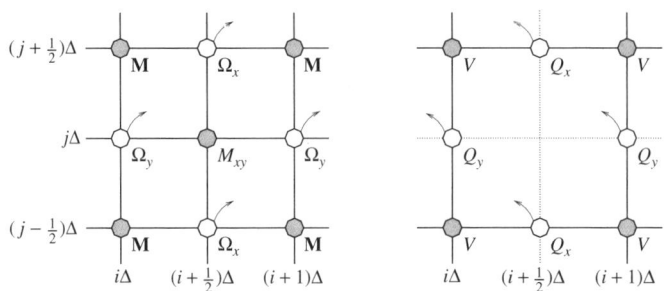

Figure 7.12: *Computational grid for the multidimensional DWN shown in Figure 7.11.*

junctions—this vector is written as **M** in Figure 7.12. Waveguide connections (of delay $T/2$) are represented by solid lines and self-loops, and sign-inversions in the signal paths are not shown. Note that at the gray dots, we will have parallel scattering junctions and at the white dots we will have series junctions; junction quantities are calculated at alternating multiples of $T/2$ seconds.

7.2.2 Boundary Termination of the Mindlin Plate

There are four common types of conditions applied at a plate boundary [126]. At a southern boundary, parallel to the x axis, the four conditions can be written as

$$q_y = m_y = m_{xy} = 0 \quad \text{Free edge} \tag{7.30a}$$

$$v = \omega_x = m_y = 0 \quad \text{Simply supported edge (1)} \tag{7.30b}$$

$$v = m_{xy} = m_y = 0 \quad \text{Simply supported edge (2)} \tag{7.30c}$$

$$v = \omega_x = \omega_y = 0 \quad \text{Clamped edge} \tag{7.30d}$$

These conditions are again lossless and of the form of (3.8). (We note that conditions on v, ω_x, and ω_y in (7.30) are usually written in terms of their time integrals w, ψ_x, and ψ_y, but the formulation above is equivalent.) The same conditions also reduce to a similar set of conditions that can be applied to the classical plate [23].

As we mentioned in Section 7.1.3, the passive boundary termination of an MDWD network, such as that shown in Figure 7.10, is not at all straightforward. Indeed, it is somewhat complicated by the fact that we must approximate all the system variables at any given grid point on the boundary. The 'wave-canceling' method [146] discussed in Section 3.10 becomes exceedingly complex when vector wave variables and reflection-free ports are involved; passivity is not easy to ensure. Termination of the DWN derived from the MDKC of Figure 7.11, operating on the computational grid of Figure 7.12, however, is simpler, because we are able to work directly with the termination of the lumped network representation. Suppose we choose our southern boundary at $y = 0$ according to Figure 7.13. The only quantities to be calculated in this arrangement will be Q_y and Ω_y, at coincident series scattering junctions, and M_{xy} at parallel scattering junctions. This arrangement is to be preferred because we do not need to worry about the termination of the vector scattering junctions at which $\mathbf{M} = [M_x, M_y]$ is calculated.

The southern boundary terminations corresponding to the four conditions (7.30) are shown in Figure 7.14. The conditions $q_y = 0$ and $\omega_y = 0$ that appear in (7.30a) and (7.30d)

7.2. PLATES

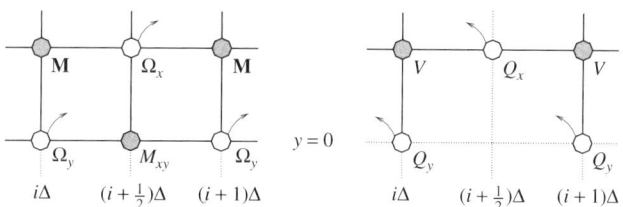

Figure 7.13: *Southern computational grid boundary at $y = 0$ for the DWN for Mindlin's system.*

can be dealt with rather simply, by terminating the boundary series junctions at which the junction currents Q_{yJ} and Ω_{yJ} are calculated in an open-circuit[3]. For these conditions, the gyrator coupling between the two subnetworks (indicated by curved arrows in Figure 7.13) may be dropped entirely. Similarly, the condition $m_{xy} = 0$ can be implemented by short-circuiting the appropriate parallel boundary junctions.

The other conditions, involving variables not calculated directly on the boundary, require a slightly more involved treatment; the analysis is similar to that performed in Section 4.4.4, and the termination problem becomes (for the most part) that of setting the self-loop immittances at the boundary junctions that cannot be trivially terminated in an open- or short-circuit. To this end, we provide the waveguide immittances at the junctions in the problem *interior* at which M_{xyJ}, Ω_{yJ}, and Q_{yJ} are calculated. From Figure 7.11, it is possible to read off these values directly. As for the DWN for the Timoshenko system in Section 7.1.2, immittances in the two overlapped networks are distinguished by a tilde ($\tilde{\ }$). For example, at a parallel junction in the five-variable grid at location $x = (i + 1/2)\Delta$, $y = j\Delta$, for i and j integer, where we calculate $M_{xyJ,i+1/2,j}$ (see Figure 7.12), there are waveguide connections to the north, south, east, and west; their admittances are defined as

$$\tilde{Y}_{y+,i+\frac{1}{2},j} = \tilde{Y}_{y-,i+\frac{1}{2},j} = \tilde{Y}_{x+,i+\frac{1}{2},j} = \tilde{Y}_{x-,i+\frac{1}{2},j} = \frac{1}{R_{15}} = \frac{1}{r_3}$$

respectively, which are simply the admittances of the multidimensional unit elements from Figure 7.11. Similarly, the four connecting waveguide impedances at series junctions at locations $x = i\Delta$, $j = i\Delta$ (at which $\Omega_{yJ,i,j}$ are calculated) will have impedances

$$\tilde{Z}_{y+,i,j} = \tilde{Z}_{y-,i,j} = R_{14} = r_2 \qquad \tilde{Z}_{x+,i,j} = \tilde{Z}_{x-,i,j} = R_{15} = r_3$$

In the other waveguide network (on the right in Figure 7.12), the connecting northward and southward impedances at the series junctions at which $Q_{yJ,i,j}$ is calculated will be

$$Z_{y+,i,j} = Z_{y-,i,j} = R_9 = r_1$$

The self-loop immittances at the three types of junctions will be

$$\tilde{Y}_{c,i+\frac{1}{2},j} = \frac{1}{R_8} = \frac{4v_0}{(D(1-v))_{i+\frac{1}{2},j}} - \frac{4}{r_3} \qquad (7.31\text{a})$$

---

[3]In keeping with the notation of Chapter 4, we have appended a 'J' to the subscript of any grid function, to indicate that it is to be calculated as a junction current or voltage.

$$\tilde{Z}_{c,i,j} = R_4 = \frac{v_0(\rho h^3)_{i,j}}{6} - 2r_2 - 2r_3 \qquad (7.31\text{b})$$

$$Z_{c,i,j} = R_2 = \frac{2v_0}{(\kappa^2 Gh)_{i,j}} - 2r_1 \qquad (7.31\text{c})$$

We assume r_1, r_2, and r_3 to be chosen according to (7.23), (7.26), and (7.27) respectively.

Free Edge

Referring to Figure 7.14(a), as mentioned previously, the conditions $q_y = 0$ and $m_{xy} = 0$ are rather simply dealt with by open- or short-circuiting the respective junctions. In this case, the gyrator connection between the two waveguide meshes can be severed at the boundary junctions, and in order to get a lossless numerical condition equivalent to $m_y = 0$ we need only set, for $j = 0$,

$$\tilde{Z}_{c,i,0} = \frac{v_0(\rho h^3)_{i,0}}{12} - r_2$$

The resulting positivity condition is less restrictive than that implied by (7.31b), and does not degrade the passivity bound from (7.24) and (7.28).

Simply Supported Edge (1)

Conditions (7.30b) are somewhat more complicated to implement than the others, because none of the variables calculated on the boundary can be zeroed out by short- or open-circuiting. From Figure 7.14(b), we can see that in addition to the gyrator coupling between the two meshes that must be maintained, we also must keep the waveguides that lie along the boundary.

In the problem interior, parallel and series junctions in the five-variable mesh are connected, through waveguides, to four neighboring junctions; on the southern boundary, however, each is connected to three—two to the east and west and one to the north. Owing to this asymmetry, we might suspect that it will be necessary to adjust the boundary waveguide impedances away from the values that they would take in the interior (which is r_3). In fact, it is possible to show that by introducing transformers, with turns ratios of $n = 2$ in these waveguides, we indeed have a lossless termination that satisfies conditions (7.30b). We must set the boundary immittances to

$$\tilde{Y}_{x^-,i+\frac{1}{2},0} = \tilde{Y}_{x^+,i+\frac{1}{2},0} = \frac{1}{2r_3} \qquad Z_{x^-,i,0} = Z_{x^+,i,0} = \frac{r_3}{2}$$

Notice that each waveguide now includes scaling factors (of 2 and 1/2) and that the impedances at either end are no longer identical, due to the transformer impedance matching. The self-loop immittances should be set according to

$$\tilde{Y}_{c,i+\frac{1}{2},0} = \frac{2v_0}{(D(1-v))_{i+\frac{1}{2},0}} - \frac{2}{r_3}$$

$$\tilde{Z}_{c,i,0} = \frac{v_0(\rho h^3)_{i,0}}{12} - r_2 - r_3$$

$$Z_{c,i,0} = = \frac{v_0}{(\kappa^2 Gh)_{i,0}} - r_1$$

7.2. PLATES

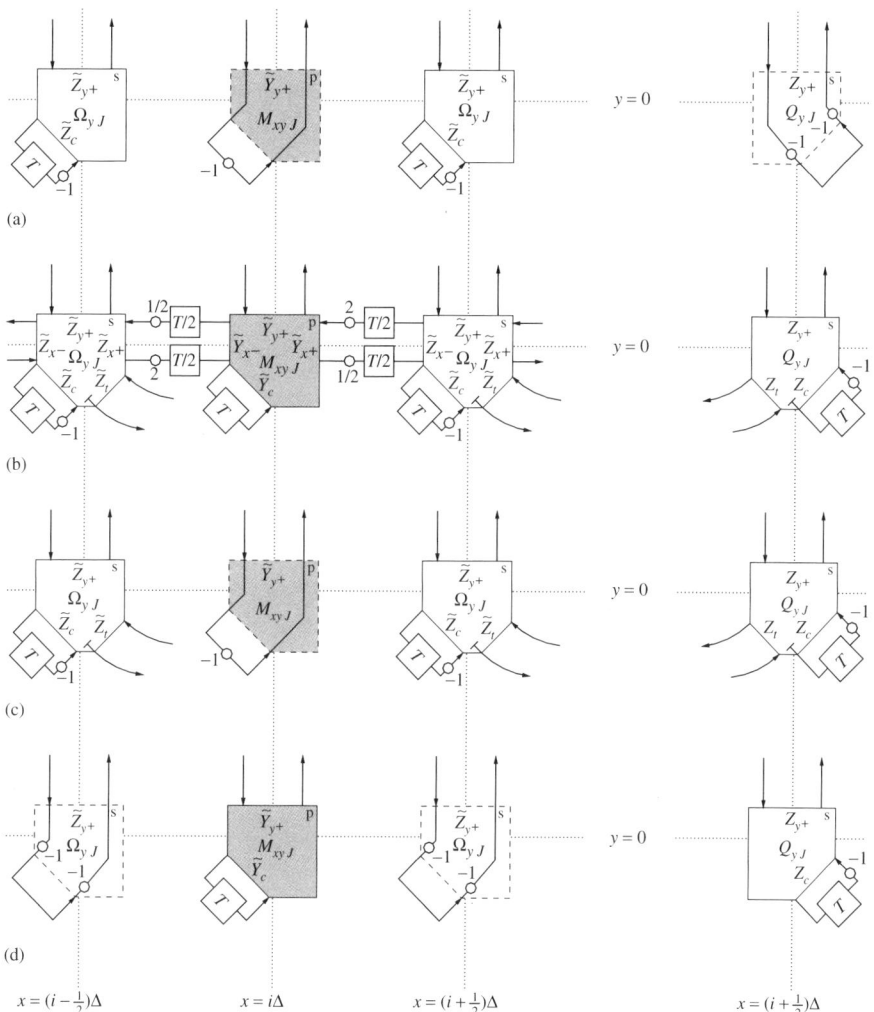

Figure 7.14: *Various lossless boundary terminations for the DWN for Mindlin's system*—(a) *free boundary;* (b) *simply supported edge (1);* (c) *simply supported edge (2);* (d) *clamped edge.*

which are precisely half the values they would take at interior junctions from (7.31) (and thus the positivity condition on these immittances is no different from the condition over interior self-loop immittances). We also mention that the gyrator coefficient, which takes on a value of $R_G = \Delta$ over the interior, should also be halved to $R_G = \Delta/2$ at the boundary junctions.

Simply Supported Edge (2)

The condition $m_{xy} = 0$ from (7.30c) can be set by short-circuiting the parallel boundary junctions. The condition $v = 0$ can be dealt with as for the preceding case, and we will again require

$$Z_{c,i,0} = \frac{v_0}{(\kappa^2 Gh)_{i,0}} - r_1$$

and the gyrator coefficient at the boundary junctions should be set to $R_G = \Delta/2$. The self-loop impedances at the series junctions in the five-variable mesh should be set, in order to ensure $m_y = 0$, as

$$\tilde{Z}_{c,i,0} = \frac{v_0(\rho h^3)_{i,0}}{12} - r_2$$

The positivity requirement on this impedance is again less restrictive than condition (7.31b) on the mesh interior.

Clamped Edge

For conditions (7.30d), we may immediately terminate the series junctions in the five-variable mesh with an open-circuit, and the gyrator coupling can be dropped entirely, as for the case of the free boundary condition. The remaining self-loop immittances should be set as

$$\tilde{Y}_{c,i+\frac{1}{2},0} = \frac{2v_0}{(D(1-v))_{i+\frac{1}{2},0}} - \frac{1}{r_3}$$

$$Z_{c,i,0} = = \frac{v_0}{(\kappa^2 Gh)_{i,0}} - r_1$$

which, as before, are less restrictive settings than those over the mesh interior, from (7.31).

7.2.3 Simulation: Mindlin's System for Plates of Uniform and Varying Thickness

For the sake of illustration, we present two DWN simulations of the vibration of a Mindlin plate. In both cases, the plate is assumed to be square, with side length 1 m and to be made of steel; the material parameters are thus $\rho = 5.38 \times 10^4 \text{kg/m}^3$, $E = 1.4 \times 10^{12} \text{N/m}^2$, and $G = 5.39 \times 10^{11} \text{N/m}^2$. κ is taken to be 5/6, and Poisson's ratio v is set to 0.3. The DWN is initialized with a transverse velocity distribution that takes the form of a single lobe of a 2D raised cosine, of radius 0.1 m, amplitude 0.0005m/s, and centered at coordinates $x = 0.3$m, $y = 0.3$m, where $x = 0$m, $y = 0$m are the coordinates of the bottom left-hand plate corner. The grid spacing is set to be $\Delta = 1$cm in the DWN in both the cases.

In the first simulation (see Figure 7.15), the plate thickness is 1 cm over the entire plate surface. Boundary conditions are of the free type, given by (7.30a), and implemented as per the DWN termination discussed in the previous section, and shown in Figure 7.14(a). In the second simulation, shown in Figure 7.16, the plate thickness is variable—over most of the plate, it is a constant 1 cm, but over the circular region outlined in black (radius 0.2 m, and centered at $x = 0.6$m, $y = 0.6$m), it rises in a raised 2D cosine distribution to a peak of 6 cm. In both simulations, snapshots of the transverse velocity distribution are

7.3. CYLINDRICAL SHELLS

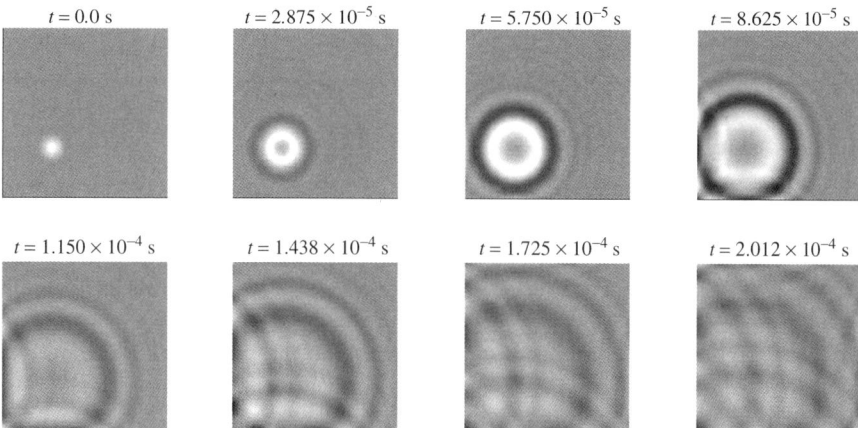

Figure 7.15: *DWN simulation of Mindlin's system for a steel plate of uniform thickness, with free edges.*

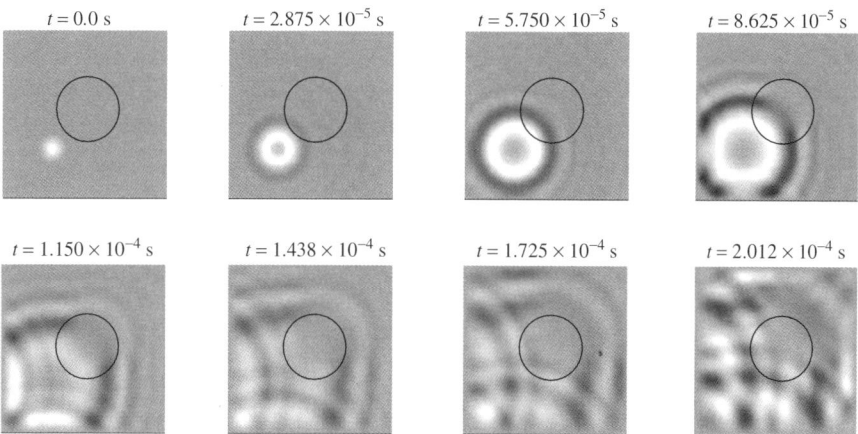

Figure 7.16: *DWN simulation of Mindlin's system for a steel plate of varying thickness. The variation is limited to the interior of the black circles. Boundary conditions are of the clamped type.*

taken every 2.875×10^{-5}s. The boundary termination is of the clamped type (7.30d) in this case, and has been enforced in the DWN according to Figure 7.14(d).

Light- and dark-colored regions correspond to positive and negative velocities respectively. The plots have been normalized and interpolated for better visibility. Notice in particular the numerical directional dependence of the propagation velocities at short wavelengths.

7.3 Cylindrical Shells

A shell is simply a plate with some curvature; it also supports wave motion, but the curvature complicates the motion considerably. As we will see, however, certain types of shell systems can also be represented by (2 + 1)D MDKCs. We will look first at the so-called cylindrical membrane shell formulation [108], then at a more modern (and elaborate) cylindrical shell formulation due to Naghdi and Cooper [46, 172].

7.3.1 The Membrane Shell

The simplest type of cylindrical shell theory is the *membrane shell* formulation of Rayleigh [108]. In this very basic theory, the shell is assumed to behave somewhat like a membrane, in that the restoring stiffness is assumed negligible. The shell is assumed to lie parallel to the x axis and has radius a. We define $\theta = a\theta'$, where θ' is the angular coordinate. This theory models the displacement of the shell from its equilibrium position; in contrast to Mindlin's plate system, however, displacements in all the three directions are modeled as a function of time, and we will write these three displacements as w_z (transverse), w_x (axial), and w_θ (tangential). In the membrane theory, these three displacements complemented by three in-surface stresses n_x (axial), n_θ (tangential), and $n_{x\theta}$ (shear) form a closed system; bending moments and transverse shear stresses are not modeled. The system can be written as

$$\rho h \frac{\partial^2 w_z}{\partial t^2} = -\frac{1}{a} n_\theta \tag{7.32}$$

$$\rho h \frac{\partial^2 w_x}{\partial t^2} = \frac{\partial n_x}{\partial x} + \frac{\partial n_{x\theta}}{\partial \theta} \tag{7.33a}$$

$$\rho h \frac{\partial^2 w_\theta}{\partial t^2} = \frac{\partial n_\theta}{\partial \theta} + \frac{\partial n_{x\theta}}{\partial x} \tag{7.33b}$$

$$n_x = \frac{Eh}{1-\nu^2} \left(\frac{\partial w_x}{\partial x} + \nu \frac{\partial w_\theta}{\partial \theta} + \frac{\nu}{a} w_z \right) \tag{7.33c}$$

$$n_\theta = \frac{Eh}{1-\nu^2} \left(\nu \frac{\partial w_x}{\partial x} + \frac{\partial w_\theta}{\partial \theta} + \frac{1}{a} w_z \right) \tag{7.33d}$$

$$n_{x\theta} = \frac{Eh}{2(1+\nu)} \left(\frac{\partial w_\theta}{\partial x} + \frac{\partial w_x}{\partial \theta} \right) \tag{7.33e}$$

The material constants E, ν, ρ, and the shell thickness h are as discussed in Section 7.2, and are now assumed to be smooth functions of x and θ. If we define the velocities v_z, v_θ, and v_x by

$$v_z = \frac{\partial w_z}{\partial t} \qquad v_\theta = \frac{\partial w_\theta}{\partial t} \qquad v_x = \frac{\partial w_x}{\partial t}$$

then we again have a symmetric hyperbolic system of the form of (3.1) in the dependent variable $\mathbf{w} = [v_z, v_x, v_\theta, n_x, n_\theta, n_{x\theta}]^T$ where the system matrices are

$$\mathbf{P} = \mathbf{P}_R = \begin{bmatrix} \mathbf{P}_R^+ & \cdot \\ \cdot & \mathbf{P}_R^- \end{bmatrix} \qquad \mathbf{A}_1 = \mathbf{A}_{R1} = \begin{bmatrix} \cdot & \cdot \\ \cdot & \mathbf{A}_{M1}^- \end{bmatrix} \qquad \mathbf{A}_2 = \mathbf{A}_{R2} = \begin{bmatrix} \cdot & \cdot \\ \cdot & \mathbf{A}_{M2}^- \end{bmatrix}$$

$$\mathbf{B} = \mathbf{B}_R = \begin{bmatrix} \cdot & \mathbf{b}_\times \\ -\mathbf{b}_\times^T & \cdot \end{bmatrix}$$

7.3. CYLINDRICAL SHELLS

where

$$\mathbf{P}_R^+ = \rho h \qquad \mathbf{P}_R^- = \begin{bmatrix} \rho h & 0 & 0 & 0 & 0 \\ 0 & \rho h & 0 & 0 & 0 \\ 0 & 0 & \frac{1}{Eh} & \frac{-v}{Eh} & 0 \\ 0 & 0 & \frac{-v}{Eh} & \frac{1}{Eh} & 0 \\ 0 & 0 & 0 & 0 & \frac{2(1+v)}{Eh} \end{bmatrix} \qquad \mathbf{b}_x = \begin{bmatrix} 0 & 0 & 0 & \frac{1}{a} & 0 \end{bmatrix}$$

and $\mathbf{A}_{M1}^-$ and $\mathbf{A}_{M2}^-$ are as defined in (7.20) and (7.21). The lower five-variable system described by $\mathbf{P}_R^-$, $\mathbf{A}_{R1}^-$, and $\mathbf{A}_{M2}^-$, when uncoupled from the one-variable system in P_R^+ is essentially equivalent to the lower subsystem in the Mindlin plate theory, except that our independent variables are now x and θ instead of x and y. (In fact, if we replace any occurrence of $h^3/12$ in $\mathbf{P}_M^-$ by h, we get exactly $\mathbf{P}_R^-$.) We thus expect the MDKC to be very similar to that of the right-hand subnetwork at top in Figure 7.10.

We again introduce current-like variables[4]

$$(i_1, i_9, i_{10}, i_{11}, i_{12}, i_{13}) = (r_1 v_z, v_x, v_\theta, r_4 n_x, r_4 n_\theta, r_5 n_{x\theta})$$

and make use of coordinates defined by (3.23) in terms of the physical coordinates $[x, \theta, t]^T$. r_1, r_4, and r_5 are, as before, positive constants that we will later use for optimization. The MDKC for the membrane shell system is shown in Figure 7.17, with circuit parameters as indicated. (We have marked the points B and B' in the figure in anticipation of a connection with the shell model in the next section.)

Optimal settings for r_4 and r_5, which follow from positivity constraints on the inductances $L_9, \ldots, L_{13}$, can be shown (through an analysis identical to that performed on the Mindlin plate system) to be

$$r_4 = \left(\frac{Eh}{1-v}\right)_{\max} \sqrt{\frac{2(\rho h)_{\min}}{\left(\frac{Eh}{1-v}\right)_{\max} + \left(\frac{Eh}{1+v}\right)_{\max}}} \qquad (7.34)$$

$$r_5 = \left(\frac{Eh}{1+v}\right)_{\max} \sqrt{\frac{2(\rho h)_{\min}}{\left(\frac{Eh}{1-v}\right)_{\max} + \left(\frac{Eh}{1+v}\right)_{\max}}} \qquad (7.35)$$

in which case we must have, for passivity

$$v_0 \geq v_R \triangleq \sqrt{\frac{\left(\frac{Eh}{1-v}\right)_{\max} + \left(\frac{Eh}{1+v}\right)_{\max}}{(\rho h)_{\min}}} \qquad (7.36)$$

The parameter r_1 is as yet unconstrained (notice that the inductance K_1 is nonnegative for any choice of $v_0 \geq 0$).

We have presented the MDKC for the membrane shell because it is an important building block in a more modern theory (to be presented shortly). It should be obvious from this MDKC that we can immediately arrive at an MDWD network, and after applying network transformations, get a multidimensional DWN as well.

---

[4]The reason for the unusual numbering of these variables will be made clear in the next section.

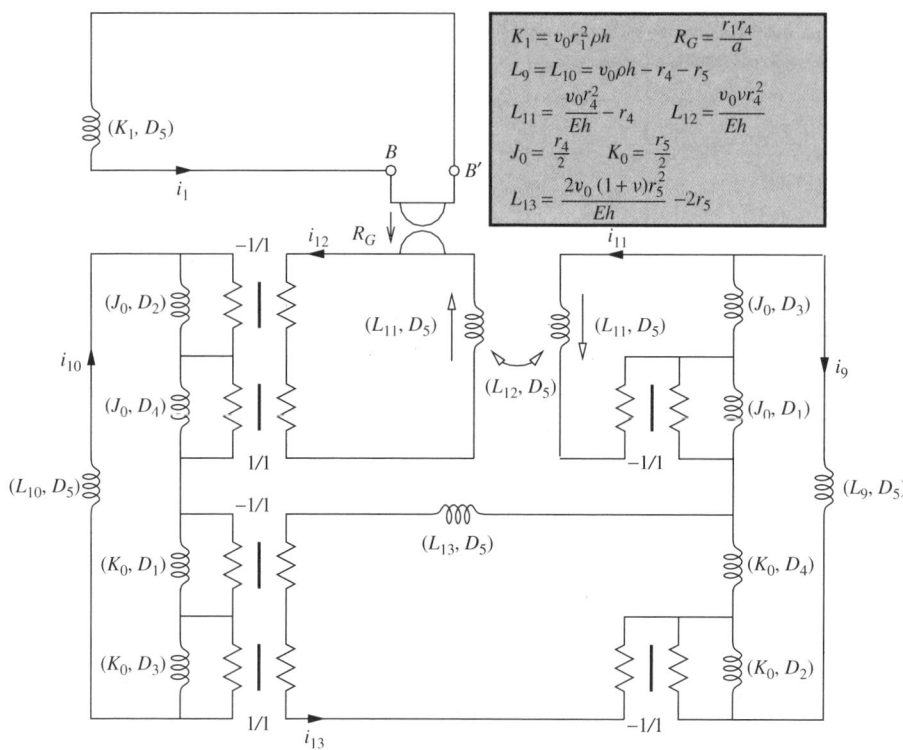

Figure 7.17: *MDKC for the cylindrical membrane shell system.*

7.3.2 The Naghdi–Cooper System II Formulation

As we mentioned, the membrane model of the cylindrical shell neglects certain important effects, and in particular, the crucial transverse shear effects. Many so-called higher-order shell theories have appeared in the literature; for a good survey of these theories, we refer to the literature [108, 111]. We have decided to focus on the Naghdi–Cooper system II shell model [46, 108, 172] because it can be simply interpreted as a passive circuit. We note in passing that not all shell theories have this property—Mirsky–Herrmann theory [108], for example, does not, and Naghdi and Cooper's system II, for example, is simplified from their proposed system I, which also does not. The problem, more specifically, is that these systems can apparently not be written in the special symmetric hyperbolic form of (3.1), for which the matrices $\mathbf{A}_1$ and $\mathbf{A}_2$ will be independent of x and θ. This property is essential here in that such dependence considerably complicates the interloop coupling in an MDKC (notice that the port-resistances of the Jaumann two-ports that realize this coupling have been constant for every system we have looked at so far).

Another reason for choosing this particular shell model is that it can be simply written as Mindlin's plate system (in coordinates x and θ instead of x and y) coupled with the membrane shell. We can write this system in the form of (3.1), where the dependent variable $\mathbf{w}$ is defined by

$$\mathbf{w} = [v_z, q_x, q_\theta, \omega_x, \omega_\theta, m_x, m_y, m_{xy}, v_x, v_\theta, n_x, n_\theta, n_{x\theta}]^T \tag{7.37}$$

7.3. CYLINDRICAL SHELLS

Here, the first eight variables are precisely those that appear in the Mindlin theory (see Section 7.2), but in cylindrical coordinates; we have changed subscripts y to θ and written v_z instead of v for the radial transverse velocity. The other five variables appear in subsystem (7.33) of the membrane shell theory (see Section 7.3.1). The system matrices are

$$\mathbf{P} = \mathbf{P}_{NC} = \begin{bmatrix} \mathbf{P}_M & \cdot \\ \cdot & \mathbf{P}_R \end{bmatrix} \quad \mathbf{A}_1 = \mathbf{A}_{NC1} = \begin{bmatrix} \mathbf{A}_{M1} & \cdot \\ \cdot & \mathbf{\bar{A}}_{M1} \end{bmatrix} \quad \mathbf{A}_2 = \mathbf{A}_{NC2} = \begin{bmatrix} \mathbf{A}_{M2} & \cdot \\ \cdot & \mathbf{\bar{A}}_{M2} \end{bmatrix}$$

$$\mathbf{B} = \mathbf{B}_{NC} = \begin{bmatrix} \mathbf{B}_M & \mathbf{B}_{NC\times} \\ -\mathbf{B}_{NC\times}^T & \cdot \end{bmatrix}$$

where $\mathbf{P}_M$, $\mathbf{A}_{M1}$, $\mathbf{\bar{A}}_{M1}$, $\mathbf{A}_{M2}$, $\mathbf{\bar{A}}_{M2}$, and $\mathbf{B}_M$ are as defined in Section 7.2, and $\mathbf{P}_R$ appears in Section 7.3.1. The coupling matrix is

$$\mathbf{B}_{NC\times} = \begin{bmatrix} 0 & 0 & 0 & 0 & 0 & 0 & 0 \\ 0 & 0 & 0 & 0 & 0 & 0 & 0 \\ 0 & 0 & 0 & 0 & 0 & 0 & 0 \\ \frac{1}{a} & 0 & 0 & 0 & 0 & 0 & 0 \\ 0 & 0 & 0 & 0 & 0 & 0 & 0 \end{bmatrix}^T$$

Note that this coupling disappears in the limit as the shell radius a becomes large (effectively leaving us with the Mindlin system). $\mathbf{B}_{NC}$ is indeed antisymmetric so we are guaranteed a lossless MDKC network; in fact, this network can be directly constructed from the two networks shown in Figures 7.10 and 7.17, by attaching terminals A and A' in the former to B and B' in the latter. The scaling parameters r_1, r_2, and r_3 can be chosen optimally according to (7.23), (7.26), and (7.27), and r_4 and r_5 can be set as in (7.34) and (7.35), giving a bound for passivity on the combined network,

$$v_0 \geq v_{NC} \triangleq \max(v_M, v_R)$$

where v_M is the bounding space step/time step ratio for the Mindlin network, from (7.29) (in cylindrical coordinates), and v_R is the same quantity for the membrane shell system, from (7.36).

Because this system can be constructed entirely by connecting subnetworks that we have already examined in detail, it seems unnecessary to show the discrete MDWD network or the alternate MDKC and its discrete form suitable for DWN implementation. The MDWD network will be exactly the combination of the 'Mindlin' system, shown at the bottom in Figure 7.10, and the MDWD network corresponding to the MDKC for the membrane shell, shown in Figure 7.17; recall that the MDKC for the membrane shell system (with a free port with terminals B and B') is identical in form to that of the uncoupled five-variable Mindlin subsystem and thus its MDWD counterpart will be of the same form as well.

For the transformed network to be used to generate a DWN, a few comments are in order. For the Mindlin system, we first applied network theoretic rules in order to arrive at a modified form, shown at the top in Figure 7.11. In this case, the transverse velocity v (renamed v_z in this section) and the bending moments m_x, m_y, and m_{xy} (renamed m_x, m_θ and $m_{x\theta}$) have been interpreted as voltages instead of currents. This transformed network can then be connected (via terminals A and A' in Figure 7.11) to a transformed form of the membrane shell system, shown in Figure 7.17; for the shell subsystem, n_x, n_θ, and $n_{x\theta}$ will be treated as voltages and v_x and v_θ as currents.

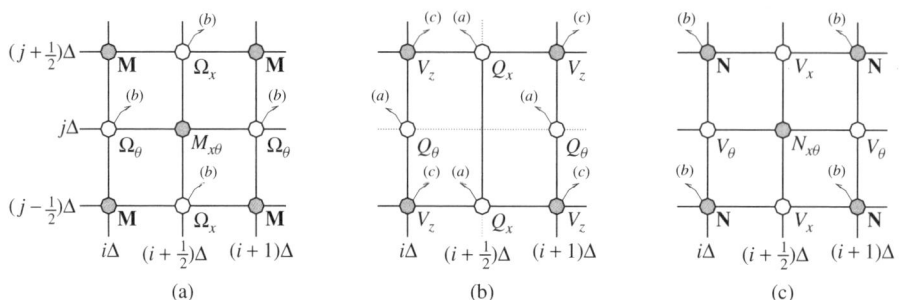

Figure 7.18: *Computational grid for the DWN for Naghdi and Cooper's system II. Grids (a) and (b) correspond to a DWN for a Mindlin-type subsystem, and are coupled to a membrane shell-type DWN operating on grid (c). Grid functions (capitalized versions of the dependent variables (7.37)) are indicated next to the grid points at which they are calculated. Gray/white coloring of grid points indicates calculation at parallel/series junctions at alternating time steps.*

After connecting these transformed subnetworks and applying the usual alternative discretization rules, we end up with a DWN that will operate on an interleaved grid as shown in Figure 7.18. Grids (a) and (b) are precisely the Mindlin grid shown in Figure 7.12, and are coupled instantaneously to a third grid (c) that adds the effect of curvature to the system.

7.4 Elastic Solids

The system defining the behavior of a (3 + 1)D linear, isotropic, elastic solid is somewhat easier to handle numerically than the (2 + 1)D plate and (1 + 1)D beam systems that are derived from it; the physics is less obscured by modeling assumptions. Numerical simulation of the full (3 + 1)D system is, of course, much more computationally expensive.

Such a medium is characterized by its density, ρ, and two material parameters λ and μ, called the *Lamé* coefficients, which describe its resilience; there are two parameters because a solid will resist compressional and shear forces to different degrees. Other elastic parameters, which we have already made use of earlier in this chapter, can be defined in terms of these two constants. Young's modulus E and Poisson's ratio ν can be written as

$$E = \frac{\mu(3\lambda + 2\mu)}{\lambda + \mu} \qquad \nu = \frac{\lambda}{2(\lambda + \mu)}$$

We remark that μ is the same as G that was used in the treatment of the Timoshenko beam (see Section 7.1), the Mindlin plate (see Section 7.2), and the Naghdi–Cooper shell model of Section 7.3.2. For the sake of generality, we allow all these parameters to be functions of x, y, and z.

The equations of motion of the solid can be written in terms of stress and displacement fields [108]. There are nine stresses: σ_{xx}, σ_{yy}, and σ_{zz} are *normal stresses* in the direction

7.4. ELASTIC SOLIDS

indicated by the double subscript, and σ_{xy}, σ_{xz}, σ_{yz}, σ_{yx}, σ_{zx}, and σ_{zy} are *shear stresses*. The displacements of a point in the medium from its equilibrium position are given by $\mathbf{d} = [w_x, w_y, w_z]^T$. If the material is assumed to be in rotational equilibrium, then we have

$$\sigma_{xy} = \sigma_{yx} \qquad \sigma_{xz} = \sigma_{zx} \qquad \sigma_{yz} = \sigma_{zy}$$

so that there are a total of six independent stresses acting at a given point in the solid.

Newton's Laws for a solid (neglecting body forces) are written as

$$\rho \frac{\partial^2 w_x}{\partial t^2} = \frac{\partial \sigma_{xx}}{\partial x} + \frac{\partial \sigma_{xy}}{\partial y} + \frac{\partial \sigma_{xz}}{\partial z} \tag{7.38a}$$

$$\rho \frac{\partial^2 w_y}{\partial t^2} = \frac{\partial \sigma_{xy}}{\partial x} + \frac{\partial \sigma_{yy}}{\partial y} + \frac{\partial \sigma_{yz}}{\partial z} \tag{7.38b}$$

$$\rho \frac{\partial^2 w_z}{\partial t^2} = \frac{\partial \sigma_{xz}}{\partial x} + \frac{\partial \sigma_{yz}}{\partial y} + \frac{\partial \sigma_{zz}}{\partial z} \tag{7.38c}$$

The stress-strain relation, or *Hooke's Law* [108] is expressed as a linear proportionality between the six stresses and spatial derivatives of the displacements (the strain):

$$\sigma_{xx} = 2\mu \frac{\partial w_x}{\partial x} + \lambda \nabla \cdot \mathbf{d} \tag{7.39a}$$

$$\sigma_{yy} = 2\mu \frac{\partial w_y}{\partial y} + \lambda \nabla \cdot \mathbf{d} \tag{7.39b}$$

$$\sigma_{zz} = 2\mu \frac{\partial w_z}{\partial z} + \lambda \nabla \cdot \mathbf{d} \tag{7.39c}$$

$$\sigma_{xy} = \mu \left(\frac{\partial w_x}{\partial y} + \frac{\partial w_y}{\partial x} \right) \tag{7.39d}$$

$$\sigma_{xz} = \mu \left(\frac{\partial w_x}{\partial z} + \frac{\partial w_z}{\partial x} \right) \tag{7.39e}$$

$$\sigma_{yz} = \mu \left(\frac{\partial w_y}{\partial z} + \frac{\partial w_z}{\partial y} \right) \tag{7.39f}$$

The systems (7.38) and (7.39) taken together are sometimes called the *Navier* system [108, 176]. By introducing the velocities defined by

$$v_x \triangleq \frac{\partial w_x}{\partial t} \qquad v_y \triangleq \frac{\partial w_y}{\partial t} \qquad v_z \triangleq \frac{\partial w_z}{\partial t}$$

it is possible to manipulate these equations into the symmetric hyperbolic form of (3.1), with $\mathbf{w} = [v_x, v_y, v_z, \sigma_{xx}, \sigma_{yy}, \sigma_{zz}, \sigma_{xy}, \sigma_{xz}, \sigma_{yz}]^T$, and

$$\mathbf{P} = \begin{bmatrix} \mathbf{P}_N^+ & \cdot \\ \cdot & \mathbf{P}_N^- \end{bmatrix} \quad \mathbf{A}_1 = \begin{bmatrix} \cdot & \mathbf{A}_{N1\times} \\ \mathbf{A}_{N1\times}^T & \cdot \end{bmatrix} \quad \mathbf{A}_2 = \begin{bmatrix} \cdot & \mathbf{A}_{N2\times} \\ \mathbf{A}_{N2\times}^T & \cdot \end{bmatrix} \quad \mathbf{A}_3 = \begin{bmatrix} \cdot & \mathbf{A}_{N3\times} \\ \mathbf{A}_{N3\times}^T & \cdot \end{bmatrix}$$

with

$$\mathbf{P}_N^+ = \mathrm{diag}(\rho, \rho, \rho) \qquad \mathbf{P}_N^- = \begin{bmatrix} \frac{1}{E} & \frac{-\nu}{E} & \frac{-\nu}{E} & 0 & 0 & 0 \\ \frac{-\nu}{E} & \frac{1}{E} & \frac{-\nu}{E} & 0 & 0 & 0 \\ \frac{-\nu}{E} & \frac{-\nu}{E} & \frac{1}{E} & 0 & 0 & 0 \\ 0 & 0 & 0 & \frac{1}{\mu} & 0 & 0 \\ 0 & 0 & 0 & 0 & \frac{1}{\mu} & 0 \\ 0 & 0 & 0 & 0 & 0 & \frac{1}{\mu} \end{bmatrix}$$

and

$$\mathbf{A}_{N1\times} = \begin{bmatrix} -1 & 0 & 0 \\ 0 & 0 & 0 \\ 0 & 0 & 0 \\ 0 & -1 & 0 \\ 0 & 0 & -1 \\ 0 & 0 & 0 \end{bmatrix}^T \quad \mathbf{A}_{N2\times} = \begin{bmatrix} 0 & 0 & 0 \\ 0 & -1 & 0 \\ 0 & 0 & 0 \\ -1 & 0 & 0 \\ 0 & 0 & 0 \\ 0 & 0 & -1 \end{bmatrix}^T \quad \mathbf{A}_{N2\times} = \begin{bmatrix} 0 & 0 & 0 \\ 0 & 0 & 0 \\ 0 & 0 & -1 \\ 0 & 0 & 0 \\ -1 & 0 & 0 \\ 0 & -1 & 0 \end{bmatrix}^T$$

Phase and Group Velocities

The characteristic polynomial relation for the Navier system, in terms of frequencies ω and wavenumber magnitude $\|\boldsymbol{\beta}\|_2 = \sqrt{\beta_x^2 + \beta_y^2 + \beta_z^2}$, will be

$$\omega^3 \left(\omega^2 - \frac{\lambda + 2\mu}{\rho} \|\boldsymbol{\beta}\|_2^2 \right) \left(\omega^2 - \frac{\mu}{\rho} \|\boldsymbol{\beta}\|_2^2 \right)^2 = 0$$

and has roots

$$\omega = 0, \quad \pm\sqrt{\frac{\lambda + 2\mu}{\rho}} \|\boldsymbol{\beta}\|_2, \quad \pm\sqrt{\frac{\mu}{\rho}} \|\boldsymbol{\beta}\|_2$$

Ignoring the nonpropagating modes with frequency $\omega = 0$ and the multiplicities of the other modes, we can see that wave propagation is dispersionless, at least for the constant-coefficient problem. There are two wave speeds,

$$\gamma_{N,P}^p = \gamma_{N,P}^g = \sqrt{\frac{\lambda + 2\mu}{\rho}} \quad \geq \quad \gamma_{N,S}^p = \gamma_{N,S}^g = \sqrt{\frac{\mu}{\rho}}$$

which are also known as the P-wave and S-wave (or compressional and shear wave) speeds [53].

For the varying-coefficient problem, the global maximum group velocity is then

$$\gamma_{N,\max}^g = \sqrt{\left(\frac{\lambda + 2\mu}{\rho}\right)_{\max}}$$

7.4. ELASTIC SOLIDS

7.4.1 Scattering Networks for the Navier System

This system was first represented using an MDKC by Nitsche [176]. Choosing current-like variables

$$(i_1, i_2, i_3, i_4, i_5, i_6, i_7, i_8, i_9) = (v_x, v_y, v_z, r_1\sigma_{xx}, r_1\sigma_{yy}, r_1\sigma_{zz}, r_2\sigma_{xy}, r_2\sigma_{xz}, r_2\sigma_{yz})$$

he derived an MDKC for the Navier system, which we have reproduced (with some minor changes) at top in Figure 7.19, where coordinates defined by the transformation matrix (3.25) have been used. Notice that again, because **P** is not diagonal, the time derivatives of the components σ_{xx}, σ_{yy}, and σ_{zz} are coupled. In the circuit representation, this is represented by a three-port coupled inductance of the form of (2.48), with a matrix inductance $\mathbf{L}_{456}$ defined by

$$\mathbf{L}_{456} = \begin{bmatrix} L_4 & -L_\times & -L_\times \\ -L_\times & L_5 & -L_\times \\ -L_\times & -L_\times & L_6 \end{bmatrix}$$

where the inductances L_4, L_5, L_6, and $L_\times$ appear in the figure.

The positivity condition on the inductances L_1, L_2, L_3, L_7, L_8, and L_9, as well as a positive definiteness condition on the matrix $\mathbf{L}_{456}$ gives, as optimal choices of the parameters r_1 and r_2,

$$r_1 = \left(\frac{E}{1-2\nu}\right)_{\max} \sqrt{\frac{\rho_{\min}}{\left(\frac{E}{1-2\nu}\right)_{\max} + 4\mu_{\max}}}$$

$$r_2 = 2\mu_{\max} \sqrt{\frac{\rho_{\min}}{\left(\frac{E}{1-2\nu}\right)_{\max} + 4\mu_{\max}}}$$

and a bound on v_0

$$v_0 \geq v_N \triangleq \sqrt{\frac{\left(\frac{E}{1-2\nu}\right)_{\max} + 4\mu_{\max}}{\rho_{\min}}}$$

When the material parameters are constant, v_N reduces to

$$v_N = \sqrt{\frac{3(\lambda + 2\mu)}{\rho}} = \sqrt{3}\gamma_{N,\max}^g$$

A modified MDKC is shown at the top in Figure 7.20, where velocities are treated as currents and stresses as voltages; as such, the coupled inductance in Figure 7.19 has become a coupled capacitance. This network may be discretized in a way very similar to the network for Maxwell's equations, as described in Section 6.4. Under the spectral mappings defined by (6.9) (with step sizes of $T_j = \Delta/2$, $j = 1, \ldots, 6$), the connecting LSI two-ports decompose into series/parallel connections of multidimensional unit elements. The one-port inductances and capacitances as well as the coupled capacitance are discretized using the trapezoid rule, with a step size of $T_7 = \Delta$. The resulting multidimensional DWN is shown at bottom in Figure 7.20 and the interleaved computational grid in Figure 7.21 (which is very similar to the grid for the DWN for Maxwell's equations, as shown in Figure 6.12).

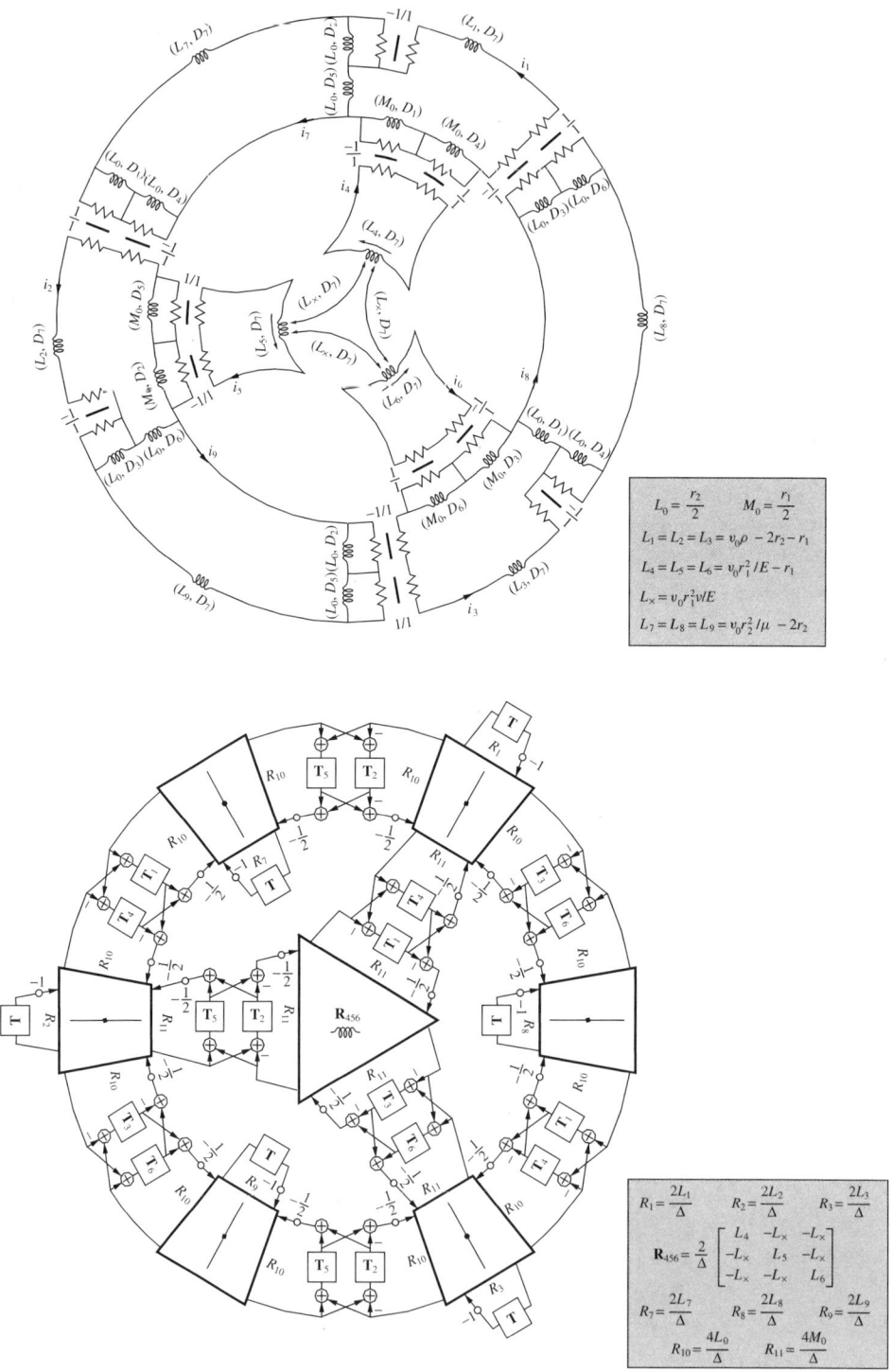

Figure 7.19: *MDKC and MDWD network for the Navier system.*

7.4. ELASTIC SOLIDS

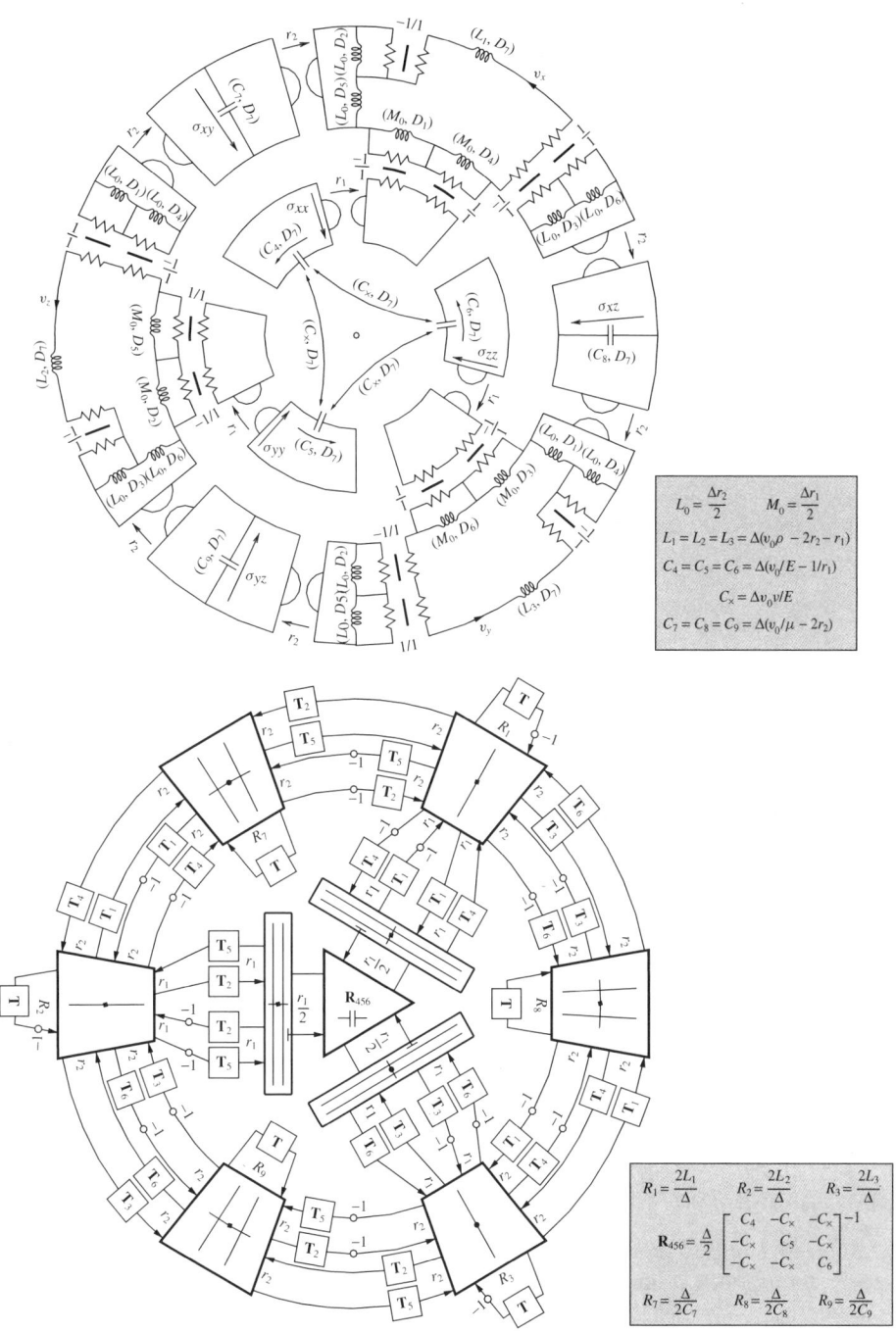

Figure 7.20: *Modified MDKC and multidimensional DWN for the Navier system.*

258   CHAPTER 7. APPLICATIONS TO VIBRATING SYSTEMS

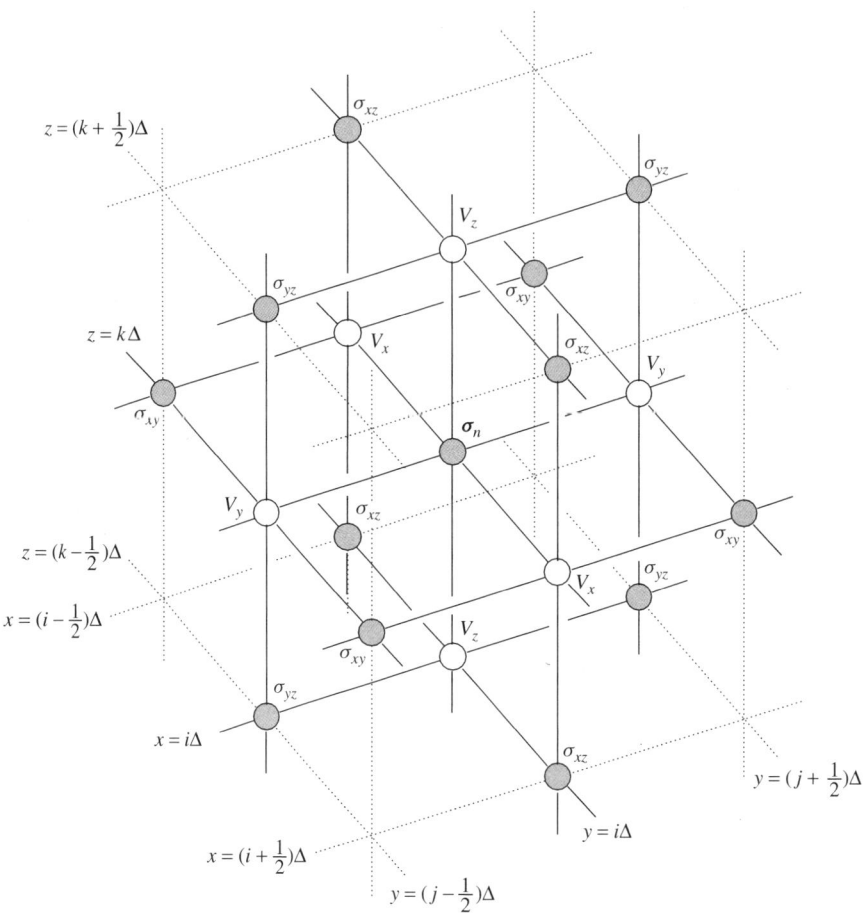

Figure 7.21: *Computational grid for the DWN for Navier's system; stresses and velocities are calculated at alternating multiples of $T/2$ and at alternating grid locations. In the DWN implementation, waveguide connections (of delay length $T/2$) between series junctions (white) and parallel junctions (gray) are shown as dark lines; waveguide sign inversions and self-loops are not shown here. At the center grid point, a vector parallel junction calculates the vector $\boldsymbol{\sigma}_n = (\sigma_{xx}, \sigma_{yy}, \sigma_{zz})$ of normal stresses.*

7.4.2 Boundary Conditions

The simplest boundary conditions for the Navier system are of the free type, that is, all stresses normal to the boundary are zero [108]. For a 'bottom' boundary $z = 0$, these conditions can be written as

$$\sigma_{xz} = \sigma_{yz} = \sigma_{zz} = 0 \qquad (7.40)$$

This condition is lossless and of the form of (3.8).

Considering the DWN shown at bottom in Figure 7.20, and the associated computational grid shown in Figure 7.21, it is easy to see that in this case it is best to arrange the grid

7.4. ELASTIC SOLIDS

such that parallel junctions (at which approximations to σ_{xz} and σ_{yz} are calculated) lie on this bottom boundary. The first two of conditions (7.40) can be ensured by short-circuiting the parallel junctions. As a result, the remaining series junctions on the boundary (at which approximations to v_z are calculated) are decoupled from the parallel junctions, and it remains only to set a self-loop impedance at these junctions so as to approximate the condition $\sigma_{zz} = 0$. We leave the determination of these self-loop impedances as an exercise to the reader.

Chapter 8

Time-varying and Nonlinear Systems

Among the more interesting developments in the field of scattering methods have been applications to time-varying and nonlinear problems. For lumped systems, in which case the goal is the numerical integration of a set of ODEs, there has been a good deal of work in areas such as cochlear modeling [107], vocal tract modeling [239], and the modeling of nonlinear circuit components [59] such as transistors [55], switching elements [204, 92] and applications to nonlinear transmission lines [60, 61, 168]. The concept of a generalized adaptor (see Section 2.2.5) with memory, as another means of approaching nonlinear circuit elements, has been explored in [208]. Another direction has been the modeling of lumped elements in musical instruments such as woodwind toneholes [273], the single reed [28], and the piano hammer [16, 15, 266], usually in conjunction with digital waveguides (see Section 4.6). More related work, in automated strategies for developing such algorithms for synthesis purposes, has appeared [209, 184, 52]. Even more recently, Fettweis has begun work in the modeling of the equations of special relativity [75, 77], as applied to particle motion.

One of the offshoots (one might say the goal) of the work of Fettweis and others on MDWD networks for distributed systems has been a set of passive numerical methods for the simulation of the nonlinear equations of fluid dynamics [21, 71, 70, 91, 74, 100, 169, 259]. These systems, also described by hyperbolic or nearly hyperbolic systems of PDEs, have circuit representations, and as such it is possible to develop numerical methods in the same way as outlined in Chapter 3; the procedure is, however, complicated by the necessarily nonlinear nature of the requisite circuit elements. (It should be of great interest to members of the TLM community that the techniques introduced in Chapter 6 may be applied to arrive at DWN-like structures for these systems.) As before, the circuit-based approach gives rise to structures made up entirely, in the lossless case, of scattering junctions connected to either digital transmission lines or wave digital elements, but new considerations arise; most importantly, the use of power-normalized waves becomes essential (thus rendering scattering junctions orthogonal). For nonlinear systems, iterative methods also play a role, as port resistances become dependent on wave variables themselves.

Wave and Scattering Methods for Numerical Simulation   Stefan D. Bilbao
© 2004 John Wiley & Sons, Ltd   ISBNs: 0-470-87017-6

Summary

In this chapter, we first show how the definitions of the linear and time-invariant circuit elements introduced in Chapter 2 have been extended to the time-varying and nonlinear cases, both in the lumped and distributed settings. We then examine linear time-varying hyperbolic systems in Section 8.2 and develop an MDKC representation for a time-varying $(1 + 1)$D transmission line model. We then turn to nonlinear systems, first in the lumped case, where we discuss systems that arise in the modeling of musical instruments in Section 8.3, as well as in applications to special relativity in Section 8.4. We then summarize the work of Fettweis and his students in the distributed problems that arise in fluid dynamics, first dealing with the useful test problem Burger's equation in Section 8.5, and then continuing with a treatment of the $(1 + 1)$D gas dynamics equations in Section 8.6. In keeping with the overall goal of unifying wave digital and digital waveguide network approaches to numerical integration, we also show how the methods of Chapter 6 can be applied to the one-dimensional gas dynamics system in order to yield a DWN-like structure. We take the opportunity to point out some connections to recent trends in the analysis of such systems involving so-called *skew-selfadjoint* forms and *entropy variables*.

8.1 Time-varying and Nonlinear Circuit Elements

The definitions of the lumped linear circuit elements mentioned in Section 2.1.4 can be extended, in the wave digital filtering context, to the time-varying [240] and nonlinear [55, 59, 92, 204, 168] cases. Particular care must be taken, as we will see, to ensure that passivity is maintained, especially when the application is to simulation of nonlinear systems.

8.1.1 Lumped Elements

The extension of the lumped resistor, defined by (2.7), is straightforward and has the same form as the linear resistor,

$$v = R(t, i)i$$

except that the resistance is now a function of both time t (in the time-varying case), and perhaps of the current i in the full nonlinear case[1]. Just as in the linear case, the instantaneous power dissipated by the resistor will be $w_{\text{inst}} = Ri^2$, and as long as R remains nonnegative for all choices of these variables, the element is passive.

Reactive elements require a slightly more involved treatment. An appropriate definition for an inductor [76] is given by

$$v = \sqrt{L(t, i)} \frac{d}{dt} \left(\sqrt{L(t, i)} i \right) \tag{8.1}$$

for a positive function L of the variables t and i. Losslessness follows immediately through multiplication of both sides of this equation by i,

$$w_{\text{inst}} = vi = \frac{d}{dt} \left(\frac{1}{2} L(t, i) i^2 \right)$$

---

[1] All the nonlinearities we will examine will be of the autonomous type, that is, without explicit dependence on the time variable.

8.1. TIME-VARYING AND NONLINEAR CIRCUIT ELEMENTS

Again, the instantaneous power absorbed is the time rate of change of positive function $\frac{1}{2}L(t,i)i^2$, again to be interpreted as a stored energy. A similar definition for a nonlinear capacitor also holds, namely,

$$i = \sqrt{C(t,i)}\frac{d}{dt}(\sqrt{C(t,i)}v) \qquad (8.2)$$

for some positive function $C(t,i)$. Such elements have appeared in the context of piano hammer modeling [16, 28, 15], and in the modeling of the relativistic mechanics of particles [75, 77]. The representations of these elements in circuit diagrams are unchanged from the linear case, but it must be borne in mind that a definition such as

$$v = L\frac{di}{dt} \qquad \text{or} \qquad v = \frac{d}{dt}(Li)$$

(in the case of the inductor) are distinct from (8.1) in the nonlinear and time-varying cases, and do not, in general, correspond to lossless or even passive elements. We note that this important consideration is often not taken into account [107].

For all these elements, wave digital discretization may be carried out as before, through the application of the trapezoid rule and the introduction of wave variables, now necessarily of the power-normalized type. The interpretation of the trapezoid rule as a bilinear transformation is obviously no longer valid in the nonlinear case. For all practical purposes, the nonlinear character of any one of these elements is essentially transferred to the port resistances of the adaptor to which it is connected; the wave digital elements themselves are identical to their linear counterparts. As long as the circuit element values remain positive, then so do the port resistances, and an adaptor scattering matrix will be forced to be orthogonal (see Section 2.2.5) and can be interpreted in terms of rotations and reflections. The problem, then, is in determining the scattering matrix, which, though orthogonal, may be dependent on the input waves; a system of nonlinear algebraic equations results. This problem is usually solved in practice using iterative methods, but existence and uniqueness of such a solution are matters that have not been broached in any detail in the literature (and should be). These systems of equations are usually localized to a small set of connected adaptors, a basic result of using wave variables.

8.1.2 Distributed Elements

It should be clear that in order to build circuit models for time-varying or nonlinear systems of PDEs, where there is a dependence on a spatial variable **x**, we will need appropriate distributed circuit elements. Nonlinear resistances, as in the lumped case, are simple to model; a voltage-current relation of the form

$$v = Ri$$

will correspond to a passive resistor as long as R is positive, regardless of its functional dependencies. The definitions of transformers and gyrators also remain unchanged in the nonlinear case; the turns ratio or gyrator coefficient may have any dependence without affecting losslessness.

A generalized definition of the MD inductor (see e.g., [71]) is similar to the lumped form:

$$v = \sqrt{L}\frac{\partial}{\partial t_j}\left(\sqrt{L}i\right) = \frac{1}{2}\left(L\frac{\partial i}{\partial t_j} + \frac{\partial(Li)}{\partial t_j}\right) \quad (8.3)$$

Here again, t_j is some coordinate defined by a transformation such as (3.16) or (3.22). The instantaneous absorbed power density will be

$$w_{\text{inst}} = vi = \sqrt{L}i\frac{\partial\sqrt{L}i}{\partial t_j} = \frac{1}{2}\frac{\partial Li^2}{\partial t_j}$$

and the element can be considered to be lossless as per the definition of (3.29) provided the stored energy flux $\mathbf{E}$ is defined to be

$$\mathbf{E} = \tfrac{1}{2}Li^2\mathbf{e}_j$$

where $\mathbf{e}_j$ is a unit vector in the direction t_j. This is the same as the definition in the linear case, from (3.36). Here, L is constrained to be positive, but may be a function (smooth) of any of the dependent or independent variables in the problem. This losslessness is reflected in the MDWD one-port; if the port resistance is chosen to be $R = 2L/T_j$ for some step-size T_j in direction t_j, then in terms of power-normalized waves $\underline{a}$ and $\underline{b}$ (see Section 2.2.2), the one-port is defined at a grid point with coordinates $\mathbf{t}$ by

$$\underline{b}(\mathbf{t}) = -\underline{a}(\mathbf{t} - \mathbf{T}_j) \quad (8.4)$$

for a vector shift $\mathbf{T}_j = T_j\mathbf{e}_j$, just as in the linear case. Again, as in the lumped case, it is essential to use power-normalized waves because passivity is not guaranteed otherwise [21]. (The reason for this should be clear from the discussion in Section 3.4.1; we cannot obtain a wave relation such as (8.4) in terms of voltage wave variables because the differential operator does not necessarily commute with the inductance.)

A generalized distributed capacitor can be similarly defined by

$$i = \sqrt{C}\frac{\partial}{\partial t_j}\left(\sqrt{C}i\right) = \frac{1}{2}\left(C\frac{\partial i}{\partial t_j} + \frac{\partial(Ci)}{\partial t_j}\right) \quad (8.5)$$

for some capacitance C, which again may have arbitrary smooth functional dependence; if C is always positive, then the capacitance is lossless.

8.2 Linear Time-varying Distributed Systems

Linear and time-varying distributed systems have not been examined in any detail in the scattering simulation literature, though time-varying WDFs [239] and DWNs [228] have both been proposed, with a focus on vocal tract modeling. Though it is true that time-variations in material parameters generally render a system nonpassive, we will show here how passive network representations may be developed for an important class of systems. Consider a system of the form

$$\frac{\partial}{\partial t}(\mathbf{P}(\mathbf{x},t)\mathbf{w}) + \sum_{j=1}^{n}\mathbf{A}_j\frac{\partial\mathbf{w}}{\partial x_j} + \mathbf{B}(\mathbf{x},t)\mathbf{w} + \mathbf{f}(\mathbf{x},t) = \mathbf{0} \quad (8.6)$$

8.2. LINEAR TIME-VARYING DISTRIBUTED SYSTEMS

which is a simple generalization of the $(n+1)$D symmetric hyperbolic form (3.1) to the case where $\mathbf{P}$ and $\mathbf{B}$ depend on both the spatial coordinates $\mathbf{x}$ and time t; $\mathbf{P}$ is assumed to be positive definite for all values of these coordinates and smoothly varying. The matrices $\mathbf{A}_j$ are again assumed to be constant and symmetric, and $\mathbf{B}$ is not required to have any particular structure. It is easy to show that in this form, it is not possible to arrive immediately at an energy condition such as (3.5). In order to put system (8.6) into more useful form, note that we can factor $\mathbf{P}$ as $\mathbf{P} = \mathbf{P}^{\frac{T}{2}}\mathbf{P}^{\frac{1}{2}}$ where $\mathbf{P}^{\frac{1}{2}}$ is some left matrix square root of $\mathbf{P}$. We can then rewrite (8.6) as

$$\mathbf{P}^{\frac{T}{2}}\frac{\partial \mathbf{P}^{\frac{1}{2}}\mathbf{w}}{\partial t} + \frac{\partial \mathbf{P}^{\frac{T}{2}}}{\partial t}\mathbf{P}^{\frac{1}{2}}\mathbf{w} + \sum_{j=1}^{n}\mathbf{A}_j\frac{\partial \mathbf{w}}{\partial x_j} + \mathbf{B}\mathbf{w} + \mathbf{f} = 0$$

Now introduce a new dependent variable $\mathbf{z}$ defined by $\mathbf{w} = e^{\kappa}\mathbf{z}$, where $\kappa = \kappa(t)$ and is assumed differentiable. Then, in terms of the new variable $\mathbf{z}$, we have

$$\mathbf{P}^{\frac{T}{2}}\frac{\partial \mathbf{P}^{\frac{1}{2}}\mathbf{z}}{\partial t} + \left(\frac{\partial \kappa}{\partial t}\mathbf{P} + \frac{\partial \mathbf{P}^{\frac{T}{2}}}{\partial t}\mathbf{P}^{\frac{1}{2}} + \mathbf{B}\right)\mathbf{z} + \sum_{j=1}^{n}\mathbf{A}_j\frac{\partial \mathbf{z}}{\partial x_j} + \tilde{\mathbf{f}} = 0$$

with $\tilde{\mathbf{f}} = e^{-\kappa}\mathbf{f}$. Assuming that this source term is zero, we can then take the inner product of this expression with $\mathbf{z}^T$ to get

$$\frac{\partial}{\partial t}\left(\frac{1}{2}\mathbf{z}^T\mathbf{P}\mathbf{z}\right) + \frac{1}{2}\sum_{j=1}^{n}\frac{\partial}{\partial x_j}\left(\mathbf{z}^T\mathbf{A}_j\mathbf{z}\right) = -\mathbf{z}^T\mathbf{Q}\mathbf{z}$$

where

$$\mathbf{Q} = \frac{\partial \kappa}{\partial t}\mathbf{P} + \frac{1}{2}\frac{\partial \mathbf{P}}{\partial t} + \frac{1}{2}\left(\mathbf{B} + \mathbf{B}^T\right)$$

If $\mathbf{Q}$ is positive semidefinite, then integrating over $\mathbb{R}^n$ gives the energy condition

$$\frac{d}{dt}\int_{\mathbb{R}^n}\frac{1}{2}\mathbf{z}^T\mathbf{P}\mathbf{z}\,d\mathbf{x} \leq 0$$

which is identical to the condition derived in Section 3.1, under the replacement of $\mathbf{w}$ with $\mathbf{z}$. As long as $\mathbf{B}$ and the time derivative of $\mathbf{P}$ are bounded, it is always possible to make a choice of κ such that $\mathbf{Q}$ is positive semidefinite. For instance, we can choose $\kappa = \kappa_0 t$, with

$$\kappa_0 \geq -\frac{1}{2}\min_{\mathbf{x}\in\mathbb{R}^n, t\geq 0}\frac{\lambda_{\min}\left(\frac{\partial \mathbf{P}}{\partial t} + \mathbf{B} + \mathbf{B}^T\right)}{\lambda_{\min}(\mathbf{P})}$$

where $\lambda_{\min}(\cdot)$ signifies 'minimum eigenvalue of.' Here, we essentially have a passivity condition in an exponentially weighted norm (if passivity is the right word to use here).

It is important to realize that a system with explicitly time-varying parameters, linear or not, is usually not passive for the simple reason that the time variations themselves imply some sort of external forcing. This important fact is not taken into consideration in many models of systems such as the vocal tract (to be discussed shortly).

8.2.1 A Time-varying Transmission Line Model

Consider a generalization of the source-free $(1+1)$D transmission line system,

$$\frac{\partial (li)}{\partial t} + \frac{\partial u}{\partial x} + ri = 0 \tag{8.7a}$$

$$\frac{\partial (cu)}{\partial t} + \frac{\partial i}{\partial x} + gu = 0 \tag{8.7b}$$

where l, c, r, and g are all smooth positive functions of x and t. Introducing the variables

$$i_1 = i e^{-\kappa_0 t} \qquad\qquad i_2 = u e^{-\kappa_0 t}/r_0$$

where r_0 is a positive constant as well as the scaled time variable $t' = v_0 t$, and transformed coordinates as per (3.19), we can rewrite this system as

$$\sqrt{L_1}\frac{\partial}{\partial t'}\left(\sqrt{L_1} i_1\right) + r_1 i_1 + \frac{r_0}{\sqrt{2}}\frac{\partial}{\partial t_1}(i_1+i_2) + \frac{r_0}{\sqrt{2}}\frac{\partial}{\partial t_2}(i_1-i_2) = 0$$

$$\sqrt{L_2}\frac{\partial}{\partial t'}\left(\sqrt{L_2} i_2\right) + r_2 i_2 + \frac{r_0}{\sqrt{2}}\frac{\partial}{\partial t_1}(i_1+i_2) + \frac{r_0}{\sqrt{2}}\frac{\partial}{\partial t_2}(i_2-i_1) = 0$$

with

$$L_1 = v_0 l - r_0 \qquad\qquad L_2 = v_0 c r_0^2 - r_0$$

and

$$r_1 = \kappa_0 l + \frac{1}{2}\frac{\partial l}{\partial t} + r \qquad r_2 = r_0^2\left(\kappa_0 c + \frac{1}{2}\frac{\partial c}{\partial t} + g\right)$$

Under the choices

$$r_0 = \sqrt{l_{\min}/c_{\min}} \qquad\qquad v_0 \geq 1/\sqrt{l_{\min} c_{\min}}$$

where now we have

$$l_{\min} = \min_{x\in\mathbb{R}, t\geq 0} l \qquad\qquad c_{\min} = \min_{x\in\mathbb{R}, t\geq 0} c$$

then L_1 and L_2 are nonnegative, and the terms involving them can be interpreted as voltages across passive inductors if power-normalized waves are employed (see Section 3.4.1 for more information on this definition of inductors). If we also choose

$$\kappa_0 \geq -\frac{1}{2}\min\left(\min_{x\in\mathbb{R},t\geq 0}\left(\frac{1}{l}\frac{\partial l}{\partial t}+r/l\right), \min_{x\in\mathbb{R},t\geq 0}\left(\frac{1}{c}\frac{\partial c}{\partial t}+g/c\right)\right) \tag{8.8}$$

then r_1 and r_2 are nonnegative and can be interpreted as passive resistances. The resulting MDKC is shown in Figure 8.1; an MDWD network can be immediately obtained through the methods discussed in Chapter 3, or network manipulations and alternative integration rules may be employed to get a DWN. A balanced form (see Section 3.11) is also possible, and gives a less strict bound on v_0, but the bound on κ_0 remains unchanged.

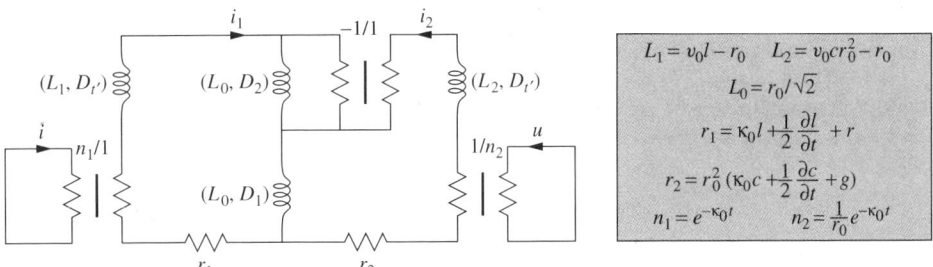

Figure 8.1: *MDKC for the time-varying $(1+1)$D transmission line system (8.7). The exponential weighting of the current variables can be viewed (formally) as a time-varying transformer coupling.*

A direct application of this MDKC to an important music or speech-synthesis problem would be the simulation of acoustic wave propagation in the vocal tract, under time-varying conditions. Such a system of PDEs is mentioned the literature in [196] and has the exact form of (8.7), with $r = g = 0$, and under the replacements

$$l \to \frac{\rho}{A} \qquad c \to \frac{A}{\rho\gamma^2} \qquad i \to v \qquad u \to p + \rho\gamma^2$$

where ρ is the air density, γ is the speed of sound, $A(x,t)$ is the cross-sectional surface area of the tube, $v(x,t)$ is the volume velocity, and $p(x,t)$ is the pressure variation. The condition (8.8) then reduces to

$$\kappa_0 \geq \max_{\mathbf{x}\in\mathbb{R}, t\geq 0} \left| \frac{\partial \ln(\sqrt{A})}{\partial t} \right|$$

If the time variation in A is slow, then κ_0 will be close to zero, and the exponential weighting will not be overly severe. The problems, for real-time synthesis applications, are that we will need to have an *a priori* estimate of the maximal time variation of the vocal tract area, and that we will apply an exponential weighting to the signal output from the scattering simulation. This exponential weighting may be viewed as a passive operation involving time-varying transformers (as shown in Figure 8.1).

8.3 Lumped Nonlinear Systems in Musical Acoustics

Musical acoustics abounds in useful test cases for scattering methods. In particular, the excitation mechanisms for nearly all Western musical instruments can be modeled as lumped elements, sometimes source driven, and necessarily nonlinear. In this short section, we examine two representative cases, the piano hammer, and the single reed.

8.3.1 Piano Hammers

A piano hammer, when in contact with a string, may be modeled, in the simplest case, as a nonlinear harmonic oscillator [39]. The string is assumed to lie along the x axis, and has transverse displacement $y_s(x, t)$ (typically described by a variant of the wave equation, with stiffness and loss terms included [94, 39, 17, 29]). The hammer, of mass m_h and at vertical position y_h, strikes the string from below, at position x_h along the string. When the hammer is in contact with the string (i.e., when $y_h \geq y_s(x_h, t)$), the hammer experiences a nonlinear stiffening spring force in the hammer felt, giving rise to the equation of motion

$$m_h \frac{d^2 y_h}{dt^2} = -K_h(y_h - y_s(x_h, t))^p$$

Here, $K_h > 0$ and $p > 0$ are constants that define the nonlinearity. Defining the new variable $y = y_h - y_s(x_h, t)$, we can rewrite this as

$$m_h \frac{d^2 y_h}{dt^2} = -K_h y^p \tag{8.9}$$

See Figure 8.2(a).

Before proceeding directly to the circuit representation for this equation, it is worth examining the stiffness nonlinearity as a one-port, in order to find a lossless capacitor representation of the form (8.2). Associating voltage with force, we have, then, a relation of the form

$$v = K_h y^p$$

Differentiating this equation, and setting $i = \frac{dy}{dt}$ gives

$$\frac{dv}{dt} = K_h p y^{p-1} i \quad \Rightarrow \quad i = \frac{1}{K_h p} \left(\frac{v}{K_h} \right)^{\frac{1-p}{p}} \frac{dv}{dt}$$

which can be written in the form of (8.2), with a capacitance defined by

$$C = \frac{2}{1+p} K_h^{-1/p} v^{\frac{1-p}{p}}$$

This reduces to $C = 1/K_h$, as expected, in the linear case, when $p = 1$. It is important to note that although the capacitance becomes unbounded in the limit as v tends to zero, the stored energy $\frac{1}{2}Cv^2$ remains bounded, and tends to zero.

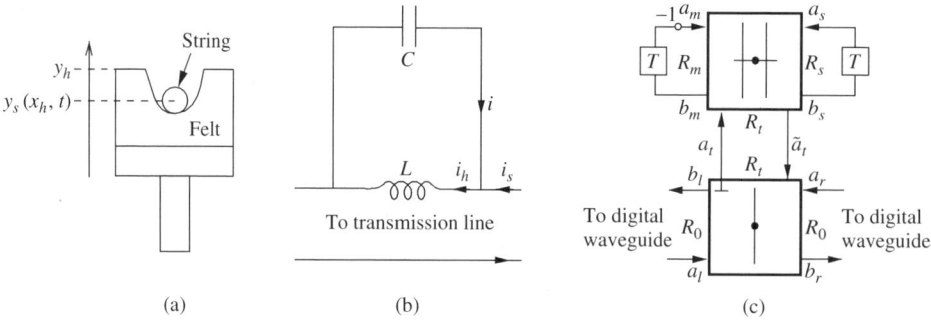

Figure 8.2: (a) *The piano hammer and string*, (b) *a two-port circuit equivalent, and* (c) *its wave digital counterpart*.

8.3. LUMPED NONLINEAR SYSTEMS IN MUSICAL ACOUSTICS

Defining the currents $i_h = \frac{dy_h}{dt}$, $i = \frac{dy}{dt}$ and $i_s = \frac{\partial y_s}{\partial t}|_{x=x_h}$, we then have conservation of current,

$$i = i_h - i_s$$

which, when combined with the equation of motion of the hammer (8.9), leads directly to a passive circuit representation, as shown in Figure 8.2(b). Here, a parallel connection of a capacitor of capacitance C (representing the felt stiffness) and an inductor of inductance $L = m_h$, (representing hammer inertia) is connected in series with a pair of lines carrying current i_s. The two resulting free ports are to be connected to transmission line-like circuit elements that model wave propagation along the string. The wave digital counterpart is shown in Figure 8.2(c), in which the port resistances of the wave digital inductor and capacitor are set to $R_m = 2L/T$ and $R_s = T/2C$. In the case of ideal string motion, the two free ports may be connected to digital waveguides of impedance Z_0 (the characteristic impedance of the string itself); the port resistances R_0 should then be set to simply $R_0 = Z_0$. Notice that as we have a direct connection of a parallel adaptor and a series adaptor, a reflection-free port will be necessary (preferably chosen at the series adaptor, and whose port resistance will be set to $R_t = 2R_0$). As discussed earlier, power-normalized variables must be used in order to ensure losslessness.

One cycle in the operation of the WD hammer consists of the following steps: first, referring to the wave quantities of Figure 8.2(c), read in the wave variables a_l and a_r incident on the series adaptor from the digital waveguides. Next, form the output wave a_t at the reflection-free port. One then has three input wave variables at the parallel adaptor: a_m from the inductor, a_s from the capacitor, and a_t. Writing the equation for the voltage at the parallel junction gives the following equation:

$$v = \frac{2}{\frac{1}{R_m} + \frac{1}{R_s} + \frac{1}{R_t}} \left(\frac{a_m}{\sqrt{R_m}} + \frac{a_s}{\sqrt{R_s}} + \frac{a_t}{\sqrt{R_t}} \right)$$

As R_s depends on v, this is a nonlinear equation that must be solved using iterative techniques. Once it has been solved, scattering may be performed, first at the parallel junction, and then at the series junction. From the orthogonality of the scattering operation, this digital model preserves energy to machine accuracy. The algorithm is initialized by setting the values in the two delay registers according to the initial position and velocity of the hammer.

This simple model describes the behavior of the hammer when it is in contact with the string; when it leaves the string, the coupling is broken and the string will vibrate freely; the uncoupling method has not been described here, but it could be introduced formally into the model through the use of a nonlinear lossless ideal diode. (This has been approached using WDFs, in the context of the modeling of electronic components, [204].) Loss in the hammer felt, also, is not modeled here, though Hunt–Crossley type loss [129] has been introduced in the WD model [16]. There exist other more involved hammer models in which felt stiffness is not merely dependent on the instantaneous depression, but on its past state [241]. It would be of great interest to know whether a circuit representation exists for such a model; indeed, nonlinear passive circuit models could well serve as a starting point for investigations into new models.

8.3.2 The Single Reed

In the case of the piano, the sound created is purely the result of free vibration; in other words, the initial conditions of the hammer and string are sufficient to describe the subsequent string motion. This is reflected by the lack of a source term in the model discussed above. For other instruments, this is generally not the case. All woodwind instruments and string instruments under bowed conditions require a continuous source of energy to produce a sound. As such, we should not expect to arrive at purely passive models, but it should be possible to localize the source term in order to bound the subsequent state of motion in terms of the supplied energy. The reed is typically modeled as a lumped nonlinearity, like the hammer, but of a quite different form. We describe the standard model here, as discussed in the literature [31, 143, 105, 272].

The reed motion is described by

$$\frac{d^2y}{dt^2} + g\frac{dy}{dt} + \omega^2 y = -\Delta p/\mu \tag{8.10}$$

Here y is the deflection of the reed from equilibrium (taken positive when the reed is bending outwards, and constrained to be greater than $-H$, where H is the slit width at equilibrium) and

$$\Delta p = p_\mathrm{m} - p \tag{8.11}$$

is the pressure difference between the pressure p_m at the mouthpiece entrance (the blowing pressure) and the pressure p at the interface with the instrument body. μ is the mass per unit area of the reed, g is a loss coefficient, and ω is the resonant frequency of the lumped reed model.

The pressure drop Δp across the mouthpiece is related to the acoustic volume flow u at the input by

$$\Delta p = \frac{\rho}{2}\left(\frac{u}{w(y+H)}\right)^2 \mathrm{sgn}(u) \tag{8.12}$$

Finally, the acoustic flow u_in at the entrance to the instrument body can be written as a difference of the two contributions u (at the entrance to the mouthpiece) and the flow induced by the motion of the reed u_r, or

$$u_\mathrm{in} = u - u_\mathrm{r} \tag{8.13}$$

u_r is defined by

$$u_\mathrm{r} = S_\mathrm{r}\frac{dy}{dt} \tag{8.14}$$

where S_r is an equivalent surface area for the reed.

The equations (8.10) to (8.14) form a system of five equations in the six variables Δp, p, u, u_r, u_in, and y; they are closed by a connection to the instrument body (assumed here to be a cylindrical tube of impedance Z_0). See Figure 8.3(a) for a graphical representation of the relevant quantities.

Before proceeding directly from the defining of equations for the single reed model (i.e., (8.10)—(8.14)) to a circuit representation, it is useful to rewrite the system in terms

8.3. LUMPED NONLINEAR SYSTEMS IN MUSICAL ACOUSTICS

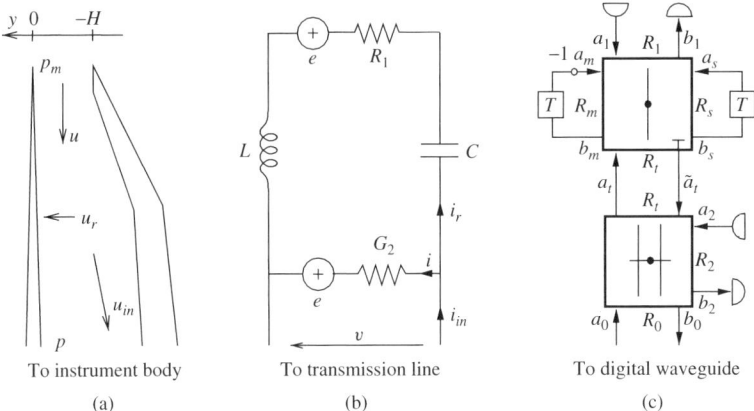

Figure 8.3: (a) *The single reed system*, (b) *a one-port circuit equivalent, and* (c) *its wave digital counterpart.*

of voltage and current variables. Defining

$$\Delta v = -\Delta p \qquad v = p \qquad e = p_m$$
$$i_r = u_r \qquad i = -u \qquad i_{in} = -u_{in}$$

and the coefficients

$$L = \frac{\mu}{S_r} \qquad C = \frac{S_r}{\mu \omega^2} \qquad R_1 = \frac{\mu g}{S_r}$$
$$R_2 = \sqrt{\frac{\rho |\Delta v|}{2}} \frac{1}{w(y+H)} \qquad R_0 = Z_0$$

the system defining the reed becomes

$$v - L\frac{di_r}{dt} - R_1 i_r - \frac{1}{C} \int_0^t i_r(t')\, dt' - e = 0$$
$$i_{in} - i - i_r = 0$$
$$e + i/R_2 = v$$

This has an immediate circuit interpretation as shown in Figure 8.3(b). Notice that there remains an open port, to be connected to the network representing the instrument body (of impedance R_0, corresponding to an acoustic tube). We also note that we have split the source into two separate contributions, each of which is paired with a resistor; thus, each source/resistor pair can be treated as a single resistive source. It is possible to arrive at a circuit with only a single source, but the topology becomes marginally more complex.

The wave digital counterpart follows directly, and is shown in Figure 8.3(c), where again, the use of power-normalized wave quantities is assumed. The port resistances corresponding to the inductor and capacitor, as indicated in the figure, are given, as usual, by

$$R_m = \frac{2L}{T} \qquad R_s = \frac{T}{2C}$$

The port resistances of the resistive sources in the series and parallel adaptor remain as R_1 and R_2.

We thus have a connection between a parallel and a series adaptor; as they are in direct (instantaneous) communication, a delay-free loop results, unless we choose one of the two interconnected ports to be reflection-free. This can be done, for the series adaptor, by setting the port resistance, according to (2.35), as

$$R_t = R_m + R_s + R_1$$

The port resistance at the adjoining port at the parallel adaptor is then also set to R_t.

The open port at the parallel adaptor is to be connected to a digital waveguide, of impedance Z_0 (with port resistance $R_0 = Z_0$). For complete details on the operation of this scattering model of the reed, we refer the reader to the literature [28]. We note that there will be, as in the case of the piano hammer, a single algebraic nonlinear equation to be solved at each step in the recursion.

A perfect power or energy balance can easily be derived by inspection of the wave digital network in Figure 8.3(c). From orthogonality of the scattering junctions, it is easy to show that, for the wave digital reed coupled to a digital waveguide with perfectly reflecting termination, we have the following power balance:

$$\frac{1}{T}(E_s(n+1) - E_s(n)) - w_i(n) + w_l(n) = 0 \qquad (8.15)$$

where $E_s(n)$ is equal to $T([a_m(n)]^2 + [a_s(n)]^2)$ plus the sum of the squares of the values stored in the digital waveguide, $w_i(n) = [a_1(n)]^2 + [a_2(n)]^2$, and $w_l(n) = [b_1(n)]^2 + [b_2(n)]^2$. $E_s(n)$ has the interpretation of a total stored energy in the reed/bore system, $w_i(n)$ that of externally supplied power, and $w_l(n)$ that of power loss. In other words, the change in stored energy per sample plus the power lost must equal the power supplied externally.

As mentioned at the beginning of this section, we have only examined here a simplified case of a more complete model. The inertial effects of air mass in the mouthpiece [272] contribute another inductor to the circuit of Figure 8.3(b). Extensions to this method could equally well be made for the case of reed beating against the mouthpiece at full closure[143] (i.e., for $y = -H$), for which a wave digital ideal diode [204] (connected across the capacitor) is an appropriate passive nonlinear device. Time variation in H (due to variations in the player's lip pressure) is also possible, but will necessarily invoke the use of another source, as such variation introduces energy (albeit very little) into the system. Yet another improvement would be a model of the reed which is itself nonlinear, that is, one in which the mass, stiffness, and damping parameters become 'effective' quantities dependent on the amount of deflection. This is an ongoing subject of research [272], but we do note that a nonlinear circuit model could well serve as a useful starting point for such investigations.

8.4 From Wave Digital Principles to Relativity Theory

by Alfred Fettweis

8.4.1 Origin of the Challenge

There are essentially three different simple expressions available by means of which an inductance can be described, that is,

8.4. FROM WAVE DIGITAL PRINCIPLES TO RELATIVITY THEORY

$$u = D(L'i), \quad u = L''Di, \quad D = d/dt \tag{8.16}$$

$$u = \sqrt{L}D(\sqrt{L}i) = \tfrac{1}{2}(D(Li) + LDi), \tag{8.17}$$

where u is the voltage and i the current. In general, these expressions are not equivalent, but the requirement $L \geq 0$ is necessary and sufficient for guaranteeing passivity (and in fact, losslessness).

In the simplest nonlinear case, that is, if the inductance depends only on its own current i, the following can be shown to hold:

$$L' = \frac{1}{2}L + \frac{1}{2i}\int_0^i L\,di, \quad L = \frac{2}{i^2}\int_0^i iL''\,di, \tag{8.18}$$

$$L'' = L + \frac{1}{2}i\frac{dL}{di}, \quad L = 2L' - \frac{2}{i^2}\int_0^i iL'\,di. \tag{8.19}$$

From (8.18) it follows that

$$L'' \geq 0 \;\Rightarrow\; L \geq 0 \;\Rightarrow\; L' \geq 0. \tag{8.20}$$

From (8.19), it further follows that none of the implication arrows in (8.20) may be reversed, as can be shown rigorously by means of the counterexamples $L = L_0/(1+x^2)$ and $L' = L_0/(1+x^2)$, respectively, where $x = i/i_0$, with L_0 and i_0 as constants. Hence, in order to have, for all i, $L \geq 0$, the requirement $L' \geq 0$ is not sufficient and $L'' \geq 0$ not necessary.

In the most general case, that is, if L depends on any of the dependent variables and/or on time, the situation is far more complicated, but for (8.17) the simple requirement $L \geq 0$ remains valid. This confirms the fundamental role of (8.17) versus either one of the expressions (8.16), and this is still true if there is more than one dimension, in which case D can be any of the relevant partial differential operators.

Despite this, the relation between force, mass, and velocity in classical relativity is not introduced as

$$\mathbf{f} = \sqrt{m}D\left(\sqrt{m}\mathbf{v}\right) = \tfrac{1}{2}\left(D(m\mathbf{v}) + \mathbf{v}Dm\right) \tag{8.21}$$

but as

$$\mathbf{f} = D\left(m'\mathbf{v}\right), \tag{8.22}$$

where we have written m' instead of m in order to conform with the notation in (8.16), m and m' being related in the same way as L and L' in (8.18) and (8.19). In fact, (8.22) is imposed as a postulate to be confirmed by experiment and is not the result of a pure deductive argument. This corresponds to placing prime emphasis on $m'\mathbf{v}$, thus on momentum, while (8.21), like (8.17), places prime emphasis on passivity, thus, on energy. The classical expression obtained for m' is

$$m' = m_0/\alpha, \quad \alpha = \sqrt{1-\beta^2}, \quad \beta = v/c, \quad v^2 = \mathbf{v}^T\mathbf{v} \tag{8.23}$$

where m_0 is a constant (also called the *rest mass*) and c is the velocity of light. The corresponding value obtained for m is rather unwieldy.

The question thus arises whether some different kind of postulate could reverse this situation. Such a postulate must be of an immediately comprehensible physical nature and must relate to aspects of energy and thus of work done. It must not affect the validity of Einsteinian kinematics, which is based on the Lorentz transformation and is the result of admirable deductive logic. The results must of course be in agreement with indisputable experimental facts.

A very brief outline of this challenging topic will be presented hereafter. Some details can be found in [75] to [79], with [78] being the most complete presentation so far, although certain aspects will be mentioned that have not yet been published.

8.4.2 The Principle of Newtonian Limit

In the context of relativistic kinematics, the first step is to consider two reference frames S and S', with S' moving with constant velocity $\mathbf{v}_0$ with respect to S. We assume $\mathbf{v}_0 = (v_0, 0, 0)^T$, and the Lorentz transformation is thus given by

$$x' = \frac{1}{\alpha_0}(x - v_0 t), \quad y' = y, \quad z' = z, \quad t' = \frac{1}{\alpha_0}\left(t - \beta_0 \frac{x}{c}\right), \tag{8.24}$$

$$\alpha_0 = \sqrt{1 - \beta_0^2}, \quad \beta_0 = v_0/c, \tag{8.25}$$

Primed and unprimed coordinates refer to S and S', respectively. The corresponding relations for the velocity of a particle are

$$v'_x = \frac{v_x - v_0}{1 - \beta_0 \beta_x}, \quad v'_y = \frac{\alpha'}{\alpha} v_y, \quad v'_z = \frac{\alpha'}{\alpha} v_z, \tag{8.26}$$

$$\beta_x = v_x/c, \quad 1 - \beta_0 \beta_x = \alpha_0 \alpha/\alpha', \tag{8.27}$$

$$\mathbf{v} = (v_x, v_y, v_z)^T, \quad v^2 = \mathbf{v}^T \mathbf{v}, \quad \beta = v/c, \quad \alpha = \sqrt{1 - \beta^2}, \tag{8.28}$$

$$\mathbf{v}' = (v'_x, v'_y, v'_z)^T, \quad v'^2 = \mathbf{v}'^T \mathbf{v}', \quad \beta' = v'/c, \quad \alpha' = \sqrt{1 - \beta'^2}. \tag{8.29}$$

We call a particle *instantaneously motionless* if in the reference frame and at the time instant under consideration its velocity is zero. We then define the principle of Newtonian limit in the following way: For a relevant quantity that can be defined in a given reference frame, say in S', under Newtonian and under relativistic conditions the ratio of the two corresponding quantities must go to unity if one approaches a time instant at which the particle is instantaneously motionless.

8.4.3 Newton's Second Law

Let us then add a subscript 1 to quantities evaluated at some arbitrary fixed time instant t_1, and thus correspondingly at t'_1. We consider a particle P of rest mass m_0 and assume S' to be such that $\mathbf{v}_0 = \mathbf{v}_1$, that is, that P is instantaneously motionless at t'_1. The principle of the Newtonian limit then yields

$$\mathbf{f}'_1 = m_0 (D'\mathbf{v}')_1, \quad D' = d/dt', \tag{8.30}$$

8.4. FROM WAVE DIGITAL PRINCIPLES TO RELATIVITY THEORY

which is also valid in classical relativity. Consider, however, the work done in S' between t'_1 and $t'_2 = t'_1 + \Delta t'$, that is,

$$\Delta W' = \int_{t'_1}^{t'_2} \mathbf{f}'^T \mathbf{v}' dt'. \tag{8.31}$$

For the corresponding quantity under Newtonian conditions, which imply $t' = t$, $\mathbf{f}' = \mathbf{f}$, and $\mathbf{v}' = \mathbf{v} - \mathbf{v}_0$, we have

$$(\Delta W')_N = \int_{t_1}^{t_2} \mathbf{f}^T (\mathbf{v} - \mathbf{v}_0) dt. \tag{8.32}$$

Both $\Delta W'$ and $(\Delta W')_N$ become zero for $\Delta t' \to 0$, but according to the principle of Newtonian limit we require

$$\lim_{\Delta t' \to 0} \left(\Delta W' / (\Delta W')_N \right) = 1, \tag{8.33}$$

and this obviously for any value of $\mathbf{f}'_1$.

From (8.24) to (8.33) one obtains, after some lengthy calculations, an expression for $\mathbf{f}$ that has precisely the form of (8.21) with

$$m = m_0/\alpha^2, \quad \alpha^2 = 1 - \beta^2, \tag{8.34}$$

thus with α as in (8.23). The alternative relativistic extension of Newton's Second Law is thus indeed given by (8.21) and (8.34). For the power absorbed one finds

$$\mathbf{v}^T \mathbf{f} = DW_k, \quad W_k = \frac{1}{2} m v^2 = m_0 c^2 \beta^2 / 2\alpha^2, \tag{8.35}$$

W_k being the kinetic energy. It is different but simpler than the expression $m_0 c^2 (1/\alpha - 1)$ from classical relativity. The total energy is then

$$W = W_0 + W_k, \quad W_0 = \text{constant},$$

the rest (internal) energy W_0 being in fact an integration constant; it is unspecified by (8.35). Since W_k can be decomposed according to $W_k = \frac{1}{2} m c^2 - \frac{1}{2} m_0 c^2$ one is tempted to set $W = \frac{1}{2} m c^2$ and $W_0 = \frac{1}{2} m_0 c^2$, but the classical $W_0 = m_0 c^2$ is also compatible with (8.35).

Observe that the quadruple

$$\begin{pmatrix} \mathbf{f} \\ \frac{1}{c} \mathbf{v}^T \mathbf{f} \end{pmatrix} = \begin{pmatrix} \mathbf{f} \\ \frac{1}{c} DW_k \end{pmatrix} \tag{8.36}$$

is identical to the four-vector originally introduced by Minkowski and thus has Lorentz invariant form.

The result given by (8.21) and (8.34) can be extended to the case of nonnegative definite symmetric matrices $\mathbf{m}_0$ and $\mathbf{m} = \mathbf{m}_0/\alpha^2$. However, even if the properties of an object are constant (as for rigid-body dynamics) these matrices need not be constant because the object may change its orientation while moving. Hence, a proper extension for the rigid case should be of the form

$$\mathbf{f} = \frac{1}{\alpha} \mathbf{m}_0^{T/2} \mathbf{D} \left(\frac{1}{\alpha} \mathbf{m}_0^{1/2} \mathbf{v} \right) \tag{8.37}$$

where $\mathbf{m}_0^{T/2} = \left(\mathbf{m}_0^{1/2} \right)^T$ and $\mathbf{m}_0 = \mathbf{m}_0^{T/2} \mathbf{m}_0^{1/2}$. The kinetic energy is now given by

$$W_k = \frac{1}{2\alpha^2} \mathbf{v}^T \mathbf{m}_0 \mathbf{v} = \frac{1}{2} \mathbf{v}^T \mathbf{m} \mathbf{v}, \tag{8.38}$$

but the Lorentz invariance mentioned in relation with (8.36) is usually no longer valid. The case of a nonrigid object requires more care.

8.4.4 Newton's Third Law and Some Consequences

Newton's First Law is purely qualitative. Modifying the Second Law as given by (8.21) and (8.34), however, can be shown to require, in the alternative theory, the modification of the Third Law ($\mathbf{f}_1 = -\mathbf{f}_2$, also in classical relativity), according to

$$\alpha_1 \mathbf{f}_1 = -\alpha_2 \mathbf{f}_2, \quad \alpha_i = \sqrt{1 - \beta_i^2}, \quad \beta_i = v_i/c, \quad i = 1, 2. \tag{8.39}$$

Here, the subscripts 1 and 2 refer to the two particles P_1 and P_2 now involved, with v_1 and v_2 their respective velocities.

If there are n particles P_ν, with rest masses $m_{\nu 0}$, velocities $\mathbf{v}_\nu$, and momenta (defined as in classical relativity)

$$\mathbf{p}_\nu = m_{\nu 0} \mathbf{v}_\nu / \alpha_\nu, \quad \alpha_\nu = \sqrt{1 - \beta_\nu^2}, \quad \beta_\nu = v_\nu/c, \quad v_\nu^2 = \mathbf{v}_\nu^T \mathbf{v}_\nu,$$

for $\nu = 1$ to n, one obtains, using (8.39),

$$D(\mathbf{p}_1 + \mathbf{p}_2 + \cdots + \mathbf{p}_n) = \mathbf{0}.$$

Hence conservation of momentum holds exactly as in classical relativity.

Consider a particle P moving with velocity $\mathbf{v}$ in a field that exerts an actual force $\mathbf{f}$ (in the sense of the alternative theory) upon P. Let $\mathbf{f}_0$ be the force that would act upon P according to classical expressions. One finds, again using (8.39),

$$\mathbf{f} = \mathbf{f}_0/\alpha, \quad \alpha = \sqrt{1 - \beta^2}, \quad \beta = v/c, \quad v^2 = \mathbf{v}^T \mathbf{v}, \tag{8.40}$$

This can be confirmed, independently of (8.39), for an electromagnetic field. Using the known rules for transforming such a field from S' to S and furthermore the Lorentz invariance mentioned for (8.36) one finds that $\mathbf{f}$ is indeed given by (8.40) with

$$\mathbf{f}_0 = q (\mathbf{E} + \mathbf{v} \times \mathbf{B}), \tag{8.41}$$

q being the charge of the particle, $\mathbf{E}$ the electric field, and $\mathbf{B}$ the magnetic induction.

8.4. FROM WAVE DIGITAL PRINCIPLES TO RELATIVITY THEORY

Obviously, (8.21), (8.34), and (8.40) together imply that classical relativity and the alternative approach lead to exactly the same behavior of particles in fields. As a further consequence of (8.40), assuming that at a given location (i.e., position and time instant) one can assign to an electromagnetic field a velocity $\mathbf{v}$, one must replace the classical expressions for the energy density w and the Poynting vector $\mathbf{S}$ by

$$w = \frac{1}{2\alpha}\left(\varepsilon\,|\mathbf{E}|^2 + \mu\,|\mathbf{H}|^2\right), \quad \mathbf{S} = \frac{1}{\alpha}\mathbf{E}\times\mathbf{H}. \tag{8.42}$$

8.4.5 Moving Electromagnetic Fields

We consider an electromagnetic field (in vacuum) using standard notation, apply the rules recalled shortly after (8.40), and adopt primed and unprimed notation as in Section 8.4.3 but write $\mathbf{v}$, α, and β instead of $\mathbf{v}_0$, α_0, and β_0. Assuming that at the location under consideration we have $\mathbf{S}' = \mathbf{0}$ one finds, for the field energy in an elementary volume $\mathrm{d}V$ of S,

$$w\,\mathrm{d}V = w'\,\mathrm{d}V' + \frac{1}{2\alpha^2}\mathbf{v}^\mathrm{T}\mathbf{m}_0\mathbf{v}\,\mathrm{d}V' \tag{8.43}$$

where, using (8.42),

$$\mathbf{S}' = \mathbf{E}'\times\mathbf{H}', \quad w' = \tfrac{1}{2}\left(\varepsilon\,|\mathbf{E}'|^2 + \mu\,|\mathbf{H}'|^2\right), \quad \mathrm{d}V = \mathrm{d}x\cdot\mathrm{d}y\cdot\mathrm{d}z,$$

$$\mathbf{m}_0 = \frac{2}{c^2}\left(2w'\mathbf{1} - \varepsilon\mathbf{E}'\mathbf{E}'^\mathrm{T} - \mu\mathbf{H}'\mathbf{H}'^\mathrm{T}\right), \quad \mathbf{1} = \text{unit matrix}$$

and so on. For a given location and a given field in S one can always achieve $\mathbf{S}' = \mathbf{0}$ by choosing $\mathbf{v}$ according to

$$2\mathbf{v}/(1 + v^2/c^2) = \mathbf{S}/w, \quad v = |\mathbf{v}| < c. \tag{8.44}$$

This suggests interpreting (impossible to do in the context of classical relativity) the field at the given location to be instantaneously motionless in S' (thus $\mathbf{v}' = \mathbf{0}$, $\alpha' = 1$), moving in S with $\mathbf{v}$ given by (8.44), and having in $\mathrm{d}V$ a rest energy and a kinetic energy given by the first and the second term in the right-hand side of (8.43). There are infinitely many choices other than (8.44) that lead to $\mathbf{S}' = \mathbf{0}$ but yield the same lateral velocity and this way also the same kinetic energy.

One is thus further tempted to associate with the kinetic energy a force density $\mathbf{f}$ related to $\mathbf{m}_0$ and $\mathbf{v}$ in a way corresponding to (8.37). To this should be added, in a pure electromagnetic field, $\mathbf{f} = (\rho\mathbf{E} + \mu\mathbf{i}\times\mathbf{H})/\alpha$, $\mathbf{i}$ being the current density (in general $\mathbf{i} \neq \rho\mathbf{v}$) in the relevant Maxwell equation, and $\mathbf{f}$ being also related to the velocity in a way as just discussed. If such results should be confirmed they open up new perspectives for examining the behavior of objects consisting of fields.

8.4.6 The Bertozzi Experiment

In 1964, Bertozzi [22] reported, among other things, results regarding calorimetric measurements of the kinetic energy W_k of fast electrons (charge e) accelerated by a field of voltage

u (notation modified to better suit our purpose) and then impacting and thus heating a target. Let W_c be the estimate of W_k as determined by calorimetry, and define $\delta = W_c/eu$. Since there are unavoidable losses (e.g., through X-ray generation) we have $W_c < W_k$, thus $\delta < 1$ if indeed $W_k = eu$ as is the case according to classical relativity. For the alternative theory we may have, as can be shown to follow from (8.39) and the presence of unavoidable secondary effects, $W_c > eu$, thus $\delta > 1$. Both test results ($eu = 1.5$ and 4.5 MeV) mentioned in [22] yield $\delta = 1.067$, thus indeed $\delta > 1$. However, it is also stated that the experiment was only 10% accurate, which would make the measured values to be not incompatible with $\delta < 1$. Obviously, higher accuracy is needed.

8.5 Burger's Equation

A simple nonlinear PDE that is often used as a model problem for fluid dynamical systems is given by the *inviscid Burger's equation* [112]:

$$\frac{\partial u}{\partial t} + u\frac{\partial u}{\partial x} = 0 \tag{8.45}$$

It is similar in form to the advection equation mentioned in Section 3.5 and, as we will see, its circuit representation is identical. The problem is assumed to be defined for $x \in \mathbb{R}$, $t \geq 0$. u can be considered to be a current, as before, through a single loop, and Kirchhoff's Voltage Law around the loop will give (8.45). The question, however, is of the type of circuit elements to be included in this loop; clearly, they must be nonlinear, and certainly reactive as well. The viscous form of Burger's equation has also been approached in the circuit context [274].

Using coordinate transformation (3.19), (8.45) can be rewritten as

$$\left(\frac{v_0 + u}{\sqrt{2}}\right)\frac{\partial u}{\partial t_1} + \left(\frac{v_0 - u}{\sqrt{2}}\right)\frac{\partial u}{\partial t_2} = 0$$

Assuming that the solution is differentiable[2], this can be rewritten as

$$\frac{1}{2}\underbrace{\left(L_1\frac{\partial u}{\partial t_1} + \frac{\partial L_1 u}{\partial t_1}\right)}_{v_1} + \frac{1}{2}\underbrace{\left(L_2\frac{\partial u}{\partial t_2} + \frac{\partial L_2 u}{\partial t_2}\right)}_{v_2} = 0 \tag{8.46}$$

where

$$L_1 = \frac{1}{\sqrt{2}}\left(v_0 + \frac{2}{3}u\right) \qquad L_2 = \frac{1}{\sqrt{2}}\left(v_0 - \frac{2}{3}u\right) \tag{8.47}$$

Thus, Burger's equation, in the form of (8.46), can be interpreted as a series combination of two nonlinear inductances, as shown in Figure 8.4(a). The resultant MDWD network,

---

[2]This is an assumption made by Fettweis et al. in all of their fluid dynamics work, and is not entirely justified, especially if discontinuities (shocks) are to be modeled. Even in the simple model of Burger's equation, these shocks can develop [112]. Over regions of continuity in the problem, these numerical methods will give the correct solution, but shock velocities, should they develop, may not be correct for this reason.

8.5. BURGER'S EQUATION

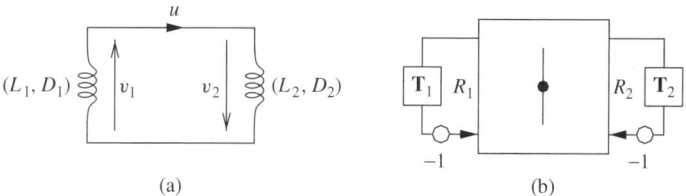

Figure 8.4: *The* $(1+1)$D *inviscid Burger's equation*—(a) *MDKC and* (b) *MDWD network.*

with port resistances

$$R_1 = \frac{2L_1}{T_1} \qquad R_2 = \frac{2L_2}{T_2} \qquad (8.48)$$

appears in Figure 8.4(b). We emphasize that this network is passive only if power-normalized wave variables are employed.

The positivity condition on these inductances now depends on the solution itself, u, and we must have

$$v_0 \geq \tfrac{2}{3} \max_{x \in \mathbb{R}, t \geq 0} |u|$$

An *a priori* estimate of $\max_{x \in \mathbb{R}, t \geq 0} |u|$ must be available; this is a consistent feature of all the circuit-based methods for the fluids systems that we will examine presently.

Let us now examine the scattering operation. First choose $T_1 = T_2 = \sqrt{2}\Delta$, so that the current grid function for the current at location $x = i\Delta$ and $t = nT$ can be written as $u_i(n)$. The two power-normalized input wave variables entering the adaptor at the same location and time step are $\underline{a}_{1,i}(n)$ $\underline{a}_{2,i}(n)$, and we have

$$u_i(n) = \frac{2}{R_{1,i}(n) + R_{2,i}(n)} \left(\sqrt{R_{1,i}(n)} \underline{a}_{1,i}(n) + \sqrt{R_{2,i}(n)} \underline{a}_{2,i}(n) \right)$$

which, from (8.47) and (8.48), and using $v_0 = \Delta/T$ can be rewritten as

$$u_i(n) = \frac{\sqrt{\Delta}}{v_0} \left(\sqrt{v_0 + \tfrac{2}{3} u_i(n)} \underline{a}_{1,i}(n) + \sqrt{v_0 - \tfrac{2}{3} u_i(n)} \underline{a}_{2,i}(n) \right) \qquad (8.49)$$

This is precisely the nonlinear algebraic equation that is to be solved (in $u_i(n)$); once $u_i(n)$ is determined then so are the port resistances, and the output wave variables $\underline{b}_{1,i}(n)$ and $\underline{b}_{2,i}(n)$ can be obtained through scattering as per (2.33). As mentioned before, it is not at all clear from the form of (8.49) whether a solution exists and is unique. We note, however, that we (and others [21, 100]) have successfully programmed simulations for the gas dynamics equations (see next section), using simple iterative methods to solve the nonlinear algebraic systems; the results would appear to be in accord with published simulation results using difference methods [234].

8.6 The Gas Dynamics Equations

The behavior of a lossless one-dimensional fluid is described by the following set of conservation equations, also known as *Euler's equations*:

$$\frac{\partial \rho}{\partial t} + \frac{\partial (\rho v)}{\partial x} = 0 \qquad \text{Conservation of mass} \qquad (8.50a)$$

$$\frac{\partial (\rho v)}{\partial t} + \frac{\partial (\rho v^2 + p)}{\partial x} = 0 \qquad \text{Conservation of momentum} \qquad (8.50b)$$

$$\frac{\partial (\rho e)}{\partial t} + \frac{\partial (\rho v e + p v)}{\partial x} = 0 \qquad \text{Conservation of energy} \qquad (8.50c)$$

where $\rho(x,t)$ is density, $v(x,t)$ is volume velocity, $p(x,t)$ is absolute pressure, and $e(x,t)$ is total energy, internal plus kinetic. The three equations are not complete without a constitutive relation among the four dependent variables. The WDF people, in their treatment of *hydrodynamics* [21], often leave out the energy equation and make an assumption of the type $p = G(\rho)$, which essentially reduces system (8.50) to a two-variable system (in ρ and v). For gas dynamics, we assume polytropic gas behavior [243]:

$$e = \frac{v^2}{2} + \frac{p}{\rho(\gamma - 1)} \qquad (8.51)$$

where $\gamma > 1$ is a constant that follows directly from thermodynamics (it is equal to the ratio of specific heats [275]).

Before proceeding any further, we mention the *scaling* of the dependent variables [21, 71]; this is done, as in the linear problems discussed in Chapters 3 and 7, in order to optimize the stability condition on the resulting network. The variables are scaled as

$$\hat{v} = \frac{v}{v_0} \qquad \hat{p} = \frac{p}{p_0} \qquad \hat{\rho} = \frac{\rho v_0^2}{p_0} \qquad (8.52)$$

The parameters v_0 and p_0 have dimensions of velocity and pressure, respectively, and nondimensionalize the system. v_0 will again become the space-step/time-step ratio in the numerical simulation routine, and p_0 plays a role similar to that of r_0 in the $(1+1)$D transmission line problem as discussed in Section 3.6 and follows directly from physical considerations.

Using the energy density definition (8.51), system (8.50) can be written in *nonconservative* [275] form (after some tedious algebraic manipulations) as

$$\begin{bmatrix} 1 & 0 & 0 \\ 0 & \hat{\rho} & 0 \\ 0 & 0 & 1 \end{bmatrix} \frac{\partial}{\partial t'} \begin{bmatrix} \hat{\rho} \\ \hat{v} \\ \hat{p} \end{bmatrix} + \begin{bmatrix} \hat{v} & \hat{\rho} & 0 \\ 0 & \hat{\rho}\hat{v} & 1 \\ 0 & \gamma\hat{p} & \hat{v} \end{bmatrix} \frac{\partial}{\partial x} \begin{bmatrix} \hat{\rho} \\ \hat{v} \\ \hat{p} \end{bmatrix} = \mathbf{0}$$

The problem here is that this system is not, in its present form, suitable for a circuit representation involving reciprocal elements[3]. Although it is not explicitly stated anywhere in the literature, the solution of Fettweis et al. has been to scale system (8.50) by left

---

[3] We note, however, that Fries [100] has obtained an MDKC directly from these conservation laws, though certain extra parameters must be introduced in order to control stability.

8.6. THE GAS DYNAMICS EQUATIONS

multiplication by the matrix diag($1/\hat{\rho}, 1, 1/(\gamma \hat{p})$) (though as previously mentioned, they work with the two-variable hydrodynamics system, or its analogues in higher dimensions). The scaled system takes the form:

$$\begin{bmatrix} \frac{1}{\hat{\rho}} & 0 & 0 \\ 0 & \hat{\rho} & 0 \\ 0 & 0 & \frac{1}{\gamma \hat{p}} \end{bmatrix} \frac{\partial}{\partial t'} \begin{bmatrix} \hat{\rho} \\ \hat{v} \\ \hat{p} \end{bmatrix} + \begin{bmatrix} \frac{\hat{v}}{\hat{\rho}} & 1 & 0 \\ 0 & \hat{\rho}\hat{v} & 1 \\ 0 & 1 & \frac{\hat{v}}{\gamma \hat{p}} \end{bmatrix} \frac{\partial}{\partial x} \begin{bmatrix} \hat{\rho} \\ \hat{v} \\ \hat{p} \end{bmatrix} = \mathbf{0} \quad (8.53)$$

While this scaling does not change smooth solutions to system (8.50), problems may occur if shocks are anticipated [243]. Such so-called *weak solutions* [243] to system (8.50) (solutions involving discontinuities that must be described using the integral formulation of (8.50)) are not necessarily preserved under such a scaling. Entropy variables, to be briefly mentioned in Section 8.6.3, allow a potential means of avoiding these difficulties.

Finally, by employing the conservation of mass equation to simplify the other equations, the scaled system (8.53) can be written in *skew-selfadjoint form* [243] as

$$\mathbf{P}\frac{\partial \mathbf{w}}{\partial t'} + \frac{\partial \mathbf{Pw}}{\partial t'} + \mathbf{A}\frac{\partial \mathbf{w}}{\partial x} + \frac{\partial \mathbf{Aw}}{\partial x} = \mathbf{0} \quad (8.54)$$

where the symmetric matrices $\mathbf{P}$ and $\mathbf{A}$ are defined by

$$\mathbf{P} = \begin{bmatrix} \frac{1}{\hat{\rho}} & 0 & 0 \\ 0 & \hat{\rho} & 0 \\ 0 & 0 & \frac{1}{\alpha \hat{p}} \end{bmatrix} \quad \mathbf{A} = \begin{bmatrix} \frac{\hat{v}}{\hat{\rho}} & 0 & 0 \\ 0 & \hat{\rho}\hat{v} & 1 \\ 0 & 1 & \frac{\hat{v}}{\alpha \hat{p}} \end{bmatrix}$$

$\mathbf{w}$ is the state, $[\hat{\rho}, \hat{v}, \hat{p}]^T$, and $\alpha = \frac{2\gamma}{\gamma-1} > 0$. $\mathbf{P}$, in addition, will be positive definite if the density and pressure are positive everywhere. The significance of this skew-selfadjoint form can be seen by taking the inner product of (8.54) with $\mathbf{w}$, in which case we get

$$\frac{1}{2}\frac{\partial \mathbf{w}^T \mathbf{Pw}}{\partial t'} + \frac{1}{2}\frac{\partial \mathbf{w}^T \mathbf{Aw}}{\partial x} = 0 \quad (8.55)$$

For the Cauchy problem (i.e., the problem defined over the entire x axis, so boundary conditions are effectively ignored), we may integrate over the domain to get

$$\frac{d}{dt'} \int_{-\infty}^{+\infty} \frac{1}{2}\mathbf{w}^T \mathbf{Pw}\, dx = 0 \quad (8.56)$$

and thus $\int_{-\infty}^{+\infty} \frac{1}{2}\mathbf{w}^T \mathbf{Pw}\, dx$ is the global conserved quantity. It can be seen, by comparison between (8.55) and (8.56) with (3.3) and (3.5) (in the lossless case), that this skew-selfadjointness property is the natural extension of symmetric hyperbolicity to the nonlinear case. It is interesting that the generalized definitions of the inductor and capacitor, as per (8.3) and (8.5), are completely commensurate; we will see this in the next section. The theory of skew-selfadjoint forms has recently seen quite a bit of activity, in particular with regard to so-called *entropy variables* [98, 115, 128, 243, 244, 245], which we will look at briefly in Section 8.6.3.

8.6.1 MDKC and MDWD Network for the Gas Dynamics Equations

It is particularly easy to see the form of the MDKC for the gas dynamics equations in the scaled form of (8.54). Applying the usual coordinate transformation (3.19), (8.54) becomes

$$\tfrac{1}{2}(LD_1\mathbf{w} + D_1L\mathbf{w}) + \tfrac{1}{2}(MD_2\mathbf{w} + D_2M\mathbf{w}) + N\mathbf{w} = 0 \qquad (8.57)$$

with

$$\mathbf{L} = \begin{bmatrix} \frac{1+\hat{v}}{\hat{\rho}} & 0 & 0 \\ 0 & \hat{\rho}(1+\hat{v})-1 & 0 \\ 0 & 0 & \frac{1+\hat{v}}{\alpha\hat{\rho}}-1 \end{bmatrix} \quad \mathbf{M} = \begin{bmatrix} \frac{1-\hat{v}}{\hat{\rho}} & 0 & 0 \\ 0 & \hat{\rho}(1-\hat{v})-1 & 0 \\ 0 & 0 & \frac{1-\hat{v}}{\alpha\hat{\rho}}-1 \end{bmatrix}$$

and

$$\mathbf{N} = \begin{bmatrix} 0 & 0 & 0 \\ 0 & D_1 + D_2 & D_1 - D_2 \\ 0 & D_1 - D_2 & D_1 + D_2 \end{bmatrix}$$

The MDKC is shown in Figure 8.5(a), where the inductances can be read directly from the entries of $\mathbf{L}$, $\mathbf{M}$, and $\mathbf{N}$. $\mathbf{L}$ and $\mathbf{M}$ represent the inductances in the three loops in directions t_1 and t_2 respectively, and $\mathbf{N}$ gives the coupling between the second and third loops (notice that it can be realized as a simple linear and shift-invariant Jaumann two-port, just as in the linear systems of Chapter 3). The first loop, with current $\hat{\rho}$ is decoupled from the other two, although the inductances in this loop are dependent on $\hat{v}$.

The MDWD network follows immediately and is shown in Figure 8.5(b). It should be kept in mind that the port resistances at the adaptors are now functions of the dependent variables (the currents in the MDKC), and thus of the wave variables themselves. In a given updating cycle, the current values of the port resistances must be determined from

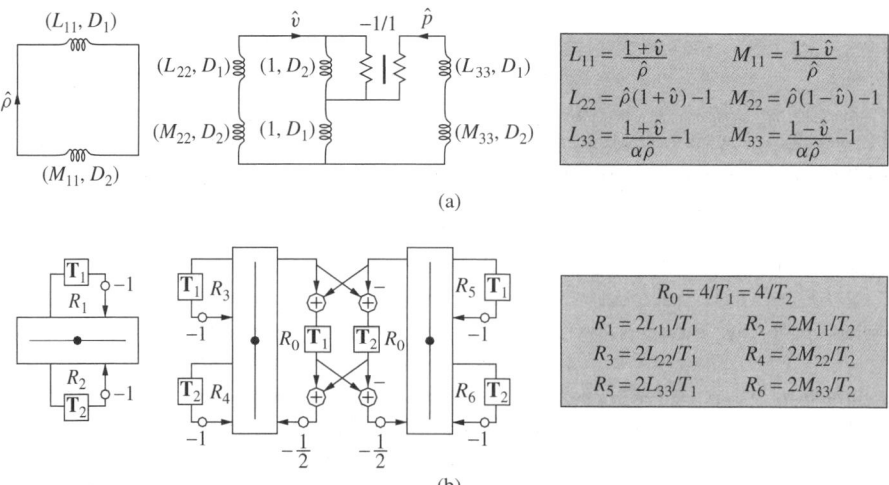

Figure 8.5: *The $(1+1)$D gas dynamics system—(a) MDKC and (b) MDWD network.*

8.6. THE GAS DYNAMICS EQUATIONS

the incoming waves. This leads to a system of coupled nonlinear algebraic equations (three, one for each adaptor) to be solved at every grid point, and at every time step.

Passivity is contingent upon the positivity of all the inductances in the network; this is essentially a condition on the positivity of the diagonal matrices **L** and **M**. Proceeding down the diagonals, this requirement on the first elements leads to the natural condition

$$1 \pm \hat{v} \geq 0 \quad \Rightarrow \quad v_0 \geq |v|_{\max}$$

where $|v|_{\max}$ is the maximum value that $|v|$ will take over the problem domain, and during the simulation period. We have also assumed that ρ remains positive, and used the definition of the scaled quantity $\hat{v}$ from (8.52). The conditions on the other elements of **L** and **M** are more strict. We get

$$\hat{\rho}(1 \pm \hat{v}) - 1 \geq 0 \quad \Rightarrow \quad 1 \pm \hat{v} \geq \frac{1}{\hat{\rho}} \quad \Rightarrow \quad v_0 \geq \frac{p_0}{\rho_{\min} v_0} + |v|_{\max} \quad (8.58a)$$

$$\frac{1 \pm \hat{v}}{\alpha \hat{p}} - 1 \geq 0 \quad \Rightarrow \quad 1 \pm \hat{v} \geq \alpha p \quad \Rightarrow \quad v_0 \geq \frac{\alpha p_{\max} v_0}{p_0} + |v|_{\max} \quad (8.58b)$$

where $\rho_{\min}$ and $p_{\max}$ are, respectively, the minimal value of ρ and the maximum value of p that will be encountered in the problem space. These quantities as well as $|v|_{\max}$ must be estimated *a priori*. It is also worth mentioning that for the above reasoning to be valid, it has been assumed that ρ and p will remain positive, and that ρ is bounded from below. Although this has not been mentioned in the literature, there does not appear to be any assurance that these assumptions will remain valid during the course of a simulation.

We still have one degree of freedom left, namely, the value of the parameter p_0. An optimal setting is easily shown to be

$$p_0 = v_0 \sqrt{\alpha p_{\max} \rho_{\min}}$$

in which case the two bounds on v_0 from (8.58) coalesce, giving

$$v_0 \geq |v|_{\max} + \sqrt{\frac{\alpha p_{\max}}{\rho_{\min}}}$$

We note that the numerical methods examined here can also be applied to fluid dynamic systems in $(2 + 1)$D [21] and $(3 + 1)$D [71]. The nonlinear algebraic systems to be solved become larger, but are still localized. Also, we mention that these numerical methods do not seem to reduce to conventional finite difference schemes along the lines of Godunov's method and its offspring [234].

8.6.2 An Alternate MDKC and Scattering Network

The use of network manipulations and alternate spectral mappings in order to derive digital waveguide networks from an MDKC has been discussed in detail in Chapter 6, and we have seen the idea applied again in Chapter 7 to elastic solid systems. This same idea can be employed in the present case as well. Consider the network of Figure 8.6(a), which is equivalent to that of Figure 8.5(a); the right-hand pair of inductors in series in the 'pressure' loop has been replaced by a gyrator closed on a parallel combination of capacitors;

notice that although we are now transforming nonlinear operators, the network transformation techniques are no different from the linear case. We thus have a powerful means of developing stable numerical methods at our disposal.

The element values L_{11}, L_{22}, M_{11}, and M_{22} are the same as before, except that they are scaled by a factor Δ, and the capacitance values will be

$$C_{33} = L_{33}\Delta/2 \qquad F_{33} = M_{33}\Delta/2$$

where L_{33} and M_{33} are as defined in Figure 8.5. (Notice that we have scaled the entire system by Δ, as in Chapter 6.)

The resulting MD digital network is shown in Figure 8.6(b). As before, the two-port $AABB'$ transforms to a pair of bidirectional delay lines under the application of the spectral mappings defined by (6.4). The other circuit elements, namely, the nonlinear inductors and capacitors, must be discretized using the trapezoid rule, and so we are left with a network that is neither an MDWD network nor a DWN, but which contains elements of both. The port resistances are determined in the usual way; for an inductance L and direction t_j, by $R = 2L/T_j$ and for a capacitance C of direction t_j, by $R = T_j/(2C)$. The port resistances of the paired multidimensional unit elements will be $R_0 = 1/\sqrt{2}$.

It is possible to choose the directional shift lengths in the one-port inductances and capacitances differently from those in the unit elements such that the network could conceivably operate in an interleaved (offset) configuration; parallel junctions that calculate p alternate with series junctions calculating v. A potential problem here is that the port resistances at the parallel junctions (say) depend on v, but v is not calculated at these grid locations; some approximation is thus necessary, but we do not pursue the matter further here.

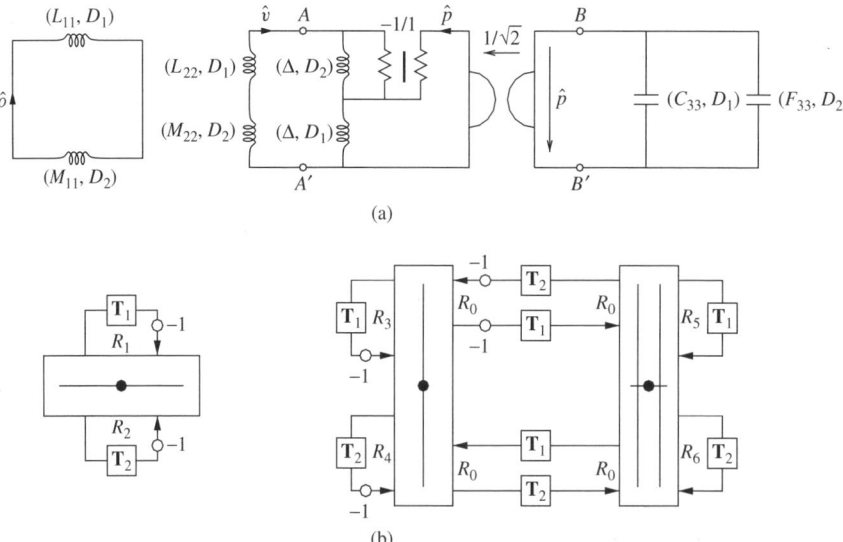

Figure 8.6: *Alternative networks for the $(1+1)$D gas dynamics system—(a) MDKC and (b) scattering network.*

8.6.3 Entropy Variables

In the opening remarks of Section 8.6, we showed how Fettweis et al. have effectively employed a skew-selfadjoint form of the gas dynamics system in order to generate a circuit model. Such forms have been the subject of a great deal of research [243]. One particular form, which makes use of the so-called *entropy variables*, would appear to be of fundamental importance because it arises from a change of variables (and not a simple scaling of the system, as was the case for the system arrived at earlier). We recap some recent results here [243].

Consider a system of conservation laws,

$$\frac{\partial \mathbf{u}}{\partial t} + \frac{\partial \mathbf{f}(\mathbf{u})}{\partial x} = \mathbf{0} \tag{8.59}$$

where the $\mathbf{f}(\mathbf{u})$ are smooth, and are, in general, nonlinear mappings. The Euler system (8.50), with $\mathbf{u} = [\rho, \rho v, \rho e]^T$ and $\mathbf{f}(\mathbf{u}) = [\rho v, \rho v^2 + p, \rho v e + p v]^T$, again complemented by the constitutive relation (8.51) is of this form. It has been noted [98, 157, 244] that (8.59) implies a further conservation law,

$$\frac{\partial \mathcal{U}}{\partial t} + \frac{\partial \mathcal{F}}{\partial x} = 0 \tag{8.60}$$

for some smooth convex scalar function $\mathcal{U}(\mathbf{u})$, and a scalar flux $\mathcal{F}(\mathbf{u})$, over any time interval over which solutions to (8.59) remain smooth. If discontinuities (shocks) develop, then (8.60) becomes an inequality ($\leq$). $\mathcal{U}$ and $\mathcal{F}$ are related by

$$\left(\frac{d\mathcal{U}}{d\mathbf{u}}\right)^T \frac{d\mathbf{f}}{d\mathbf{u}} = \left(\frac{d\mathcal{F}}{d\mathbf{u}}\right)^T$$

It has been shown [245] that system (8.59) is symmetrized through left multiplication by the Hessian $\mathbf{P}_\mathcal{U}$ of $\mathcal{U}$,

$$\mathbf{P}_\mathcal{U} = \frac{d^2 \mathcal{U}}{d\mathbf{u}^2}$$

so that we have

$$\mathbf{P}_\mathcal{U} \frac{\partial \mathbf{u}}{\partial t} + \mathbf{A}_\mathcal{U} \frac{\partial \mathbf{u}}{\partial x} = \mathbf{0} \tag{8.61}$$

where $\mathbf{A}_\mathcal{U}$ and $\mathbf{P}_\mathcal{U}$ are symmetric, and, in addition, $\mathbf{P}_\mathcal{U}$ is positive definite (a result of the convexity requirement on $\mathcal{U}$). This nonlinear system is of the same form as the (linear) symmetric hyperbolic system (3.1) discussed in Section 3.1, and possesses many similar properties; this form, however, cannot be easily approached through MD circuit methods. Furthermore, *weak solutions* (i.e., solutions involving discontinuities) will not be preserved under such a scaling [243]. This is the same defect as that of Fettweis's MDKC for the Euler system, as discussed in Section 8.6.

It was later shown that (8.59) can also symmetrized with respect to a new variable $\mathbf{z}$, defined by

$$\mathbf{z} = \frac{d\mathcal{U}}{d\mathbf{u}}$$

In this case, symmetrization is carried out through a variable change and not a scaling, so weak solutions are indeed preserved. If, furthermore, the flux **f** is homogeneous [243], it can be shown that there is also a skew-selfadjoint form of (8.59). The gas dynamics system (8.50) can be written in skew-selfadjoint form as

$$\mathbf{P}_{\mathcal{U}} \frac{\partial \mathbf{z}}{\partial t} + \frac{\partial \mathbf{P}_{\mathcal{U}} \mathbf{z}}{\partial t} + \mathbf{A}_{\mathcal{U}} \frac{\partial \mathbf{z}}{\partial x} + \frac{\partial \mathbf{A}_{\mathcal{U}} \mathbf{z}}{\partial x} = \mathbf{0} \tag{8.62}$$

if $\mathcal{U}$ is chosen as

$$\mathcal{U} = -(p\rho^\alpha)^{1/\alpha+\gamma} \qquad \alpha > 0$$

which is closely related to the physical entropy of the system [243]. The new variables **z** are referred to as *entropy variables*.

System (8.62) is of the same form as (8.54) (though note that we have neglected to perform the variable scalings), but it is written in terms of entropy variables **z** as opposed to nonconservative variables **w**. Applying coordinate transformation (3.19), we can get the system into the form

$$\tfrac{1}{2} (\mathbf{L}_{\mathcal{U}} D_1 \mathbf{z} + D_1 \mathbf{L}_{\mathcal{U}} \mathbf{z}) + \tfrac{1}{2} (\mathbf{M}_{\mathcal{U}} D_2 \mathbf{z} + D_2 \mathbf{M}_{\mathcal{U}} \mathbf{z}) = \mathbf{0} \tag{8.63}$$

where

$$\mathbf{L}_{\mathcal{U}} = v_0 \mathbf{P}_{\mathcal{U}} + \mathbf{A}_{\mathcal{U}} \qquad\qquad \mathbf{M}_{\mathcal{U}} = v_0 \mathbf{P}_{\mathcal{U}} - \mathbf{A}_{\mathcal{U}}$$

Both $\mathbf{L}_{\mathcal{U}}$ and $\mathbf{M}_{\mathcal{U}}$ will be positive definite if v_0 is chosen sufficiently large. Though system (8.63) would appear to be in the correct form for an MDKC representation, there are certain difficulties.

First, the matrices $\mathbf{L}_{\mathcal{U}}$ and $\mathbf{M}_{\mathcal{U}}$, unlike **L** and **M** from (8.57), are not diagonal. Fettweis was able to take advantage of the fact that in terms of the variables **w**, all interloop couplings are linear and shift-invariant (the coupling matrix **N** is a constant), so nonlinearities can be well-isolated. This is no longer the case here. It is of course possible to write down an MDKC corresponding to (8.63)—the entries of $\mathbf{L}_{\mathcal{U}}$ and $\mathbf{M}_{\mathcal{U}}$ become inductances directly. But positive definiteness of $\mathbf{L}_{\mathcal{U}}$ and $\mathbf{M}_{\mathcal{U}}$ does not imply that the off-diagonal entries will be positive, so our MDKC will not necessarily be a concretely passive representation.

We might attempt to avoid this by treating (8.63) as a simple series combination of two vector inductors, of inductances $\mathbf{L}_{\mathcal{U}}$ and $\mathbf{M}_{\mathcal{U}}$. In analogy with definitions of the coupled inductances in Section 2.2.7, it is certainly possible to define lossless nonlinear coupled inductances by

$$\mathbf{v}_1 = \tfrac{1}{2} (\mathbf{L}_{\mathcal{U}} D_1 \mathbf{z} + D_1 \mathbf{L}_{\mathcal{U}} \mathbf{z})$$

$$\mathbf{v}_2 = \tfrac{1}{2} (\mathbf{M}_{\mathcal{U}} D_2 \mathbf{z} + D_2 \mathbf{M}_{\mathcal{U}} \mathbf{z})$$

and the resulting MDKC is essentially identical to that of Figure 3.6, for the simple linear advection equation, except that the current is now **z**. It is difficult to introduce wave variables, however, because power-normalization is not straightforward in the nonlinear vector case; for an inductor of positive definite vector inductance **L**, the relationship analogous to (8.3) does not hold, that is,

$$\tfrac{1}{2} (\mathbf{L} D_j \mathbf{i} + D_j (\mathbf{L} \mathbf{i})) \neq \mathbf{L}^{T/2} D_j \left(\mathbf{L}^{1/2} \mathbf{i} \right) \tag{8.64}$$

8.6. THE GAS DYNAMICS EQUATIONS

The second form of the inductor (which is distinct, and also lossless) on the right of (8.64), involving some left square root $\mathbf{L}^{T/2}$ of $\mathbf{L}$, is that which would be essential for power-normalization because then we would be able to write

$$\mathbf{L}^{-T/2}\mathbf{v} = D_j(\mathbf{L}^{1/2}\mathbf{i})$$

and then define power-normalized vector wave variables by

$$\underline{\mathbf{a}} = \tfrac{1}{2}\left(\mathbf{R}^{-T/2}\mathbf{v} + \mathbf{R}^{1/2}\mathbf{i}\right)$$

$$\underline{\mathbf{b}} = \tfrac{1}{2}\left(\mathbf{R}^{-T/2}\mathbf{v} - \mathbf{R}^{1/2}\mathbf{i}\right)$$

where $\mathbf{R}^{T/2}$ is the left square root of some positive definite matrix port resistance $\mathbf{R}$. Making the usual choice of $\mathbf{R} = 2\mathbf{L}/T_j$ (or rather $\mathbf{R}^{T/2} = \sqrt{2/T_j}\mathbf{L}^{T/2}$), we would then arrive at the familiar wave relationship of (3.39) in terms of the vector waves $\underline{\mathbf{b}}$ and $\underline{\mathbf{a}}$. (The two inductor definitions of (8.64) do, however, coincide if the nonlinearity is confined to the diagonal elements of $\mathbf{L}$.) We note that Fettweis and Bose [33] have begun to approach the problem of vector, or multiport inductors in the time-varying case; the analysis is by no means simple, and it is not at all clear which results may be extended to the fully nonlinear case.

It would be of fundamental interest to know whether a passive MDKC for general nonlinear systems of the form of (8.63) (and its analogues in higher dimensions), amenable to wave digital discretization in fact exists. In such an MDKC or MDWD network, the global conserved quantity would have the interpretation of an entropy, which can be thought of as a generalized form of energy [98].

Chapter 9

Concluding Remarks

We begin by addressing the general questions of Section 1.2 and then pose some new ones.

9.1 Answers

- *To what types of systems can multidimensional wave digital and digital waveguide network simulation approaches be applied?*

Let us first confine our attention to numerical methods that result from the passive discretization of a multidimensional Kirchhoff circuit (MDKC) representation; all the methods discussed in this book are of this form, except for the multi-grid methods of Section 5.4, and the type I and II DWNs for the transmission line and parallel-plate problems (see Section 4.3.6 and Section 4.4.2 respectively). We can then rephrase our question as: *For which types of systems do there exist passive MDKC representations?*

In general, MDKC representations follow directly only for $(n+1)$D symmetric hyperbolic systems of the form of (3.1), and we repeat this definition here:

$$\mathbf{P}\frac{\partial \mathbf{w}}{\partial t} + \sum_{k=1}^{n} \mathbf{A}_k \frac{\partial \mathbf{w}}{\partial x_k} + \mathbf{B}\mathbf{w} + \mathbf{f} = \mathbf{0} \qquad (9.1)$$

Recall that the coupling matrices $\mathbf{A}_k$ are constrained to be symmetric and constant, so all the material parameter variation is confined to the symmetric positive-definite matrix coefficient $\mathbf{P}$ of the time derivatives, and the coefficient $\mathbf{B}$ of the constant-proportional terms. (Indeed, $\mathbf{B}$ can be *time-varying* without affecting the network representation of any system in which it appears. We examined the more general case of time variation of $\mathbf{P}$, which occurs in, for example, time-varying vocal tract models [196], in Section 8.2.) For passivity, we generally require that the symmetric part of $\mathbf{B}$ be positive semidefinite. The *symmetry* of the $\mathbf{A}_k$ implies a network representation involving only *reciprocal* reactive MD-circuit elements[1]. In fact, the structure of the $\mathbf{A}_k$ directly specifies the MD network topology; if

---

[1] Recall that a reciprocal circuit element is one whose impedance matrix (more generally, its hybrid matrix) is Hermitian.

Wave and Scattering Methods for Numerical Simulation   Stefan D. Bilbao
© 2004 John Wiley & Sons, Ltd   ISBNs: 0-470-87017-6

any of the $\mathbf{A}_k$ contains a nonzero entry in the (p,q)th position, then we will necessarily have some coupling between the pth and qth loops in the network representation. The constancy of the $\mathbf{A}_k$ has important implications for the energetic analysis of the system, as we saw in Section 3.1, and also ensures that all interloop couplings can be accomplished through the use of linear and shift-invariant (LSI) coupling elements (generally the Jaumann or lattice two-ports introduced in Section 3.6.2 for MDWD networks, or the alternative hybrid form of Section 6.1.2 for DWNs). This is crucial, because the spatial derivative information is concentrated in these coupling elements; discretization of a reactive element is usually only passive if its defining parameters (usually inductances) are independent of the integration directions. (The main exceptions here are the reactive elements discussed at the end of Section 3.4.1 and again in Section 8.1.2; such elements can be used in the time-varying case, as mentioned above, or in nonlinear problems which were discussed at length in Chapter 8.) In fact, all the linear and time-invariant MD-passive systems we have examined have networks that can be decomposed into two networks as in Figure 9.1. The subnetwork on the left is made up of a set of time inductors (or capacitors) with spatially varying inductances (or capacitances) and perhaps resistances as well, and can be discretized through the trapezoid rule in time. In the expanded signal flow graphs, these always give rise to *self-loops*. The right-hand subnetwork is linear and shift-invariant, and contains all the spatial derivative terms, in the form of Jaumann or lattice connections for MDWD networks, or the hybrid form for DWNs, discussed in Chapter 6. It can be discretized by the MD trapezoid rule (or alternative discretization rules for DWNs), and the full discrete network will be passive if the subnetworks are MD-passive separately.

A few other features are worthy of comment. First, although the $\mathbf{P}$ matrix is diagonal for many of the systems we have looked at, it is *not* so for any of the vibrating elastic systems in $(2+1)$D and $(3+1)$D of Chapter 7 (nor will it be for, say, Maxwell's equations in anisotropic media, or mutually inductive coupled transmission lines, or systems in general curvilinear coordinates). The nondiagonal elements of $\mathbf{P}$ are modeled in an MDKC as mutual inductances or capacitances, and subsequently require vector-scattering junctions, as discussed in Section 2.2.7. This leads to more complex 'block' scattering matrices, but passivity is not compromised. Second, consider the $\mathbf{B}$ matrix that is not constrained to be of any particular form. If the symmetric part of $\mathbf{B}$ is positive semidefinite, then it always implies *loss*, and can be realized as a purely resistive coupling network. The antisymmetric part of $\mathbf{B}$, which does not produce loss, gives rise to *dispersion* (as for the Timoshenko beam in Section 7.1, the Mindlin plate in Section 7.2 and the shell models of Section 7.3), and corresponds to lossless gyrator couplings among the circuit loops. In general, if the $\mathbf{B}$ matrix is not sparse, the resulting resistive and gyrator couplings in the MDKC may considerably complicate the resulting discrete network (i.e., various reflection-free ports will be required).

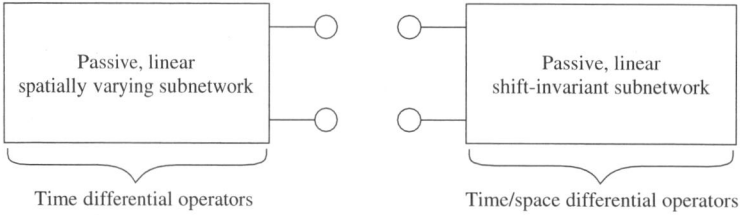

Figure 9.1: *Network decomposition of passive symmetric hyperbolic systems.*

Third, the system (9.1) forms a subclass of what are called *strongly hyperbolic* systems, for which the initial value problem is well-posed [112]. A general strongly hyperbolic system can be also written in the form of (9.1), but the matrices $\mathbf{A}_k$ are not constrained to be symmetric (any real linear combination of the $\mathbf{A}_k$ must, however, possess a set of real distinct eigenvalues). If the $\mathbf{A}_k$ cannot be simultaneously symmetrized by some change of variables, then it is clear that the set of reactive reciprocal circuit elements does not suffice for a passive circuit representation, even though energy estimates of the type as discussed in Section 3.1 may be available. It would be quite interesting to know how a network could be designed for such strongly, but not symmetric hyperbolic systems. Furthermore, system (9.1) is not even the most general form of symmetric hyperbolic system. The matrices $\mathbf{A}_k$ may be functions of space or time variables, though the simple energetic analysis of Section 3.1 becomes more involved; passivity does not immediately follow. Fourth, scattering methods may be extended to cover certain parabolic or borderline- parabolic systems as well (through a hyperbolic approximation [26, 50, 274]). Fifth, for nonlinear systems, the natural extension of symmetric hyperbolicity that leads to circuit representations is *skew self-adjointness*. We examined one such extension in Section 8.6.

- *What features do the two methods share, and what distinguishes them?*

The most basic difference between the two methods is, as we emphasized early on in Chapter 1, that a DWN can always be viewed as a large network of scattering junctions, connected to one another *portwise* by bidirectional delay lines; a MDWD network is better thought of as the discrete image of a multidimensional Kirchhoff circuit representation. Though a signal flow graph for a MDWD network follows immediately, it cannot be decomposed into a collection of discrete transmission lines, or bidirectional delay lines. Both the methods, however, operate using exclusively scattering and shifting operations—it may be useful to flip to Figure 6.1 which shows the expanded signal flow graphs for the DWN and MDWD networks for the (1 + 1)D transmission line equations. Notice that the portwise connectivity of the MDWD network is always lost in these (and all) expanded forms; for certain adaptors, the two wave signal paths entering the 'port' are not connected to the two terminals of another single port. We have seen, of course, in Chapter 6, that many DWN forms operating on regular grids can also be derived from MDWD circuit representations; DWNs are more general in the sense that they may be constructed completely *locally*, and in irregular arrangements (recall the general discussion of DWNs in Section 1.1.3), without using an MDKC.

We have also seen, in Chapter 4, that DWNs are, in general, equivalent to simple two-step centered finite difference schemes of the FDTD variety. In Section 3.8, we showed that MDWD simulation methods also correspond to finite difference schemes, but they are, in general, multistep methods. In both cases, the calculations have been rearranged (using wave variables) in a *one-step* form. If the underlying system is lossless, and power-normalized wave variables are employed, then at any time step, the discrete network recursion involves an *orthogonal transformation* (scattering) applied to the entire set of wave variables stored in memory, followed by a *permutation* (shifting) operation. Lossless-ness, and thus stability is ensured in a very direct way. The two methods possess distinct spectral properties that we examined briefly in the case of the (1 + 1)D transmission line in Section 4.3.8.

There is another more subtle distinction. For almost all first-order PDE systems that follow from the physical laws, the state variables can be separated into two types that are *dual* to one another. To be more precise, in all the systems of the form of (9.1) which we have examined in this book, the time derivative of a variable of one type is always related to spatial derivatives of variables of the other type; for example, voltages and currents are dual variables in the transmission line and parallel-plate problems, as are electric and magnetic field components described by Maxwell's equations (6.7), and so on. This duality implies a certain structure in the coupling matrices $\mathbf{A}_k$. In the MDKCs developed by Fettweis et al., however, this structure is ignored, and all variables are treated generally as *currents*. In the resulting numerical methods, this usually implies that all the dependent variables are computed *together* at the same spatial locations, and at all time steps. (As we mentioned in Section 3.8, however, it is possible to design MDWD simulation methods operating on 'checkerboard' grids; the variables are all computed together, but at locations interleaved in time and space.) For the DWNs discussed in Chapters 4 and 7, we have made use of this duality in order to design networks in which the two sets of variables are computed at alternating time instants and spatial locations. One set is interpreted as *current-like*, and the other as *voltage-like*. The resulting alternations are generally indicated by gray/white junction colorings in the signal flow graphs. (It is worth calling attention, at this point, to the analogous distinction between the so-called 'expanded' and 'condensed' node TLM formulations [44].)

- *What are their relative advantages?*

The answers to the above question have several practical implications.

Because any DWN can always be written in the form of a large network of port-wise connected scattering junctions, passive (and thus stable) boundary termination is straightforward. For the transmission line (see Section 4.3.9), the parallel-plate problem (see Section 4.4.4), beams (see Section 7.1.3) and plates (see Section 7.2.2) many useful lossless boundary conditions can be effected through the use of simple short- or open-circuit terminations, or self-loops connected to the scattering junctions that lie on the boundary. In fact, the termination of any boundary junction with an *all-pass* (more generally *bounded real*) scattering one-port will *always* yield a numerical scheme that is guaranteed lossless (more generally passive), though the physical interpretation of such an arbitrary termination may not be obvious. It is worth emphasizing that the ease with which passivity (or exact losslessness, if so desired) can be ensured in the presence of boundary conditions is one great benefit of a scattering formulation. It can be quite difficult to ensure the stability of a boundary condition applied to a finite difference scheme. The same is not true of MDWD networks; because portwise connectivity is lost in the expanded signal flow graph, passive termination is no longer simple or straightforward. We indicated in Section 3.10 and Section 7.1.3 the severe difficulties inherent in the 'wave-canceling' termination approach described in the literature [146], and looked at a possible theoretical foundation for passive distributed network termination in Section 3.10.1.

An MDKC is always mapped, via spectral transformations or integration rules, to a discrete time and space MDWD image network. As mentioned above, the resulting numerical method always operates on a regular grid in some coordinate system. It is thus impossible, through the approach of Fettweis et al., to arrive at structures that operate on irregular grids. As we have seen in Section 5.4, because DWNs may be constructed locally, multigrid methods become a possibility—DWNs of differing densities or in different coordinate systems may be simply connected to one another in such a way that passivity is maintained across the interface. Such interfaces can be designed so as to be locally consistent with the underlying model problem; as a result, numerical reflection vanishes as the grid spacings become small. The applications to numerical integration over irregular problem domains should be self-evident.

- *Can they be unified in a formal way?*

The clear answer to this question, as shown in Chapter 6, is *yes*. A passive MDKC representation of a system of PDEs is no more than that—a representation. It illustrates, however, how the system can be decomposed into simpler elements, each of which is passive in its own right, and immediately suggests a stable numerical scheme. (In the opinion of this author, the development of the TLM method has been hampered by the lack of such a representation, or any direct link between the PDE system and the scattering method, leaving the algorithm designer with very little to go on besides experience and intuition.) Each element can be discretized through the application of one or a set of spectral mappings, or multidimensional integration rules, in such a way that this passivity is preserved (this is discussed in Section 3.4.3) in a discrete network, which can then be directly implemented as a computer simulation routine. The MD trapezoid rule, or bilinear transform, which gives rise to wave digital networks, is one such rule but is by no means the only way of proceeding; in multiple dimensions, the family of such mappings is large and diverse. DWNs are the result of the application of another member of this family. As we have seen, especially in Chapter 7, DWNs are now applicable to any system that has been approached using MDWD networks (we have covered most of them in this book). It would appear, then, justifiable to lump these techniques together as simply 'wave' or 'scattering' methods.

As indicated in Figure 9.2, the unification of these two methods also implies their membership within a larger class of passive numerical methods; indeed, *any* set of passive spectral mappings or integration rules may be applied to *any* circuit representation of a system of PDEs, and the result is necessarily a stable numerical method built of the same basic scattering and shifting operations as the DWN or an MDWD network. We note, however, that for a given circuit representation of a system of PDEs, it is not at all obvious which spectral mappings should be applied in order to give rise to a useful structure. It should be possible to elucidate the link to a certain degree; does a particular network topology imply a particular integration rule or mapping? This is important, because for a given system of PDEs, there is not a single MDKC representation; any rules or transformations from classical network theory can be used to manipulate the MDKC into an infinite number of new topologies, each of which, upon discretization, gives rise to a distinct numerical method. This being said, the MD circuit framework (or rather, language) does at least give us a set of guiding principles.

9.2 Questions

The answers given above are necessarily incomplete, and, of course, give rise to yet more questions.

- *What about TLM?*

TLM methods have not been covered in this book; as it is nearly certain that the majority of the readers of this book will be familiar with TLM and not MDWD networks, this might

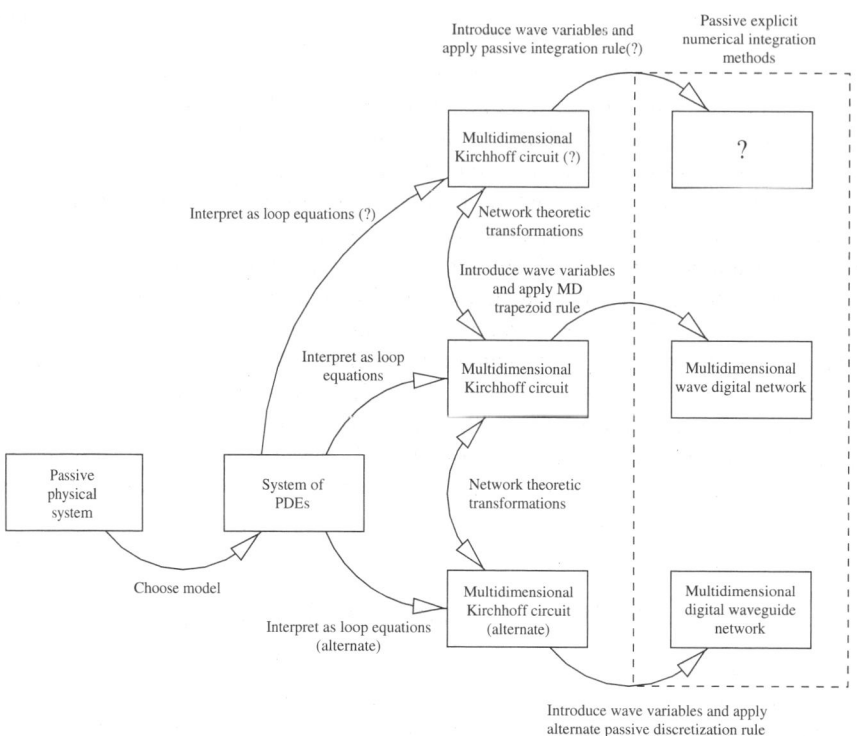

Figure 9.2: *The generalized family of passive numerical methods.*

seem to be a major oversight. To confuse the target audience even more, we have introduced DWNs, a family of scattering methods that bear, in many ways, a very strong resemblance to TLM structures. One reason for this is that there already exist comprehensive books on the subject of TLM. The more important reason, however, is that Fettweis's multidimensional circuit representation approach rules out the standard development of TLM in terms of lumped elements. Indeed, the point of this book is that there is a different (and, we feel, more powerful) framework for arriving at TLM-like structures. For various reasons, the DWN is a closer match to the methods of Fettweis than TLM; chief among these are the reliance on elementary scattering operations (scattering operations in TLM are usually constrained to be only orthogonal), the use of power-normalized waves, and, more generally, a similarity in notation, due to the common roots of DWNs and WDFs in signal processing theory.

On the other hand, TLM has developed in many interesting directions that we have not been able to discuss in this book. Many different types of structures have been proposed, in particular those for which the dependent variables are not interleaved. Given that we have shown certain DWNs can be derived from MDKCs, just as MDWD networks are, it would be interesting to know whether the various TLM structures can be arrived at in a similar way. A compact MD circuit representation would empower an algorithm designer enormously, and would almost certainly make a useful tool for designing new structures that are potentially more efficient and which may have better numerical properties.

- *What is the difference between stability and passivity?*

9.2. QUESTIONS

Upon setting a scattering method directly alongside its finite difference counterpart, a curious observation one can make is that the condition under which the finite difference scheme is stable (in the sense of von Neumann [238]) is *not* the same as the condition that the network be made up exclusively of passive elements. Because in explicit schemes for hyperbolic systems, the difference scheme coefficients and network element values are usually parameterized by v_0, the space step/time step ratio, the question is often one of the *range* of values of v_0 for which a given scheme is stable or passive. As we have seen, a distinction between passivity and stability manifests itself in various ways in many very different settings. We saw, for example, in Section 4.3.6, that several different waveguide networks for the transmission line problem, though all equivalent in infinite-precision arithmetic to the same simple centered difference approximation, are passive over quite different ranges of v_0, depending on material parameter variation. Even more striking examples will be seen in Appendix A, in the case of the triangular scheme for the $(2 + 1)$D wave equation, and in particular for the so-called 'interpolated' difference schemes for the wave equation in $(2 + 1)$D and $(3 + 1)$D; these are rudimentary constant-coefficient difference schemes, and yet the difference between the stability condition and the passivity condition for the equivalent waveguide mesh is already quite complex; for the other mesh structures examined in Appendix A, von Neumann stability and passivity imply one another. Other instances appear throughout this book. The question is one of network topology—that is, there are many network topologies corresponding to a given difference scheme, and though the stability bound on v_0 will be the same for all of them, the bounds for passivity will be, in general, distinct (see Section A.2.3 and Section A.2.4 for some interesting examples). In sum, passivity is a sufficient, but not necessary condition for numerical stability; the distinction is essentially that between concretely and abstractly passive networks (see Section 1.1.1). It may well be, however, that it is always possible to find a particular topology such that these conditions imply one another. It would be of great interest to make sense of the 'gray area' between the two conditions, and to be able to answer a more precise question, namely: *Is there any stable difference scheme that can* not *be written in a scattering form?*

- Given that scattering methods are equivalent to finite difference schemes, why should one not use a finite difference scheme directly?

This is perhaps the most difficult question to answer; in fact, it almost surely cannot be answered, given the great number of factors that play a role in the design of numerical simulation algorithms. It is, however, important to at least address it.

We have shown that scattering methods, whether they are derived from wave digital principles or DWNs, are always equivalent to finite difference schemes of some type. This is an obvious observation, but bears being repeated. How, then, are they different? One interesting way of viewing this question is in terms of the *Lax–Richtmyer Equivalence Theorem*, which, with a few technical caveats, can be stated as follows: *A finite difference scheme for a given model problem is convergent if and only if it is both consistent with the model problem, and stable.* Finite difference schemes are usually designed in the following way: find a consistent approximation to a model problem, usually by replacing partial derivatives by difference operators, and then show that the scheme is stable (somehow). For TLM design, the approach has traditionally been the opposite: build a network out of elements that are known to behave in a stable manner when combined (such as orthogonal scattering junctions and simple transmission line-like elements), and *then* attempt to prove consistency with a given model problem. The MDKC/MDWD approach essentially assures both consistency and stability at once, at least for the Cauchy problem, through the use of a passive circuit representation of the model system itself.

One of the great difficulties in designing a finite difference scheme is in ensuring that it remains stable when boundary conditions are applied. For scattering methods, the MD representation is incomplete is this case (though we have outlined the beginning of a possible line of attack on this problem in Section 3.10.1), and it is simpler to fall back on the lumped network representation; it is immediately clear that terminating boundary junctions with passive elements must leave the entire network passive as a whole, yielding a stable simulation routine. But how can they be chosen consistently with the boundary conditions in the model problem? In this book, we have 'crossed over' in a sense by proving consistency through comparison with finite difference boundary conditions, bringing us back to the question above. Not quite, though, as we now have a useful means at our disposal for *proving* the stability of a terminated finite difference scheme; if the scheme, including boundary conditions, can be written as a passive network, then it must be stable. This could prove to be an alternative means of approaching the stability problem to more traditional approaches such as GKSO theory [112, 113, 148, 181]. In addition, the network representation immediately provides us with a huge family of stable boundary conditions (in essence, *any* termination by lumped passive elements will do), which do not depend at all on the geometry of the network. Thus scattering methods could well be used as a powerful design technique, even if they are ultimately implemented in terms of finite differences. When time-varying and nonlinear problems are approached, the numerical methods arrived at through network representations appear to be entirely novel.

We feel that the emphasis on passivity (and thus stability) in scattering methods is a crucial one; it is reassuring for an algorithm designer to know that a numerical method is stable by construction. On the other hand, it is also true that proponents of scattering methods have more to learn from straight-ahead finite difference practitioners than they may realize; there are tools that have been developed in the finite difference context which also allow for energy-based stability guarantees (the *energy method* [112, 199] being chief among them). Certainly there are relative strengths and weaknesses in both points of view. Interdisciplinary research, both among the various scattering method camps, and with the mainstream simulation community can lead only to improvement on all fronts.

Appendix A

Finite Difference Schemes for the Wave Equation

In this appendix, we reexamine the finite difference schemes corresponding to the waveguide meshes discussed in Chapter 4 and the first part of Chapter 5, in the special case for which the underlying model problem is lossless, source-free and does not exhibit any material parameter variation. In this case, these finite difference schemes will solve the wave equation given by

$$\frac{\partial^2 u}{\partial t^2} = \gamma^2 \nabla^2 u \qquad (A.1)$$

in either $(2+1)$D or $(3+1)$D, depending on the type of mesh. Here, γ is the *wave speed*, and ∇^2 is the *Laplacian*. These schemes will be linear and shift-invariant, and as such, it is possible to analyze them in the frequency domain, through what is called *von Neumann analysis* [238]; this appendix is intended partly as a short primer on this subject.

Some of the material in this appendix has appeared previously [30].

Summary

Appendix A serves a dual purpose. First, it is intended as a review of the basics of the spectral or *von Neumann* analysis of finite difference schemes; this analysis is quite powerful and revealing if the underlying physical model problem is linear and shift-invariant. We pay close attention to the *numerical stability* conditions that can be arrived at through a straightforward application of these spectral methods. For this review portion of the appendix, we depend primarily on the excellent text by Strikwerda [238]. (For the reader with no prior exposure to the analysis of finite difference methods, Section A.1 could well serve as point of departure, before jumping directly into network and scattering theory in Chapter 2.) We then systematically revisit all of the large variety of forms of DWN for the two- and three-dimensional wave equation, applying this spectral analysis to the equivalent difference schemes. The comparison of the von Neumann numerical stability conditions with the passivity conditions on the associated mesh structures yields somewhat surprising results for the so-called triangular (Section A.2.3) and interpolated mesh (Section A.2.2, Section A.3.3) structures; indeed, the conditions do not coincide in these cases, leaving us with some fundamental and puzzling questions about the nature of this discrepancy. In addition, we also introduce some techniques for rigorously analyzing certain vector-type schemes (the hexagonal scheme, in Section A.2.4 and the tetrahedral scheme, in Section A.3.4), and look at a theoretical means of obtaining optimally direction-independent numerical dispersion properties for certain schemes for which we have free parameters at our disposal (the interpolated schemes in Section A.2.2 and Section A.3.3). We also evaluate the relative memory requirements and computational efficiencies of the various schemes, and provide numerical dispersion error plots for all the schemes.

A.1 Von Neumann Analysis of Difference Schemes

In this section, we summarize the basics of von Neumann analysis provided by Strikwerda [238]. Consider the $(N+1)$D real-valued grid function $U_{\mathbf{m}}(n)$, defined for integer n and for $\mathbf{m} = [m_1, \ldots, m_N] \in \mathbb{Z}^N$, the set of all integer N-tuples. Such a grid function will be used, in a finite difference scheme, as an approximation to the continuous solution $u(\mathbf{x}, t)$ to some problem, at the location $\mathbf{x} = \mathbf{m}\Delta$, and at time $t = nT$, where Δ is the grid spacing and T is the time step. Here, and henceforth in this appendix, we have assumed that the grid spacing is uniform in all the spatial coordinates, and that the spatial domain is unbounded. As in the main body of the book, we define the space step/time step ratio to be

$$v_0 \triangleq \frac{\Delta}{T}$$

The spatial Fourier transform of $U_{\mathbf{m}}(n)$ is defined by

$$\hat{U}_{\boldsymbol{\beta}}(n) = \frac{1}{(2\pi)^{N/2}} \sum_{\mathbf{m} \in \mathbb{Z}^N} e^{-j\Delta \mathbf{m} \cdot \boldsymbol{\beta}} U_{\mathbf{m}}(n) \Delta^N$$

and is a periodic function of $\boldsymbol{\beta} = [\beta_1, \ldots, \beta_N]^T$, a vector of spatial wavenumbers. The transform can be inverted by

$$U_{\mathbf{m}}(n) = \frac{1}{(2\pi)^{N/2}} \int_{[-\pi/\Delta, \pi/\Delta]^N} e^{j\Delta \mathbf{m} \cdot \boldsymbol{\beta}} \hat{U}_{\boldsymbol{\beta}}(n) \, d\beta_1 \, d\beta_2 \ldots d\beta_N$$

where $\boldsymbol{\beta} \in [-\pi/\Delta, \pi/\Delta]^N$ refers to the hypercube enclosed by the intervals $-\pi/\Delta \leq \beta_k \leq \pi/\Delta$, for $k = 1, \ldots, N$.

If, for a given grid spacing Δ, we define the discrete spatial l_2 norm of $U_{\mathbf{m}}(n)$ by

$$\|U(n)\|_2 = \left(\sum_{\mathbf{m} \in \mathbb{Z}^N} U_{\mathbf{m}}^2(n) \Delta^N \right)^{1/2}$$

A.1. VON NEUMANN ANALYSIS OF DIFFERENCE SCHEMES

and the corresponding spectral L_2 norm of $\hat{U}_\beta(n)$ by

$$\|\hat{U}(n)\|_2 = \left(\int_{[-\pi/\Delta, \pi/\Delta]^N} |\hat{U}_\beta(n)|^2 \, d\beta_1 \, d\beta_2 \ldots d\beta_N \right)^{1/2}$$

then, if $U_\mathbf{m}(n)$ and $\hat{U}_\beta(n)$ are in l_2 and L_2, respectively, *Parseval's relation* gives

$$\|U(n)\|_2 = \|\hat{U}(n)\|_2$$

A.1.1 One-step Schemes

Consider the following *one-step* explicit difference scheme, which relates values of the grid function $U_\mathbf{m}(n+1)$ to values at the previous time step:

$$U_\mathbf{m}(n+1) = \sum_{\mathbf{k} \in \mathbb{K}} \alpha_\mathbf{k} U_{\mathbf{m}-\mathbf{k}}(n)$$

where $\mathbb{K}$ is some subset of $\mathbb{Z}^N$, and the parameters $\alpha_\mathbf{k}$ are constants; it is initialized by setting $U_\mathbf{m}(0)$ equal to some function $U_{\mathbf{m},0}$ (assumed to be in l_2). Taking the spatial Fourier transform of this recursion gives

$$\hat{U}_\beta(n+1) = \left(\sum_{\mathbf{k} \in \mathbb{K}} \alpha_\mathbf{k} e^{-j\Delta \mathbf{k} \cdot \beta} \right) \hat{U}_\beta(n)$$

$$= G_\beta \, \hat{U}_\beta(n) \qquad (A.2)$$

G_β so defined is called the *spectral amplification factor* for a one-step finite difference scheme. (A.2) implies we have, in particular, that

$$\hat{U}_\beta(n) = G_\beta^n \hat{U}_{\beta,0} \qquad (A.3)$$

where $\hat{U}_{\beta,0}$ is the spatial Fourier transform of the initial condition $U_{\mathbf{m},0}$. (A.3) further implies that

$$\|\hat{U}(n)\|_2 \leq \left(\max_\beta |G_\beta| \right)^n \|\hat{U}_0\|_2$$

and finally, through Parseval's relation, that

$$\|U(n)\|_2 \leq \left(\max_\beta |G_\beta| \right)^n \|U_0\|_2$$

If the $\alpha_\mathbf{k}$ which define the difference scheme are independent of the grid spacing and the time step, then such a difference scheme is called *stable* if

$$\max_\beta |G_\beta| \leq 1$$

The l_2 norm of the solution to the difference equation will thus not increase as the simulation progresses.

A.1.2 Multistep Schemes

Multistep methods can be treated in a very similar way. An explicit M-step method is defined by

$$U_{\mathbf{m}}(n+1) = \sum_{r=1}^{M}\sum_{\mathbf{k}\in\mathbb{K}_r} \alpha_{\mathbf{k}} U_{\mathbf{m}-\mathbf{k}}(n+1-r)$$

for constant coefficients $\alpha_{\mathbf{k}}$ defined over subsets $\mathbb{K}_r$ of $\mathbb{Z}^N$. Taking the Fourier transform of this recursion gives

$$\hat{U}_{\beta}(n+1) = \sum_{r=1}^{M}\sum_{\mathbf{k}\in\mathbb{K}_r} \alpha_{\mathbf{k}} e^{-j\Delta \mathbf{k}\cdot\boldsymbol{\beta}} \hat{U}_{\beta}(n+1-r) \tag{A.4}$$

A simple way of examining (A.4) is to look for solutions of the form $\hat{U}_{\beta}(q) = G_{\beta}^q \hat{U}_{\beta}(0)$. This gives the *amplification polynomial equation*

$$G_{\beta}^M = \sum_{r=1}^{M}\sum_{\mathbf{k}\in\mathbb{K}_r} \alpha_{\mathbf{k}} e^{-j\Delta \mathbf{k}\cdot\boldsymbol{\beta}} G_{\beta}^{M-r}$$

the solutions of which, $G_{\beta,\nu}$, $\nu = 1, \ldots, M$ must be bounded by unity for stability (though in general, this is not sufficient, as we will show presently for a special case).

A particular form of the amplification polynomial equation, which will appear frequently in our subsequent treatment of finite difference schemes for the wave equation, is that of a simple two-step centered difference approximation, namely,

$$G_{\beta}^2 + B_{\beta} G_{\beta} + 1 = 0 \tag{A.5}$$

for some real function B_{β}. This expression has solutions

$$G_{\beta,\pm} = \frac{1}{2}\left(-B_{\beta} \pm \sqrt{B_{\beta}^2 - 4}\right) \tag{A.6}$$

which will be bounded by (and in fact equal to) unity in magnitude if we have $|B_{\beta}| \leq 2$ for all β. Furthermore, if $|B_{\beta}| > 2$ for some β, then we will necessarily have an amplification factor with magnitude greater than one at that frequency. For any β for which $G_{\beta,\pm}$ are not equal, we can write

$$\hat{U}_{\beta}(n+1) = \frac{G_{\beta,-}\hat{U}_{\beta,0} - \hat{U}_{\beta,1}}{G_{\beta,-} - G_{\beta,+}} G_{\beta,+}^{n+1} + \frac{G_{\beta,+}\hat{U}_{\beta,0} - \hat{U}_{\beta,1}}{G_{\beta,+} - G_{\beta,-}} G_{\beta,-}^{n+1}$$

where $\hat{U}_{\beta,0}$ and $\hat{U}_{\beta,1}$ are the spatial frequency spectra of the two grid functions (at time steps $n = 0$ and $n = 1$) used to initialize the two-step method. It is easy to show that the l_2 norm of $U_{\mathbf{m}}(n)$ can be bounded in terms of the norms of the initial conditions if the spectral amplification factors are distinct and bounded by 1 in magnitude at all wavenumbers.

It is important to realize, however, that the condition that these roots $G_{\beta,\pm}$ be bounded by unity is necessary, but not sufficient to ensure no growth in the l_2 norm of the solution; this point has not been addressed in the finite difference treatment of waveguide meshes. In fact, as shown in Section 4.3.4, the simple centered difference approximation to the wave equation admits linearly growing solutions.

This behavior can be examined in the spectral domain as we will now show. Notice that the solutions (A.6) of the amplification polynomial equation for the two-step scheme can coincide if, and only if, at some frequency $\beta = \beta_0$, $B_{\beta_0} = \pm 2$, in which case we have $G_{\beta_0,+} = G_{\beta_0,-} = \mp 1$. The evolution of the particular spatial frequency component at frequency β_0 can be written as

$$\hat{U}_{\beta_0}(n) = (\mp 1)^n \hat{U}_{\beta_0,0} + n(\mp 1)^{n-1} \left(\hat{U}_{\beta_0,1} \pm \hat{U}_{\beta_0,0} \right)$$

We can thus expect some linear growth at any such frequency β_0 if we do not properly initialize the algorithm so as to cancel the linearly growing part of the solution. It also follows that in employing such a method, one may need to be particularly careful when applying an excitation that contains such frequency components, and that nonlinear signal quantization may pump energy into such modes, even if none is originally present there.

Strikwerda does not classify such linear growth as unstable, because the wave equation itself admits, in addition to traveling wave solutions, a solution that grows linearly with time[1]. For the physical modeling of musical instruments and acoustic spaces, however (the problems to which finite difference schemes of the form to be discussed shortly are often applied), such solutions are nonphysical and definitely not acceptable. These comments concerning this mild linear instability apply to schemes in unbounded domains; when boundary conditions are present, further analysis will be required.

In order to simplify the analysis of these schemes, we mention that for difference schemes for the wave equation, it is often possible to write

$$B_\beta = -2\lambda^2 F_\beta - 2 \qquad (A.7)$$

where $\lambda^2 \triangleq \gamma^2/v_0^2$ and F_β is independent of λ. In this case, the stability condition can be rewritten as

$$\max_\beta |B_\beta| \leq 2 \qquad \Longleftrightarrow \qquad \max_\beta |\lambda^2 F_\beta + 1| \leq 1$$

This new condition on F_β is easier to analyze: we first require

$$\max_\beta F_\beta \leq 0 \qquad (A.8)$$

and if (A.8) holds, we get a further bound on λ, namely,

$$\lambda \leq \sqrt{\frac{-2}{\min_\beta F_\beta}} \qquad \Longrightarrow \qquad T \leq \frac{\Delta}{\gamma} \sqrt{\frac{-2}{\min_\beta F_\beta}} \qquad (A.9)$$

Thus, the stability of these schemes can be simply analyzed in terms of the global maximum and minimum of F_β.

For certain schemes (in particular, the interpolated schemes to be discussed in Sections A.2.2 and A.3.3), the function F_β depends on several parameters. Condition (A.8) tells us the range of parameters over which our scheme is stable, and over the stability region, condition (A.9) gives us a maximum time step T in terms of the grid spacing Δ.

---

[1] $u = t$, for instance, satisfies (A.1).

A.1.3 Vector Schemes

For two of the schemes that we will examine (hexagonal and tetrahedral), it will be necessary to analyze a vectorized system of difference equations. In general, the analysis of vector forms is considerably more difficult; the typical approach will invoke the *Kreiss Matrix Theorem* [238], which is a set of equivalent conditions that can be used to check the boundedness of a particular amplification matrix. In the general vector case, we will be analyzing the evolution of a q-element vector $\hat{\mathbf{U}}_\beta(n) = [\hat{U}_{1,\beta}(n), \ldots, \hat{U}_{q,\beta}(n)]^T$ of spatially Fourier-transformed functions of β. The L_2 norm is defined by

$$\|\hat{\mathbf{U}}(n)\|_2 = \left(\int_{[\pi/\Delta, \pi/\Delta]^N} \hat{\mathbf{U}}_\beta^*(n) \hat{\mathbf{U}}_\beta(n) \, d\beta_1 \, d\beta_2 \ldots d\beta_N \right)^{1/2}$$

where * denotes transpose conjugation.

The schemes for the wave equation that we will examine, however, have a relatively simple form. The column vector of grid spatial frequency spectra $\hat{\mathbf{U}}_\beta(n)$ satisfies an equation of the form

$$\hat{\mathbf{U}}_\beta(n+1) + \mathbf{B}_\beta \hat{\mathbf{U}}_\beta(n) + \hat{\mathbf{U}}_\beta(n-1) = \mathbf{0} \tag{A.10}$$

for some Hermitian matrix function of β, $\mathbf{B}_\beta$. Because $\mathbf{B}_\beta$ is Hermitian, we may write $\mathbf{B}_\beta = \mathbf{J}_\beta^* \mathbf{\Lambda}_\beta \mathbf{J}_\beta$, for some unitary matrix $\mathbf{J}_\beta$, and a real diagonal matrix $\mathbf{\Lambda}_\beta$ containing the eigenvalues of $\mathbf{B}_\beta$. As such, we may change variables via $\hat{\mathbf{V}}_\beta(n) = \mathbf{J}_\beta \hat{\mathbf{U}}_\beta(n)$, to get

$$\hat{\mathbf{V}}_\beta(n+1) + \mathbf{\Lambda}_\beta \hat{\mathbf{V}}_\beta(n) + \hat{\mathbf{V}}_\beta(n-1) = \mathbf{0} \tag{A.11}$$

The system thus decouples into a system of scalar two-step spectral update equations; because $\hat{\mathbf{U}}_\beta(n)$ and $\hat{\mathbf{V}}_\beta(n)$ are related by a unitary transformation, we have $\|\hat{\mathbf{U}}(n)\|_2 = \|\hat{\mathbf{V}}(n)\|_2$, and we may apply stability tests to the uncoupled system (A.11). We thus require that the eigenvalues of $\mathbf{B}_\beta$, namely $\Lambda_{\beta,j}$ for $j = 1, \ldots, q$, which are the elements on the diagonal of $\mathbf{\Lambda}_\beta$, all satisfy

$$\max_\beta |\Lambda_{\beta,j}| \leq 2 \tag{A.12}$$

At frequencies β_0 for which any of the eigenvalues satisfies (A.12) with equality, we may again have the same problem with mild linear growth in the solution.

A.1.4 Numerical Phase Velocity

For a given amplification factor G_β, the *numerical phase velocity* at frequency β is defined by

$$v_{\beta,\text{phase}} = \left| \frac{\log(G_\beta / |G_\beta|)}{j \|\beta\|_2 T} \right|$$

where $\|\beta\|_2$ is the Euclidean norm of the vector β. This expression gives the speed of propagation for a plane wave of wave number β according to the numerical scheme for which G_β is an amplification factor. For the wave equation model problem, the speed of any plane wave solution will simply be γ but the numerical phase velocity will in general be different, and in particular, wave speeds will be directionally dependent to a certain degree, depending on the type of scheme used. For all these schemes, the numerical phase velocity for at least one of the amplification factors will approach the correct physical velocity near the spatial DC frequency, by *consistency* of the numerical scheme with the wave equation[2].

A.2 Finite Difference Schemes for the (2 + 1)D Wave Equation

Waveguide meshes of rectilinear [267], interpolated rectilinear [213], triangular [213, 269], and hexagonal [269] forms have all been applied to solve the (2 + 1)D wave equation. Though they have often been written as scattering forms, they can also be recast as finite difference schemes. There are quite a few computational issues that arise, which serve to distinguish between these difference schemes. Among them are the density of grid points, the possibility of decomposing a given scheme into more computationally efficient subschemes, the operation count, spectral characteristics, the ease with which boundary conditions can be implemented, as well as the maximum allowable time step. It is, of course, impossible to say which is best, without knowing problem specifics. The following is intended partly as a catalog as well as an indication of certain features that probably deserve more attention, and in particular, the distinction between passivity and stability, which becomes apparent in the cases of the triangular and interpolated meshes.

All the schemes to be discussed here are explicit; though implicit schemes are not nearly as attractive for hyperbolic problems (such as the wave equation) as for parabolic problems, they may have some advantages, especially in terms of the reduction of numerical dispersion. For more information on such schemes as applied to the wave equation in one and two spatial dimensions, we refer to [27, 38].

It is worthwhile introducing two new quantities at this point. In addition to Δ, the 'nearest-neighbor' grid spacing or interjunction spacing, T the time step, v_0, which will always be equal to Δ/T, and $\lambda = \gamma/v_0$, we also define ρ_S, the *computational density* of a particular scheme S to be the number of grid points at which the difference scheme is operative, per unit volume and per unit time. Thus, if the N-dimensional volume of the spatial domain $\mathcal{D}$ of a particular problem is $|\mathcal{D}|$ and the total time over which we wish to obtain a solution is $\mathcal{T}$, then the total number of grid point calculations that will need to be made will be $|\mathcal{D}|\mathcal{T}\rho_S$. Similarly, we can define the *add density* σ_S to be $A_S\rho_S$ if scheme S requires A_S adds in order to update at any given grid point. A multiply density could be defined similarly, though we will not, for reasons of space, do so here.

A.2.1 The Rectilinear Scheme

The finite difference scheme corresponding to a rectilinear mesh is obtained by applying centered differences to the wave equation over a rectangular grid with indices i and j (which

---

[2]Regrettably, a full discussion of consistency of difference schemes would take us too far afield, and we refer to the literature [238] for a full exposition. The idea, grossly speaking, is that for a stable difference scheme, consistency is our guarantee that the numerical solution to the difference scheme converges to the solution of the continuous model problem as the grid spacing and time step are decreased. It is usually checked via a Taylor expansion of the difference scheme.

refer to points with spatial coordinates $x = i\Delta$ and $y = j\Delta$). The difference scheme, given originally as (4.54), is

$$U_{i,j}(n+1) + U_{i,j}(n-1) = \lambda^2 \Big(U_{i+1,j}(n) + U_{i-1,j}(n) + U_{i,j+1}(n) + U_{i,j-1}(n)\Big)$$
$$+ \Big(2 - 4\lambda^2\Big) U_{i,j}(n)$$
(A.13)

and the amplification polynomial equation is of the form (A.5), with

$$B_\beta = -2\Big(1 + \lambda^2 \big(\cos(\beta_x \Delta) + \cos(\beta_y \Delta) - 2\big)\Big)$$

for $\boldsymbol{\beta} = [\beta_x, \beta_y]^T$. From (A.7), we thus have

$$F_\beta = \cos(\beta_x \Delta) + \cos(\beta_y \Delta) - 2$$

and we have

$$\max_\beta F_\beta = 0 \qquad \min_\beta F_\beta = -4$$

Condition (A.8) is thus satisfied, and condition (A.9) gives the bound

$$\lambda \leq \frac{1}{\sqrt{2}} \qquad \text{(for stability)}$$

which implies that the amplification factor $|G_{\beta,\pm}| = 1$ for such values of λ. Because $\lambda = \gamma/v_0$, this bound is the same as the bound for passivity of the associated mesh scheme, given in (4.64). The amplification factors, however, are distinct at all spatial frequencies only for $\lambda < 1/\sqrt{2}$. If $\lambda = 1/\sqrt{2}$, then the factors are degenerate for $\beta_x = \beta_y = 0$ and for $\beta_x = \beta_y = \pm\pi/\Delta$, and we are then in the situation discussed in Section A.1.2 in which linear growth of the solution may occur. This is an important special case, because it corresponds to the standard finite difference scheme for the rectilinear waveguide mesh (i.e., the realization without self-loops). The waveguide mesh implementation does not allow such growth at these frequencies[3].

As far as assessing the computational requirements of the finite difference scheme, first consider the case $\lambda < 1/\sqrt{2}$. Five adds are required at each grid point in order to update. Given that $T = \Delta/v_0$, we can write the computational and add densities for the scheme as

$$\rho_{\text{rect}} = \frac{v_0}{\Delta^3} \qquad \sigma_{\text{rect}} = \frac{5v_0}{\Delta^3} \qquad \text{for} \qquad v_0 > \sqrt{2}\gamma$$

For $\lambda = 1/\sqrt{2}$, however, scheme (A.13) simplifies to

$$U_{i,j}(n+1) + U_{i,j}(n-1) = \tfrac{1}{2}\Big(U_{i+1,j}(n) + U_{i-1,j}(n) + U_{i,j+1}(n) + U_{i,j-1}(n)\Big) \quad \text{(A.14)}$$

---

[3] As an example of such growth at the spatial DC frequency, consider initializing the scheme (A.14) using $U_{i,j}(0) = 1$ for $i + j$ even and $U_{i,j}(1) = -1$ for $i + j$ odd. Then we will have $U_{i,j}(n) = 2n - 1$, for $i + j + n$ even. It is simple to show that a waveguide implementation does not allow us to choose bounded wave variable initial conditions that yield these values for $U_{i,j}(0)$ and $U_{i,j}(1)$.

A.2. FINITE DIFFERENCE SCHEMES FOR THE (2 + 1)D WAVE EQUATION

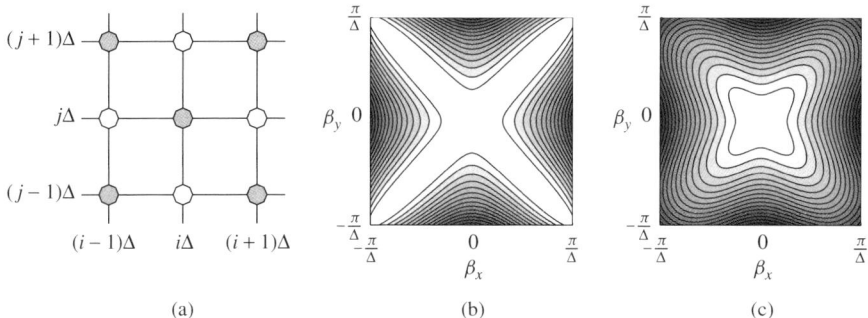

Figure A.1: *The rectilinear scheme (A.13)—(a) grid, of spacing Δ, where gray/white coloring indicates a subgrid decomposition possible when $\lambda = 1/\sqrt{2}$. (b) $v_{\beta,\text{phase}}/\gamma$ for $\lambda = 1/\sqrt{2}$. Contour lines are drawn, representing successive deviations of 2% from the ideal value of 1, which is obtained at spatial DC. (c) $v_{\beta,\text{phase}}/\gamma$ away from the stability bound, for $\lambda = 1/2$.*

which may be operated on alternating grids. That is, $U_{i,j}(n)$ need only be calculated for $i + j + n$, even (or odd). The computational and add densities, for $\lambda = 1/\sqrt{2}$ are then

$$\rho^s_{\text{rect}} = \frac{v_0}{2\Delta^3} \qquad \sigma^s_{\text{rect}} = \frac{2v_0}{\Delta^3} \qquad \text{for} \qquad v_0 = \sqrt{2}\gamma$$

where we note that the reduced scheme (A.14) requires only four adds for updating at a given grid point; in addition, the multiplies by 1/2 may be accomplished in a fixed-point implementation by simple bit-shifting operations. The increased efficiency of this scheme must be weighed against the fact that because grid density is reduced, the scheme is now applicable over a smaller range of spatial frequencies. The numerical phase velocities of the schemes, at the stability limit, and away from it, at $\lambda = 1/2$, are plotted in Figure A.1. It is interesting to note that away from the stability limit, the numerical dispersion is somewhat less directionally dependent; this important factor may be useful from the point of view of frequency-warping techniques [213], which may be used to reduce numerical dispersion effects for schemes that are relatively directionally independent. This idea has been discussed in the waveguide mesh context (where self-loops will be present) in [237].

A.2.2 The Interpolated Rectilinear Scheme

This scheme, like the standard rectilinear scheme, is defined over a grid with indices i and j, for points with $x = i\Delta$ and $y = j\Delta$. Updating, in this case, at a given point, requires access to values of the grid function at the previous time step at nearest-neighbor grid points to the north, east, west, and south, as well as those to the northeast, northwest, southeast, and southwest, which are more distant by a factor of $\sqrt{2}$. The scheme is referred to as 'interpolated' [213] because it is derived as an approximation to a hypothetical (and nonrealizable) multidirectional difference scheme with minimally directionally dependent numerical dispersion. (It is perhaps more useful to think of the scheme as interpolating between two rectilinear schemes operating on grids with a relative angle of 45°.) The difference scheme will have the form

$$U_{i,j}(n+1) + U_{i,j}(n-1)$$
$$= \lambda^2 a \Big(U_{i,j+1}(n) + U_{i,j-1}(n) + U_{i+1,j}(n) + U_{i-1,j}(n) \Big)$$
$$+ \lambda^2 b \Big(U_{i+1,j+1}(n) + U_{i+1,j-1}(n) + U_{i-1,j+1}(n) + U_{i-1,j-1}(n) \Big) \quad \text{(A.15)}$$
$$+ \lambda^2 c U_{i,j}(n)$$

for constants a, b, and c which satisfy the constraints

$$a + 2b = 1 \qquad 4a + 4b + c = \frac{2}{\lambda^2} \quad \text{(A.16)}$$

for consistency with the wave equation. If $b = 0$, we get the standard rectilinear scheme, and if $a = 0$, we get a rectilinear scheme operating on a grid of spacing $\sqrt{2}\Delta$, which is rotated by 45° with respect to that of the standard scheme. This general form was put forth in [213], and the free parameter a may be adjusted to give a less directionally dependent numerical phase velocity; it may thus be used in conjunction with frequency-warping methods for reducing dispersion error. In general, the interpolated scheme cannot be decomposed into mutually exclusive subschemes.

It is possible to examine the stability of this method as in the previous case. We again have an amplification polynomial equation of the form of (A.5), with

$$B_\beta = -2\lambda^2 \Big(a(\cos(\beta_x \Delta) + \cos(\beta_y \Delta)) + (1-a)\cos(\beta_x \Delta)\cos(\beta_y \Delta) - 1 - a \Big) - 2$$

and thus

$$F_\beta = a\big(\cos(\beta_x \Delta) + \cos(\beta_y \Delta)\big) + (1-a)\cos(\beta_x \Delta)\cos(\beta_y \Delta) - 1 - a$$

Note that F_β is *multilinear* [4] in $\cos(\beta_x \Delta)$ and $\cos(\beta_y \Delta)$, so that any extrema must occur at the corners of the region in the spatial frequency plane defined by $|\beta_x| = |\beta_y| = 0$,

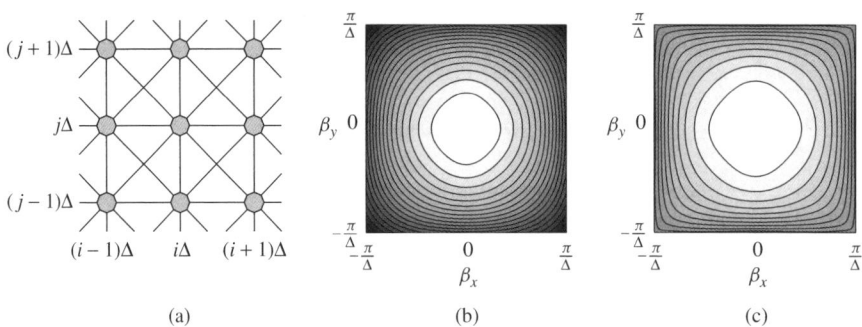

Figure A.2: *The interpolated rectilinear scheme* (A.15)—(a) *numerical grid and connections.* (b) $v_{\beta,\text{phase}}/\gamma$ *for $a = 0.62$ at the 'passivity' bound, $\lambda = 1/\sqrt{1+a}$.* (c) $v_{\beta,\text{phase}}/\gamma$ *for $a = 0.62$, at the stability bound, for $\lambda = 1/\sqrt{2a}$.*

A.2. FINITE DIFFERENCE SCHEMES FOR THE (2 + 1)D WAVE EQUATION

π/Δ. Thus, we need evaluate F_β only for $\boldsymbol{\beta}^T = [\beta_x, \beta_y] = [0, 0]$, $[\pi/\Delta, 0]$, $[0, \pi/\Delta]$, and $[\pi/\Delta, \pi/\Delta]$:

$$F_{\boldsymbol{\beta}^T=[0,0]} = 0 \qquad F_{\boldsymbol{\beta}^T=[\pi/\Delta,0]} = F_{\boldsymbol{\beta}^T=[0,\pi/\Delta]} = -2 \qquad F_{\boldsymbol{\beta}^T=[\pi/\Delta,\pi/\Delta]} = -4a$$

The global maximum of F_β is nonpositive (and thus condition (A.8) is satisfied) only if $a \geq 0$. The global minimum of F_β, over this range of a will then be

$$\min_\beta F_\beta = \begin{cases} -2, & 0 \leq a \leq \frac{1}{2} \\ -4a, & a \geq \frac{1}{2} \end{cases}$$

and the stability bound on λ will be

$$\lambda \leq \begin{cases} 1, & 0 \leq a \leq \frac{1}{2} \\ \frac{1}{\sqrt{2a}}, & a \geq \frac{1}{2} \end{cases} \qquad \text{(for von Neumann stability)} \qquad \text{(A.17)}$$

It is interesting to look at the interpolated scheme from a waveguide mesh point of view (see Chapter 4 for details). At each grid point we will have a nine-port parallel scattering junction: four connections are made to neighboring points to the north, south, east, and west, through unit-delay bidirectional delay lines of admittance Y_a; four more connections are made to the points to the northeast, southeast, northwest, and southwest using waveguides of admittance Y_b, and there will be a self-loop of admittance Y_c. If the junction voltage is written as $U_{i,j}(n)$, then the difference scheme corresponding to this waveguide mesh will be exactly (A.15), with

$$\lambda^2 a = \frac{2Y_a}{Y_J} \qquad \lambda^2 b = \frac{2Y_b}{Y_J} \qquad \lambda^2 c = \frac{2Y_c}{Y_J}$$

where the junction admittance Y_J (assumed positive) will be given by

$$Y_J = 4Y_a + 4Y_b + Y_c$$

The passivity condition will then be a condition on the positivity of Y_a, Y_b, and Y_c. From the previous discussion, we already require $a \geq 0$, so this ensures that $Y_a \geq 0$. Requiring $Y_b \geq 0$ is equivalent to requiring $b \geq 0$; from the first of constraints (A.16), this is true only for $a \leq 1$. Requiring $Y_c \geq 0$ is equivalent to requiring finally, from the second of constraints (A.16), that

$$\lambda \leq \frac{1}{\sqrt{1+a}}, \qquad 0 \leq a \leq 1 \qquad \text{(for passivity)}$$

The difference between the constraints for stability from (A.17) and the passivity constraint above is striking; these bounds are graphed in Figure A.3.

This is not the last time that we will find a discrepancy between von Neumann stability of a scheme and passivity of the related mesh structure; it will come up again in the following section during a discussion of the triangular scheme, and in Section A.3.3 when we look at the (3 + 1)D interpolated scheme. It is interesting to note that for a given value of a, with $0 \leq a \leq 1$, the numerical dispersion properties can always be improved if we are willing to forgo passivity (and a mesh implementation). We have plotted the numerical

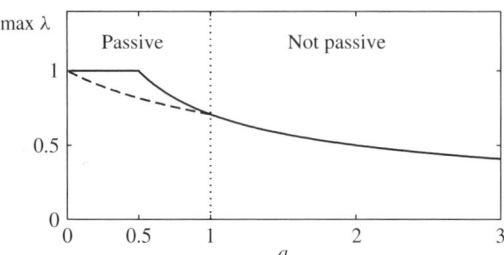

Figure A.3: *Stability bounds for the interpolated rectilinear scheme, as a function of the free parameter a. The solid line indicates the maximum value of λ for a given value of a, and the dashed line the maximum value of λ allowed in a passive waveguide mesh implementation. Note that there is a passive realization only for $0 \leq a \leq 1$.*

phase velocities of this scheme for $a = 0.62$, at both the stability limit and the passivity limit in Figure A.2.

Finally, we mention that the computational and add densities for this scheme will be, in general,

$$\rho_{\text{interp}} = \frac{v_0}{\Delta^3} \qquad \sigma_{\text{interp}} = \frac{10 v_0}{\Delta^3}$$

over the range of v_0 allowed by the stability constraint (A.17). For the scheme at the passivity bound (for $\lambda = 1/\sqrt{1+a}$, with $0 < a < 1$), we have

$$\rho_{\text{interp}}^p = \frac{\gamma \sqrt{1+a}}{\Delta^3} \qquad \sigma_{\text{interp}}^p = \frac{9\gamma \sqrt{1+a}}{\Delta^3}$$

We recall that for $a = 0$ or $a = 1$, at the stability limit, we again have the standard rectilinear scheme for which a grid decomposition is possible; this was discussed in the previous section.

Optimally Direction-independent Numerical Dispersion

Although the choice of the free parameter a which gives a maximally direction-independent numerical dispersion profile has been made in the past through computerized optimization procedures [213], we note here that it is possible to make a theoretical choice as well, based on a Taylor series expansion of the spectrum.

The spectral amplification factors for the interpolated scheme can be written in terms of the function B_β, or, equivalently, in terms of the function F_β. It should be clear, then, that if F_β is directionally independent, then so are the amplification factors and thus the numerical phase velocity (see Section A.1.4) as well. Ideally, we would like F_β to be a function of the spectral radius $\|\beta\|_2 = (\beta_x^2 + \beta_y^2)^{1/2}$ alone. Now examine the Taylor expansion of F_β about $\beta = 0$:

$$F_\beta = -\Delta^2 \|\beta\|_2^2 + \Delta^4 \left(\frac{1}{4!} \left(\beta_x^4 + \beta_y^4 \right) + \frac{1-a}{4} \beta_x^2 \beta_y^2 \right) + O(\Delta^6)$$

The directionally independent $O(\Delta^2)$ term reflects the fact that the scheme is consistent with the wave equation; higher-order terms in general show directional dependence. The

A.2. FINITE DIFFERENCE SCHEMES FOR THE (2 + 1)D WAVE EQUATION

choice of $a = 2/3$, discussed also in [252], however, gives

$$F_\beta = -\Delta^2 \|\beta\|_2^2 + \frac{1}{4!}\Delta^4 \|\beta\|_2^4 + O(\Delta^6) \qquad \text{for} \quad a = 2/3$$

and the directional dependence is confined to higher-order powers of Δ. Thus, for this choice of a, the numerical scheme is maximally direction-independent about spatial DC. Note that this value of a does fall within the required bounds for a passive waveguide mesh implementation. The value of 0.62 (for which the numerical dispersion profile is plotted in Figure A.2), which is very close to 2/3, was chosen by visual inspection of dispersion profiles for various values of a.

A.2.3 The Triangular Scheme

The simplest difference scheme that can be used to solve the wave equation on a triangular grid, and which corresponds to the waveguide mesh discussed in Section 5.1.1 in the constant-coefficient case, is given by

$$U_{i,j}(n+1) + U_{i,j}(n-1) = \tfrac{2}{3}\lambda^2 \Big(U_{i,j+2}(n) + U_{i,j-2}(n) + U_{i+1,j+1}(n) + U_{i+1,j-1}(n)$$

$$+ U_{i-1,j+1}(n) + U_{i-1,j-1}(n) \Big)$$

$$+ 2\left(1 - 2\lambda^2\right) U_{i,j}(n) \qquad (A.18)$$

for a grid defined by points at indices (i, j), for integer i and j such that $i + j$ is even. These coordinates refer to grid points at locations $x = \sqrt{3}i\Delta/2$ and $y = j\Delta/2$, so that a given grid point is equidistant from its six neighbors. This arrangement is shown in Figure A.4(a) and can be considered to be a rectilinear grid under a coordinate transformation; we refer to [261] for a discussion of the range of allowable spatial frequencies for such a grid. Difference schemes for the advection equation in $(2 + 1)$D operating on triangular grids, and the resulting numerical dispersion effects are discussed in detail in [290].

In this case, we will again have an amplification polynomial of the form (A.5), with

$$B_\beta = -2\left(1 + \frac{2}{3}\lambda^2\left(\cos(\beta_y\Delta) + 2\cos\left(\frac{\beta_y\Delta}{2}\right)\cos\left(\frac{\sqrt{3}\beta_x\Delta}{2}\right) - 3\right)\right)$$

$$F_\beta = \frac{2}{3}\left(\cos(\beta_y\Delta) + 2\cos\left(\frac{\beta_y\Delta}{2}\right)\cos\left(\frac{\sqrt{3}\beta_x\Delta}{2}\right) - 3\right)$$

Because F_β is not multilinear (see Section A.2.2) in the cosines, finding the extrema is not as simple as in the interpolated case—one can proceed either through some tedious algebra, change to stretched rectilinear coordinates, in which F_β becomes multilinear again, or make use of a computer. In any case, these extrema can be shown to be

$$\max_\beta F_\beta = 0 \qquad \min_\beta F_\beta = -3$$

and thus, from (A.9),

$$\lambda \leq \sqrt{\tfrac{2}{3}} \qquad \text{(for stability)}$$

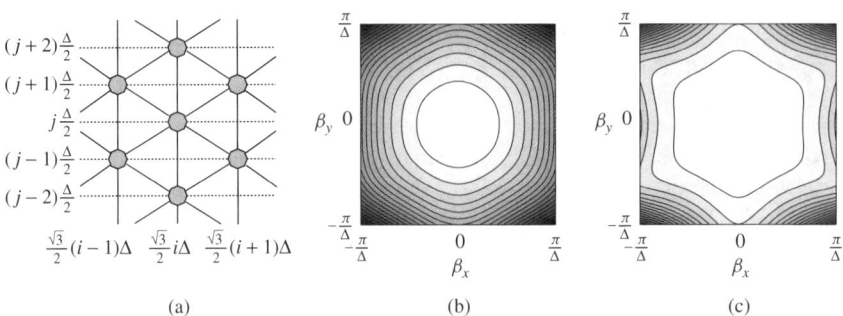

Figure A.4: *The triangular scheme (A.18)—(a) numerical grid and connections. (b) $v_{\beta,\text{phase}}/\gamma$ for the scheme at the passivity bound, $\lambda = 1/\sqrt{2}$. (c) $v_{\beta,\text{phase}}/\gamma$ at the stability bound, for $\lambda = \sqrt{2/3}$.*

This is surprising because the bound for passivity, from (5.5), of the triangular mesh is $\lambda \leq 1/\sqrt{2}$. That is to say, for a given interjunction spacing of Δ, a triangular waveguide mesh, of the type mentioned in Section 5.1.1, is concretely passive for time steps T with $T \leq \Delta/(\sqrt{2}\gamma)$. The corresponding difference equation, namely (A.18), is *stable* (in the sense of von Neumann), for $T \leq \sqrt{2}\Delta/(\sqrt{3}\gamma)$. The waveguide mesh can of course operate in a nonpassive mode for $1/\sqrt{2} < \lambda \leq \sqrt{2/3}$ (where we will require negative self-loop immittances, and will not have a simple positive definite energy measure for the network in terms of the wave quantities). The numerical dispersion characteristics of the scheme at the two bounds are considerably different, and are plotted in Figure A.4(b) and (c); the phase velocities are near the correct physical velocity over a much wider range of spatial frequencies at the stability bound, though the dispersion is also more directional.

The question that arises here is of the distinction between passive and stable numerical methods (this was also seen for the mesh for the transmission line equations in Section 4.3.6, as well as in the previous section on the interpolated rectilinear scheme). Is it always possible to find a passive realization of a stable numerical method? The discussion on the hexagonal mesh will help answer this question. To this end, we note that at the stability limit, we can rewrite B_β as

$$B_\beta = 2(1 - \tfrac{2}{9}|\psi_\beta|^2) \qquad \text{for} \qquad \lambda = \sqrt{\tfrac{2}{3}}$$

for a function ψ_β whose squared magnitude is given by

$$|\psi_\beta|^2 = 1 + 4\cos^2\left(\frac{\beta_y \Delta}{2}\right) + 4\cos\left(\frac{\beta_y \Delta}{2}\right)\cos\left(\frac{\sqrt{3}\beta_x \Delta}{2}\right)$$

The spectral amplification factors at the stability limit will then be, from (A.6),

$$G_{\beta,\pm} = -1 + \tfrac{2}{9}|\psi_\beta|^2 \pm \tfrac{2}{3}|\psi_\beta|\left(\tfrac{1}{9}|\psi_\beta|^2 - 1\right)^{\tfrac{1}{2}} \qquad (A.19)$$

For $\lambda = \sqrt{2/3}$ (its limiting value), the triangular scheme has the same potential for instability as the rectilinear scheme. Linear growth may occur for this scheme at the seven spatial frequency pairs

$$\beta^T = [0, 0], \quad [0, \pm 4\pi/3\Delta], \quad [2\pi/\sqrt{3}\Delta, \pm 2\pi/3\Delta], \quad [-2\pi/\sqrt{3}\Delta, \pm 2\pi/3\Delta]$$

The computational and add densities for the triangular scheme in general, and at the stability ($\lambda = \sqrt{2/3}$) and passivity bounds ($\lambda = 1/\sqrt{2}$) will be

$$\rho_{tri} = \frac{2v_0}{\sqrt{3}\Delta^3} \qquad \sigma_{tri} = \frac{14v_0}{\sqrt{3}\Delta^3}$$

$$\rho_{tri}^s = \frac{\sqrt{2}\gamma}{\Delta^3} \qquad \sigma_{tri}^s = \frac{7\sqrt{2}\gamma}{\Delta^3}$$

$$\rho_{tri}^p = \frac{2\sqrt{2}\gamma}{\sqrt{3}\Delta^3} \qquad \sigma_{tri}^p = \frac{4\sqrt{6}\gamma}{\Delta^3}$$

Here, we have taken into account the fact that at the passivity bound, we require one less add per point (in the waveguide mesh implementation, the self-loop disappears). We also mention that the triangular difference scheme is doubly pathological in the sense that not only do its passivity and stability regimes not coincide (and aside from the interpolated rectilinear schemes, it is the only scheme examined in this appendix that exhibits this behavior), but it also cannot be decomposed into even/odd mutually exclusive subschemes, as can all the other schemes to be discussed here (again, excepting the interpolated scheme). It seems reasonable to conjecture that these two 'symptoms' are related (somehow).

A.2.4 The Hexagonal Scheme

The hexagonal scheme is different from those previously discussed in that updating is not the same at every point on the grid. Indeed, one-half the grid points have a 'mirror-image' orientation with respect to the other half, as shown in Figure A.5(a). For this reason, we will take special care in the analysis of this system; first suppose that we have two grid functions $U_1(n)$ and $U_2(n)$ defined over the two subgrids (labeled 1 and 2, in Figure A.5). We index these two grid functions as $U_{1,i,j}(n)$ and $U_{2,i+2,j}(n)$, for i and j integer such that $i = 3m$, for integer m, and $j + i/3$ is even. $U_{1,i,j}(n)$ will serve as an approximation to some continuous function u_1 at the point ($x = \Delta i/2$, $y = \sqrt{3}j\Delta/2$, $t = nT$), and $U_{2,i+2,j}(n)$ will approximate a function u_2 at a point with coordinates ($x = \Delta i/2 + \Delta$, $y = \sqrt{3}j\Delta/2$, $t = nT$). As before, the distance between any grid point and its nearest neighbors (three in this case) is Δ. The difference scheme for the hexagonal waveguide mesh can then be written as the system

$$U_{1,i,j}(n+1) + U_{1,i,j}(n-1) = \tfrac{4}{3}\lambda^2 \Big(U_{2,i+2,j}(n) + U_{2,i-1,j+1}(n) + U_{2,i-1,j-1}(n) \Big)$$
$$+ 2\left(1 - 2\lambda^2\right) U_{1,i,j}(n) \qquad \text{(A.20a)}$$

$$U_{2,i+2,j}(n+1) + U_{2,i+2,j}(n-1) = \tfrac{4}{3}\lambda^2 \Big(U_{1,i,j}(n) + U_{1,i+3,j+1}(n) + U_{1,i+3,j-1}(n) \Big)$$
$$+ 2\left(1 - 2\lambda^2\right) U_{2,i+2,j}(n) \qquad \text{(A.20b)}$$

Consistency of (A.20) with the wave equation is not immediately apparent. We can check it as follows. First expand (A.20) in a Taylor series in terms of the continuous functions u_1 and u_2 to get

$$\left(T^2\frac{\partial^2}{\partial t^2} + 4\lambda^2\right)u_1 = \lambda^2\left(4 + \Delta^2\nabla^2\right)u_2$$

$$\left(T^2\frac{\partial^2}{\partial t^2} + 4\lambda^2\right)u_2 = \lambda^2\left(4 + \Delta^2\nabla^2\right)u_1$$

to $O(\Delta^4, T^4)$. This system can then be reduced to

$$\left(T^2\frac{\partial^2}{\partial t^2} + 4\lambda^2\right)^2 u = \lambda^4\left(4 + \Delta^2\nabla^2\right)^2 u$$

where u is either of u_1 or u_2. Discarding higher-order terms in T and Δ gives the wave equation.

In terms of the spatial Fourier spectra of the grid functions U_1 and U_2, we may write the differencing system (A.20) in the vector form of (A.10) with

$$\hat{U}_\beta = \begin{bmatrix} \hat{U}_{1,\beta} \\ \hat{U}_{2,\beta} \end{bmatrix} \qquad B_\beta = \begin{bmatrix} -2(1 - 2\lambda^2) & -\frac{4}{3}\lambda^2\psi_\beta \\ -\frac{4}{3}\lambda^2\psi_\beta^* & -2(1 - 2\lambda^2) \end{bmatrix}$$

where

$$\psi_\beta = e^{j\beta_x\Delta} + 2e^{-j\beta_x\Delta/2}\cos(\sqrt{3}\beta_y\Delta/2)$$

Because B_β is Hermitian, we can change variables so that the system is in the form of (A.11), with

$$\Lambda_\beta = \begin{bmatrix} -2(1 - 2\lambda^2) + \frac{4}{3}\lambda^2|\psi_\beta| & 0 \\ 0 & -2(1 - 2\lambda^2) - \frac{4}{3}\lambda^2|\psi_\beta| \end{bmatrix}$$

The necessary stability condition, from (A.12) will then be

$$\max_\beta | - 2(1 - 2\lambda^2) \pm \tfrac{4}{3}\lambda^2|\psi_\beta|| \leq 2 \qquad (A.21)$$

It is easy to check that $|\psi_\beta|$ takes on a maximum of 3 when $\beta_x = \beta_y = 0$, and is minimized for $\beta_x = 0$, $|\beta_y| = 4\pi/(3\sqrt{3}\Delta)$ and for $|\beta_x| = 2\pi/3$, $|\beta_y| = 2\pi/(3\sqrt{3}\Delta)$, where it takes on the value 0. It is then easy to show that we require $\lambda \leq 1/\sqrt{2}$ in order to satisfy (A.21). This coincides with the passivity bound, from (5.4).

An analysis of numerical dispersion is more complex in the vector case. Beginning from the uncoupled system defined by Λ_β, whose upper and lower diagonal entries we will call $\Lambda_{\beta,1}$ and $\Lambda_{\beta,2}$ respectively, we can see that we will thus have two pairs of spectral amplification factors, one for each uncoupled scalar equation. These will be given by

$$G_{\beta,1,\pm} = \frac{1}{2}\left(-\Lambda_{\beta,1} \pm \sqrt{\Lambda_{\beta,1}^2 - 4}\right) \qquad G_{\beta,2,\pm} = \frac{1}{2}\left(-\Lambda_{\beta,2} \pm \sqrt{\Lambda_{\beta,2}^2 - 4}\right)$$

A.2. FINITE DIFFERENCE SCHEMES FOR THE (2 + 1)D WAVE EQUATION

It is useful to check the values of the amplification factors at the spatial DC frequency, and at the stability bound, where we have $\Lambda_{\beta,1} = 2$, $\Lambda_{\beta,2} = -2$. At this frequency, the spectral amplification factors take on the values

$$G_{\beta=0,1,\pm} = -1 \qquad G_{\beta=0,2,\pm} = 1 \qquad (A.22)$$

Clearly, the pair of spectral amplification factors $G_{\beta=0,2,\pm}$ correctly represents wave propagation at spatial DC, but the factors $G_{\beta=0,1,\pm}$ will be responsible for *parasitic oscillations* [238] in the hexagonal scheme; they will not, in general, be overly problematic, since the energy allowed into such modes must vanish as the grid spacing Δ is decreased; this is a result of the consistency of the numerical scheme (A.20) with the wave equation, as was shown earlier in this subsection. In order to clarify this point, it is useful to examine the diagonalizing transformation defined by $\mathbf{J}_\beta$, which takes the Fourier-transformed hexagonal scheme in the form of (A.10) in the variable $\hat{\mathbf{U}}_\beta$, to that of (A.11) in $\hat{\mathbf{V}}_\beta$. At $\beta = 0$, and for $\lambda = 1/\sqrt{2}$, we have

$$\mathbf{B}_{\beta=0} = \begin{bmatrix} 0 & -2 \\ -2 & 0 \end{bmatrix} \qquad \mathbf{\Lambda}_{\beta=0} = \begin{bmatrix} 2 & 0 \\ 0 & -2 \end{bmatrix} \qquad \mathbf{J}_{\beta=0} = \frac{1}{\sqrt{2}} \begin{bmatrix} -1 & 1 \\ 1 & 1 \end{bmatrix}$$

and thus $\hat{V}_{1,\beta=0} = (-\hat{U}_{1,\beta=0} + \hat{U}_{2,\beta=0})/\sqrt{2}$ and $\hat{V}_{2,\beta=0} = (\hat{U}_{1,\beta=0} + \hat{U}_{2,\beta=0})/\sqrt{2}$. Because scheme (A.20) is consistent with the wave equation, for any reasonable choice of initial conditions, we must have that $\hat{U}_{1,\beta=0} \approx \hat{U}_{2,\beta=0}$, as Δ becomes small. Thus $\hat{V}_{1,\beta=0}$, the component of the numerical solution whose spectral amplification is governed by the parasitic factor $G_{\beta=0,1,\pm}$, must vanish in this limit as well.

The computational and add densities, for the general scheme (A.20) and at the stability limit for $\lambda = 1/\sqrt{2}$, will be given by

$$\rho_{\text{hex}} = \frac{4v_0}{3\sqrt{3}\Delta^3} \qquad \sigma_{\text{hex}} = \frac{16v_0}{3\sqrt{3}\Delta^3}$$

$$\rho_{\text{hex}}^s = \frac{2\sqrt{2}\gamma}{3\sqrt{3}\Delta^3} \qquad \sigma_{\text{hex}}^s = \frac{2\sqrt{2}\gamma}{\sqrt{3}\Delta^3}$$

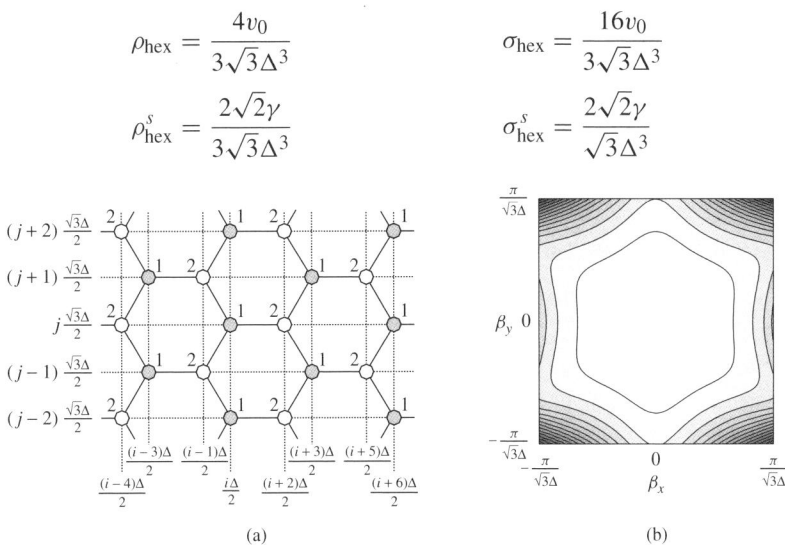

Figure A.5: *The hexagonal scheme (A.20)—(a) numerical grid and connections, in which gray/white coloring of points indicates a division into mutually exclusive subschemes at the stability bound. (b) $v_{\beta,\text{phase}}/\gamma$ for the scheme at the passivity bound, $\lambda = 1/\sqrt{2}$, for the dominant mode.*

As in the rectilinear scheme, we have used the fact that the hexagonal scheme decouples into two independent subschemes at the stability limit.

One other point is worthy of comment. Consider again the vector equation that describes the time evolution of the spatial spectra for the hexagonal scheme, which, in diagonalized form, is exactly (A.11). At the stability limit, then, for $\lambda = 1/\sqrt{2}$, we will have

$$\Lambda_\beta = \begin{bmatrix} \frac{2}{3}|\psi_\beta| & 0 \\ 0 & -\frac{2}{3}|\psi_\beta| \end{bmatrix}$$

Let us examine the second uncoupled subsystem. From (A.22), the spectral amplification factors will then be

$$G_{\beta,2,\pm} = \tfrac{1}{3}|\psi_\beta| \pm \left(\tfrac{1}{9}|\psi_\beta|^2 - 1\right)^{\frac{1}{2}}$$

It is of interest to see the effect of the amplification factors after *two* time steps; these will simply be the squares of $G_{\beta,2,\pm}$, which are

$$G_{\beta,2,\pm}^2 = -1 + \tfrac{2}{9}|\psi_\beta|^2 \pm \tfrac{2}{3}|\psi_\beta|\left(\tfrac{1}{9}|\psi_\beta|^2 - 1\right)^{\frac{1}{2}} \tag{A.23}$$

The important point here is that the *two-step* spectral amplification factors for scheme (A.20) are identical to the one-step factor for the triangular scheme with grid spacing $\sqrt{3}\Delta$ at its own stability limit; these factors were given in (A.19). This is perhaps not surprising, given that, from Figure A.5(a), it is clear that either of the two subgrids for the hexagonal scheme forms a triangular grid of spacing $\sqrt{3}\Delta$. What is surprising is that a triangular waveguide mesh at the stability limit is *not a concretely passive structure* (see previous section). That is to say, it will still operate stably (in the von Neumann sense) but will require negative self-loop immittances. Thus, a hexagonal waveguide mesh, at its passivity/stability bound can be seen as a *passive realization* of the stable difference scheme on a triangular grid. The question as to whether there is always a passive realization for any stable difference scheme remains open[4].

A.2.5 Note on Higher-order Accuracy

All the schemes we have presented here are second-order accurate. In the last 20 years, there has been intense research into difference schemes that possess greater accuracy, at the expense of larger computational stencils. This problem has been approached in various ways. One, perhaps the most conventional, is to look for schemes that are of greater accuracy in the sense of truncation error (i.e., in maximizing accuracy in the limit of vanishing wavenumbers, or around the spatial DC). One interesting way of approaching this is by means of what is called the 'modified equation' approach [256, 257, 221], which relies on delicate cancellations of time and space truncation errors. Another approach is to sacrifice the truncation error measure in favor of a more even distribution of dispersion over the entire range of wavenumbers [291, 289]. Then, of course, there are what are known as spectral methods, which can be thought of as a limiting case of high accuracy as the computational

---

[4] We consider this to be the single most important issue raised in this book.

stencil becomes large; this is a huge field of its own—see the recent books by Trefethen [254] and Fornberg [97] for an overview of, and introduction to spectral methods (in these cases, more properly, pseudospectral methods).

Higher-order accuracy in scattering methods such as DWNs is an open problem; though we have shown that it is possible to obtain higher-order spatial accuracy within a scattering structure (see Section 3.12 and Section 6.3), time accuracy would appear to be limited to second order.

A.3 Finite Difference Schemes for the (3 + 1)D Wave Equation

We now look at several difference schemes that solve the wave equation in (3 + 1)D, and in particular, schemes that operate on a rectilinear grid; all the schemes that have appeared in the DWN literature are of this type. We will pay special attention to the interpolated scheme, for which the requirements for stability and passivity become even more distinct than they were in the (2 + 1)D case (see Section A.2.2).

A.3.1 The Cubic Rectilinear Scheme

This is the simplest scheme for the (3 + 1)D wave equation. The grid points, indexed by i, j, and k are located at coordinates $(x, y, z) = (i\Delta, j\Delta, k\Delta)$. The finite difference scheme is written as

$$U_{i,j,k}(n+1) + U_{i,j,k}(n-1) = \lambda^2 \Big(U_{i+1,j,k}(n) + U_{i-1,j,k}(n) + U_{i,j+1,k}(n) + U_{i,j-1,k}(n)$$

$$+ U_{i,j,k+1}(n) + U_{i,j,k-1}(n)\Big)$$

$$+ \left(2 - 6\lambda^2\right) U_{i,j,k}(n) \qquad (A.24)$$

If the grid points are located at the corners of a cubic lattice, then updating the scheme requires access to the grid function at the six neighboring corners; see Figure A.6(a). The stability analysis is very similar to that of the (2 + 1)D rectilinear scheme, except that we now have a 3-tuple of spatial frequencies, $\boldsymbol{\beta} = [\beta_x, \beta_y, \beta_z]^T$. The amplification polynomial equation is again of the form of (A.5), with

$$B_\beta = -2\left(1 + \lambda^2 \left(\cos(\beta_x \Delta) + \cos(\beta_y \Delta) + \cos(\beta_x \Delta) - 3\right)\right)$$

and thus

$$F_\beta = \cos(\beta_x \Delta) + \cos(\beta_y \Delta) + \cos(\beta_x \Delta) - 3$$

Because F_β is multilinear in the cosines, it is simple to show that

$$\max_\beta F_\beta = 0 \qquad \min_\beta F_\beta = -6$$

and so, from (A.9),

$$\lambda \leq \tfrac{1}{\sqrt{3}} \qquad \text{(for von Neumann stability)}$$

APPENDIX A. FINITE DIFFERENCE SCHEMES FOR THE WAVE EQUATION

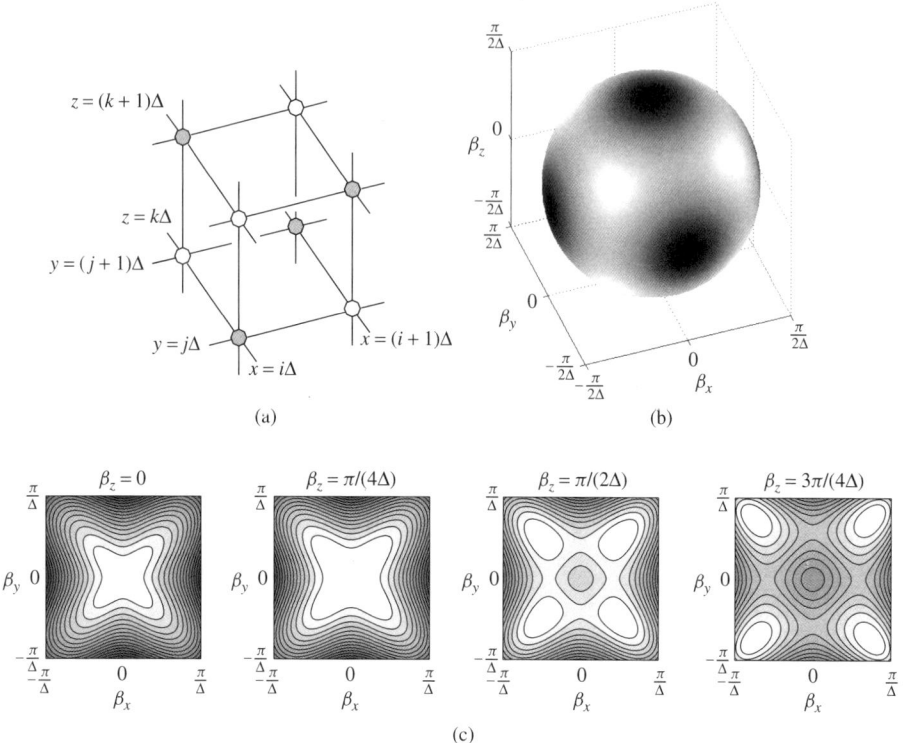

Figure A.6: *The cubic rectilinear scheme* (A.24)—(a) *numerical grid and connections, where gray/white coloring of points indicates a division into mutually exclusive subschemes at the stability bound.* (b) $v_{\beta,\text{phase}}/\gamma$ *for the scheme at the stability bound* $\lambda = 1/\sqrt{3}$, *for a spherical surface with* $\|\boldsymbol{\beta}\|_2 = \pi/(2\Delta)$—*the shading is normalized over the surface so that white corresponds to no dispersion error, and black to the maximum error over the surface (which is 7% in this case).* (c) *Contour plots of* $v_{\beta,\text{phase}}/\gamma$ *for various cross sections of the space of spatial frequencies* $\boldsymbol{\beta}$; *contours indicate successive deviations of 2% from the ideal value of 1, which is obtained at spatial DC.*

When $\lambda = 1/\sqrt{3}$, the amplification factors become degenerate and linear growth of the solution may occur for $\beta_x = \beta_y = \beta_z = 0$ and for $|\beta_x| = |\beta_y| = |\beta_z| = \pi/\Delta$. The computational and add densities are

$$\rho_{\text{cub}} = \frac{v_0}{\Delta^4} \qquad \sigma_{\text{cub}} = \frac{7v_0}{\Delta^4}$$

for $v_0 > \sqrt{3}\gamma$, and

$$\rho^s_{\text{cub}} = \frac{\gamma}{2\Delta^4} \qquad \sigma^s_{\text{cub}} = \frac{3\gamma}{\Delta^4}$$

at the stability limit $v_0 = \sqrt{3}\gamma$. At this limit, the scheme may, like the $(2+1)$D scheme, be divided into two mutually exclusive subschemes. See Figure A.6(b) and (c) for plots of the numerical dispersion properties of the cubic rectilinear scheme.

A.3.2 The Octahedral Scheme

The grid for an octahedral scheme is constructed from two superimposed rectilinear grids; if the points of the first grid are located at cube corners, then the points of the second will occur at the centers of the cubes defined by the first. The relevant difference scheme on an octahedral grid can be written as

$$U_{i,j,k}(n+1) + U_{i,j,k}(n-1) = \tfrac{3}{4}\lambda^2 \big(U_{i-1,j+1,k+1}(n) + U_{i+1,j+1,k+1}(n)$$
$$+ U_{i-1,j-1,k+1}(n) + U_{i+1,j-1,k+1}(n)$$
$$+ U_{i-1,j-1,k-1}(n) + U_{i-1,j+1,k-1}(n) \quad (A.25)$$
$$+ U_{i+1,j+1,k-1}(n) + U_{i+1,j-1,k-1}(n)\big)$$
$$+ \big(2 - 8\lambda^2\big) U_{i,j,k}(n)$$

for i, j, and k, which are either all even or all odd integers. Now, we have taken the spacing between nearest neighbors to be Δ, so the indices i, j, and k refer to a point with coordinates $x = i\Delta/\sqrt{3}$, $y = j\Delta/\sqrt{3}$, and $z = k\Delta/\sqrt{3}$. The amplification polynomial equation is again of the form (A.5), with

$$B_\beta = -2\left(1 + 3\lambda^2 \left(\cos\left(\frac{\beta_x \Delta}{\sqrt{3}}\right) \cos\left(\frac{\beta_y \Delta}{\sqrt{3}}\right) \cos\left(\frac{\beta_x \Delta}{\sqrt{3}}\right) - 1\right)\right)$$

and

$$F_\beta = 3\left(\cos\left(\frac{\beta_x \Delta}{\sqrt{3}}\right) \cos\left(\frac{\beta_y \Delta}{\sqrt{3}}\right) \cos\left(\frac{\beta_x \Delta}{\sqrt{3}}\right) - 1\right)$$

and it is again easy to determine that

$$\max_\beta F_\beta = 0 \qquad \min_\beta F_\beta = -6$$

which are the same as the bounds in the cubic rectilinear case. We again have that

$$\lambda \leq \tfrac{1}{\sqrt{3}} \qquad \text{(for von Neumann stability)}$$

Thus, the stability bound coincides with the passivity bound for the mesh implementation. For $\lambda = 1/\sqrt{3}$, instabilities may appear at any spatial frequency triplets $\beta = [\beta_x, \beta_y, \beta_z]^T$ where each component is either 0 or $\pm\sqrt{3}\pi/\Delta$.

The computational and add densities are given by

$$\rho_{\text{oct}} = \frac{3\sqrt{3}\nu_0}{4\Delta^4} \qquad \sigma_{\text{oct}} = \frac{27\sqrt{3}\nu_0}{4\Delta^4} \qquad \text{for} \qquad \nu_0 > \sqrt{3}\gamma$$

$$\rho_{\text{oct}}^s = \frac{9\gamma}{8\Delta^4} \qquad \sigma_{\text{oct}}^s = \frac{9\gamma}{\Delta^4} \qquad \text{for} \qquad \nu_0 = \sqrt{3}\gamma$$

At the stability limit, the scheme can be divided into two mutually exclusive subschemes; plots of numerical dispersion are shown in Figure A.7(b) and (c). It is interesting to note

318 APPENDIX A. FINITE DIFFERENCE SCHEMES FOR THE WAVE EQUATION

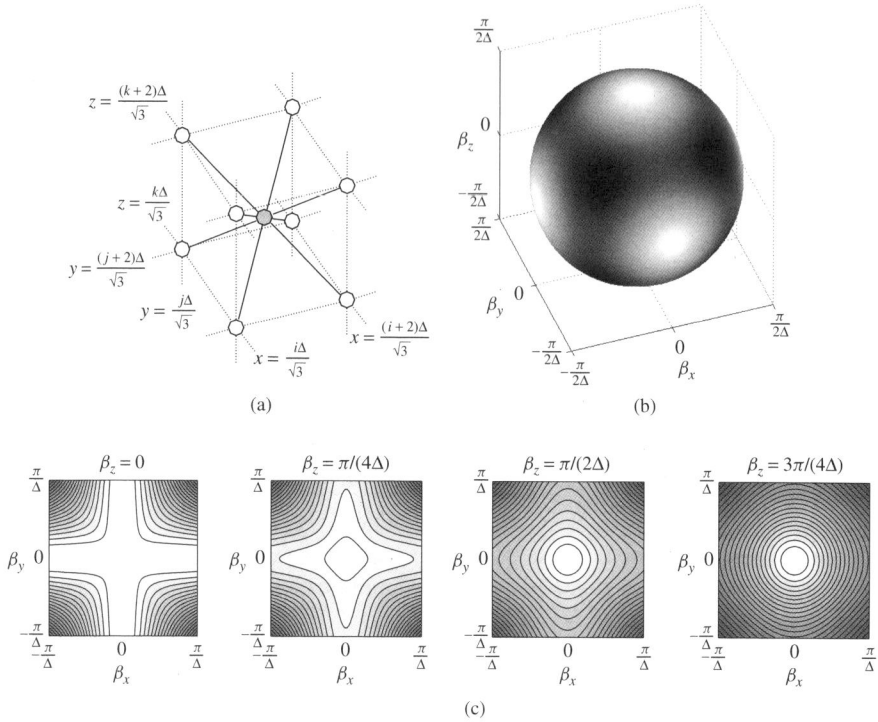

Figure A.7: *The octahedral scheme (A.25)—(a) numerical grid and connections, in which gray/white coloring of points indicates a division into mutually exclusive subschemes at the stability bound. (b) $v_{\beta,\text{phase}}/\gamma$ for the scheme at the stability bound $\lambda = 1/\sqrt{3}$, for a spherical surface with $\|\boldsymbol{\beta}\|_2 = \pi/(2\Delta)$—the shading is normalized over the surface so that white corresponds to no dispersion error, and black to the maximum error over the surface (which is 5% in this case). (c) Contour plots of $v_{\beta,\text{phase}}/\gamma$ for various cross-sections of the space of spatial frequencies $\boldsymbol{\beta}$; contours indicate successive deviations of 2% from the ideal value of 1, which is obtained at spatial DC.*

that there is no dispersion error along the six axial directions; this should be compared with the cubic rectilinear scheme, for which wave propagation is dispersionless along the diagonal directions (there are eight such directions).

A.3.3 The $(3+1)$D Interpolated Rectilinear Scheme

In the interest of achieving a more uniform numerical dispersion profile in $(3+1)$D, it is of course possible to define an interpolated scheme [210, 214], in the same way as was done in $(2+1)$D in Section A.2.2. We will again have a two-step scheme, and updating at a given grid point is performed with reference to, at the previous time step, the grid point at the same location as well as the 26 nearest neighbors: the six points a distance Δ away, 12 points at a distance of $\sqrt{2}\Delta$, and eight points that are $\sqrt{3}\Delta$ away (see Figure A.9(a)). We present here a complete analysis of the relevant stability conditions, as well as the conditions under which a waveguide mesh implementation exists. We also look at a means of minimizing directional dependence of numerical dispersion.

A.3. FINITE DIFFERENCE SCHEMES FOR THE (3 + 1)D WAVE EQUATION

Like the cubic rectilinear and octahedral schemes, this scheme will be defined over a rectilinear grid indexed by i, j, and k and will have the general form

$$U_{i,j,k}(n+1) + U_{i,j,k}(n-1) = \lambda^2 a\Big(U_{i+1,j,k}(n) + U_{i-1,j,k}(n) + U_{i,j+1,k}(n)$$
$$+ U_{i,j-1,k}(n) + U_{i,j,k+1}(n) + U_{i,j,k-1}(n)\Big)$$
$$+ \lambda^2 b\Big(U_{i+1,j+1,k}(n) + U_{i+1,j-1,k}(n) + U_{i-1,j+1,k}(n)$$
$$+ U_{i-1,j-1,k}(n) + U_{i+1,j,k+1}(n) + U_{i-1,j,k+1}(n)$$
$$+ U_{i,j+1,k+1}(n) + U_{i,j-1,k+1}(n) + U_{i+1,j,k-1}(n)$$
$$+ U_{i-1,j,k-1}(n) + U_{i,j+1,k-1}(n) + U_{i,j-1,k-1}(n)\Big)$$
$$+ \lambda^2 c\Big(U_{i+1,j+1,k+1}(n) + U_{i+1,j+1,k-1}(n)$$
$$+ U_{i-1,j-1,k+1}(n) + U_{i-1,j-1,k-1}(n)$$
$$+ U_{i+1,j-1,k+1}(n) + U_{i+1,j-1,k-1}(n)$$
$$+ U_{i-1,j+1,k+1}(n) + U_{i-1,j+1,k-1}(n)\Big)$$
$$+ \lambda^2 d U_{i,j}(n) \tag{A.26}$$

In order for scheme (A.26) to satisfy the wave equation, we require the constants a, b, c, and d to satisfy the constraints

$$c = \frac{1 - a - 4b}{4} \qquad\qquad d = \frac{2}{\lambda^2} - 4a - 4b - 2 \tag{A.27}$$

and a family of difference schemes parameterized by a, b, and λ results.

The stability analysis of this scheme proceeds along the same lines as that of the $(2+1)$D scheme, though as we shall see, the stability condition on the parameters a and b is considerably more complex. As before, we have an amplification polynomial of the form of (A.5), now with

$$F_\beta = a\big(\cos(\beta_x \Delta) + \cos(\beta_y \Delta) + \cos(\beta_z \Delta)\big)$$
$$+ 2b\big(\cos(\beta_x \Delta)\cos(\beta_y \Delta) + \cos(\beta_x \Delta)\cos(\beta_z \Delta) + \cos(\beta_y \Delta)\cos(\beta_z \Delta)\big)$$
$$+ (1 - a - 4b)\cos(\beta_x \Delta)\cos(\beta_y \Delta)\cos(\beta_z \Delta) - 2a - 2b - 1$$

where, as before, $B_\beta = -2\lambda^2 F_\beta - 2$. Because F_β is again multilinear in the three cosines, its extrema can only occur at the eight corners of the spectral 'cube.' These extrema are

$$F_{\beta^T=[0,0,0]} = 0$$
$$F_{\beta^T=[\pi/\Delta,0,0]} = F_{\beta^T=[0,\pi/\Delta,0]} = F_{\beta^T=[0,0,\pi/\Delta]} = -2$$
$$F_{\beta^T=[\pi/\Delta,\pi/\Delta,0]} = F_{\beta^T=[\pi/\Delta,0,\pi/\Delta]} = F_{\beta^T=[\pi/\Delta,\pi/\Delta,0]} = -4a - 8b$$
$$F_{\beta^T=[\pi/\Delta,\pi/\Delta,\pi/\Delta]} = -4a + 8b - 2$$

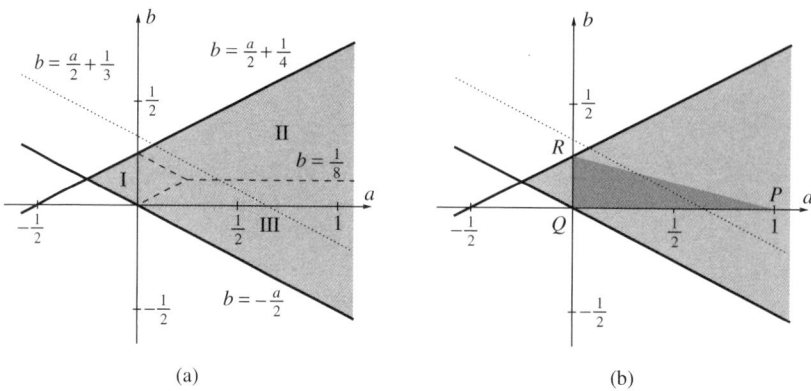

Figure A.8: (a) *Stability region, in gray, for the interpolated rectilinear scheme, plotted in the (a, b) plane. This region can be divided into three subregions, labeled I, II, and III separated by dashed lines, over which different stability conditions on λ apply. In region I, we must have $\lambda \leq 1$, in region II $\lambda \leq 1/\sqrt{2a+4b}$, and in region III $\lambda \leq 1/\sqrt{2a-4b+1}$. The dotted line indicates choices of a and b for which numerical dispersion is optimally direction independent.* (b) *The subset of stable schemes for which a passive waveguide mesh implementation exists is shown in dark gray. Over this region, we require $\lambda \leq 1/\sqrt{2a+2b+1}$. This bound is more strict than the stability conditions mentioned above in the same region. We also remark that this interpolated scheme reduces to other simpler schemes under particular choices of a and b. At point P, we have the cubic rectilinear scheme (see Section A.3.1), at point Q, we have the octahedral scheme (see Section A.3.2), and at point R, we have what might be called a 'dodecahedral' scheme. Notice, in particular, that none of these schemes is optimally direction-independent (i.e., P, Q, and R do not lie on the dotted line).*

The nonpositivity requirement on F_β then amounts to requiring that these extreme values be nonpositive. The resulting stability region in the (a, b) plane is shown in gray in Figure A.8(a).

Assuming that a and b fall in this region, we must now find the values of λ which satisfy (A.9). The minimum value of F_β depends on a and b in a nontrivial way; referring to Figure A.8(a), the stability domain can be divided into three regions, and in each there is a different closed form expression for the upper bound on λ. These bounds are given explicitly in the caption to Figure A.8(a).

In order to examine the directional dependence of the dispersion error, we may expand F_β in a Taylor series about $\boldsymbol{\beta} = \mathbf{0}$, as was done in the (2 + 1)D case. We have

$$F_\beta = -\Delta^2 \|\boldsymbol{\beta}\|_2^2 + \Delta^4 \left(\frac{1}{4!} \left(\beta_x^4 + \beta_y^4 + \beta_z^4 \right) \right.$$
$$\left. + \frac{1-a-2b}{4} \left(\beta_x^2 \beta_y^2 + \beta_x^2 \beta_z^2 + \beta_y^2 \beta_z^2 \right) \right) + O(\Delta^6)$$

which implies that

$$F_\beta = -\Delta^2 \|\boldsymbol{\beta}\|_2^2 + \Delta^4 \tfrac{1}{4!} \|\boldsymbol{\beta}\|_4^4 + O(\Delta^6) \qquad \text{for } b = -a/2 + 1/3$$

A.3. FINITE DIFFERENCE SCHEMES FOR THE (3 + 1)D WAVE EQUATION

and the dispersion error is directionally independent to fourth order. This special choice of the parameters a and b is plotted as a dotted line in Figure A.8(a). It is well worth comparing this optimization method with the computer-based techniques applied to the same problem [214].

The computational and add densities for the scheme will be

$$\rho_{\text{3Dinterp}} = \frac{v_0}{\Delta^4} \qquad \sigma_{\text{3Dinterp}} = \frac{27 v_0}{\Delta^4}$$

Considerable computational savings are possible if any of a, b, c, or d is zero.

Finally, we remark that the $(3+1)$D interpolated scheme can be realized as a waveguide mesh, in which, at any given junction, we will have four types of waveguide connections: those of admittances Y_a, Y_b, and Y_c are connected to the neighboring junctions located at grid points at distances Δ, $\sqrt{2}\Delta$, and $\sqrt{3}\Delta$ away respectively, and a self-loop of admittance Y_d is also connected to every junction. We end up with exactly difference scheme (A.26), with

$$\lambda^2 a = \frac{2Y_a}{Y_J} \quad \lambda^2 b = \frac{2Y_b}{Y_J} \quad \lambda^2 c = \frac{2Y_c}{Y_J} \quad \lambda^2 d = \frac{2Y_d}{Y_J} \quad Y_J = 6Y_a + 12Y_b + 8Y_c + Y_d$$

The passivity condition is then a positivity condition on these admittances, and thus on the parameters a, b, c, and d. Recalling the expression for c in terms of a and b from (A.27), we must have

$$a \geq 0 \quad b \geq 0 \quad b \leq \frac{1-a}{4}$$

This region is shown, in dark gray, in Figure A.8(b). The positivity condition on d (expressed in terms of a, b, and λ as per (A.27)) gives the bound on λ, which is

$$\lambda \leq \sqrt{\frac{1}{2a + 2b + 1}} \qquad \text{(for passivity)}$$

A.3.4 The Tetrahedral Scheme

The tetrahedral scheme in $(3+1)$D [269] is somewhat similar to the hexagonal scheme in $(2+1)$D, in that the grid is divided evenly into two sets of points at which updating is performed using 'mirror-image' stencils. It is different, however, because grid points can easily be indexed with reference to a regular cubic lattice; the hexagonal scheme operates on a rectangular grid in stretched or transformed coordinates. In fact, a tetrahedral scheme can be obtained directly from an octahedral scheme simply by removing half of the grid points it employs; as such, any given grid point in the tetrahedral scheme has four nearest neighbors. As usual, we assume the nearest-neighbor grid spacing to be Δ. See Figure A.10(a) for a representation of the numerical grid.

As for the hexagonal scheme, we will view this as a vector scheme operating on two distinct subgrids, labeled 1 and 2 in Figure A.10(a). The two grid functions $U_{1,i,j,k}(n)$ and $U_{2,i+1,j+1,k+1}(n)$ are defined for integer i, j, and k all even such that $(i+j+k)/2$ is also even. $U_{1,i,j,k}$ will be used to approximate a continuous function u_1 at the point with coordinates $x = i\Delta/\sqrt{3}$, $y = j\Delta/\sqrt{3}$, and $z = k\Delta/\sqrt{3}$, and $U_{2,i+1,j+1,k+1}$ approximates u_2 at

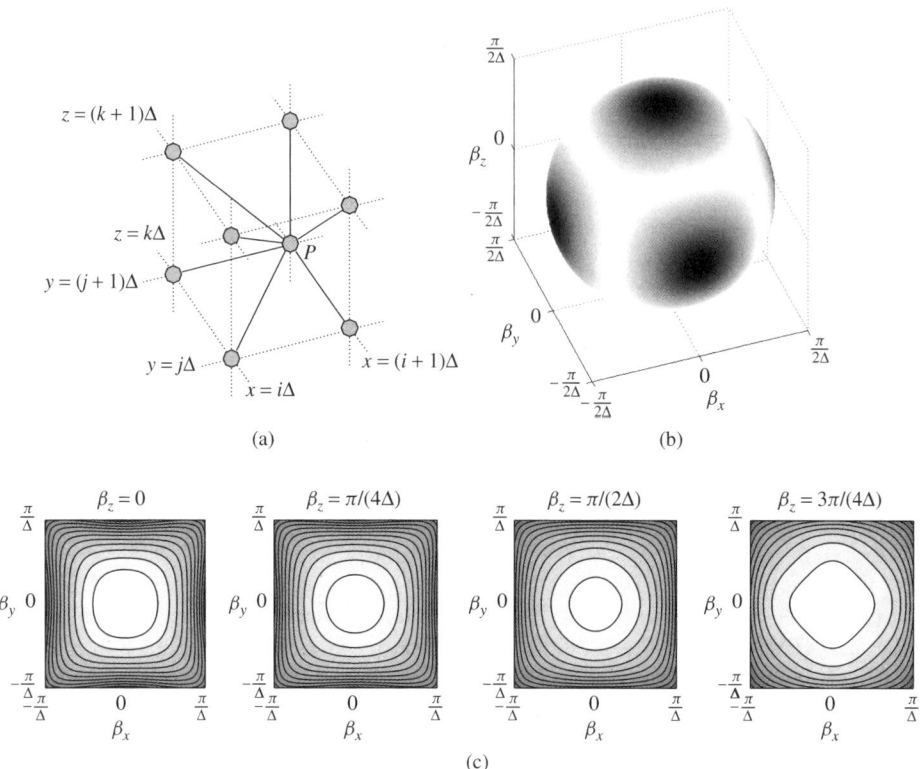

Figure A.9: *The $(3+1)$D interpolated rectilinear scheme (A.26)—(a) numerical grid and connections, from a central grid point (labeled P) to its neighbors in one octant. (b) $v_{\beta,\text{phase}}/\gamma$ for the scheme with $a = 0.42$ and $b = 0.1233$ at the stability bound $\lambda = 0.8617$, for a spherical surface with $\|\boldsymbol{\beta}\|_2 = \pi/(2\Delta)$—the shading is normalized over the surface so that white and black refer to minimal and maximal dispersion error, respectively. Here, unlike for the cubic rectilinear and octahedral schemes, there are no dispersionless directions. The variation in the numerical phase velocity is, however, quite small, ranging from 96.81 to 97.32% of the correct wave speed. (c) Contour plots of $v_{\beta,\text{phase}}/\gamma$ for various cross-sections of the space of spatial frequencies $\boldsymbol{\beta}$; contours indicate successive deviations of 2% from the ideal value of 1, which is obtained at spatial DC.*

coordinates $x = (i+1)\Delta/\sqrt{3}$, $y = (j+1)\Delta/\sqrt{3}$, and $z = (k+1)\Delta/\sqrt{3}$. The numerical scheme can then be written as

$$U_{1,i,j,k}(n+1) + U_{1,i,j,k}(n-1)$$
$$= \tfrac{3}{2}\lambda^2 \Big(U_{2,i+1,j+1,k+1}(n) + U_{2,i+1,j-1,k-1}(n)$$
$$+ U_{2,i-1,j-1,k+1}(n) + U_{2,i-1,j+1,k-1}(n) \Big)$$
$$+ 2\left(1 - 3\lambda^2\right) U_{1,i,j,k}(n) \tag{A.28a}$$

A.3. FINITE DIFFERENCE SCHEMES FOR THE (3 + 1)D WAVE EQUATION 323

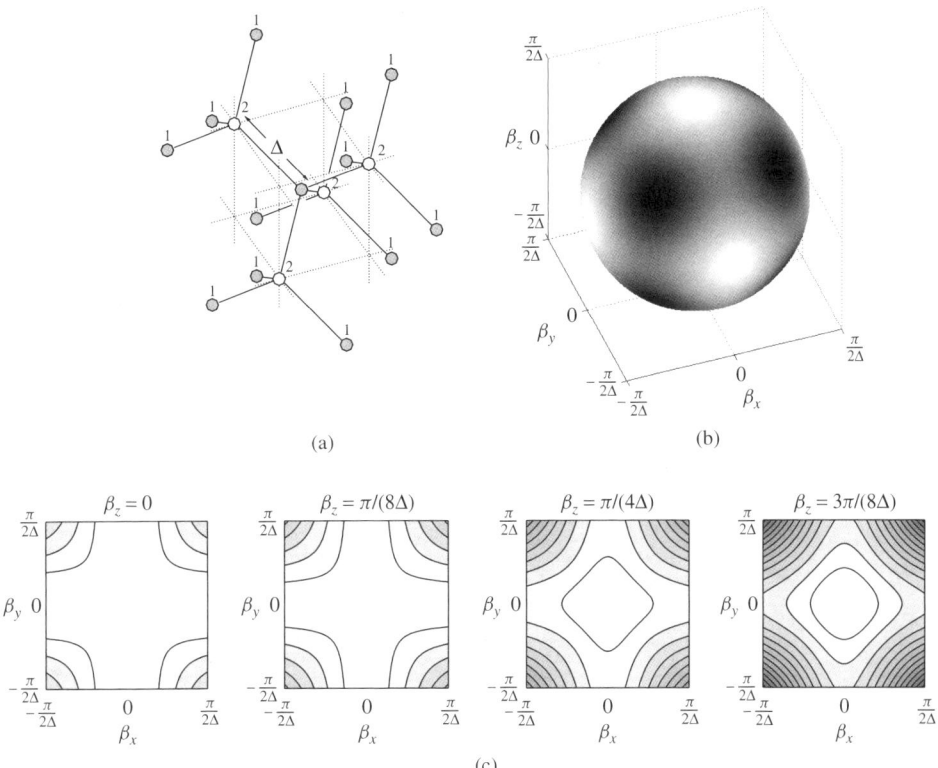

Figure A.10: *The tetrahedral scheme (A.28)—(a) numerical grid and connections, in which gray/white coloring of points indicates a division into mutually exclusive subschemes at the stability bound. The scheme can be indexed similarly to the octahedral scheme (see Figure A.7). The two subgrids with mutually inverse orientations are labeled 1 and 2. (b) $v_{\beta,\text{phase}}/\gamma$ for the scheme at the stability bound $\lambda = 1/\sqrt{3}$, for a spherical surface with $\|\beta\|_2 = \pi/(2\Delta)$—the shading is normalized over the surface so that white corresponds to no dispersion error, and black to the maximum error over the surface (which is 6% in this case). (c) Contour plots of $v_{\beta,\text{phase}}/\gamma$ for various cross sections of the space of spatial frequencies β; contours indicate successive deviations of 2% from the ideal value of 1, which is obtained at spatial DC. Here we have only plotted spatial frequencies $|\beta_x|$, $|\beta_y|$, and $|\beta_z|$ all less than $\pi/(2\Delta)$.*

$$U_{2,i+1,j+1,k+1}(n+1) + U_{2,i+1,j+1,k+1}(n-1)$$
$$= \tfrac{3}{2}\lambda^2 \Big(U_{1,i,j,k}(n) + U_{1,i,j+2,k+2}(n)$$
$$+ U_{1,i+2,j+2,k}(n) + U_{1,i+2,j,k+2}(n)\Big)$$
$$+ 2\left(1 - 3\lambda^2\right) U_{2,i+1,j+1,k+1}(n) \qquad \text{(A.28b)}$$

As for the hexagonal scheme, we may check consistency of this system with the wave equation by treating the grid functions as samples of continuous functions u_1 and u_2, and expanding (A.28) in terms of partial derivatives; both grid functions updated according to this scheme will approximate the solution to the wave equation on their respective grids.

Determining the stability condition proceeds as for the hexagonal scheme; taking spatial Fourier transforms of (A.28) gives a vector spectral update equation of the form (A.10), with $\mathbf{B}_\beta$ given by

$$\mathbf{B}_\beta = \begin{bmatrix} -2(1 - 3\lambda^2) & -\frac{3}{2}\lambda^2 \psi_\beta \\ -\frac{3}{2}\lambda^2 \psi_\beta^* & -2(1 - 3\lambda^2) \end{bmatrix}$$

with

$$\psi_\beta = 2\left(e^{j\Delta\beta_x/\sqrt{3}} \cos(\Delta(\beta_y + \beta_z)/\sqrt{3}) + e^{-j\Delta\beta_x/\sqrt{3}} \cos(\Delta(\beta_y - \beta_z)/\sqrt{3})\right)$$

$\mathbf{B}_\beta$ is again Hermitian, and has eigenvalues

$$\Lambda_{\beta,1} = -2\left(1 - 3\lambda^2\right) + \tfrac{3}{2}\lambda^2 |\psi_\beta|$$

$$\Lambda_{\beta,2} = -2\left(1 - 3\lambda^2\right) - \tfrac{3}{2}\lambda^2 |\psi_\beta|$$

The stability condition can thus be written as

$$\left| -2\left(1 - 3\lambda^2\right) \pm \tfrac{3}{2}\lambda^2 |\psi_\beta| \right| \leq 2 \tag{A.29}$$

ψ_β can be shown to take on a maximum of 4 and a minimum of 0, and it then follows that (A.29) will be satisfied if and only if $\lambda \leq 1/\sqrt{3}$, the same bound as obtained for the cubic rectilinear and octahedral schemes. The bound is the same as the bound for passivity of a tetrahedral mesh, as discussed in Section 5.2. We note that as for these other schemes, the grid permits a subdivision into mutually exclusive subschemes at this stability limit (see Figure A.10(a)). By a simple comparison with the hexagonal scheme, we can obtain the four spectral amplification factors

$$G_{\beta,1,\pm} = \frac{1}{2}\left(-\Lambda_{\beta,1} \pm \sqrt{\Lambda_{\beta,1}^2 - 4}\right) \qquad G_{\beta,2,\pm} = \frac{1}{2}\left(-\Lambda_{\beta,2} \pm \sqrt{\Lambda_{\beta,2}^2 - 4}\right)$$

and it is easy to see that parasitic modes (characterized by the amplification factors $G_{\beta,1,\pm}$) will be present in the tetrahedral scheme, due to the nonuniformity of updating on the numerical grid. The numerical dispersion characteristics of the dominant modes with amplification factors $G_{\beta,2,\pm}$ are shown in planar and spherical cross-sections in Figure A.10(b) and (c).

The computational and add densities of this scheme, in general, are

$$\rho_{\text{tetr}} = \frac{3\sqrt{3}v_0}{8\Delta^4} \qquad \sigma_{\text{tetr}} = \frac{15\sqrt{3}v_0}{8\Delta^4}$$

for $v_0 > \sqrt{3}\gamma$, and

$$\rho_{\text{tetr}}^s = \frac{9\gamma}{16\Delta^4} \qquad \sigma_{\text{tetr}}^s = \frac{9\gamma}{4\Delta^4}$$

at the stability limit $v_0 = \sqrt{3}\gamma$.

Appendix B

Eigenvalue and Steady State Problems

by Gunnar Nitsche

B.1 Introduction

In the previous chapters, MDWD and DWN methods for the numerical solution of time-dependent partial differential equations (PDEs) have been presented. The detailed time domain behavior of the solution, however, is sometimes not of primary concern for linear and time-invariant systems, and one is more interested in global information such as the eigenvalues and eigenfunctions of a system or its steady state response to a sinusoidal excitation. An approach to solving problems of this kind is presented in this appendix.

Elliptic PDEs can be considered to be a special case of hyperbolic PDEs by searching solutions of the latter that are static, that is, those that are not time-dependent, or more generally, which exhibit a sinusoidal dependence on time (steady state solution). Conventional numerical methods such as finite elements directly discretize elliptic PDEs. This, however, is not possible for MDWD and DWN methods, so we must first discretize the full time-dependent hyperbolic PDE and then look for steady state solutions of the discrete model.

The algorithm presented here addresses both the eigenvalue and steady state problems mentioned above, thus widening the scope of MDWD and DWN methods to include problems that cannot be approached directly. The main advantages of the algorithm are as follows:

- Easy application: based only on the procedure performing one time step.

- Efficiency: number of operations depends roughly linearly on the number of sampling points (spatial + temporal).

- Accuracy: single precision is usually sufficient to achieve relative errors $\ll 10^{-4}$.

- Generality: any stable time domain model can be treated, even in the general lossy case.

- Both eigenvalue and steady state problems can be solved effectively by the same approach.

This chapter is fairly compact. For more details, please refer to [175, 176, 177]. A related approach, but for symmetric (equivalent to lossless) matrices, is given in [47].

B.2 Abstract Time Domain Models

For all the methods discussed in the main body of the book, the state of the system at some time instant t_k may be represented by a set of state variables, either by samples of the field quantities (in FDTD) or by certain wave quantities (in DWNs and MDWD methods). It is useful to represent these by a column vector $\mathbf{x}(t_k)$. (Note that in the case of FDTD $\mathbf{x}(t_k)$ contains two steps worth of field samples at time instants t_k and $t_{k-\frac{1}{2}}$.) For the analysis of the eigensystem, no sources will be present, so that the transition from the kth time step to the next can be written as

$$\mathbf{x}(t_{k+1}) = \mathbf{S}\mathbf{x}(t_k), \qquad t_k = kT, \tag{B.1}$$

where T is the time step. The matrix $\mathbf{S}$ is usually large (e.g. $>100\,000$ unknowns) but sparse and it is assumed that there is a subroutine performing the matrix-vector multiplication (B.1) in such a way that the sparsity of $\mathbf{S}$ is exploited. In some cases, for example, when applying MDWD-based methods, the complete time step (B.1) may consist of up to four partial steps, that is, $\mathbf{S}$ is implemented as product of up to four different matrices.

B.3 Typical Eigenvalue Distribution of a Discretized PDE

In practice, one is often not interested in all the eigenvalues, but only a few relevant ones corresponding to low frequency modes. The discrete model gives satisfactory accuracy only for sufficiently low frequencies, and there is usually no practical reason for analyzing the higher modes of a system. The key observation that leads to an efficient algorithm for the calculation of the relevant eigenvalues is that the eigenvalue distribution of a time domain model corresponding to a physical system is not arbitrary, but exhibits a special structure.

The main property of the distribution is that there are few low-frequency eigenvalues and many higher-frequency ones; this does not usually depend on the problem under consideration. In general, the number of modes $z = e^{j\omega}$ in the frequency interval $0 \leq \omega \leq \omega_g$ can be expected to be approximately proportional to ω_g^d for $\omega_g \ll 1$, where d is the number of spatial dimensions. So, if ω_g is divided by μ, the number of modes with frequency less than ω_g is divided by μ^d, that is, by the same factor as the total number of eigenvalues is multiplied by if the number of sampling points in every spatial direction is multiplied by μ.

B.4 Excitation and Filtering

A state vector consisting of a linear combination of purely low-frequency eigenvectors can be constructed as follows. First excite all eigenmodes with approximately equal power by taking a random start vector $\mathbf{x}_0$ and then filter the low frequency modes out of the sequence of successive time steps

$$\mathbf{x}_k = \mathbf{x}(t_k) = \mathbf{S}^k \mathbf{x}_0 \,. \tag{B.2}$$

A finite impulse response (FIR) low-pass filter is ideally suited for this purpose since only one output sample is needed. Applying such a filter with coefficients h_k, $k = 0, 1, \ldots, n-1$, is equivalent to forming a linear combination of the $\mathbf{x}_k$, that is,

$$\mathbf{y}_0 = \sum_{k=0}^{n-1} h_k \mathbf{x}_k = \sum_{k=0}^{n-1} h_k \mathbf{S}^k \mathbf{x}_0 \,. \tag{B.3}$$

Since the first computed sample, $\mathbf{y}_0$, contains the complete state information, no further output samples are needed. The vector $\mathbf{y}_0$ will be nearly free of eigenvectors that correspond to eigenvalues of $\mathbf{S}$ lying in the stopband of the filter. We use a class of nonrecursive equiripple low-pass filters designed by the well-known Parks-McClellan algorithm [183, 195] with the design parameters

$$\omega_s = \frac{\pi}{M}, \quad M = 2, 3, \ldots, 20, \quad \omega_p = 0.75\,\omega_s, \quad A_s = 70\,\text{dB}, \quad A_p = 1.5\,\text{dB}, \tag{B.4}$$

where ω_s is the stopband edge frequency, ω_p the passband edge frequency, A_s the minimum attenuation in the stopband and A_p the maximum attenuation in the passband.

The result $\mathbf{y}_0$ in (B.3) is used as a start vector $\mathbf{x}_0$ for a second filtering pass, such that the total stopband attenuation becomes 140 dB, that is, the full dynamic range of single precision floating point numbers.

B.5 Partial Similarity Transform

Let $\mathbf{y}$ be the vector resulting from the second filtering step. The next step of the algorithm is to construct a (usually) much smaller matrix $\mathbf{T}$ in such a way that the relevant eigenvalues of $\mathbf{S}$ are well approximated by the eigenvalues of $\mathbf{T}$. This relationship between the matrices $\mathbf{S}$ and $\mathbf{T}$ will be called *partial similarity*.

Owing to the filtering, the spectrum of the sequence

$$\tilde{\mathbf{y}}_k = \mathbf{S}^k \mathbf{y} \tag{B.5}$$

is low-pass limited with an edge frequency of $\omega_s = \pi/M$, so it can be downsampled by a factor of M by using only every Mth state vector $\mathbf{y}_k$ of the sequence $\tilde{\mathbf{y}}_k$,

$$\mathbf{y}_k = \left(\mathbf{S}^M\right)^k \mathbf{y} \,. \tag{B.6}$$

$$\mathbf{q}_1 := \frac{\mathbf{y}}{\sqrt{\mathbf{y}^T\mathbf{y}}}$$

for $k := 1$ to *termination condition*

$\quad\mathbf{w} := \mathbf{S}^M\mathbf{q}_k$

$\quad$for $i := 1$ to k

$\quad\quad t_{ik} := \mathbf{w}^T\mathbf{q}_i$

$\quad\quad \mathbf{w} := \mathbf{w} - t_{ik}\mathbf{q}_i$

$\quad$end

$\quad t_{k+1,k} := \sqrt{\mathbf{w}^T\mathbf{w}}$

$\quad \mathbf{q}_{k+1} := \dfrac{\mathbf{w}}{t_{k+1,k}}$

end

Figure B.1: *Algorithm using modified Gram-Schmidt orthogonalization to perform a partial similarity transform.*

An orthogonal basis for the subspace spanned by the sequence $\mathbf{y}_k$ can be found by applying the so-called modified Gram-Schmidt orthogonalization procedure to the vectors $\mathbf{y}_k$ [156]. This gives the algorithm in Figure B.1, which recursively constructs matrices $\mathbf{Q}_k$ and $\mathbf{T}_k$ obeying

$$\mathbf{S}^M\mathbf{Q}_k = \mathbf{Q}_k\mathbf{T}_k + t_{k+1,k}\mathbf{q}_{k+1}\mathbf{e}_{k+1}^T, \tag{B.7}$$

where

$$\mathbf{Q}_k = [\mathbf{q}_1, \ldots, \mathbf{q}_k], \quad \mathbf{Q}_k^T\mathbf{Q}_k = \mathbf{I}_k, \quad \mathbf{T}_k = \begin{bmatrix} t_{11} & t_{12} & \cdots & & \cdots & t_{1k} \\ t_{21} & t_{22} & \ddots & & & \vdots \\ 0 & \ddots & \ddots & \ddots & & \vdots \\ \vdots & \ddots & \ddots & \ddots & t_{k-1,k-1} & t_{k-1,k} \\ 0 & \cdots & 0 & & t_{k,k-1} & t_{kk} \end{bmatrix}, \tag{B.8}$$

and $\mathbf{e}_{k+1}$ is the column vector with its first k elements being zero and its $(k+1)$th element being one. Since $\mathbf{q}_{k+1}$ is orthogonal to the $\mathbf{q}_i$, $i = 1, 2, \ldots, k$, we have

$$\mathbf{T}_k = \mathbf{Q}_k^T\mathbf{S}^M\mathbf{Q}_k, \quad \text{and} \quad \mathbf{S}^M\mathbf{Q}_k \cong \mathbf{Q}_k\mathbf{T}_k \tag{B.9}$$

which is satisfied optimally in the least-squares sense with error norm $t_{k+1,k}$ [156]. Equation (B.9) can be considered to be an approximate partial similarity transform, since if $\mathbf{u}$ is an eigenvector of $\mathbf{T}_k$ with eigenvalue λ according to

$$\mathbf{T}_k\mathbf{u} = \lambda\mathbf{u}, \tag{B.10}$$

then
$$S^M Q_k u \cong Q_k T_k u = \lambda Q_k u, \quad (B.11)$$

so v is an approximate eigenvector of S with eigenvalue z (as long as $v \neq 0$),

$$Sv \cong zv, \quad \text{where} \quad v = Q_k u \quad \text{and} \quad z = \lambda^{1/M}. \quad (B.12)$$

The definition of z by (B.12) is unique since the prior low-pass filtering ensures that z is the solution of (B.12) with smallest phase angle, that is, lowest frequency. The size of the upper Hessenberg matrix T_k is in a range where a full eigenanalysis is feasible, for example, by applying the standard EISPACK routine HQR2 [223]. Eigenvalues and eigenvectors of S can easily be derived from those of T_k by using (B.12). Note that up to this step exclusively real vectors and matrices occur in the algorithm. A final verification of the solution is necessary to check the accuracy of the calculated modes and to reject spurious modes.

B.6 Steady State Problems

Consider the case where the time-domain model (B.1) is sinusoidally excited according to

$$x(t_{k+1}) = Sx(t_k) + \text{Re}\, f_0 e^{jk\omega_0}, \quad (B.13)$$

where $\text{Re}\, f_0 e^{jk\omega_0}$ denotes the real part of $f_0 e^{jk\omega_0}$, f_0 is a known complex constant vector and ω_0 is the normalized angular frequency. The problem is to find the steady state solution of (B.13), that is,

$$x(t_k) = \text{Re}\, x_0 e^{jk\omega_0} \quad \text{for all } k, \quad (B.14)$$

where x_0 is an unknown complex constant vector. The case $\omega_0 = 0$ is clearly included. Equations (B.13) and (B.14) form a linear system of equations that can be transformed into an eigenvalue problem to ensure the applicability of the algorithm described in sections B.4 and B.5. Such a problem can be constructed from (B.13) by including the exciting oscillator in an extended system $\tilde{S}$. A sinusoidal oscillator of frequency ω_μ can be modeled by

$$y_\mu(t_{k+1}) = \Omega_\mu y_\mu(t_k), \quad \text{where} \quad y_\mu = \begin{pmatrix} y_{R,\mu} \\ y_{I,\mu} \end{pmatrix}, \quad y_\mu(0) = \begin{pmatrix} 1 \\ 0 \end{pmatrix}, \quad (B.15)$$

$$\text{and} \quad \Omega_\mu = \begin{pmatrix} \cos\omega_\mu & -\sin\omega_\mu \\ \sin\omega_\mu & \cos\omega_\mu \end{pmatrix}. \quad (B.16)$$

Generalizing the approach to take into account a simultaneous excitation by m different frequencies and substituting the $y_\mu(t_k)$ into (B.13) yields

$$x(t_{k+1}) = Sx(t_k) + \sum_{\mu=1}^{m} \text{Re}\, f_\mu e^{jk\omega_\mu} = Sx(t_k) + \sum_{\mu=1}^{m} [f_{R,\mu}, f_{I,\mu}] y_\mu(t_k), \quad (B.17)$$

where $\mathbf{f}_{R,\mu} = \operatorname{Re} \mathbf{f}_\mu$ and $\mathbf{f}_{I,\mu} = -\operatorname{Im} \mathbf{f}_\mu$. The above equations can be written as

$$\tilde{\mathbf{x}}(t_{k+1}) = \tilde{\mathbf{S}} \tilde{\mathbf{x}}(t_k), \tag{B.18}$$

where

$$\tilde{\mathbf{S}} = \left[\begin{array}{c|ccc} \mathbf{S} & \mathbf{f}_{R,1}\,\mathbf{f}_{I,1} & \cdots & \mathbf{f}_{R,m}\,\mathbf{f}_{I,m} \\ \hline & \Omega_1 & & 0 \\ 0 & & \ddots & \\ & 0 & & \Omega_m \end{array} \right] \quad \text{and} \quad \tilde{\mathbf{x}} = \begin{bmatrix} \mathbf{x} \\ \mathbf{y}_1 \\ \vdots \\ \mathbf{y}_m \end{bmatrix}. \tag{B.19}$$

The algorithm for the eigenanalysis described in Sections B.4 and B.5 is applicable if (B.1) is replaced by (B.18). The start vector should no longer be chosen randomly but the oscillators should be initialized according to (B.15) and the vector $\mathbf{x}$ in (B.19) should be initialized with zero. By construction, $\tilde{\mathbf{S}}$ is forced to have the eigenvalues $e^{\pm j\omega_\mu}$, and the corresponding eigenvectors give the solution of the steady state problem. Clearly, the frequencies ω_μ must lie in the passband of the low-pass filter. The possibility of calculating the responses to several different excitations in one program run makes the approach very efficient.

It is quite unusual to solve a linear equation system by means of an equivalent eigenvalue problem. However, in this special case the eigenvalues are known and the problem reduces to finding the corresponding eigenvectors, which could be done by applying the so-called inverse iteration [47] for every frequency. In practice, however, doing so seems not to be very attractive because of the excellent performance of the HQR2-algorithm.

B.7 Generalization to Multiple Eigenvalues

Under certain symmetries, multiple eigenvalues may occur, where for physically relevant PDEs it can be assumed that there are as many linearly independent eigenvectors as the multiplicity of the eigenvalue. The excitation (B.2) finds only one eigenvector in such a situation.

In order to get all l eigenvectors corresponding to an l-fold eigenvalue, l linearly independent excitations are required. This can be accomplished by performing the excitation and filtering procedure of section B.4 l times, since random vectors are almost surely linearly independent. However, the partial similarity transform has to be adapted.

Let $\mathbf{y}_k, k = 1, \ldots, l$, be the l results of the second filtering according to (B.3). Firstly, these are orthogonalized (see Figure B.2(a)), and then a partial similarity transform is calculated (see Figure B.2(b)). The resulting matrix $\mathbf{T}$ has l subdiagonals, so it is no longer upper Hessenberg. Thus, the EISPACK routine RG [223] for real general matrices has to be applied instead of HQR2.

The support of complex and generalized eigenvalue problems is also possible [176], but outside the scope of this chapter.

for $k := 1$ to l
    for $i := 1$ to $k - 1$
        $r_{ik} := \mathbf{q}_i^T \mathbf{y}_k$
        $\mathbf{y}_k := \mathbf{y}_k - r_{ik} \mathbf{q}_i$
    end
    $r_{kk} := \sqrt{\mathbf{y}_k^T \mathbf{y}_k}$
    $\mathbf{q}_k := \dfrac{\mathbf{y}_k}{r_{kk}}$
end

(a)

for $k := 1$ to *termination condition*
    $\mathbf{w} := \mathbf{S}^M \mathbf{q}_k$
    for $i := 1$ to $k + l - 1$
        $t_{ik} := \mathbf{q}_i^T \mathbf{w}$
        $\mathbf{w} := \mathbf{w} - t_{ik} \mathbf{q}_i$
    end
    $t_{k+l,k} := \sqrt{\mathbf{w}^T \mathbf{w}}$
    $\mathbf{q}_{k+l} := \dfrac{\mathbf{w}}{t_{k+l,k}}$
end

(b)

Figure B.2: (a) *Orthogonalization of the l start vectors, and* (b) *partial similarity transform for up to l–fold eigenvalues.*

```
Number of sampling points:   30 * 25 * 22 = 16 500
Number of state variables:   102 920
Downsampling factor:         10
Partial similarity transform: 35 steps
Memory used for similarity transform:   15 290 kBytes

Results

   Mode    Frequency    Absolute value       Error

    1      0.05346466     0.9910821       8.729978e-07
    2      0.05833124     0.9909683       1.700687e-06
    3      0.06298642     0.9908357       1.536901e-06
    4      0.07015239     0.9918733       3.160058e-05
    5      0.07058950     0.9915792       8.485109e-05

Number of time steps:   684
```

Figure B.3: *Eigenanalysis of an MDWD model for Maxwell's equations (lossy cavity).*

B.8 Numerical Example

As an example, an MDWD model for Maxwell's equations in three spatial dimensions is analyzed (based on Figure 6.10, with some additional loss added to L_1, L_2, and L_3). The corresponding physical problem is a lossy cuboidal cavity. The model provides the full time domain solution of an arbitrary excitation, but this is usually more detailed information than required for designing a resonating cavity. The designer is generally more interested in the lowest-frequency eigenvalues and corresponding eigenvectors.

The results of the eigenvalue analysis are given in Figure B.3. Though the number of state variables is very large (>100 000) and loss is present, the five lowest frequency eigenmodes can be computed quite quickly and accurately. The absolute values of the eigenvalues correspond to the quality of the resonator. The total execution time is dominated by the amount necessary to compute the time steps.

Bibliography

[1] S. Abarbanel and D. Gottlieb. A mathematical analysis of the PML method. *Journal of Computational Physics*, 134(2):357–363, 1 July 1997.

[2] S. Abarbanel and D. Gottlieb. On the construction and analysis of absorbing layers in CEM. *Applied Numerical Mathematics*, 27(4):331–340, August 1998.

[3] J. Abel, T. Smyth, and J. O. Smith III. A simple, accurate wall loss filter for acoustic tubes. In *Proceedings of the COST-G6 Digital Audio Effects Conference*, pages 254–258, London, UK, September 2003.

[4] R. Abraham. *Linear and Multilinear Algebra*. W. A. Benjamin, Inc., New York, USA, 1966.

[5] S. Akhtarzad and P. B. Johns. Solution of Maxwell's equations in three space dimensions and time by the T.L.M. method of numerical analysis. *Proceedings of the IEE*, 122(12):1344–1348, December 1975.

[6] K. Balemarthy and S. Bass. General linear boundary conditions in MD wave digital simulations. In *Proceedings of the IEEE International Symposium on Circuits and Systems*, volume 1, pages 73–76, Seattle, Washington, USA, April–May 1995.

[7] M. Banerjee. On the vibration of skew plates of variable thickness. *Journal of Sound and Vibration*, 63(3):377–383, 8 April 1979.

[8] B. Bank. Modelling the longitudinal vibration of piano strings. In *Proceedings of the Stockholm Musical Acoustics Conference*, pages 143–146, Stockholm, Sweden, August 2003.

[9] S. Bass. Sampling grid properties in wave digital PDE simulations. Technical Report CSE-TR-26-94, University of Notre Dame, 1994.

[10] S. Basu and A. Fettweis. On the factorization of scattering transfer matrices for multidimensional lossless two-ports. *IEEE Transactions on Circuits and Systems*, CAS-32(9):925–934, September 1985.

[11] S. Basu and A. Fettweis. On synthesizable multidimensional lossless two-ports. *IEEE Transactions on Circuits and Systems*, 35(12):1478–1486, December 1988.

[12] S. Basu and A. Zerzghi. Multidimensional digital filter approach for numerical solution of a class of PDEs of the propagating wave type. In *Proceedings of the IEEE International Symposium on Circuits and Systems*, volume 5, pages 74–77, Monterey, California, USA, May–June 1998.

[13] V. Belevitch. Summary of the history of circuit theory. *Proceedings of the IRE*, 50:848–855, May 1962.

[14] V. Belevitch. *Classical Network Theory*. Holden Day, San Francisco, California, USA, 1968.

[15] J. Bensa, S. Bilbao, R. Kronland-Martinet, and J. O. Smith III. The wave digital piano hammer: A passive formulation. In *Proceedings of the 144th meeting of the Acoustical Society of America*, Cancun, Mexico, December 2002. CD-ROM.

[16] J. Bensa, S. Bilbao, R. Kronland-Martinet, and J. O. Smith III. A power-normalized non-linear lossy piano hammer. In *Proceedings of the Stockholm Musical Acoustics Conference*, pages 365–368, Stockholm, Sweden, August 2003.

[17] J. Bensa, S. Bilbao, R. Kronland-Martinet, and J. O. Smith III. The simulation of piano string vibration: from physical models to finite difference schemes and digital waveguides. *Journal of the Acoustical Society of America*, 114(2):1095–1107, 2003.

[18] J.-P. Berenger. A perfectly matched layer for the absorption of electromagnetic waves. *Journal of Computational Physics*, 114(2):185–200, October 1994.

[19] J.-P. Berenger. Three-dimensional perfectly matched layer for the absorption of electromagnetic waves. *Journal of Computational Physics*, 127(2):363–379, September 1996.

[20] D. P. Berners. *Acoustics and Signal Processing Techniques for Physical Modelling of Brass Instruments*. PhD thesis, Department of Electrical Engineering, Stanford University, 1999.

[21] R. Bernhardt and D. Dahlhaus. Numerical integration of the Euler equations by means of wave digital filters. In *Proceedings of the IEEE International Conference on Acoustics, Speech, and Signal Processing*, volume 6, pages 1–4, Adelaide, Australia, April 1994.

[22] W. Bertozzi. Speed and energy of relativistic electrons. *American Journal of Physics*, 32:551–555, 1964.

[23] D. Beskos, editor. *Boundary Element Analysis of Plates and Shells*. Springer-Verlag, New York, USA, 1991.

[24] S. Bilbao. Digital waveguide networks as multidimensional wave digital filters. In *Proceedings of the COST G-6 Conference on Digital Audio Effects*, pages 49–54, Verona, Italy, December 2000.

[25] S. Bilbao. Digital waveguide networks in inhomogeneous media. In *Proceedings of the COST G-6 Conference on Digital Audio Effects*, pages 249–253, Verona, Italy, December 2000.

[26] S. Bilbao. *Wave and Scattering Methods for the Numerical Integration of Partial Differential Equations*. PhD thesis, Department of Electrical Engineering, Stanford University, 2001.

[27] S. Bilbao. Stability of parameterized finite difference schemes for the wave equation. *Numerical Methods for Partial Differential Equations*, 2004. In press.

[28] S. Bilbao, J. Bensa, and R. Kronland-Martinet. The wave digital reed: a passive formulation. In *Proceedings of the COST-G6 Digital Audio Effects Conference*, pages 225–230, London, UK, September 2003.

[29] S. Bilbao, J. Bensa, R. Kronland-Martinet, and J. O. Smith III. From the physics of piano strings to digital waveguides. In *Proceedings of the International Computer Music Conference*, pages 45–48, Göteborg, Sweden, 2002.

[30] S. Bilbao and J. O. Smith III. Finite difference schemes for the wave equation: stability, passivity and numerical dispersion. *IEEE Transactions on Acoustics, Speech, and Signal Processing*, 11(3):255–266, May 2003.

[31] G. Borin, G. De Poli, and D. Rocchesso. Elimination of delay-free loops in discrete-time models of nonlinear acoustic systems. *IEEE Transactions on Speech and Audio Processing*, 8:597–606, 2000.

[32] N. K. Bose. *Applied Multidimensional Systems Theory*. Van Nostrand Reinhold, New York, USA, 1982.

[33] N. K. Bose and A. Fettweis. Skew-symmetry in the equivalent representation problem of a time-varying multiport inductor. In *Proceedings of the IEEE International Symposium on Circuits and Systems*, volume 3, pages 662–665, Bangkok, Thailand, May 2003.

[34] A. Bossavit and L. Kettunen. Yee-like schemes on a tetrahedral mesh, with diagonal lumping. *International Journal of Numerical Modelling*, 12:129–142, January–April 1999.

[35] A. Bruckstein and T. Kailath. An inverse scattering framework for several problems in signal processing. *IEEE ASSP Magazine*, 4(1):6–20, January 1987.

[36] K. P. Bube and R. Burridge. The one-dimensional inverse problem of reflection seismology. *SIAM Review*, 25(4):497–559, 1983.

[37] R. Cacoveanu, P. Saguet, and F. Ndagijimana. TLM method: a new approach for the central node in polar meshes. *Electronics Letters*, 31(4):297–298, 16 February 1995.

[38] A. Chaigne. On the use of finite differences for musical synthesis. Application to plucked stringed instruments. *Journal d'Acoustique*, 5(2):181–211, 1992.

[39] A. Chaigne and A. Askenfelt. Numerical simulations of struck strings. I. A physical model for a struck string using finite difference methods. *Journal of the Acoustical Society of America*, 95(2):1112–1118, February 1994.

[40] D. Chandrasekharaiah. Thermoelasticity with second sound: a review. *Applied Mechanics Review*, 39(3):355–376, March 1986.

[41] Y. Chen and R. Mittra. Finite-difference time-domain algorithm for solving Maxwell's equations in rotationally symmetric geometries. *IEEE Transactions on Microwave Theory and Technology*, 44(6):832–836, 1996.

[42] Z. Chen, M. M. Ney, and W. J. R. Hoefer. A new finite difference time-domain formulation and its equivalence with the TLM symmetrical condensed node. *IEEE Transactions on Microwave Theory and Technology*, MTT-39(12):2160–2168, December 1991.

[43] D. Cheng. *Field and Wave Electromagnetics*, page 438. Addison-Wesley, USA, second edition, 1990.

[44] C. Christopoulos. *The Transmission-Line Modelling Method*. IEEE Press, New York, USA, 1995.

[45] P. R. Cook. *Identification of Control Parameters in an Articulatory Vocal Tract Model with Applications to the Synthesis of Singing*. PhD thesis, Department of Electrical Engineering, Stanford University, 1990.

[46] R. Cooper and M. Naghdi. Propagation of non-axially symmetric waves in elastic cylindrical shells. *Journal of the Acoustical Society of America*, 29(12):1365–1373, December 1957.

[47] J. K. Cullum and R. A. Willoughby. *Lanczos Algorithms for Large Symmetric Eigenvalue Computations, Vol. 1: Theory, Progress in Scientific Computing* 3. Birkhäuser, Boston, 1985.

[48] A. Czyzewski and B. Kostek. Waveguide modelling of the panpipes. In *Proceedings of the Stockholm Musical Acoustics Conference*, volume 1, pages 259–262, Stockholm, Sweden, August 2003.

[49] G. Dahlquist. A special stability problem for linear multistep methods. *BIT*, 3:27–43, 1963.

[50] D. de Cogan. *Transmission Line Matrix (TLM) Techniques for Diffusion Applications*. Gordon and Breach Science Publishers, Amsterdam, The Netherlands, 1998.

[51] C. de Vaal and R. Nouta. Suppression of parasitic oscillations in floating point wave digital filters. In *Proceedings of the IEEE International Symposium on Circuits and Systems*, pages 1018–1022, New York, USA, May 1978.

[52] G. DeSanctis, A. Sarti, and S. Tubaro. Automatic synthesis strategies for object-based dynamical physical models in musical acoustics. In *Proceedings of the COST-G6 Digital Audio Effects Conference*, pages 219–224, London, UK, September 2003.

[53] W. Elmore and M. Heald. *Physics of Waves*. McGraw-Hill, New York, USA, 1969.

[54] P. Enders and D. de Cogan. TLM for diffusion: the artefact of the standard initial conditions and its elimination with an abstract TLM suite. *International Journal of Numerical Modelling*, 14:107–114, 2001.

[55] M. Erbar and E.-H. Horneber. Models for transmission lines with connecting transistors based on wave digital filters. *International Journal of Circuit Theory and Applications*, 23:395–412, July–August 1995.

[56] K. Erickson and A. N. Michel. Stability analysis of fixed-point digital filters using computer generated Lyapunov functions—Part II: Wave digital filters and lattice digital filters. *IEEE Transactions on Circuits and Systems*, CAS–32(2):132–142, February 1985.

[57] C. Erkut. *Aspects in Analysis and Model-Based Sound Synthesis of Plucked String Instruments*. PhD thesis, Laboratory of Acoustics and Audio Signal Processing, Helsinki University of Technology, 2002.

[58] C. Eswarappa and W. J. R. Hoefer. Bridging the gap between TLM and FDTD. *IEEE Microwave and Guided Wave Letters*, 43(8):4–6, August 1995.

[59] T. Felderhoff. Simulation of nonlinear circuits with period-doubling and chaotic behavior by wave digital filter principles. *IEEE Transactions on Circuits and Systems*, 41(7):485–489, July 1994.

[60] T. Felderhoff. Jacobi's method for massive parallel wave digital filter algorithm. In *Proceedings of the IEEE International Symposium on Circuits and Systems*, volume 2, pages 1621–1624, Atlanta, Georgia, USA, May 1996.

[61] T. Felderhoff. A new wave description for nonlinear elements. In *Proceedings of the IEEE International Conference on Acoustics, Speech, and Signal Processing*, volume 3, pages 221–224, Atlanta, Georgia, USA, May 1996.

[62] A. Fettweis. Digital filters related to classical structures. *AEU: Archive für Elektronik und Übertragungstechnik*, 25:79–89, February 1971. (See also U.S. Patent 3,967,099, 1976, now expired.).

[63] A. Fettweis. Pseudopassivity, sensitivity, and stability of wave digital filters. *IEEE Transactions on Circuit Theory*, 19(6):668–673, November 1972.

[64] A. Fettweis. Reciprocity, inter-reciprocity and transposition in wave digital filters. *International Journal of Circuit Theory and Applications*, 1:323–337, 1973.

[65] A. Fettweis. On sensitivity and roundoff noise in wave digital filters. *IEEE Transactions on Acoustics, Speech, and Signal Processing*, ASSP–22(5):383–384, October 1974.

[66] A. Fettweis. Principles of multidimensional wave digital filtering. In J. K. Aggarwal, editor, *Digital Signal Processing*, pages 261–282. Western Periodicals, North Hollywood, California, USA, 1978.

[67] A. Fettweis. Multidimensional circuit and systems theory. In *Proceedings of the IEEE International Symposium on Circuits and Systems*, pages 951–957, Montreal, Canada, May 1984.

[68] A. Fettweis. Wave digital filters: theory and practice. *Proceedings of the IEEE*, 74(2):270–327, February 1986.

[69] A. Fettweis. The role of passivity and losslessness in multidimensional digital signal processing—new challenges. In *Proceedings of the IEEE International Symposium on Circuits and Systems*, volume 1, pages 112–115, Singapore, June 1991.

[70] A. Fettweis. Discrete passive modelling of physical systems described by PDEs. In *Proceedings of EUSIPCO-92, Sixth European Signal Processing Conference*, volume 1, pages 55–62, Brussels, Belgium, August 1992.

[71] A. Fettweis. Discrete passive modelling of viscous fluids. In *Proceedings of the IEEE International Symposium on Circuits and Systems*, pages 1640–1643, San Diego, California, USA, May 1992.

[72] A. Fettweis. Multidimensional wave digital filters for discrete-time modelling of Maxwell's equations. *International Journal of Numerical Modelling*, 5:183–201, August 1992.

[73] A. Fettweis. Multidimensional wave digital principles: from filtering to numerical integration. In *Proceedings of the IEEE International Conference on Acoustics, Speech, and Signal Processing*, volume 6, pages 173–181, Adelaide, Australia, April 1994. IEEE Press.

[74] A. Fettweis. Improved wave-digital approach to numerically integrating the PDEs of fluid dynamics. In *Proceedings of the IEEE International Symposium on Circuits and Systems*, volume 3, pages 361–364, Scottsdale, Arizona, USA, May 2002.

[75] A. Fettweis. The wave-digital method and some of its relativistic implications. *IEEE Transactions on Circuits and Systems I: Fundamental Theory and Applications*, CAS-49(6):862–868, 2002.

[76] A. Fettweis. Robust numerical integration using wave digital concepts. In *Proceedings of the 5th DSPS Educators Conference*, pages 23–32, Tokyo, Japan, September 2003.

[77] A. Fettweis. Wave digital concepts and relativity theory. In D. Liu and P. Antsaklis, editors, *Stability and Control of Dynamical Systems with Applications*, pages 3–22. Birkhäuser, Boston, Massachusetts, USA, 2003.

[78] A. Fettweis. Losslessness in nonlinear Kirchhoff circuits and in relativity theory, Problems of Nonlinear Analysis in Engineering Systems, 9(3)(19):141–163, 2003.

[79] A. Fettweis. Nonlinear Kirchhoff circuits and relativity theory. *AEÜ International Journal of Electronics*, 58(1):21–29, 2004.

[80] A. Fettweis and S. Basu. On improved representation theorems for multidimensional lossless bounded matrices. *International Journal of Circuit Theory and Applications*, 19(5):453–457, September–October 1991.

[81] A. Fettweis and T. Leickel. On floating-point implementations of modified wave digital filters. In *Proceedings of the IEEE International Symposium on Circuits and Systems*, volume 4, pages 1812–1815, San Diego, California, USA, May 1992.

[82] A. Fettweis, T. Leickel, M. Bolle, and U. Sauvagerd. Realization of filter banks by means of wave digital filters. In *Proceedings of the IEEE International Symposium on Circuits and Systems*, volume 3, pages 2013–2016, New Orleans, Louisiana, USA, May 1990.

[83] A. Fettweis, H. Levin, and A. Sedlmeyer. Wave digital lattice filters. *International Journal of Circuit Theory and Applications*, 2(2):203–211, June 1974.

[84] A. Fettweis and K. Meerkötter. Suppression of parasitic oscillations in wave digital filters. *IEEE Transactions on Circuits and Systems*, CAS–22(3):239–246, March 1974.

[85] A. Fettweis and K. Meerkötter. On adaptors for wave digital filters. *IEEE Transactions on Acoustics, Speech, and Signal Processing*, ASSP-23(6):516–525, December 1975.

[86] A. Fettweis and K. Meerkötter. On parasitic oscillations in digital filters under looped conditions. *IEEE Transactions on Circuits and Systems*, CAS–24(9):475–481, September 1975.

[87] A. Fettweis and G. Nitsche. Numerical integration of partial differential equations by means of multidimensional wave digital filters. In *Proceedings of the IEEE International Symposium on Circuits and Systems*, volume 2, pages 954–957, New Orleans, Louisiana, USA, May 1990.

[88] A. Fettweis and G. Nitsche. Massively parallel algorithms for numerical integration of partial differential equations. In E. Deprettere and A.-J. van der Veen, editors, *Algorithms and Parallel VLSI Architectures*, volume B: Proceedings, pages 475–484. Elsevier, Amsterdam, The Netherlands, 1991.

[89] A. Fettweis and G. Nitsche. Numerical integration of partial differential equations using principles of multidimensional wave digital filters. *Journal of VLSI Signal Processing*, 3(1–2):7–24, June 1991.

[90] A. Fettweis and G. Nitsche. Transformation approach to numerically integrating PDEs by means of WDF principles. *Multidimensional Systems and Signal Processing*, 2(2):127–159, May 1991.

[91] A. Fettweis and G. A. Seraji. New results in numerically integrating PDEs by the wave digital approach. In *Proceedings of the IEEE International Symposium on Circuits and Systems*, volume 5, pages 17–20, Orlando, Florida, USA, May–June 1999.

[92] A. Fiedler and H. Grotstollen. Simulation of power electronic circuits with principles used in wave digital filters. *IEEE Transactions on Industrial Applications*, 33(1):49–57, January–February 1997.

[93] H. D. Fischer. Wave digital filters for numerical integration. *ntz-Archiv*, 6:37–40, February 1984.

[94] N. Fletcher and T. Rossing. *The Physics of Musical Instruments*. Springer-Verlag, New York, USA, 1991.

[95] F. Fontana and D. Rocchesso. Physical modelling of membranes for percussive instruments. *Acustica United with Acta Acustica*, 84:529–542, May–June 1998.

[96] F. Fontana and S. Serafin. Modelling Savart's trapezoidal violin using a digital waveguide mesh. In *Proceedings of the Stockholm Musical Acoustics Conference*, pages 51–53, Stockholm, Sweden, August 2003.

[97] B. Fornberg. *A Practical Guide to Pseudospectral Methods*. Cambridge Monographs on Applied and Computational Mathematics, Cambridge, UK, 1995.

[98] K. Friedrichs and P. Lax. Systems of conservation equations with a convex extension. In *Proceedings of the National Academy of Sciences of the U.S.A.*, pages 1686–1688, 1971.

[99] M. Fries. Multidimensional reactive elements on curvilinear coordinate systems and their MDWDF discretization. In *Proceedings of the IEEE International Conference on Acoustics, Speech, and Signal Processing*, volume 6, pages 9–12, Adelaide, Australia, April 1994. IEEE Press.

[100] M. Fries. Simulation of one-dimensional Euler flow by means of multidimensional wave digital filters. In *Proceedings of the IEEE International Symposium on Circuits and Systems*, volume 6, pages 9–12, London, UK, May–June 1994.

[101] M. Fries and A. Schrick. MDWDF-Verfahren und TLM-Methode zur Integration der 2D Maxwell-Gleichungen. In *Tagungsband ITG-Diskussionssitzung Neue Anwendungen theoretischer Konzepte in der Elektrotechnik*, pages 137–144, Berlin, Germany, 1995. In German.

[102] M. Fusco. FDTD algorithm in curvilinear coordinates. *IEEE Transactions on Antennas and Propagation*, 38(1):76–89, January 1990.

[103] M. Fusco, M. Smith, and L. Gordon. A three-dimensional FDTD algorithm in curvilinear coordinates. *IEEE Transactions on Antennas and Propagation*, 39(10):1463–1466, October 1991.

[104] P. Garabedian. *Partial Differential Equations*. Chelsea Publishing Company, New York, USA, second edition, 1986.

[105] B. Gazengel, J. Gilbert, and N. Amir. Time domain simulation of single reed wind instruments. From the measured impedance to the synthesis signal. Where are the traps? *Acta Acustica*, 3:445–472, 1995.

[106] Y. Genin. An algebraic approach to A-stable linear multistep-multiderivative integration formulas. *BIT*, 14(4):382–406, 1974.

[107] C. Giguere and P. Woodland. A computational model of the auditory periphery for speech and hearing. I. Ascending path. *Journal of the Acoustical Society of America*, 95(1):331–342, January 1994.

[108] K. Graff. *Wave Motion in Elastic Solids*. Dover Publications, New York, USA, 1975.

[109] A. H. Gray, Jr. and J. D. Markel. Digital lattice and ladder filter synthesis. *IEEE Transactions on Audio and Electroacoustics*, AU-21:491–500, December 1973.

[110] A. H. Gray, Jr. and J. D. Markel. A normalized digital filter structure. *IEEE Transactions on Acoustics, Speech, and Signal Processing*, ASSP-23:268–277, June 1975.

[111] J. Greenspon. Vibrations of a thick-walled cylindrical shell—comparison of the exact theory with approximate theories. *Journal of the Acoustical Society of America*, 32(5):571–578, May 1960.

[112] B. Gustaffson, H.-O. Kreiss, and J. Oliger. *Time Dependent Problems and Difference Methods*. John Wiley & Sons, New York, USA, 1995.

[113] B. Gustaffson, H.-O. Kreiss, and A. Sundstrom. Stability theory of difference approximations for mixed initial boundary value problems. II. *Mathematics of Computation*, 26(119):649–686, 1972.

[114] C. Harris and C. Crede, editors. *Shock and Vibration Handbook*. McGraw-Hill, New York, USA, second edition, 1976.

[115] A. Harten. On the symmetric form of systems of conservation laws. *Journal of Computational Physics*, 49:151–164, January 1983.

[116] G. Hemetsberger. Stability verification of multidimensional Kirchoff circuits by suitable energy functions. In *Proceedings of the IEEE International Conference on Acoustics, Speech, and Signal Processing*, volume 6, pages 13–16, Adelaide, Australia, April 1994.

[117] G. Hemetsberger and R. Hellfajer. Approach to simulating acoustics in supersonic flow by means of multidimensional vector-WDFs. In *Proceedings of the IEEE International Symposium on Circuits and Systems*, pages 73–76, London, UK, May–June 1994. IEEE Press.

[118] J. L. Herring and C. Christopoulos. Solving electromagnetic field problems using a multiple grid transmission-line modelling method. *IEEE Transactions on Antennas and Propagation*, 42(12):1654–1658, December 1994.

[119] J. L. Herring and C. Christopoulos. Multigrid transmission-line modelling method for solving electromagnetic field problems. *Electronics Letters*, 27(20):1794–1795, 26 September 1991.

[120] C. Hirsch. *Numerical Computation of Internal and External Flows*. John Wiley & Sons, Chichester, UK, 1988.

[121] W. J. R. Hoefer. The transmission line matrix method—theory and applications. *IEEE Transactions on Microwave Theory and Technology*, 33:882–893, 1985.

[122] W. J. R. Hoefer. *The Electromagnetic Wave Simulator*. John Wiley & Sons, Chichester, UK, 1991.

[123] R. Holland. Finite-difference solution of Maxwell's equations in generalized non-orthogonal coordinates. *IEEE Transactions on Nuclear Science*, NS-30(6):4589–4591, December 1983.

[124] R. Horn and C. Johnson. *Matrix Analysis*. Cambridge University Press, Cambridge, UK, 1985.

[125] F. Hu. On absorbing boundary conditions for linearized Euler equations by a perfectly matched layer. *Journal of Computational Physics*, 129(1):201–219, November 1996.

[126] H. Huang. *Static and Dynamic Analyses of Plates and Shells*. Springer-Verlag, Berlin, Germany, 1989.

[127] P. Huang, S. Serafin, and J. O. Smith III. Modelling high-frequency modes of complex resonators using a digital waveguide mesh. In *Proceedings of the COST G-6 Conference on Digital Audio Effects*, pages 269–272, Verona, Italy, December 2000.

[128] T. J. R. Hughes, L. Franca, and M. Mallet. A new finite element formulation for computational fluid dynamics: I. Symmetric forms of the Euler and Navier-Stokes equations and the second law of thermodynamics. *Computer Methods in Applied Mechanics and Engineering*, 54:223–234, 1986.

[129] K. Hunt and F. Crossley. Coefficient of restitution interpreted as damping in vibroimpact. *ASME Journal of Applied Mechanics*, 42:440–445, June 1975.

[130] D. Jaffe and J. O. Smith III. Extensions of the Karplus-Strong plucked string algorithm. *Computer Music Journal*, 7(2):56–68, Summer 1983.

[131] P. B. Johns. The solution of inhomogeneous waveguide problems using a transmission-line matrix. *IEEE Transactions on Microwave Theory and Technology*, MTT-22:209–215, March 1974.

[132] P. B. Johns. A simple, explicit and unconditionally stable routine for the solution of the diffusion equation. *International Journal of Numerical Methods in Engineering*, 11:1307–1328, 1977.

[133] P. B. Johns. On the relationship between TLM and finite-difference methods for Maxwell's equations. *IEEE Transactions on Microwave Theory and Technology*, MTT-35(1):60–61, January 1987.

[134] P. B. Johns. A symmetrical condensed node for the TLM method. *IEEE Transactions on Microwave Theory and Technology*, MTT-35(4):370–377, April 1987.

[135] P. B. Johns and R. L. Beurle. Numerical solution of 2-dimensional scattering problems using a transmission-line matrix. *Proceedings of the IEE*, 118:1203–1208, September 1971.

[136] S. Kaliski and L. Solarz. *Vibrations and Waves*. Elsevier, New York, USA, 1992.

[137] M. Karjalainen. Block-compiler: Efficient simulation of acoustic and audio systems. In *Preprints of AES 114th Convention*, Amsterdam, The Netherlands, May 2003.

[138] M. Karjalainen. Time-domain physical modelling and real-time synthesis using mixed modelling paradigms. In *Proceedings of the Stockholm Musical Acoustics Conference*, volume 1, pages 393–396, Stockholm, Sweden, August 2003.

[139] M. Karjalainen and J. O. Smith III. Body modelling techniques for string instrument synthesis. In *Proceedings of the International Computer Music Conference*, pages 232–239, Hong Kong, August 1996.

[140] M. Karjalainen, V. Välimäki, and P. Esquef. Efficient modelling and synthesis of bell-like sounds. In *Proceedings of the COST-G6 Digital Audio Effects Conference*, pages 181–186, Hamburg, Germany, September 2002.

[141] K. Karplus and A. Strong. Digital synthesis of plucked-string and drum timbres. *Computer Music Journal*, 7(2):43–55, Summer 1983.

[142] J. L. Kelly and C. C. Lochbaum. Speech synthesis. In *Proceedings of the Fourth International Congress on Acoustics*, pages 1–4, Copenhagen, Denmark, 1962. Paper G42.

[143] J. Kergomard. *Elementary Considerations on Reed-instrument Oscillations*. Springer, New York, USA, 1995.

[144] T. Koga. Synthesis of finite passive n-ports with prescribed two-variable reactance matrices. *IEEE Transactions on Circuit Theory*, CT-13:31–52, March 1966.

[145] H. Krauss. Wave digital simulation of transmission lines with arbitrary initial potential and current distributions. In *1996 IEEE Digital Signal Processing Workshop Proceedings*, pages 195–198, Loen, Norway, September 1996. IEEE Press.

[146] H. Krauss and R. Rabenstein. Application of multidimensional wave digital filters to boundary value problems. *IEEE Signal Processing Letters*, 2(7):183–201, July 1995.

[147] H. Krauss, R. Rabenstein, and M. Gerken. Simulation of wave propagation by multidimensional wave digital filters. *Simulation Practice and Theory*, 4:361–382, November 1996.

[148] H.-O. Kreiss. Initial boundary value problems for hyperbolic systems. *Communications on Pure and Applied Mathematics*, 23:277–298, 1970.

[149] G. Kron. Equivalent circuit of the field equations of Maxwell. *Proceedings of the IRE*, 32(5):284–299, May 1944.

[150] M. Krumpholz, C. Huber, and P. Russer. A field-theoretical comparison of FDTD and TLM. *IEEE Transactions on Microwave Theory and Technology*, 43(8):1935–1950, August 1995.

[151] M. Krumpholz and P. Russer. A field-theoretical derivation of TLM. *IEEE Transactions on Microwave Theory and Technology*, 42(9):1660–1667, September 1994.

[152] W. Ku and S. Ming. Floating-point coefficient sensitivity and roundoff noise of recursive digital filters realized in ladder structures. *IEEE Transactions on Circuits and Systems*, CAS–22(12):927–936, December 1975.

[153] P. Kundu. *Fluid Mechanics*. Academic Press, San Diego, California, USA, 1990.

[154] J. Kuttler and V. Sigillito. Vibrational frequencies of clamped plates of variable thickness. *Journal of Sound and Vibration*, 86(2):181–189, 22 January 1983.

[155] T. Laakso, V. Välimäki, M. Karjalainen, and U. Laine. Splitting the unit delay—tools for fractional delay filter design. *IEEE Signal Processing Magazine*, 13(1):30–60, January 1996.

[156] C. L. Lawson and R. J. Hanson. *Solving Least Squares Problems*. Prentice Hall, Englewood Cliffs, 1974.

[157] P. Lax. *Shock Waves and Entropy*. Academic Press, New York, USA, 1971.

[158] T. Leickel and A. Fettweis. Efficient digital-signal-processor realization of multirate filter banks using wave digital filters. In *Proceedings of the IEEE International Symposium on Circuits and Systems*, volume 4, pages 1812–1815, San Diego, California, USA, May 1992.

[159] P. Lennarz and W. Drews. Design of circularly symmetric 2-D wave digital filters. In *Proceedings of EUSIPCO-83, Second European Signal Processing Conference*, pages 199–202, Erlangen, Germany, September 1983.

[160] P. Lennarz and L. Hofmann. Computer realization of two-dimensional wave digital filters. In *Proceedings of the 1978 European Conference on Circuit Theory and Design*, pages 360–364, Lausanne, Switzerland, September 1978.

[161] K. Liew, C. Wang, Y. Xiang, and S. Kitipornchai. *Vibration of Mindlin Plates: Programming the p-version Ritz Method*. Elsevier, Amsterdam, The Netherlands, first edition, 1998.

[162] X. Liu and L. Bruton. A new three-port adaptor suitable for floating-point arithmetic and/or DSP implementations. In *Proceedings of the IEEE International Conference on Acoustics, Speech, and Signal Processing*, volume 6, pages 17–20, Adelaide, Australia, April 1994.

[163] X. Liu and A. Fettweis. Multidimensional digital filtering by using parallel algorithms based on diagonal processing. *Multidimensional Systems and Signal Processing*, 1(1):51–66, March 1990.

[164] R. H. MacNeal. An asymmetrical finite difference network. *Quarterly of Applied Mathematics*, 9(3):295–310, October 1953.

[165] J. D. Markel and A. H. Gray, Jr. *Linear Prediction of Speech Signals*. Springer-Verlag, New York, USA, 1976.

[166] W. Mathis. Recent developments in numerical integration of differential equations. *International Journal of Numerical Modelling*, 7:99–125, 1994.

[167] K. Meerkötter. Incremental pseudopassivity of wave digital filters. In *Proceedings of EUSIPCO-80, First European Signal Processing Conference*, pages 27–31, Lausanne, Switzerland, 1980.

[168] K. Meerkötter and T. Felderhoff. Simulation of nonlinear transmission lines by wave digital filter principles. In *Proceedings of the IEEE International Symposium on Circuits and Systems*, volume 2, pages 875–878, San Diego, California, USA, May 1992.

[169] B. G. Mertzios and F. N. Koumboulis. Analysis and numerical integration of nonlinear systems using MD-passive circuits. In *Proceedings of the IEEE International Symposium on Circuits and Systems*, volume 5, pages 25–28, Orlando, Florida, USA, May–June 1999.

[170] J. Morente, G. Gimenez, J. Ponti, and M. Khalladi. Dispersion analysis for a TLM mesh of symmetrical condensed nodes with stubs. *IEEE Transactions on Microwave Theory and Technology*, 43(2):452–456, 1995.

[171] D. Murphy and D. Howard. 2-D digital waveguide mesh topologies in room acoustic modelling. In *Proceedings of the COST G-6 Conference on Digital Audio Effects*, pages 211–216, Verona, Italy, December 2000.

[172] M. Naghdi and R. Cooper. Propagation of elastic waves in cylindrical shells, including the effects of transverse shear and rotatory inertia. *Journal of the Acoustical Society of America*, 28(1):56–63, January 1956.

[173] Y. Naka, H. Ikuno, M. Nishimoto, and A. Yata. FD-TD method with PMLs ABC based on the principles of multidimensional wave digital filters for discrete-time modelling of Maxwell's equations. *Transactions of the Institute of Electronics, Information and Communication Engineers: Electronics*, 81(2):305–314, February 1998.

[174] J. S. Nielsen and W. J. R. Hoefer. Generalized dispersion analysis and spurious modes of 2-d and 3-d TLM formulations. *IEEE Transactions on Microwave Theory and Technology*, 41:1375–1384, 1993.

[175] G. Nitsche. Solving eigenvalue and steady-state problems of partial differential equations (PDEs) using time-domain models. In *Proceedings Workshop on Advanced Algorithms and their Realizations*, Bonas, France, 1991.

[176] G. Nitsche. *Numerische Lösung partieller Differentialgleichungen mit Hilfe von Wellendigitalfiltern*. PhD thesis, Ruhr-Universität Bochum, 1993. In German.

[177] G. Nitsche. Solving eigenvalue and steady-state problems using time-domain models. *International Journal of Numerical Modelling*, 7:85–97, 1994.

[178] R. Nouta. The Jaumann structure in wave-digital filters. *International Journal of Circuit Theory and Applications*, 2(2):163–174, June 1974.

[179] A. V. Oppenheim and R. Schafer. *Digital Signal Processing*. Prentice Hall, Englewood Cliffs, New Jersey, USA, 1975.

[180] A. V. Oppenheim and R. Schafer. *Discrete-time Signal Processing*. Prentice Hall, Englewood Cliffs, New Jersey, USA, 1989.

[181] S. Osher. Stability of difference approximations of dissipative type for mixed initial boundary value problems. I. *Mathematics of Computation*, 23:335–340, 1969.

[182] H. Ozaki and T. Kasami. Positive real functions of several variables and their application to variable networks. *IRE Transactions on Circuit Theory*, CT-7:251–260, September 1960.

[183] T. W. Parks and J. H. McClellan. A program for the design of linear phase finite impulse response digital filters. *IEEE Transactions on Audio Electroacoustics*, AU-20:195–199, August 1972.

[184] F. Pedersini, A. Sarti, S. Tubaro, and R. Zattoni. Towards the automatic synthesis of nonlinear wave digital models for musical acoustics. In *Proceedings of EUSIPCO-98, Ninth European Signal Processing Conference*, volume 4, pages 2361–2364, Rhodes, Greece, 1998.

[185] P. Penfield. *Tellegen's Theorem and Electrical Networks*. MIT Press, Cambridge, Massachusetts, USA, 1970.

[186] J. Peng and C. Balanis. A generalized reflection-free domain-truncation method: transparent absorbing boundary. *IEEE Transactions on Antennas and Propagation*, 46(7):1015–1022, July 1998.

[187] P. Petropoulos, L. Zhao, and A. Cangellaris. A reflectionless sponge layer absorbing boundary condition for the solution of Maxwell's equations with high-order staggered finite difference schemes. *Journal of Computational Physics*, 139(1):184–208, 1 January 1998.

[188] N. Porcaro, W. Putnam, P. Scandalis, D. Jaffe, J. O. Smith III, and T. Stilson. Synthbuilder: A graphical real-time synthesis, processing and performance system. In *Proceedings of the International Computer Music Conference*, pages 61–62, Banff, Canada, September 1995.

[189] J. Porti and J. Morentz. TLM method and acoustics. *International Journal of Numerical Modelling*, 14:171–183, 2001.

[190] J. Proakis. *Digital Signal Processing*. Prentice Hall, Englewood Cliffs, New Jersey, USA, third edition, 1996.

[191] S. Pulko, A. Mallik, R. Allen, and P. B. Johns. Automatic time-stepping in TLM routines for the modelling of thermal diffusion processes. *International Journal of Numerical Modelling*, 3:127–136, 1990.

[192] R. Rabenstein. A signal processing approach to the digital simulation of multidimensional continuous systems. In *Proceedings of EUSIPCO-86, Third European Signal Processing Conference*, volume 2, pages 665–668, The Hague, The Netherlands, September 1986.

[193] R. Rabenstein and L. Trautmann. Solution of vector partial differential equations by transfer function models. In *Proceedings of the IEEE International Symposium on Circuits and Systems*, volume 5, pages 21–24, Orlando, Florida, USA, May–June 1999.

[194] R. Rabenstein and L. Trautmann. Solution of vector partial differential equations by transfer function models. In *Proceedings of the IEEE International Symposium on Circuits and Systems*, volume 1, pages 407–410, Geneva, Switzerland, May 2000.

[195] L. R. Rabiner, J. H. McClellan, and T. W. Parks. FIR digital filter design techniques using weighted chebyshev approximation. *Proceedings IEEE* 63:595–610, April 1975.

[196] L. Rabiner and R. Schafer. *Digital Processing of Speech Signals*. Prentice Hall, Englewood Cliffs, New Jersey, USA, 1978.

[197] S. S. Rao. *Mechanical Vibrations*. Addison-Wesley, USA, second edition, 1990.

[198] P. Regalia. The digital all-pass filter: A versatile signal-processing building block. *Proceedings of the IEEE*, CAS-34(1):19–37, January 1987.

[199] R. Richtmyer and K. Morton. *Difference Methods for Initial-Value Problems*. Krieger Publishing Company, Malabar, Florida, USA, second edition, 1994.

[200] D. J. Riley and C. D. Turner. Interfacing unstructured tetrahedron grids to structured-grid FDTD. *IEEE Microwave and Guided Wave Letters*, 5(9):284–286, September 1995.

[201] D. Rocchesso. Maximally diffusive yet efficient feedback delay networks for artificial reverberation. *IEEE Signal Processing Letters*, 4(9):252–255, September 1997.

[202] D. Rocchesso and J. O. Smith III. Circulant and elliptic feedback delay networks for artificial reverberation. *IEEE Transactions on Speech and Audio Processing*, 5(1):51–63, January 1997.

[203] D. Rocchesso and J. O. Smith III. Generalized digital waveguide networks. *IEEE Transactions on Speech and Audio Processing*, 11(3):242–254, May 2003.

[204] M. Roitman and P. S. R. Diniz. Simulation of non-linear and switching elements for transient analysis based on wave digital filters. *IEEE Transactions on Power Delivery*, 11(4):2042–2048, October 1996.

[205] M. Roseau. *Vibrations in Mechanical Systems*. Springer-Verlag, Berlin, West Germany, 1984.

[206] Z. Sacks, D. Kingsland, R. Lee, and J. Lee. A perfectly matched absorber for use as an absorbing boundary condition. *IEEE Transactions on Antennas and Propagation*, 43(12):1460–1463, December 1995.

[207] A. Saleh and P. Blanchfield. Analysis of acoustic radiation patterns of array transducers using the TLM method. *International Journal of Numerical Modelling*, 3:39–56, 1990.

[208] A. Sarti and G. DePoli. Generalized adaptors with memory for nonlinear wave digital structures. In *Proceedings of EUSIPCO-96, Seventh European Signal Processing Conference*, volume 3, pages 1773–1776, Trieste, Italy, 1996.

[209] A. Sarti and G. DePoli. Toward nonlinear wave digital filters. *IEEE Transactions on Signal Processing*, 47(6):1654–1658, 1999.

[210] L. Savioja. Improving the three-dimensional digital waveguide mesh by interpolation. In *Proceedings of the Nordic Acoustical Meeting (NAM'98)*, pages 265–268, Stockholm, Sweden, September 1998.

[211] L. Savioja, J. Backman, A. Järvinen, and T. Takala. Waveguide mesh method for low-frequency simulation of room acoustics. In *Proceedings of the 15th International Conference on Acoustics*, pages 637–640, Trondheim, Norway, June 1995.

[212] L. Savioja, T. Rinne, and T. Takala. Simulation of room acoustics with a 3-D finite-difference mesh. In *Proceedings of the International Computer Music Conference*, pages 463–466, Århus, Denmark, September 1994.

[213] L. Savioja and V. Välimäki. Reducing the dispersion error in the digital waveguide mesh using interpolation and frequency-warping techniques. *IEEE Transactions on Speech and Audio Processing*, 8(2):184–194, March 2000.

[214] L. Savioja and V. Välimäki. Interpolated 3-D digital waveguide mesh with frequency warping. In *Proceedings of the IEEE International Conference on Acoustics, Speech, and Signal Processing*, pages 3345–3348, Salt Lake City, Utah, USA, May 2001.

[215] R. Scaramuzza and A. J. Lowery. Hybrid symmetrical condensed node for the TLM method. *Electronics Letters*, 26(23):1947–1949, 1990.

[216] G. P. Scavone. *An Acoustic Analysis of Single-Reed Woodwind Instruments with an Emphasis on Design and Performance Issues and Digital Waveguide Techniques*. PhD thesis, Department of Music, Stanford University, 1997.

[217] T. Schetelig and R. Rabenstein. Simulation of three-dimensional sound propagation with multidimensional wave digital filters. In *Proceedings of the IEEE International Conference on Acoustics, Speech, and Signal Processing*, volume 6, pages 3537–3540, Seattle, Washington, USA, May 1998.

[218] S. Serafin, F. Avanzini, and D. Rocchesso. Bowed string simulation using an elasto-plastic friction model. In *Proceedings of the Stockholm Musical Acoustics Conference*, volume 1, pages 95–98, Stockholm, Sweden, August 2003.

[219] S. Serafin and J. Kojs. The voice of the dragon: a physical model of a rotating corrugated tube. In *Proceedings of the COST-G6 Digital Audio Effects Conference*, pages 259–263, London, UK, September 2003.

[220] S. Serafin, C. Wilkerson, and J. O. Smith III. Modelling bowl resonators using circular waveguide networks. In *Proceedings of the COST-G6 Digital Audio Effects Conference*, pages 117–121, Hamburg, Germany, September 2002.

[221] G. R. Shubin and J. B. Bell. A modified equation approach to constructing fourth order methods for acoustic wave propagation. *SIAM Journal of Scientific and Statistical Computing*, 8:135–51, 1987.

[222] N. Simons and E. Bridges. Equivalence of propagation characteristics for the transmission line matrix method and finite difference time domain methods in two dimensions. *IEEE Transactions on Microwave Theory and Technology*, 39:354–357, 1991.

[223] B. T. Smith et al. *Matrix Eigensystem Routines - EISPACK Guide*. Lecture Notes in Computer Science 6. Springer, New York 1976.

[224] J. O. Smith III. *Techniques for Digital Filter Design and System Identification with Application to the Violin*. PhD thesis, Department of Electrical Engineering, Stanford University, 1983.

[225] J. O. Smith III. A new approach to digital reverberation using closed waveguide networks. In *Proceedings of the International Computer Music Conference*, Vancouver, Canada, September 1985. Appears in Technical Report STAN-M-39, pages 1–7, Center for Computer Research in Music and Acoustics (CCRMA), Department of Music, Stanford University.

[226] J. O. Smith III. Efficient simulation of the reed-bore and bow-string mechanisms. *Music Applications of Digital Waveguides*, pages 29–34. Technical Report STAN-M-39, Center for Computer Research in Music and Acoustics (CCRMA), Department of Music, Stanford University, 1987.

[227] J. O. Smith III. Elimination of limit cycles and overflow oscillations in time-varying lattice and ladder digital filters. *Music Applications of Digital Waveguides*, pages 47–78. Technical Report STAN-M-39, Center for Computer Research in Music and Acoustics (CCRMA), Department of Music, Stanford University, 1987.

[228] J. O. Smith III. *Music Applications of Digital Waveguides*. Technical Report STAN-M-39, Center for Computer Research in Music and Acoustics (CCRMA), Department of Music, Stanford University, 1987.

[229] J. O. Smith III. Waveguide digital filters. *Music Applications of Digital Waveguides*, pages 108–181. Technical Report STAN-M-39, Center for Computer Research in Music and Acoustics (CCRMA), Department of Music, Stanford University, 1987.

[230] J. O. Smith III. Physical modelling using digital waveguides. *Computer Music Journal*, 16(4):74–91, 1992.

[231] J. O. Smith III. *Principles of Digital Waveguide Models of Musical Instruments*, pages 417–466. Kluwer Academic Publishers, Boston, Massachusetts, USA, 1998.

[232] T. Smyth, J. Abel, and J. O. Smith III. The estimation of birdsong control parameters using maximum likelihood and minimum action. In *Proceedings of the Stockholm Musical Acoustics Conference*, pages 413–416, Stockholm, Sweden, August 2003.

[233] P. P. M. So, C. Eswarappa, and W. J. R. Hoefer. Parallel and distributed TLM computation with signal processing for electromagnetic field modelling. *International Journal of Numerical Modelling*, 8(3–4):169–185, May–August 1995.

[234] G. Sod. A survey of several finite difference methods for systems of nonlinear hyperbolic conservation laws. *Journal of Computational Physics*, 27(1):1–31, April 1978.

[235] I. Sokolnikoff. *Tensor Analysis Theory and Applications to Geometry and Mechanics of Continua*. John Wiley & Sons, New York, USA, second edition, 1964.

[236] V. Soyfer. *Lysenko and the Tragedy of Soviet Science*. Rutgers University Press, New Brunswick, New Jersey, USA, 1994.

[237] S. Stoffels. Full mesh warping techniques. In *Proceedings of the COST G-6 Conference on Digital Audio Effects*, pages 223–228, Verona, Italy, December 2000.

[238] J. Strikwerda. *Finite Difference Schemes and Partial Differential Equations*. Wadsworth and Brooks/Cole Advanced Books and Software, Pacific Grove, California, USA, 1989.

[239] H. Strube. Time-varying wave digital filters and vocal-tract models. In *Proceedings of the IEEE International Conference on Acoustics, Speech, and Signal Processing*, volume 2, pages 923–926, Paris, France, May 1982.

[240] H. Strube. Time-varying wave digital filters for modelling analog systems. *IEEE Transactions on Acoustics, Speech, and Signal Processing*, ASSP-30(6):864–868, December 1982.

[241] A. Stulov. Hysteretic model of the grand piano hammer felt. *Journal of the Acoustical Society of America*, 97:2577–2585, 1995.

[242] S. Summerfeld, T. Wicks, and S. Lawson. Wave digital filters using short signed digit coefficients. *Proceedings of the IEE*, 143(5):259–266, October 1996.

[243] E. Tadmor. Skew-selfadjoint form for systems of conservation laws. *Journal of Mathematical Analysis and Applications*, 103:428–442, 1984.

[244] E. Tadmor. A minimum entropy principle in the gas dynamics equations. *Applied Numerical Mathematics*, 2:151–164, October 1986.

[245] E. Tadmor. Entropy functions for symmetric systems of conservation laws. *Journal of Mathematical Analysis and Applications*, 122:355–359, March 1987.

[246] A. Taflove. *Computational Electrodynamics*. Artech House, Boston, Massachusetts, USA, 1995.

[247] A. Taflove. *Advances in Computational Electrodynamics*. Artech House, Boston, Massachusetts, USA, 1998.

[248] L. Taxen. Polyphase filter banks using wave digital filters. *IEEE Transactions on Acoustics, Speech, and Signal Processing*, 29(3):423–428, June 1981.

[249] W. Thomson. *Theory of Vibrations with Applications*. Prentice Hall, Upper Saddle River, New Jersey, USA, fourth edition, 1993.

[250] T. Tolonen, V. Välimäki, and M. Karjalainen. Modelling of tension modulation nonlinearity in plucked strings. *IEEE Transactions in Speech and Audio Processing*, 8:300–310, May 2000.

[251] B. Tongue. *Principles of Vibration*. Oxford University Press, New York, USA, 1996.

[252] L. Trefethen. Group velocity in finite difference schemes. *SIAM Review*, 24:113–136, 1982.

[253] L. Trefethen. Instability of finite difference models for hyperbolic initial boundary value problems. *Communications on Pure and Applied Mathematics*, 37:329–367, 1984.

[254] L. Trefethen. *Spectral Methods in Matlab*. SIAM, Philadelphia, Pennsylvania, USA, 2000.

[255] V. Trenkic and C. Christopoulos. Theory of the symmetrical super-condensed node. *IEEE Transactions on Microwave Theory and Technology*, 43(6):1342–1348, 1995.

[256] J. Tuomela. A note on high-order schemes for the one-dimensional wave equation. *BIT*, 36(1):158–65, 1995.

[257] J. Tuomela. On the construction of arbitrary order schemes for the many-dimensional wave equation. *BIT*, 36(1):158–165, 1996.

[258] E. Turkel and A. Yefet. Absorbing PML boundary layers for wave-like equations. *Applied Numerical Mathematics*, 27(4):533–557, August 1998.

[259] T. Utsunomiya and A. Fettweis. Discrete modelling of plasma equations with ion motion using technique of wave digital filters. In *Proceedings of the IEEE International Conference on Acoustics, Speech, and Signal Processing*, volume 6, pages 21–24, Adelaide, Australia, April 1994.

[260] P. P. Vaidyanathan. A unified approach to orthogonal digital filters and wave digital filters, based on LBR two-pair extraction. *IEEE Transactions on Circuits and Systems*, CAS–32(7):673–686, July 1985.

[261] P. P. Vaidyanathan. *Multirate Systems and Filter Banks*, page 288. Prentice Hall, Englewood Cliffs, New Jersey, USA, 1993.

[262] V. Välimäki. *Discrete-Time Modelling of Acoustic Tubes Using Fractional Delay Filters*. PhD thesis, Helsinki University of Technology, Faculty of Electrical Engineering, Laboratory of Acoustics and Audio Signal Processing, Espoo, Finland, 1995.

[263] V. Välimäki and M. Karjalainen. Implementation of fractional delay waveguide models using all-pass filters. In *Proceedings of the IEEE International Conference on Acoustics, Speech, and Signal Processing*, pages 8–12, Detroit, Michigan, USA, May 1995.

[264] V. Välimäki, M. Karjalainen, and T. Laakso. Modelling of woodwind bores with finger holes. In *Proceedings of the International Computer Music Conference*, pages 32–39, Tokyo, Japan, September 1993.

[265] V. Välimäki, T. Tolonen, and M. Karjalainen. Plucked-string synthesis algorithms with tension modulation nonlinearity. In *Proceedings of the IEEE International Conference on Acoustics, Speech, and Signal Processing*, volume 2, pages 977–980, Phoenix, Arizona, USA, March 1999.

[266] S. A. Van Duyne, J. R. Pierce, and J. O. Smith III. Travelling wave implementation of a lossless mode-coupling filter and the wave digital hammer. In *Proceedings of the International Computer Music Conference*, pages 411–418, Århus, Denmark, September 1994.

[267] S. A. Van Duyne and J. O. Smith III. Physical modelling with the 2D digital waveguide mesh. In *Proceedings of the International Computer Music Conference*, pages 40–47, Tokyo, Japan, September 1993.

[268] S. A. Van Duyne and J. O. Smith III. A simplified approach to modelling dispersion caused by stiffness in strings and plates. In *Proceedings of the International Computer Music Conference*, pages 407–410, Århus, Denmark, September 1994.

[269] S. A. Van Duyne and J. O. Smith III. The 3D tetrahedral digital waveguide mesh with musical applications. In *Proceedings of the International Computer Music Conference*, pages 9–16, Hong Kong, August 1996.

[270] G. Verkroost and H. Butterweck. Suppression of parasitic oscillations in wave digital filters and related structures by means of controlled rounding. In *Proceedings of the IEEE International Symposium on Circuits and Systems*, pages 628–629, Munich, West Germany, April 1976.

[271] R. Voelker and R. Lomax. A finite difference transmission line matrix method incorporating a nonlinear device. *IEEE Transactions on Microwave Theory and Technology*, 38:302–312, 1990.

[272] M. Van Walstijn. *Discrete-Time Modelling of Brass and Reed woodwind Instruments with Application to Musical Sound Synthesis*. PhD thesis, Faculty of Music, University of Edinburgh, 2002.

[273] M. Van Walstijn and G. Scavone. The wave digital tonehole model. In *Proceedings of the International Computer Music Conference*, pages 465–468, Berlin, Germany, August 2000.

[274] X. Wang and S. Bass. The wave digital method and its use in a PIM chip array. Technical Report TR-97-11, University of Notre Dame, 1997.

[275] R. Warming, R. Beam, and B. Hyett. Diagonalization and simultaneous symmetrization of the gas-dynamic matrices. *Mathematics of Computation*, 29(132):1037–1045, October 1975.

[276] W. Wegener. Design of wave digital filters with very short coefficient word lengths. In *Proceedings of the IEEE International Symposium on Circuits and Systems*, pages 473–476, Munich, West Germany, April 1976.

[277] J. Wegner and J. Haddow. Linear thermoelasticity, second sound and the entropy inequality. *Wave Motion*, 18(1):67–77, October 1993.

[278] L. Weinberg. *Network Analysis and Synthesis*. McGraw-Hill, New York, USA, 1962.

[279] J. Wlodarczyk. New multigrid interface for the TLM method. *Electronics Letters*, 32(12):1111–1112, June 1996.

[280] M. R. Wohlers. *Lumped and Distributed Passive Networks; A Generalized and Advanced Viewpoint*. Academic Press, New York, USA, 1969.

[281] F. Xiao and H. Yabe. Numerical dispersion relation for FDTD method in general curvilinear coordinates. *IEEE Microwave and Guided Wave Letters*, 7(2):48–50, February 1997.

[282] C. Q. Xu. Accommodating lumped linear boundary conditions in the wave digital simulations of PDE systems. Master's thesis, University of Notre Dame, 1996. Available as Technical Report CSE-TR-26-94.

[283] C. Q. Xu, S. Bass, and X. Wang. Accommodating boundary conditions in the wave digital simulations of PDE systems. In *Proceedings of the 40th Midwest Symposium on Circuits and Systems*, volume 2, pages 694–697, 1997. IEEE Press.

[284] C. Q. Xu, S. Bass, and X. Wang. Accommodating linear and nonlinear boundary conditions in wave digital simulations of PDE systems. *Journal of Circuits, Systems and Computers*, 7(6):563–597, December 1997.

[285] A. E. Yagle. Fast algorithms for estimation and signal processing: an inverse scattering formulation. *IEEE Transactions on Acoustics, Speech, and Signal Processing*, 37(6):957–959, 1989.

[286] K. S. Yee. Numerical solution of initial boundary value problems involving Maxwell's equations in isotropic media. *IEEE Transactions on Antennas and Propagation*, 14:302–307, 1966.

[287] B. Yegnanarayana. Design of recursive group delay filters by autoregressive modelling. *IEEE Transactions on Acoustics, Speech, and Signal Processing*, ASSP-30:632–637, August 1982.

[288] L. Zhao and C. Cangellaris. A general approach for the development of unsplit-field time domain implementations of perfectly matched layers for FDTD grid truncation. *IEEE Microwave and Guided Wave Letters*, 6(5):209–211, May 1996.

[289] D. W. Zingg. Comparison of high-accuracy finite-difference methods for linear wave propagation. *SIAM Journal of Scientific Computing*, 22(2):476–502, 2000.

[290] D. W. Zingg and H. Lomax. Finite-difference schemes on regular triangular grids. *Journal of Computational Physics*, 108:306–313, 1993.

[291] D. W. Zingg, H. Lomax, and H. M. Jurgens. High-accuracy finite-difference schemes for linear wave propagation. *SIAM Journal of Scientific Computing*, 17:328–346, 1996.

[292] R. Ziolkowski. Time-derivative Lorentz material model-based absorbing boundary condition. *IEEE Transactions on Antennas and Propagation*, 45(10):1530–1535, October 1997.

Index

A-stability, 26
absorbing boundary conditions, 104
abstract passivity, 3
acoustic tubes
    junctions between, 7–8
    musical instrument bodies as, 270
    wave propagation in, 5–7
adaptor, 41, *see also* scattering junction
    conservation of power at, 43
    power-normalized form of, 41
    scattering matrices for, 42–43
    two-port, 20
    vector form of, 46
    with reflection-free port, 42
add density, 303
admittance
    at junction, 123
        matrix form of, 125
    matrix, 28
    of acoustic tube, 6
    of digital waveguide, 120
advection equation, 74
    MDKCs and MDWD networks for, 75–79
all-pass filter
    as termination in a DWN, 146, 292
    fractional delay lines and, 164
    Kelly–Lochbaum model and, 11
amplification polynomial, 300
artificial reverberation, 166, 203

balanced form
    for parallel-plate system, 107
    for Timoshenko beam system, 233–235
    for transmission line system, 105–107
    gyrator in, 39
Bass, S., 61
beam dynamics, *see* Timoshenko beam system
beam moment of inertia, 224
Berenger, J.-P., 104
Beurle, R. L., xi
bidirectional delay line, *see* digital waveguide
bilinear transform, 33–35
Bose, N. K., 287
boundary conditions
    difficulties with implementation of in MDWD networks, 100–101, 230–231, 292
    for Mindlin plate system, 242
    for Navier system, 258–259
    for parallel-plate system, 158
    for symmetric hyperbolic systems, 57–58
    for Timoshenko beam system, 230
    for transmission line system, 99–100, 144–145
    stability and, 296
boundary networks, 100–103
bounded real matrix, 37
bulk modulus, 181
Burger's equation, 278
    MDKC and scattering network for, 278–280

capacitance, 31
    coupled, *see* coupled capacitance
    of parallel-plate pair, 147

capacitance *(continued)*
  of transmission line, 79
  positivity of and passivity, 4
capacitor, 31
  generalized definition of, 263
  multidimensional, 70, 264
  vector form of, *see* coupled capacitance
  wave digital, 21, 38
CFL condition, *see* Courant–Friedrichs–Lewy condition
checkerboard grids, 214, 292
circuit elements
  memoryless, 33, 71
  multidimensional, 22, 68–71, 263–264
  N-port, *see* N-port
  non-energic, 40
  non-reciprocal, *see* non-reciprocal circuit elements
  one-port, *see* one-port
  reactive, 31
  reciprocal, *see* reciprocal circuit elements
  two-port, *see* two-port
circulant delay networks, 166
circulator, 32, 40
  magnitude truncation and, 45
computational density, 303
concrete passivity, 3
conductance
  at port, *see* port conductance
  shunt, of transmission line, 79
Cook, P., 165
coordinate transformations
  grid generation and, 60–65, 109, 113
  subgrid decomposition and, 62
coupled capacitance, 46–48
  as used in MDKCs for vibrating systems, 240, 255
coupled inductance, 46–48
  as used in MDKCs for vibrating systems, 238, 255
  generalized definitions of, 286–287

Courant–Friedrichs–Lewy condition
  explicit finite difference schemes and, 54
  magic time step and, 149
  MDWD networks and, 84
  self-loops and, 136
  space-step/time-step ratio and, 61
current source, 31
  wave digital, 38

delay-free loops, 35, 38, 73
  reflection-free ports and, 42
delay-free path, 20, 80
dielectric constant, 218
digital filters
  all-pass, *see* all-pass filter
  Kelly–Lochbaum model and, 11–12
  ladder/lattice forms of, 26, 116
  orthogonal, 26
  parasitic oscillations and, 45
digital waveguide
  admittance of, 120
  as bidirectional delay line, 118–119
  impedance of, 119
  notational ambiguity and, 121–122
  unit element and, 40
  vector form of, 124
digital waveguide networks, xi–xii, 2, 115, *see also* waveguide mesh
  as lumped networks, 115, 205, 207
  boundary termination of, 144–146, 158–161, 231–232, 242–246, 292
  finite difference schemes and, 17, 115–116, 291
  for Maxwell's equations, 218–222
  for Mindlin plate system, 240–242
  for Naghdi–Cooper system II, 251–252
  for parallel-plate system, 153–156, 174–180
  for Timoshenko beam system, 228–230
  for transmission line system, 135–139
  higher-order accuracy and, 214–217, 315

INDEX 357

initialization of, 162–164
interfaces between, 186–203
interleaved forms of, 133–135
Kelly–Lochbaum model and, 12
MDWD networks and, 115, 205–206, 291
multidimensional representations of, 205–206, 293
music and audio applications of, 164–167
numerical dispersion in, 143–144
scalar/vector coupling in, 125–126
subgrid decomposition in, 156–158
TLM and, xii, 2, 117, 149, 151, 156
unstructured, 166, 203
directional dispersion, 172
dispersion, 223
DWN, *see* digital waveguide networks

elastic solid dynamics, *see* Navier system
elliptic systems, 54, 325–330
embeddings, 63–65
energy flow, 65
  in MDKCs, 64, 83
  in multidimensional inductor, 68
energy method, 51, 93, 296
entropy variables, 285–287
Euler's equations, 181, 280
Euler–Bernoulli beam equation, 224
expanded node, 117

FDTD, *see* finite difference time domain method
feedback delay networks, 166
Fettweis, A., xi–xiii, 24–27, 33, 35, 40, 54, 60, 61, 65, 82, 100, 208, 214, 217, 261, 262, 278, 281, 285–287, 292, 294
finite difference schemes, *see also* finite difference time domain method
  add density of, 303
  computational density of, 303
  consistency of, 17, 303
  DWNs and, 17, 115–116, 291
  for parallel-plate system, 147–148, 152–153

for $(1+1)$D wave equation, 129
for $(2+1)$D wave equation, 16–17, 149, 303–314
for $(3+1)$D wave equation, 315–324
for Maxwell's equations, 223
for transmission line system, 90–93, 127–129, 141
higher-order accurate, 108–113, 214–217, 314–315
implicit, 303
in general curvilinear coordinates, 184
interpolated, 295, 305–309, 318–321
Kelly–Lochbaum model and, 17–18
MDWD networks and, 90–96
multi-step, 300–301
numerical phase velocity and, 302–303
one-step, 299
scattering methods and, 2, 295–296
symbols for, 95
upwind, 77
vector, 302
finite difference time domain method, 117, 148, *see also* finite difference schemes
  DWNs and, 291
  for Maxwell's equations, 222
  TLM and, 117–118
finite volume time domain method, 203
Fourier transform, 298
fractional delay lines, 164
frequency warping, 33, 305
  numerical dispersion correction and, 166, 306
Fries, M., 183
frozen-coefficient system, 59

gas dynamics equations, 280
  MDKCs and scattering networks for, 282–285
  nonconservative form of, 280–281
  skew-selfadjoint forms of, 281–282, 286

GKSO theory, 116, 296
Godunov's method, 283
grid decimation, 12, 128, *see also*
  interleaved grids
  in (2 + 1)D, 148, 156
grid function, 127, 298
group velocity, 59
gyrator, 31–32
  as dualizer, 32
  as used in alternate MDKCs, 211, 215, 240
  as used in networks for vibrating systems, 227, 240
  asymmetry in PDE systems and, 57, 106, 226, 290
  in MDKCs for balanced forms, 106–108, 234
  multidimensional, 71, 263
  wave digital, 39
gyrator coefficient, 32

harmonic oscillator, 21, 48–50
  nonlinear, 267
Hemetsberger, G., 65
hexagonal waveguide mesh, 170–172, 311, 314
higher-order accuracy *see* finite difference schemes, higher-order accurate
Hooke's Law, 253
Householder reflection, 43, 117
Huygens' Principle, 16–18
hybrid forms, 209
hybrid matrix, 28, 289
  for multidimensional unit element, 209
  for series/parallel connections, 210
hydrodynamics, 280, 281

ideal diode, 269, 272
impedance
  at junction, 123
    matrix form of, 125
  characteristic, 104
  for one-port circuit elements, 31
  line, 105
  matrix, 28
    for lattice two-port, 82
  Hermiticity of, 32
    positive realness of, 29, 67
  multidimensional, 67
  of digital waveguide, 119
  reflectance and, 37
incremental pseudopassivity, 37, 44
inductance, 31
  coupled, *see* coupled inductance
  mutual, 47
  of parallel-plate pair, 147
  of transmission line, 79
  positivity of and passivity, 4
inductor, 31
  generalized definition of, 70, 262–264
  multidimensional, 68–70, 264
  multidimensional wave digital, 69, 264
  vector form of, *see* coupled inductance
  wave digital, 21, 38
initial conditions
  for advection equation, 74
  for parallel-plate system, 163–164
  for transmission line system, 162
  Fourier transforms of, 299
  implementation of in DWNs, 162–164
  implementation of in MDWD networks, 96–99
input wave, *see* wave variables, definition of
interleaved grids, 91, 113
  DWNs and, 133
  FDTD and, 117, 148
  holes in, 161
  MDWD networks and, 92
interpolated difference schemes *see* finite difference schemes, interpolated
inverse scattering, 12, 26

Jaumann form, 81
  discretization of, 110
  significance of linear and shift-invariance of, 250, 290
Johns, P. B., xi, 118

junction admittance, 123
junction impedance, 123

Karplus-Strong algorithm, xii, 146
Kelly–Lochbaum model, 4–12, 18
    DWNs and, 132
    equivalent finite difference scheme for, 18
Kirchhoff's Current Law, 30
Kirchhoff's Voltage Law, 20, 30
    acoustical analogue of, 13
Krauss, H., 93, 99
Kreiss Matrix Theorem, 302
Kron, G., 117
Krumpholz, M., 118

Lagrangian mechanics, 98
Lamé coefficients, 252
lattice two-port, 81
Lax–Richtmyer Equivalence Theorem, 86, 93, 295
linear predictive coding (LPC), 12, 132
Liu, X., 60, 61
Lorentz materials, 104
lossless bounded real, 37
losslessness
    multidimensional, 23, 66
    numerical stability and, 18
    of acoustic tube junction, 9
    of boundary conditions, 292
    of inductor and capacitor, 31
    of multidimensional inductor, 69
    of N-port, 29
    of parallel and series connections, 30
    orthogonal transformations and, 291

MacNeal, R., 188
magic time step, 129, 135, 149, 151
magnetic permeability, 217
magnitude truncation, 44
matching layer *see* digital waveguide networks, interfaces between
Maxwell's equations, 217–218
    MDKCs and scattering networks for, 218–222
    as symmetric hyperbolic system, 217–218
    coordinate transformations and, 64
    dual variable structure in, 292
    FDTD and, 117, 148
    phase and group velocities for, 218
MD circuit element *see* multidimensional circuit elements
MD-losslessness, *see* losslessness, multidimensional
MD-passivity, *see* passivity, multidimensional
MDKC, *see* multidimensional Kirchhoff circuit
MDWD networks *see* multidimensional wave digital networks
MDWDF, *see* multidimensional wave digital filters
membrane shell system, 248
    as component of Naghdi–Cooper system II, 250
    as symmetric hyperbolic system, 248–249
    MDKC for, 249
memoryless circuit elements, 33, 71
Mindlin plate system, 235–236
    as symmetric hyperbolic system, 236–237
    boundary conditions for, 242
    group velocities of, 237–238
    MDKCs and scattering networks for, 238–247
multigrid methods, 187, 292
multistep schemes, 300–301
    grid interfaces and, 194
    MDWD networks and, 91–92, 291
multidimensional circuit elements, 22, 68–71, 263–264
multidimensional Kirchhoff circuit, 23, 53
    alternate forms of and DWNs, 205, 293
    energetic analysis of, 83
    for advection equation, 75, 77
    for Burger's equation, 279
    for gas dynamics equations, 282, 284
    for Maxwell's equations, 218

multidimensional Kirchhoff circuit
    (continued)
    for membrane shell system, 249
    for Mindlin system, 238–240
    for Naghdi–Cooper system II, 251
    for parallel-plate system, 87, 212–213
    for time-varying transmission line system, 266
    for Timoshenko beam system, 227, 228
    for transmission line system, 80–81, 210–211, 215
    nonuniqueness of, 293
    strongly hyperbolic systems and, 105, 291
    symmetric hyperbolic systems and, 289–291
multidimensional unit element, 207–208
    hybrid form for, 209
    series/parallel connections and, 211
multidimensional wave digital filters, 53–54
multidimensional wave digital networks, xi–xiii, 2, 22–24, 53–54
    as finite difference schemes, 54, 89–93, 291
    boundary termination of, 99–100, 292
    DWNs and, 115, 205–206, 291
    expanded signal flow graphs for, 76, 78, 206
    for advection equation, 75–76, 78
    for Burger's equation, 279–280
    for gas dynamics equations, 282–283
    for Mindlin plate system, 240
    for Naghdi–Cooper system II, 251
    for parallel-plate system, 88
    for time-varying transmission line system, 266
    for Timoshenko beam system, 227–228
    for transmission line system, 83–86, 93–96
    initialization of, 96–99
    numerical dispersion in, 93–96
    parasitic modes and, 93–96
musical sound synthesis, xii, 164–166, 201, 261, 267–272

N-port, 27
    energy balance at, 29, 66
    multidimensional, 65
    power absorbed at, 28, 29
    stored energy in, 29
Naghdi–Cooper system II, 250–251
    MDKCs and scattering networks for, 251–252
Navier system, 252–253
    as symmetric hyperbolic system, 253–254
    boundary conditions for, 258–259
    MDKCs and scattering networks for, 255–258
    phase and group velocities for, 254
Navier–Stokes equations, 54
network, 27
Nitsche, G., xii, 22, 46, 54, 60, 61, 64, 65, 70, 82, 100, 183, 214, 217, 223, 224, 226, 325
non-energic circuit elements, 40
non-reciprocal circuit elements, 32, 57, 208
    asymmetry in PDE systems and, 106
    strongly hyperbolic systems and, 105
    vibrating systems and, 223
numerical dispersion
    correction of, 165, 306
    directional dependence of, 303
        interpolated schemes and, 305–306, 308–309, 320–321
    in DWNs, 143–144
    in MDWD networks, 93–96
numerical phase velocity, 95, 143, 302–303
numerical reflection, 192

octahedral waveguide mesh, 181, 317
offset sampling, 62, 92, 107, 207
one-dimensional wave equation *see* wave equation, in (1 + 1)D

one-port, 19, 31
  multidimensional, 22, 68–71
  port resistance of, 19
  vector, 45
  wave digital, 37–39
one-sided difference approximation, 145, 159
one-way wave equation, *see* advection equation
open-circuit, 31
  as boundary termination, 100, 144, 158, 161, 243, 292
  impedance of, 31
  wave digital, 38
orthogonal filters, 26
output wave, *see* wave variables, definition of

P- and S-waves, 254
parabolic systems, 54, 117, 291
parallel connection, 13, 30
  of two circuit elements, 20
  scattering equations for, 14, 41–42, 123
  vector form of, 46
parallel-plate system, 86, 152
  alternate MDKC for, 212–214
  as symmetric hyperbolic system, 86
  balanced form of, 108
  boundary conditions for, 158
  DWNs for, 153–156, 170–180, 214
  finite difference schemes for, 147–148, 152–153, 174
  in radial coordinates, 173
  initial conditions for, 163–164
  lossless, source-free form of, 146–147, 212
  MDKC and MDWD networks for, 87–89
  phase and group velocities for, 87
  TE/TM modes and, 86, 148
parasitic modes
  at waveguide mesh interfaces, 198
  in hexagonal scheme for $(2+1)$D wave equation, 313
  in MDWD networks, 93–96
    initialization and, 98
  in tetrahedral scheme for $(3+1)$D wave equation, 324
PARCOR coefficients, 12
passivity
  abstract vs. concrete, 3–4, 140, 295
  in frequency domain, 29
  multidimensional, 65–67
    in frequency domain, 66–67
    integral vs. differential definitions of, 66
  of N-port, 29
  of one-port circuit elements, 31
  of wave digital harmonic oscillator, 49–51
  positive realness of immittances and, 29
  pseudo-, *see* pseudopassivity
  structural, 45
  vs. stability, 1, 10, 51, 77, 295
    higher-order accurate DWNs and, 112
    in DWNs for transmission line system, 140
    in interpolated difference schemes, 307–308
    in triangular scheme for $(2+1)$D wave equation, 310, 314
perfectly matched layers, 104–105
phase velocity, 59
  numerical, *see* numerical phase velocity
piano hammer-string interaction, 267–269
plate dynamics, *see* Mindlin system
Poisson's ratio, 235, 252
polytropic gas, 280
port, 27
port conductance, 35
  at reflection-free port, 42
port resistance, 19, 35
  at reflection-free port, 42
  choice of for wave digital elements, 37–39
positive realness, 29
  bounded realness and, 37
  in multiple dimensions, 67
  spectral mappings and, 33, 73

power
    active, 29
    conservation of at adaptor, 43
    conservation of at scattering
        junction, 8–9, 15, 124
    instantaneous absorbed, 28
    multidimensional, 65, 67
    pseudo-, *see* pseudopower
    total complex absorbed, 29
pseudopassivity, 33, 35, 36
pseudopower, 36

QUARL, 32, 40, 208

Rabenstein, R., 54, 99
reactive circuit elements, 31
realizability, 13, 118
reciprocal circuit elements, 57
    impedance matrices for, 289
    symmetric hyperbolic systems and, 289
rectilinear waveguide mesh
    (2 + 1)D, 15–17, 149–151, 304–305
        interpolated, 307–308
    (3 + 1)D, 181
        interpolated, 321
reflection coefficient, 20, 116
    at junction between acoustic tubes, 8
    boundedness of, 10
    PARCOR coefficient and, 12
    quantization of, 10
reflection-cancelling waves, 231
reflection-free port, 42
    in MDWD networks for balanced
        forms, 106, 108, 234
    in piano hammer model, 269
    in scattering networks for vibrating
        systems, 227, 229, 240
    in single reed model, 272
resistance, 31
    at port, *see* port resistance
    of transmission line, 79
    positivity of and passivity, 4
resistive source, 38–39
    in DWNs, 142
    in single reed model, 271

resistor, 31
    generalized definition of, 262
    multidimensional, 71, 263
    wave digital, 38
Rocchesso, D., 124, 166

scattering junction, 13–15, 122–124
    at interface corner, 190–192
    at mesh interface, 187–188, 194
    conservation of power at, 8, 15, 124
    notational ambiguity and, 119
    vector forms of, 125
scattering matrix, 37
    at junction between acoustic tubes, 8
    for adaptor, 42–43
    for lattice two-port, 82
second-sound theory, 54
self-loops, 136
    boundary termination and, 145, 292
    material parameter variation and, 207
    physical interpretation of, 136–137
semi-implicit approximation, 141, 153
series connection, 30
    scattering equations for, 41–42, 123
    vector form of, 46
series/parallel connection, 209–210
shear modulus, 225
shell dynamics, *see* membrane shell
    system, Naghdi–Cooper
    system II
shocks, 278, 281, 285
short-circuit, 31
    as boundary termination, 100, 145, 158, 231
    impedance of, 31
    wave digital, 38
single reed model, 270–272
skew-selfadjointness, 281–282, 291
slow-wave structure, 151
Smith, J. O., xii, 115, 118, 124, 164, 166, 170, 181
spectral amplification factor, 143, 299
    matrix, 95
spectral mappings, 25, 26, 33–35
    multidimensional, 71–74, 211, 213, 216, 293

spectral methods, 314
staggered grid, *see* interleaved grids
strongly hyperbolic systems, 56, 57, 104, 105, 291
Strube, H., 54
structural passivity, 45
symbol, 94
symmetric hyperbolic systems, 55–60
    boundary conditions and, 57–58
    circuit representations and, 289–291
    dual variable structure in, 291
    energetic properties of, 56–57
    perfectly matched layers and, 104–105
    propagation velocities in, 58–60
    skew-selfadjointness and, 281
    time-varying generalization of, 264–265

T-junction, 80, 81
Taflove, A., 117
TE/TM modes
    Maxwell's equations and, 218
    parallel-plate system and, 86, 117, 146, 148
telegrapher's equations, *see* transmission line system
Tellegen's Theorem, 30, 60, 71
tetrahedral waveguide mesh, 181, 324
thin plate vibration, 235
three-dimensional wave equation *see* wave equation, in $(3+1)$D
time-varying systems, 264–267
    Kelly–Lochbaum model, 10
    power-normalized waves and, 124
    symmetric hyperbolic, 289
Timoshenko beam system, 224–225
    as symmetric hyperbolic system, 225–226
    balanced form of, 233–234
    boundary conditions for, 230
    group velocities of, 226
    MDKCs and scattering networks for, 226–235
TLM, *see* transmission line matrix method, 293

transformer, 31–32
    as used in boundary termination, 159–160, 244
    as used in higher-order accurate networks, 216
    in time-varying systems, 267
    multidimensional, 71, 263
    wave digital, 39
transmission line matrix method, xi, 25, 117–118, 293–295
    DWNs and, xii, 2, 16, 149, 151, 156
    expanded and condensed nodes in, 292
    finite difference schemes and, xiii
    lumped networks and, 100
    Maxwell's equations and, 217
    multigrid methods and, 187
    stubs in, 137
transmission line system, 79, 141
    alternate MDKCs for, 211, 215
    as symmetric hyperbolic system, 79
    balanced form of, 106
    boundary conditions for, 99–100, 144–145
    distortionless, 85
    DWNs for, 135–140, 142–143, 206–207, 212
    finite difference schemes for, 90–93, 141
    generalized to multiple dimensions, 182
    initial conditions for, 162
    lossless, source-free form of, 127, 210
    MDKCs and MDWD networks for, 80–86, 206–207
    phase and group velocities for, 79
    time-varying generalization of, 266
    with constant coefficients, and simplified networks, 85
transparent absorbing boundary, 104
trapezoid rule, 19, 34
    bilinear transform and, 34
    generalized form of, 70, 90
    in multiple dimensions, 23, 69, 72, 90

trapezoid rule *(continued)*
  vector form of, 47
travelling wave
  in uniform acoustic tube, 6
  solution to $(1+1)$D wave equation, 120, 127
  solution to advection equation, 75
  solution to transmission line system, 85
Trefethen, L., 58
triangular waveguide mesh, 172–173, 310, 314
turns ratio, 31, 39
two-dimensional wave equation *see* wave equation, in $(2+1)$D
two-port, 31–32
  adaptor, 20
  hybrid forms for, 209
  wave digital, 39–40

unit element, 40
  digital waveguide and, 119
  multidimensional form of, *see* multidimensional unit element
upwind difference methods, 77–79

Van Duyne, S., xii, 115, 170, 181
vocal tract modelling, 4–11, 165, 267, *see also* Kelly–Lochbaum model
  transmission lines and, 127
voltage source, 31
von Neumann analysis, 93, 112, 297–303

wave digital filters, xi, 2, 25–26
  lumped nonlinearities and, 165
  other filter structures and, 26
wave equation
  in $(1+1)$D, 120, 127
    finite difference schemes for, 129
    travelling wave solution to, 120, 127
  in $(2+1)$D, 17, 297
    finite difference schemes for, 149, 303–314

  waveguide meshes for, 15–17, 149–151
  in $(3+1)$D, 181, 297
    finite difference schemes for, 315–324
    waveguide meshes for, 180–182
  one-way, *see* advection equation
wave variables
  definition of, 19, 35
  in DWNs, 119
  in multiple dimensions, 69
  power-normalized, 8, 35–36
    role in time-varying and nonlinear systems and, 261
  quantization of, 43–45
  vector form of, 45–46
waveguide mesh, 165, *see also* digital waveguide networks
  $(2+1)$D rectilinear, 15–17, 156–158, 304–305
    interpolated, 307–308
  $(3+1)$D rectilinear, 181
    interpolated, 321
  grid decomposition in, 173
  hexagonal, 170–172, 311, 314
  in general curvilinear coordinates, 185–186
  in radial coordinates, 173–180
    connection with a rectilinear mesh and, 198–201
  octahedral, 181, 317
  tetrahedral, 181, 324
  TLM and, 16
  triangular, 172–173, 310, 314
WDF, *see* wave digital filters
Webster's horn equation, 18
well-posedness, 104, 291
  of strongly hyperbolic systems, 56

Yee's method, 117, 206, 222, *see also* finite difference time domain method
Yee, K. S., 117, 148
Young's modulus, 224, 252